Physical Geography and Biodiversity

Physical Geography and Biodiversity

Edited by Carlos Wyatt

Syrawood Publishing House
New York

Published by Syrawood Publishing House,
750 Third Avenue, 9th Floor,
New York, NY 10017, USA
www.syrawoodpublishinghouse.com

Physical Geography and Biodiversity
Edited by Carlos Wyatt

International Standard Book Number: 978-1-68286-414-2 (Hardback)

Cataloging-in-publication Data

Physical geography and biodiversity / edited by Carlos Wyatt.
p. cm.
Includes bibliographical references and index.
ISBN 978-1-68286-414-2
1. Animal ecology. 2. Wildlife management. 3. Biodiversity. 4. Wildlife conservation.
5. Wildlife habitat improvement. I. Wyatt, Carlos.
QL82 .W55 2017
591.7--dc23

Printed in China.

TABLE OF CONTENTS

PREFACE

This book on physical geography and biodiversity deals with the study of terrestrial weather systems and its relation to natural habitats and feeding patterns of animals. The latest research in this field has yielded data that is capable of highlighting areas of potential risk as well as diversity-rich regions. The need for greater technical input in order to aid the study of the various ecological systems has led to the development of geographical information systems, databases and computer models. Topics included in this text discuss the global impact of species extinction, climate change and animal migration. This book is a compilation of chapters that discuss the most vital concepts and emerging trends in the field of physical geography. It will serve as a valuable source of reference for graduate and post graduate students.

The information contained in this book is the result of intensive hard work done by researchers in this field. All due efforts have been made to make this book serve as a complete guiding source for students and researchers. The topics in this book have been comprehensively explained to help readers understand the growing trends in the field.

I would like to thank the entire group of writers who made sincere efforts in this book and my family who supported me in my efforts of working on this book. I take this opportunity to thank all those who have been a guiding force throughout my life.

Editor

1

Strontium Isotopes and the Reconstruction of the Chaco Regional System: Evaluating Uncertainty with Bayesian Mixing Models

Brandon Lee Drake*, Wirt H. Wills, Marian I. Hamilton, Wetherbee Dorshow

Department of Anthropology, University of New Mexico, Albuquerque, New Mexico, United States of America

Abstract

Strontium isotope sourcing has become a common and useful method for assigning sources to archaeological artifacts. In Chaco Canyon, an Ancestral Pueblo regional center in New Mexico, previous studies using these methods have suggested that significant portion of maize and wood originate in the Chuska Mountains region, 75 km to the East. In the present manuscript, these results were tested using both frequentist methods (to determine if geochemical sources can truly be differentiated) and Bayesian methods (to address uncertainty in geochemical source attribution). It was found that Chaco Canyon and the Chuska Mountain region are not easily distinguishable based on radiogenic strontium isotope values. The strontium profiles of many geochemical sources in the region overlap, making it difficult to definitively identify any one particular geochemical source for the canyon's pre-historic maize. Bayesian mixing models support the argument that some spruce and fir wood originated in the San Mateo Mountains, but that this cannot explain all $^{87}Sr/^{86}Sr$ values in Chaco timber. Overall radiogenic strontium isotope data do not clearly identify a single major geochemical source for maize, ponderosa, and most spruce/fir timber. As such, the degree to which Chaco Canyon relied upon outside support for both food and construction material is still ambiguous.

Editor: John P. Hart, New York State Museum, United States of America

Funding: The University of New Mexico Department of Anthropology provided support that assisted with the initial databasing of strontium isotope data. This material is based in part upon work supported by the National Science Foundation Graduate Research Fellowship Program under Grant No. DGE-0903444. The funders had no role in the study design, data collection and analysis, decision to publish, or preparation of the manuscript.

Competing Interests: The authors have declared that no competing interests exist.

* E-mail: b.lee.drake@gmail.edu

Introduction

During the Bonito Phase (ca. AD 860 to 1140) in Chaco Canyon, New Mexico (Figure 1), a series of dispersed subsistence farming communities coalesced around the construction of communal stone buildings called "Great Houses," of which Pueblo Bonito is the most famous (Figure 2). Often referred to as the "Chaco Phenomenon," the social changes that occurred during the Bonito Phase are widely viewed as characteristic of emergent social complexity. Bonito Phase food production was based on the cultivation of maize, beans and squash, and possibly non-domesticates such as sunflowers, and there is evidence for water control features designed to direct seasonal runoff into fields [1]. These crops were grown in Chaco despite constraints imposed by aridity and high elevation, but there is considerable debate among specialists about whether the canyon could have produced enough food to support "monumental" Great House construction efforts or even sustain a relatively small residential population [2]. If the local economy could not underwrite the labor requirements of construction or meet the basic physiological needs of a small canyon residential community, then an obvious conclusion is that labor and food were brought into Chaco from somewhere else [3]. The Bonito Phase is famous for evidence of long-distance exchange in materials, especially pottery, turquoise, macaws, cacao, and timber [4]–[6] and therefore it is reasonable to think that food could have moved through these same exchange systems.

Over the past decade a series of studies based on strontium isotope analysis of maize (*Zea mays*) have attempted to identify specific source areas (farming locations) outside of Chaco [2] [7]–[10]. Researchers have concluded that at least some maize was imported from distances of up to 75 km. Diamond (2005) has argued that reliance on such long-distance exchange networks was an unsustainable economic strategy that set up conditions for the rapid collapse and abandonment of Chaco society during a major 12th century drought [10]. Emergent complex societies in the past are typically associated with evidence for long-distance exchange [11]; the demonstration of long-distance transport for construction materials and food would imply a high level of complexity in Chaco Canyon [12]. General conditions may have been wetter at the time [13], with Great House construction occurring during wet periods [14].

However, the following study illustrates that radiogenic strontium isotope ($^{87}Sr/^{86}Sr$) values obtained from Chaco maize do not discriminate statistically between the $^{87}Sr/^{86}Sr$ values of purported distant geographical sources and $^{87}Sr/^{86}Sr$ values of sources within Chaco Canyon itself [7]. A second set of strontium isotope analyses designed to identify possible long-distance sources of timber used to build Great Houses [12][15] showed a similar lack of clear discrimination among potential sources of ponderosa pine (*Pinus ponderosa*), the primary construction species. These same studies, however, indicated potential timber sources in the canyon

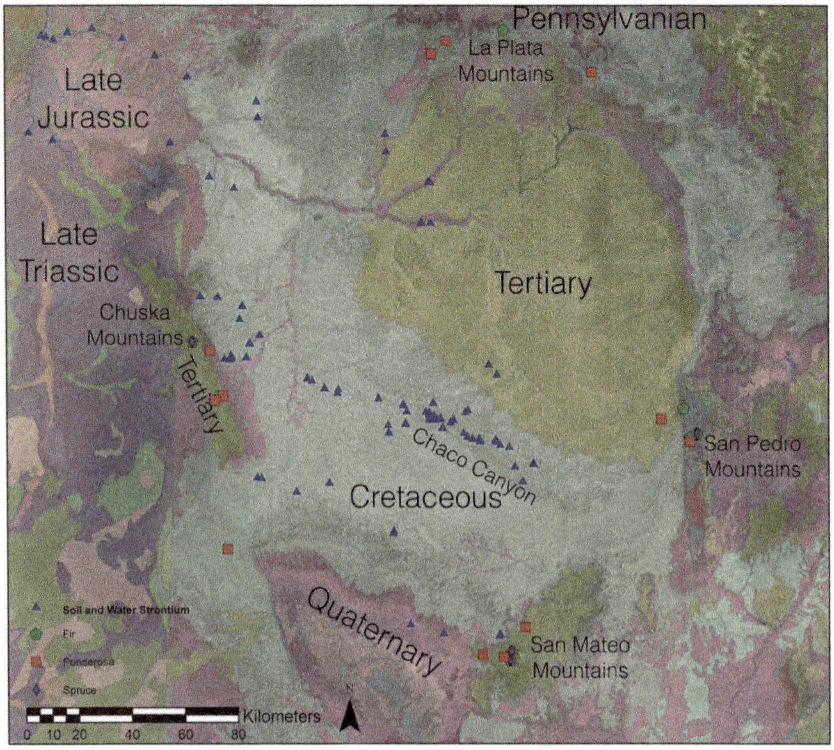

Figure 1. Location of Chaco Canyon (center point) in the San Juan Basin, surrounding upland areas and location of strontium samples associated with forested zones. A simplified geology (colored based on age) shows the relation of strontium sampling locations to their parent geology.

drainage system. Building on recent models showing that the prehistoric agricultural potential of the canyon has been previously underestimated [16]–[17], it is suggested here that the importation of costly basic resources during the Bonito Phase has not yet been proven, despite popular assumptions that Great House society was based on such transactions [10].

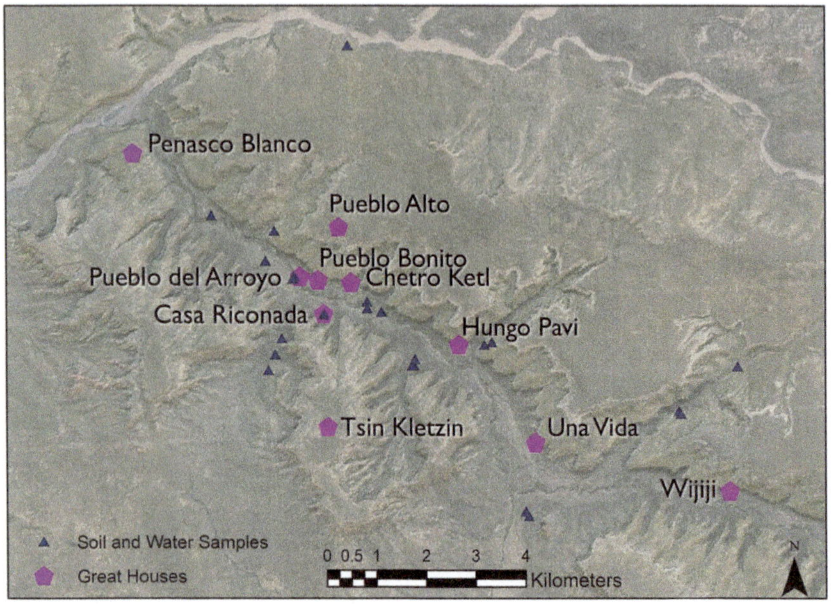

Figure 2. Chaco Canyon Great Houses, local tributaries and strontium sample locations associated with maize sourcing studies [9].

Background

Scholars propose that in the 11th century AD, Chaco saw the development of a meta-community of Great Houses as well as dozens of smaller residential buildings hierarchically organized as a way to effectively manage social relations, food production, and regional exchange [18]–[25]. In turn, the concentration of Great Houses in the canyon (Figure 2) is usually seen as the sociopolitical center of a larger regional network of farming communities outside Chaco, each containing at least one Great House [26]. Most Great House architecture in Chaco was built during the 11th century A.D., with construction declining precipitously after AD 1100, and ending by the late AD 1200s or earlier [27] [28].

The San Juan basin is a mixture of Pennsylvanian, Triassic, Jurassic, Cretaceous, Tertiary, and Quaternary sedimentary formations (Figure 1, Figure S1, S2, S3) [29]. Today, Chaco Canyon itself holds few advantages for dry-farming. The canyon's high elevation (1800 to 1900 masl), aridity, b-seasonal precipitation, and cold winters make cultivation a risky endeavor. Paleoenvironmental studies suggest that the canyon vegetation has changed little over the past 2000 years, leading many researchers to speculate that food was brought into the canyon from elsewhere. Studies in the possible sources of construction beams and maize were spurred on by the recognition that the major mountain ranges near the Colorado Plateau had differing geochemistries [12], chief among these being radiogenic strontium isotope ratios. Evidence for a substantial presence of spruce and fir in Chaco Canyon during the period of construction of the Great Houses is limited [30]–[33]. The long distance transport of tree beams had been hypothesized before [34][35], and $^{87}Sr/^{86}Sr$ values could serve as valid test. Strontium atoms themselves have the same outer electron valence shell structure as calcium, and can be substituted for calcium in plant and animal tissues. Radiogenic (^{87}Sr) to non-radiogenic (^{86}Sr) strontium isotope ratios are unique across many geological terrains, and therefore measuring them can be useful in sourcing various materials. Radiogenic strontium (^{87}Sr) forms as radioactive rubidium decays. The amount of ^{87}Sr in any particular substrate is a function of the age of the rock as well as the original abundance of strontium and rubidium. When set as a ratio with non-radiogenic strontium, variation in abundance is normalized [36]. Bedrock weathering is the primary input for soil $^{87}Sr/^{86}Sr$ [36] [37] but alluvial and aeolian deposition can also influence ratios, along with differential weathering of minerals. However, differences between the radiogenic strontium isotope ratios of bedrock and soil are generally minimal [36] [38], though the ratio does vary by depth [39]. Plants acquire the $^{87}Sr/^{86}Sr$ value of the soils that they grow in, and numerous studies have confirmed that a plant's $^{87}Sr/^{86}Sr$ value is consistently a less variable, more averaged version of the local soil's [39] [40].

A notable contribution to this effort has been a long-term study examining the $^{87}Sr/^{86}Sr$ values from maize found in Chaco Canyon and its potential regional sources [2] [7] [8]. These studies analyzed synthetic soil-water samples collected from locations throughout the southern Colorado Plateau (including 48 in Chaco Canyon) and 36 maize cobs from nine Chaco sites, including seven from Pueblo Bonito. Twenty-one of the maize cobs from four sites all dated to the late 12th century AD. The specimens from Pueblo Bonito ranged from the late 9th to early 13th centuries, while six samples from other sites dated to the 15th, 18th and 19th centuries ([7], Table 1). Although Benson and colleagues (2009) concluded that none of the analyzed maize came from sources in Chaco Canyon, they also recognized significant shortcomings in their comparative database stemming from geological complexity at the regional level. The geological variation observed by Benson's study group likely reflects the regional dominance of Cretaceous

Table 1. Comparison of medians across maize by provenience (columns) and potential sources (rows).

	Aztec Soil	Chuska Slope	Lobo Mesa	Northeastern San Juan River	Red Mesa	La Plata	Salmon Ruin	Pre 1140 Maize	Post 1140 Maize	Historic Maize
Chaco Watershed	−0.0003*	0.0002*	−0.0006**	0.0006**	0.0002	0.0004*	−0.0008**	**−0.0002**	−0.0009**	−0.0012**
Aztec Soil		0.0005**	−0.0003	0.0009**	0.0005*	0.0008*	−0.0005*	0.0001	−0.0006	−0.0009*
Chuska Slope			−0.0008**	0.0003**	0.0001	0.0002	−0.0011**	−0.0004**	−0.0011**	−0.00145**
Lobo Mesa				0.0011**	0.0008*	0.0010*	**−0.0003**	0.0004	−0.0003	**−0.0006**
Northwestern San Juan River					**−0.0003**	−0.0001	−0.0014*	−0.0008**	−0.0014**	−0.0018**
Red Mesa						0.0002	−0.0011*	**−0.0004**	−0.0011**	−0.0014*
La Plata							−0.0013*	−0.0006*	−0.0013**	−0.0016*
Salmon Ruin								**0.0006**	0	−0.0003
Pre 1140 Maize									**−0.0007**	**−0.0010**
Post 1140 Maize										−0.0003

Significance codes: 0.001 '**' 0.01 '*'.
Items in bold indicate that no significant difference between medians was found.

sandstone formations (Figure 1) combined with a variety of erosional and depositional processes (including aeolian mixing), all of which can lead to extremely patchy geochemical properties in potential cultivation sites.

Benson's efforts to identify geographic sources for maize found in Chaco have been paralleled by radiogenic strontium isotope studies designed to locate sources for wood recovered from Chaco Great Houses. More than 80% of the primary wooden beams in Great Houses were ponderosa pine (*Pinus ponderosa*), which grew in Chaco in the early 20th century A.D., albeit only a few isolated trees [30]. However, many of the construction elements in Great Houses included Douglas fir (*Pseudotsuga menziesii*), spruce (*Picea spp.*) and even aspen (*Populus tremuloides*) species, which did not grow in Chaco during the late Holocene [31]–[33]. The ten largest ruins in Chaco would have required ca. 45,000 high elevation trees (2500 m and above today), based on species proportions recovered from excavations [34]. Archaeologists have argued that over time, Chaco builders would have depleted any locally available ponderosa, forcing them to seek this species in distant mountains, up to 80 km distant [34] [12] [28]. Spruce and fir in the region primarily occur in high-altitude sub-alpine montane and mesic woodlands in the Chuska, San Mateo, San Pedro, and La Plata Mountains while Ponderosa occupies the slopes of these ranges (Figure S4). These mountains ranges are at least 70+ kilometers away from Chaco Canyon Great Houses. However, spruce, fir, and ponderosa are sporadically available in mesic and riparian mixed conifer woodlands closer to the canyon (Figure S5) [41]. These trees typically grow on north-facing slopes near arroyos and other sources of periodic water. These woodlands occur along the paths to most nearby mountain ranges (Figure S6, S7, S8, S9). It is unlikely that these trees grew directly in Chaco Canyon based on macrobotanicals found in packrat middens in the canyon [33], though only two of these samples fall within a few centuries of the construction period.

Following a pioneering investigation which demonstrated that trees growing on different substrates in the San Juan Basin produced chemically distinct signatures [34], English, et al. (2001) analyzed radiogenic strontium isotopes from *Picea sp.* (spruce), *Abies sp.* (fir), *Pinus ponderosa* (ponderosa), water, and rocks at three areas where high elevation trees grow now. They compared these values to 52 archaeological specimens of spruce and fir from six Great Houses and concluded the archaeological wood came from the San Mateo Mountains (80+ km to the south) and the Chuska Mountains (75+ km to the west) [12]. Because these species correspond to modern elevations above 2500 m, the analysis did not include other possible geographic sources. Today, isolated stands of spruce, fir, and ponderosa exist in riparian and mesic mixed-conifer forests in the San Juan Basin [41]. Historically, isolated fir and ponderosa were both available in the canyon itself [27]. If conditions were previously cooler and wetter, these isolated stands could potentially have existed in larger quantities across the San Juan Basin [13] [14] and provided a more attractive source for timber than distant mountain ranges.

However, a subsequent attempt to use radiogenic strontium isotopes for sourcing ponderosa pine was unable to establish clear connections between potential source areas and archaeological specimens recovered from canyon Great Houses. Reynolds et al (2005) compared $^{87}Sr/^{86}Sr$ values for 62 living ponderosa from 19 geological source areas with 53 archaeological ponderosa from Pueblo Bonito, Chetro Ketl, and Pueblo del Arroyo (Figure 1). The results were plotted as means and the geological sample distribution compared visually to the archaeological distribution. Data were insufficient to allow clear identification of the particular origins of any of the Chaco timbers, although the authors suggested that the ponderosa pine data indicated a shift to more northern and western sources, specifically the La Plata or San Juan Mountains, ca. 150 km away. They also noted that they could not rule out the possibility that half the archaeological beams grew in Chaco itself, based on isotopic values from water samples in the canyon [15].

As geochemical data are variable due to geologic process or instrumental error, there must be some consideration of uncertainty in the original geochemical source data before attempting to make an argument for a specific geospatial source [42]. The present study uses both frequentist inference and Bayesian mixing models to assess the distinction between potential geochemical sources for the maize and trees found in Chaco Canyon by characterizing variation in both source samples and archaeological specimens as probability statements and distributions.

The sample of potential sources for spruce, fir and ponderosa was expanded beyond only high-altitude regions to include the Chaco Watershed and Aztec ruins based on riparian and mesic mixed conifer forest located in the area (Figures S4, S5, S6, S7, S8, S9). The sample of potential sources for maize includes the Chaco Watershed. This was excluded in earlier analysis as environmental conditions were not considered suitable for maize cultivation [7]–[10]. However, recent work indicates that maize cultivation was possible [16]–[17], as water control features and field houses are present in the canyon.

The large quantity of available radiogenic strontium data not only provides a way to get a more holistic view of the prehistoric economy of Chaco Canyon, but also a large data set to assess the utility of Bayesian inference into geochemical sourcing questions.

Methods

Strontium data reported by English et al. (2001) and Reynolds et al. (2005) were used to define high-altitude geochemical sources for timber, while strontium data reported by Benson et al. (2003), Benson et al. (2009), and Benson et al. (2010) were used to define lower-elevation geochemical sources for maize cultivation in the immediate Chaco region and the broader San Juan Basin (Tables S9, S10, S11, S12) [12] [2] [15] [7]. Potential maize geochemical sources were defined using soil and synthetic water $^{87}Sr/^{86}Sr$ values ($n = 154$) and potential timber geochemical sources were defined using geologic and modern tree $^{87}Sr/^{86}Sr$ values ($n = 223$). Soil and water samples grouped by Benson et al. (2009) as either Chaco Canyon or the Upper Rio Chaco region were combined to form a single strontium isotope ratio set because the canyon is located in the hydrological center of the samples identified as the Upper Rio Chaco (Figure 2). This sample is referred to in this manuscript as the Chaco watershed. $^{87}Sr/^{86}Sr$ values for Chaco maize were grouped into temporal cultivation periods (pre-1140, post-1140, and historic). Values for archeological timbers were grouped by taxon (spruce, ponderosa, and fir). NBS SRM-987 values were reported for each study (English et al (2001) = 0.7102453+/−0.000012 [12]; Reynolds et al (2005) [15] = 0.7102680+/−0.00004; Benson et al (2003) [2] = 0.710276+/−0.000016; Benson et al (2009) [7] = 0.71029+/−0.00002) and $^{87}Sr/^{86}Sr$ values were normalized to 0.71029 per Benson et al (2009) (Tables S9–S12).

Frequentist statistics can be used to determine whether or not $^{87}Sr/^{86}Sr$ measures of central tendency for potential sources, maize, or timber species are distinguishable. First, kernel density estimates and plots were created in order to visualize the probability distributions of $^{87}Sr/^{86}Sr$ values for sources, timber, and maize. Kernel density estimates were formed with the "stats" package in the R programming language (R 3.0.2) [43][44] using a

Gaussian smoothing kernel with a smoothing bandwidth equal to the standard deviation of each geochemical source and the Great House construction timber. These plots function similarly to histograms, but can be "smoothed" by assuming a normal distribution for each data point in order to depict a more curve-like probability density function for the variable of interest. For each geochemically defined source, timber taxon, and temporal maize cultivation group, 552 equally spaced points were generated with these smoothing parameters. The resulting kernel density estimate plots were used to produce individual $^{87}Sr/^{86}Sr$ probability curves for each geochemically defined source, each timber taxon, and each temporally defined maize cultivation groups. In addition to qualitative assessments of kernel density plots, simple Wilcoxon Mann-Whitney tests, which allow for comparing medians between data sets that are not normally distributed or have unequal variances, were run to determine if means were significantly different.

As a supplement to frequentist methods, mixing models can be used to assess possible source attributions [45]. Mixing models are most commonly used in ecology for assessing the contributions of different prey items to a predator's diet [46] [47], but they are well suited for geochemical source analysis as well. In the traditional approach, after measuring N isotopic elements, basic mixing models estimate the proportional contribution of N+1 sources to a mixture by solving a system of N+1 linear equations. A clear limitation to this method is that the number of sources considered in a model is capped by the number of isotopes measured [46] [48]. The use of Bayesian inference techniques allows a release from this restriction; theoretically any number of sources could be considered regardless of the number of isotopic elements [49] [50] [51]. A Bayesian approach also allows the incorporation of prior information into the model and more ways to investigate uncertainty and variability within potential sources. In this study, the R package SIAR (Stable Isotope Analysis in R) [49] [52] was used to build Bayesian mixing models to estimate the relative contribution of geochemical sources to the radiogenic strontium signatures in Chaco maize and timber. Using SIAR, each potential geochemical source was defined as a mean and standard deviation of measured $^{87}Sr/^{86}Sr$ from the soil, water, or modern trees in that locality. Archeological samples were defined using radiogenic $^{87}Sr/^{86}Sr$ values grouped by cultivation period (for maize: pre-1140, post-1140, and historic) or by taxonomic group (for timbers: spruce, ponderosa, and fir).

The Dirichlet distribution (a generalization of the more commonly used Beta distribution) was used as a prior for the proportional contribution of each source and no other information was included in the prior. Each source was treated as independent with an equal prior probability (12.5%), but the total proportions were required to sum to one [49]. Priors for all samples were kept as equal (12.5%) so that posterior probabilities primarily reflected the data following procedures used by Parnell et al. [49]. Isotopic data for each timber and maize sample were inputted, and 200,000 Markov Chain Monte Carlo simulations with 50,000 burn-ins were run to determine the proportionate contribution of each geochemical source to the maize and timbers of Chaco Canyon. Definitive source attributions should show a high proportionate contribution for one source over the others, while ambiguous source attributions should show an even spread of posterior probabilities.

One limitation of the mixing model software is that all data is assumed to follow a normal distribution [45] [49] [50]. To our knowledge, software does not yet exist to run Bayesian mixing models without this assumption. Shapiro-Wilkes tests for normality (Table S1) indicate this is a potential issue with the Chuska Mountains, San Pedro Mountains, and Aztec soils. To test the effects of these influences, a reduced model of the data were run for ponderosa with the Chuska Mountains corrected for normality (removing data points CHM-214b and CHM-214d) and for Pre-1140 A.D. maize (removing data point CHCU43684A). Removing these data limit the generalizability of these models, but do allow for a comparison between the posterior probabilities of data coerced to be normal to the full data set. A third robustness test included a sequential increase of prior probabilities by 3.5% from 12.5% to 93% for the Chuska Mountains with the posterior probability assessed relative to the prior probability.

Results

Maize Sources

Our kernel density plots yielded a series of overlapping probability curves that reveal considerable uncertainty in source attribution for maize. $^{87}Sr/^{86}Sr$ values from Chaco Canyon maize radiocarbon-dated to Pueblo I and II periods (pre-1140 AD) largely overlap with the $^{87}Sr/^{86}Sr$ values of the Chaco watershed source, as well as many of the other defined sources (Figure 3a, 3b). Additionally, this subset of maize data has a bimodal distribution. A second, smaller peak appears outside the $^{87}Sr/^{86}Sr$ value of any known geochemical source within the study area. This indicates that there must be some yet-undefined geochemical source influencing this maize sample; it cannot be all sourced within the known geochemistry of the sampled areas.

Wilcoxon Mann-Whitney tests showed no statistical different in the $^{87}Sr/^{86}Sr$ values for maize predating 1140 A.D. and the Chaco watershed, Aztec soils, Chuska Slope soils, Lobo Mesa soils, or Salmon Ruin soil sources. Shapiro Wilkes tests for normality indicate that the Aztec soil, pre-1140 A.D., and post-1140 A.D. maize are non-normal to the 99% confidence level.

In the Bayesian mixing model results, Salmon Ruin had the highest mean source posterior probability for post-1140 A.D. maize (Figure 4a, Table 2), although it was not a large difference with respect to probability for the Aztec soil, Chaco watershed soil, or Lobo Mesa soil sources.

$^{87}Sr/^{86}Sr$ values from maize grown in the Pueblo III period (post-1140 AD) partially overlapped with canyon drainage $^{87}Sr/^{86}Sr$ values, but some cobs had ratios outside those obtained from the Chaco watershed. These outlying ratios did not correspond to any other sources in this dataset (Figure 3a, 3b), suggesting an unknown or un-sampled source. It is certain that some post-1140 AD maize cobs were from the Navajo occupation of the canyon, which post-dates the acquisition of horses and consequently could easily have been transported from farming locales outside Chaco proper. This later maize had a small statistically significant difference with all potential sources except the Salmon Ruin area (Table 1). In the Bayesian mixing models, it had the highest probability of association with Salmon Ruin and Lobo Mesa soil sources, although again the differences are not great (Figure 3a, Table 2).

To determine the effects of non-normality on the conclusion for Pre-1140 A.D. maize, a sample (CHCU43684A) was removed from this data set and the model re-run without it. Mean posterior probabilities were not substantially different as a consequence of a violation of the normality assumption for this data set (Table S3). A second run of the model was performed on data rounded to the 4th decimal place, and neither median difference was significance (Table S4) or posterior probabilities (Table S5) were strongly affected by this treatment.

Maize dated to the historic era (post 1492 A.D.) were similar to pre-1140 A.D. maize in that no single source was dominant

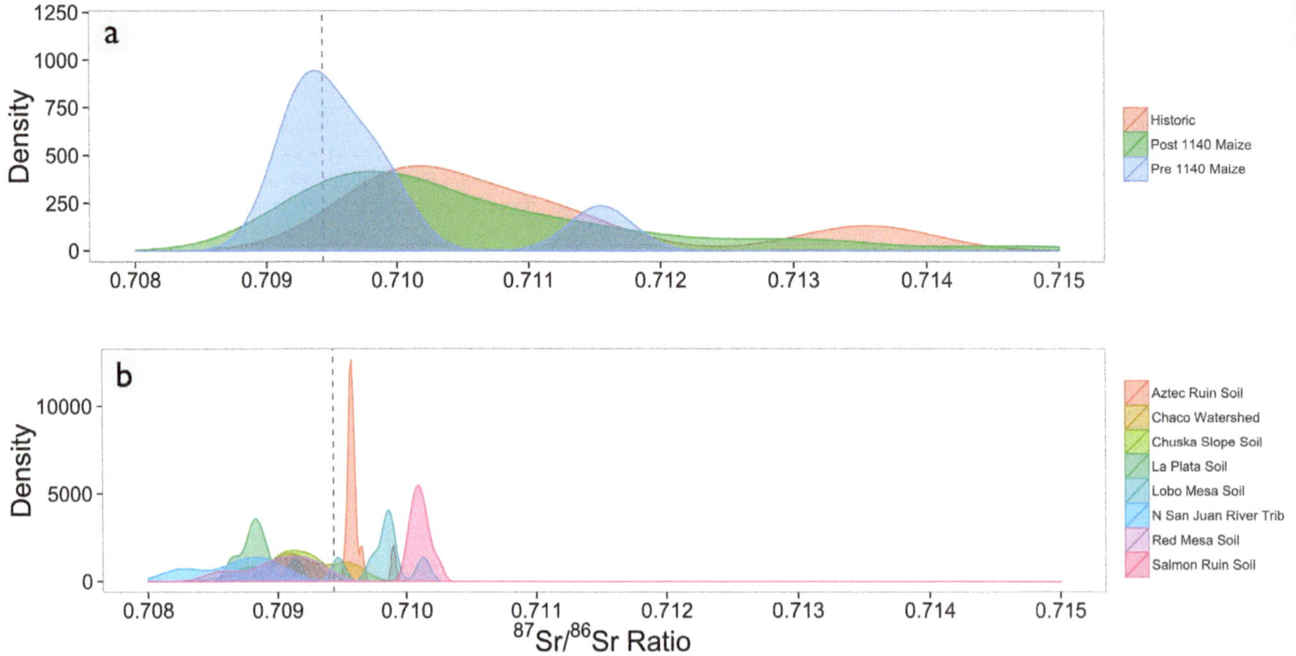

Figure 3. Kernel density distributions of historical, pre-1140 A.D., and post-1140 A.D. maize in Chaco Canyon (a) and soil and water $^{87}Sr/^{86}Sr$ sources (b). The dotted line represents the strontium isotope ratio for a single ponderosa tree stump found in the West Court of Pueblo Bonito (JPB-99). Maize from both time periods overlaps with strontium isotope ratios from the Chaco watershed, though maize dating after 1140 A.D. has more cobs with higher values than found within the canyon and in the selected source data set.

(Figure 4c, Figure 5a–c, Table 2). Overall, the Monte Carlo simulations did not support any definitive geochemical sources associated with Chaco maize from any period (Table 2, Figure 4a–c, Figure 5a–c).

Timber Sources

Wilcoxon Mann-Whitney tests indicated that there was no significant difference in medians for the Chaco watershed region and the Chuska Mountains at the 99% confidence level (Table 3), meaning these sources may not be sufficiently geochemically distinguishable. This complicates arguments for the Chuska Mountains as a major source for timber or maize [2]. The Bayesian mixing model indicated that some spruce and fir from Great Houses was likely from the San Mateo range (Figure 6a–f, Figure 7, Table 4), though some values fell within the range of the Chuska and Chaco watershed regions. It is important to note that the Bayesian model is not a confirmation that San Mateo is a source for construction timbers, only that it is the most likely given the available $^{87}Sr/^{86}Sr$ values. Although Chaco is a marginal location for ponderosa and fir today, and well outside the modern range of spruce, if trees that grow in Chaco Canyon mirror the strontium isotope values from the Chuska Mountains and, in part, the La Plata Mountains, it may be impossible to distinguish these two/three geochemical sources based on strontium isotope data alone. Therefore, the geochemical source attribution for timber data is also unclear.

To determine the effects of non-normality on the conclusion for the Chuska source, two samples (CHM-214b and CHM-214d) were removed from this data set and the model re-run without it. Mean posterior probabilities were not substantially different as a consequence of a violation of the normality assumption for this data set (Table S6). A second run of the model was performed on data rounded to the 4th decimal place, and neither median was difference significance (Table S7) or posterior probabilities (Table S8) were strongly affected by this treatment. A sequential increase of prior probabilities increased posterior probabilities for the Chuska Mountains timber source for spruce, however in all cases the posterior was lower than the prior (Figure S10).

All code and analyses are available in the supporting information R code S1. All data is available in the supplementary material (Tables S9, S10, S11, S12). In addition, instructions are offered for modifying the code to use the same approach with new data.

Discussion

Both qualitative interpretation using kernel density plots and quantitative source attributions using Bayesian mixing models indicate that geochemical sources for both maize and timber are not as distinguishable as indicated in the past, with the exception of the San Mateo Mountains. Data do not support the argument that maize was exclusively imported from outside Chaco Canyon, and do not support the Chuska Mountains as an unambiguous source for timber. There are no dominant or unambiguous geochemical, and thus geospatial, sources for Chaco maize or timber indicated in the radiogenic strontium isotope data. The overlapping kernel density plots for $^{87}Sr/^{86}Sr$ values of the Chaco watershed, maize cobs, and construction timbers from Great Houses, in addition to the Chuska Mountains and Chuska Slope soils, all share peaks near the $^{87}Sr/^{86}Sr$ value of the ponderosa stump (JPB-99) found in Pueblo Bonito (Figure 8). It is very likely that at least some of the geochemical sources of construction timber and maize are the same. $^{87}Sr/^{86}Sr$ values obtained from soil and water sources in the Chaco watershed are consistent with values from archaeological timbers from the major Great Houses (Figure 3a, 3b), but water samples from Werritos Rincon and South Gap (canyon tributaries) are also similar to the mean value

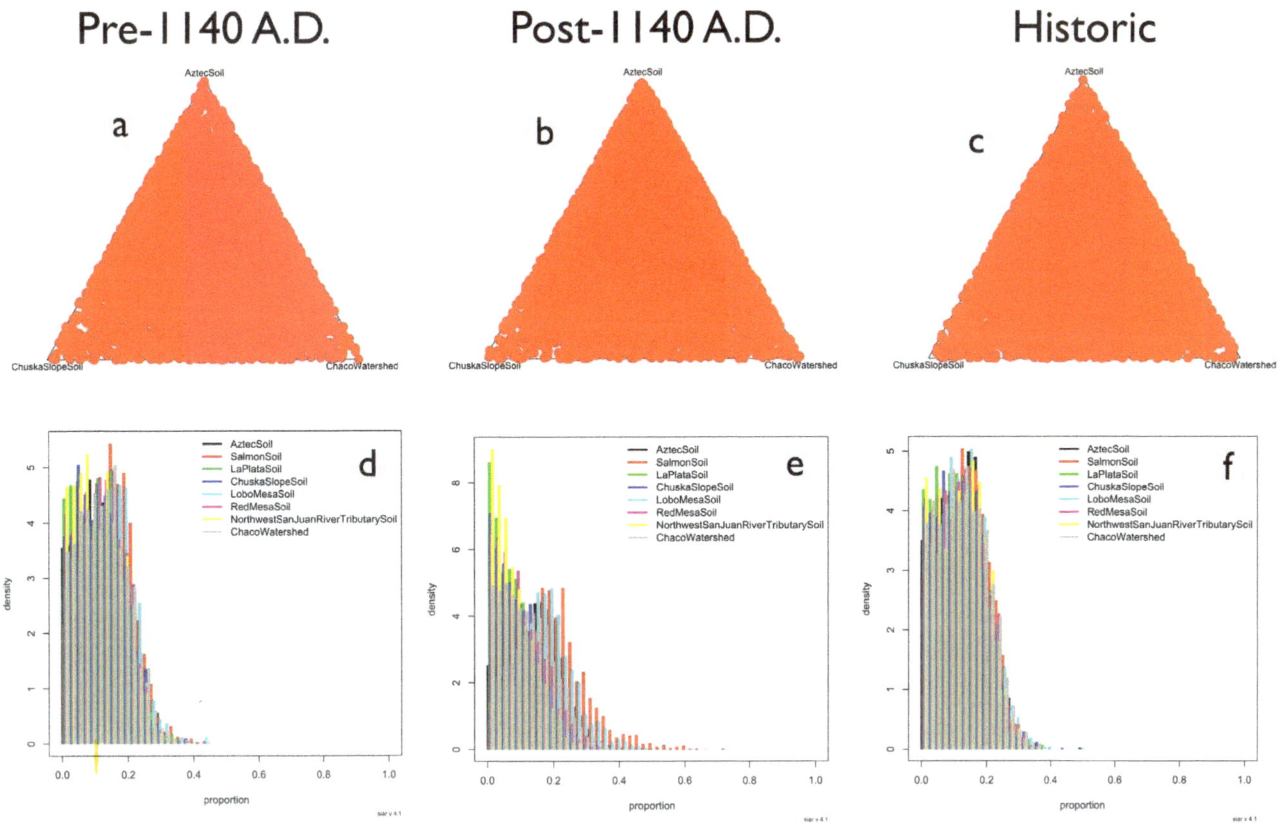

Figure 4. Ternary and proportion plots illustrating source mean posterior probabilities generated from Markov-Chain Monte Carlo simulations of strontium isotope data. There is no clear pattern for pre-1140 A.D. (a), post-1140 A.D. (b), and historic maize (c), indicating that no source attribution is more likely than others based on this model. Mean posterior probability proportions of geochemical source contributions for pre-1140 maize (d), post-1140 maize (e), and historic maize (f) are also displayed.

for the Chuska sources, the lower mode for the La Plata source, and the Pueblo Bonito stump (JPB-99).

The results of robustness testing, which included removing data points to coerce the Chuska Mountains timber source and pre-1140 A.D. maize and rounding data to the 4th decimal place, did not substantially alter results or conclusions. This indicates that the Bayesian mixing model as employed in the SIAR package for R is robust to violations in the normality assumption, at least as far as the present study is concerned. These results mirror those of Parnell et al. 2010 [49] who found that SIAR was robust to violations of its core assumptions.

It is important to note how the Bayesian mixing models should, and should not, be interpreted. The posterior probability of the Bayesian mixing models should not be regarded as a hypothesis test for a given source, but rather as a statement of uncertainty regarding that source's contribution to the observed data in ecofacts from Chaco Canyon. Each source was assigned a prior probability of 12.5%. The posterior probability reflects the

Table 2. Mean posterior probability proportions of geochemical source contributions to observed variation in Chaco maize.

	Pre 1140 A.D. Maize	Post 1140 A.D. Maize	Historic Maize
Chaco Watershed	0.127	0.145	0.128
Aztec Soil	0.131	0.114	0.135
Chuska Slope	0.120	0.098	0.119
Northwestern San Juan	0.110	0.076	0.121
Lobo Mesa	0.134	0.177	0.132
Red Mesa	0.119	0.094	0.122
La Plata	0.117	0.085	0.119
Salmon Ruin	0.143	0.211	0.125

For all maize samples, the strontium shows uniform mean posterior probabilities across all possible sources. This indicates a high degree of uncertainty when attempting to use strontium to source maize samples.

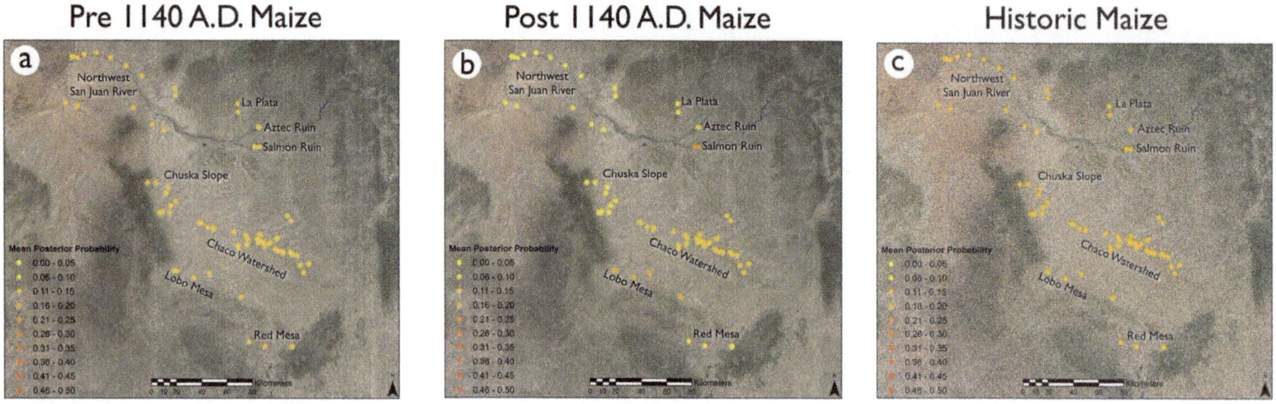

Figure 5. Map of mean posterior probabilities for source contribution in the San Juan Basin for pre-1140 A.D. maize (a), post-1140 A.D. maize (b), and historic maize (c).

assessments of the likelihood of each source contributing to the variation observed in Chaco Canyon ecofacts in light of the strontium data from selected sources. That the posterior probability for the San Mateo Mountains, 47.5% (Table 4), is so much higher than the prior reflects the higher confidence of that range being a source for at least some of the spruce and fir. However, for ponderosa, the posterior probability, 14.5%, is not much higher than the prior; this indicates that strontium data does not present strong evidence for that range being a large proportional contributor to observed $^{87}Sr/^{86}Sr$ variation relative to other potential sources. This does not rule out San Mateo as a source for ponderosa, it only indicates that strontium data do not present a compelling case for this argument. This same interpretation extends to Chaco Canyon and the Chuska Mountains as sources for spruce, fir, or ponderosa; the posterior probabilities for proportional contribution are not substantially higher than the priors. Increasing the priors does not change this interpretation (Figure S10), as posterior probabilities remain lower than weighted

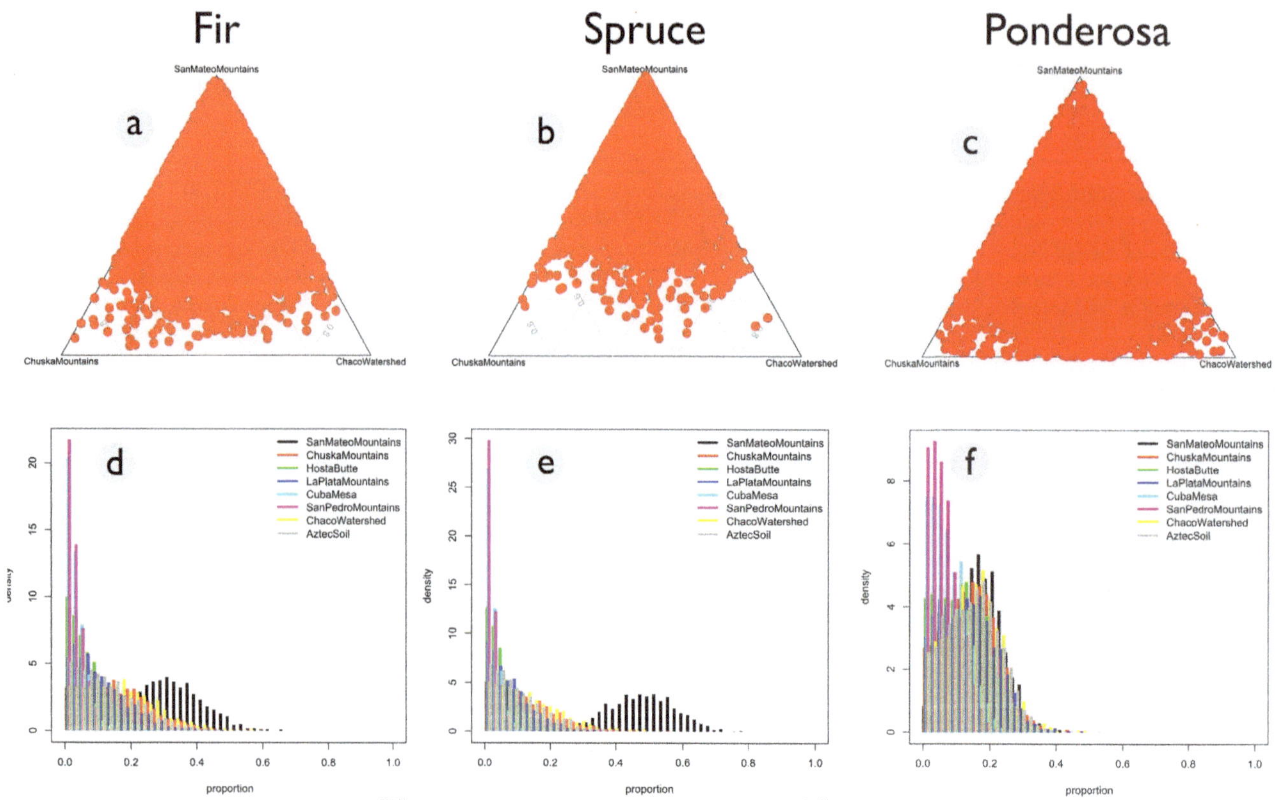

Figure 6. Ternary and proportion plots illustrating source posterior probabilities generated from Markov-Chain Monte Carlo simulations of strontium isotope data. Both Spruce and Fir (a & b) indicate San Mateo is a more likely source than either the Chuska Mountains or isolated stands near Chaco. There is no clear pattern for Ponderosa pine (c), indicating that no source attribution is more likely than others based on this model. Mean posterior probability proportions of geochemical source contributions for fir (d), spruce (e) and ponderosa (f) are also displayed.

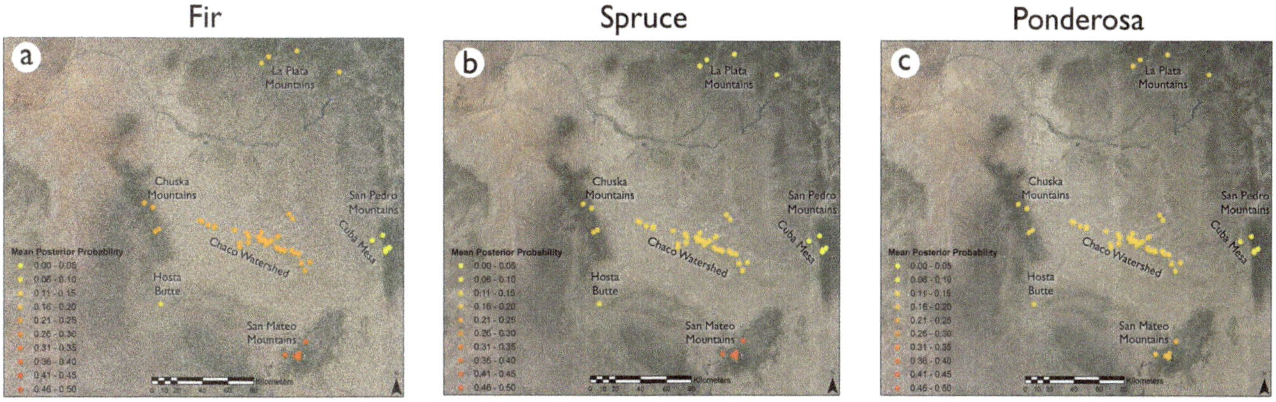

Figure 7. Map of mean posterior probabilities for source contribution in the San Juan Basin for fir (a), spruce (b), and ponderosa (c).

priors. Likewise, for almost all maize sources for all time periods analyzed in the present manuscript, posterior probabilities for potential sources were not substantially higher than the priors used. The only exception to this is Salmon Ruins and possibly Lobo Mesa for the maize grown after 1140 A.D. but before the historic period. As noted earlier, posterior probabilities for source contribution to observed geochemical variability do not rule out potential sources, but they do address which sources are more or less likely to contribute to observed radiogenic strontium isotope ratio variation in ecofacts. It is difficult to make any clear arguments about geochemical sourcing for most maize or timber in Chaco Canyon in light of the currently available radiogenic strontium isotope data.

The clearest possible source attribution that emerged from the mixing model analysis is the San Mateo range for spruce and fir trees (Figure 6–7, Table 4). However, while the Bayesian approach indicated San Mateo was a likely source for some spruce and fir, Mann-Whitney tests showed significant differences between the medians of San Mateo $^{87}Sr/^{86}Sr$ values and the species means from construction timber in Chaco Canyon. These conflicting results highlight the limitations of frequentist methods when inferring geochemical sources. A geochemical source can neither be ruled out nor confirmed based solely on a shared mean/median with another geochemical data set, because an object can be collected from a limited portion of the source and thus have a significantly different $^{87}Sr/^{86}Sr$ value from its parent population. A comparison of means can indicate if a particular geochemical source is likely, but it cannot rule one out. However, a comparison of means can be highly useful in determining if geospatially distinct sources can be differentiated geochemically. For example, the Chuska range is distinct from the Chaco watershed as a geospatial source (Figure 1), but geochemically, they overlap almost completely (Figure 8, Table 3). The Bayesian mixing model approach illustrated in this manuscript offers a more probabilistic assessment for relative source contributions. Frequentist methods are appropriate when addressing if one geochemical source can be distinguished from another; Bayesian methods can be appropriate when assessing the probability of a given geochemical source explaining variation in a set of objects.

Maize data are particularly ambiguous; posterior probabilities do not substantially deviate from the priors used for source contribution, with the exception of Salmon Ruins for Post-1140 A.D. samples (Table 2). The central problem is that variation in maize $^{87}Sr/^{86}Sr$ values from all time periods overwhelms variation in geochemical sources (Figure 3). This phenomenon is interesting for two reasons. First it indicates that not all radiogenic strontium isotope variability has been documented in the San Juan Basin despite dedicated sampling and effort, a result identified by Benson [7]–[10]. Secondly, that historical maize largely matches the distribution of post-1140 A.D. maize and has a similarly large range as pre-1140 A.D. maize. As the ancestral pueblo were the primary occupants in the pre-historic period and the Navajo for the historic period [26], this raises interesting questions regarding long-distance transportation of maize. If, as argued in the past [7]–[10], maize was transported long distances to Chaco Canyon during the Great House period as based on $^{87}Sr/^{86}Sr$ values, why would historic maize follow the same pattern? Was historic maize imported from the same locations by a different ethnic group hundreds of years after Chaco's role as a ceremonial center had ended? Unfortunately, the strontium data are too ambiguous to facilitate a straightforward interpretation of provenance for existing maize samples.

The inconclusive results are not surprising because the San Juan Basin consists largely of sedimentary sandstones and shales that contribute through aeolian and alluvial processes to soil development [53]. Chaco Canyon is 32 km in length and includes an 11,000 sq. km watershed that funnels sediment into the canyon and its tributaries. The geological formations that contribute to the Chaco watershed (primarily the Mesa Verde and Meneffee formations) are widespread across the San Juan Basin (Figure 1). Consequently, it seems likely that canyon and basin soils would be relatively homogenized with respect to strontium and therefore it is difficult to differentiate from any particular source in the larger region [54] [55]. Based on available radiogenic strontium isotopes samples, Chaco Canyon and its tributaries are equally likely to represent source areas for ponderosa as far more distant parts of the San Juan Basin. Taken in isolation, radiogenic strontium isotope ratios from the Chuska Mountains region is consistent with data from Chaco Canyon, but the same statement can be made for a much broader area across the entire Colorado plateau that, in wetter conditions, could have supported more forests in and near riparian and mesic mixed conifer forests present in the basin.

The results of quantitative analysis indicate considerable ambiguity regarding geochemical sources for Chaco canyon maize and tree lumber. While some spruce and fir trees can likely be sourced to the San Mateo range, there is no clear evidence for a primary ponderosa geochemical source. A few ponderosa stands may be sourced to the San Pedro range due to a high $^{87}Sr/^{86}Sr$ value [15], but it is unlikely that the San Pedro range was a major source for timber. Like ponderosa, maize from Chaco Canyon also

Table 3. Comparison of medians across multiple potential timber sources.

	Chuska Mountains	San Mateo Mountains	Cuba Mesa	Hosta Butte	La Plata Mountains	San Pedro Mountains	Great Houses Ponderosa	Great Houses Spruce	Great Houses Fir
Chaco Watershed	−0.0002	0.0016**	−0.0049**	−0.0017**	−0.0006**	−0.0052**	−0.0003**	0.0007**	−0.0002
Chuska Mountains		0.0018**	−0.0046**	−0.0015**	−0.0003*	−0.0050**	−0.0001*	0.0009**	0.0001
San Mateo Mountains			−0.0064**	−0.0032**	−0.0021**	−0.0068**	−0.0019**	−0.0009**	−0.0017**
Cuba Mesa				0.0032*	0.0043**	−0.0003	0.0045**	0.0055**	0.0047**
Hosta Butte					0.0011	−0.0035**	0.0014*	0.0023**	0.0015**
La Plata Mountains						−0.0047**	0.0002	0.0012**	0.0004*
San Pedro Mountains							0.0049**	0.0059**	0.0051**
Great Houses Ponderosa								0.001**	0.0002**
Great Houses Spruce									−0.0008*

Significance codes: 0.001 '**' 0.01 '*'.
Items in bold indicate that no significant difference between medians was found, potentially indicating that these sources cannot be quantitatively distinguished by strontium alone.

does not have any clear geochemical sources (Figure 4a, Figure 5a–c). The data presented here do not contradict arguments that some construction elements may have originated from the San Mateo Mountains [12] [15], but there is no geochemical evidence that the Chuska Mountains were a major source for construction timbers. Similarly, the Chuska slope does not appear to be any more likely as a source for maize compared to more local alternatives (Figure 4a–c, Figure 5a, Table 2). The range of $^{87}Sr/^{86}Sr$ values for the Chuska Mountains and Chuska slope soils overlaps considerably with values of soils from the Chaco watershed (Figure 8), and there is not a significant difference between their means (Table 3). As such, isolated ponderosa stands near Chaco Canyon and a dense ponderosa forest in the Chuska Mountains will have the same strontium isotope ratio. Therefore, $^{87}Sr/^{86}Sr$ values cannot be used to distinguish between the Chuska and Chaco regions. For maize, Chuska slope soils and the Chaco watershed can be differentiated generally, but the majority of pre-1140 A.D. maize falls within the overlapping $^{87}Sr/^{86}Sr$ values that characterize both potential sources (Figure 3a,b).

Recent modeling of potential agricultural productivity in Chaco Canyon [16] [17] indicates that there was more arable land than recognized in previous studies of Bonito Phase food production. Without a clear indication from geochemical analyses that extramural sources of maize or ponderosa pine were important during the Bonito Phase, the case for long-distance transport of food and wood has not been demonstrated. Therefore there is no unambiguous evidence that the economic foundation of Chaco society was predicated on such transactions. Some high elevation wood was brought to the canyon, at least in part including the San Mateo Mountains, but it is unknown how these trees were procured or whether they were economically significant. For maize in particular, an argument for long-distance transport is tenuous due to the close correspondence of $^{87}Sr/^{86}Sr$ values of historical maize and $^{87}Sr/^{86}Sr$ values of post-1140 A.D. maize

Table 4. Mean posterior probability proportions of geochemical source contributions to observed variation in Chaco timber.

	Spruce	Fir	Ponderosa
Chuska Mountains	0.121	0.153	0.145
Chaco Watershed	0.125	0.159	0.145
Aztec Soil	0.100	0.126	0.143
San Mateo Mountains	0.475	0.319	0.168
San Pedro Mountains	0.021	0.030	0.063
La Plata Mountains	0.078	0.108	0.135
Hosta Butte	0.056	0.072	0.124
Cuba Mesa	0.025	0.033	0.076

For most potential timber geochemical sources, mean posterior proportion probabilities show little difference from prior probabilities. This indicates a high degree of uncertainty when attempting to use strontium to source timber samples. Only the San Mateo source emerges as a higher probability source for spruce (0.475) and fir (0.319).

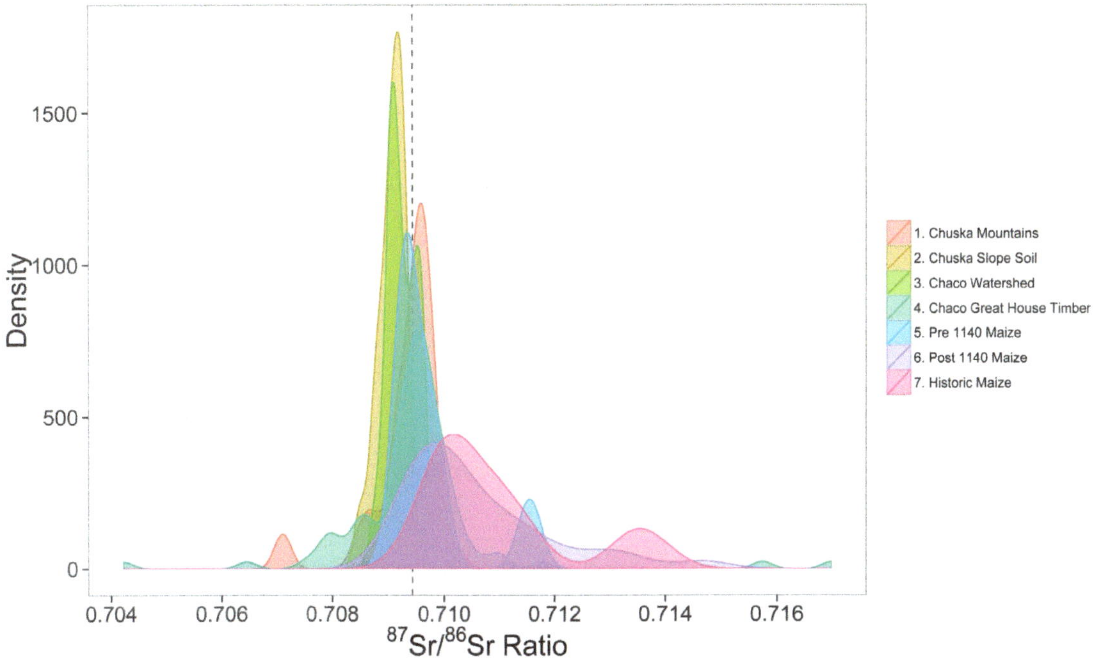

Figure 8. Correspondence of means between soils, maize, and trees in Chaco Canyon and the Chuska region. The red line represents the strontium isotope ratio for a Ponderosa tree stump found in the West Plaza of Pueblo Bonito (JPB-99). There is considerable overlap in strontium isotope ratios in the Chaco and Chuska regions.

(Figure 3a), as well as the overlap in $^{87}Sr/^{86}Sr$ values from maize from all periods have with Chaco watershed $^{87}Sr/^{86}Sr$ values (Figure 3b).

The concept of a "source" in geochemistry studies is complex and varies considerably in spatial scale, from individual specimens or samples to geographic regions, and therefore sampling designs that match data to questions are critical to the success of any sourcing enterprise [56]. In using chemical data for sourcing, an important distinction must be drawn between geochemical sources and geospatial sources [42]. Potential sources for Bonito Phase maize may lie outside the canyon and it is possible that ponderosa pine was transported long distances to Chaco. Recent analysis suggests that the northern portion of the Chaco watershed may be a maize source; this would be supported in part by the present analysis. However, while some chert artifacts do originate for the Chuska Mountains region [57], the radiogenic strontium isotope evidence does not support this region as a likely source for either maize or construction timbers. San Mateo is better supported as a support by this analysis; obsidian has been demonstrated to have come, in part, from Mt. Taylor in the San Mateo Range [58]. However, the connection between timber acquisition and lithic procurement is ambiguous. As Chaco Canyon was a source of long-distance trade, it is difficult to weight the influences of one type of artifact upon another. The present re-evaluation of radiogenic strontium isotope data does not rule out some long-distance transport of high elevation tree species, though it best supports models in which these come from the south, not the west or north. The case for long-distance acquisition in maize and ponderosa pine is ambiguous and additional research is needed to determine possible relationships between Great Houses the procurement of these costly resources.

Supporting Information

Figure S1 Geological Labels.

Figure S2 Sediment Age of Deposition.

Figure S3 Sediment Type.

Figure S4 San Juan Basin Vegetation Cover.

Figure S5 Chaco Area Vegetation Cover.

Figure S6 Chaco/Chuska Mountains Vegetation Cover.

Figure S7 Chaco/San Mateo Mountains Vegetation Cover.

Figure S8 Chaco/San Pedro Mountains Vegetation Cover.

Figure S9 Chaco/La Plata Mountains Vegetation Cover.

Figure S10 Difference between Posterior and Prior probabilities for sequentially weighted priors for the Chuska Mountains as a potential source of spruce.

Table S1 Shapiro-Wilkes Test for normality of sources.

Table S2 Shapiro-Wilkes Test for tested data.

Table S3 Posterior probabilities of maize data coerced to normality through removal of data points (Pre-1140 A.D. Maize, CHCU43684A) and full model results.

Table S4 Comparison of medians across maize by provenience (columns) and potential sources (rows). Items in bold indicate that no significant difference between medians was found. Original $^{87}Sr/^{86}Sr$ data were rounded to the 4th decimal place.

Table S5 Proportion of the 200,000 run simulations in which each maize sample (columns) fell within the mean +/− one standard deviation of each source (rows). For all maize samples, the strontium shows uniform proportions across all possible sources. This indicates a high degree of uncertainty when attempting to use strontium to source maize samples. Original $^{87}Sr/^{86}Sr$ data were rounded to the 4th decimal place.

Table S6 Posterior probabilities of source data coerced to normality through removal of data points (Chuska Mountains, CHM-214b and CHM-214d) and full model results.

Table S7 Comparison of medians across multiple potential timber sources. Items in bold indicate that no significant difference between medians was found, potentially indicating that these sources cannot be quantitatively distinguished by strontium alone. Original $^{87}Sr/^{86}Sr$ data were rounded to the 4th decimal place.

Table S8 Proportion of the 200,000 run simulations in which each timber sample (columns) fell within the mean +/− one standard deviation of each source (rows). For all timber samples, the strontium shows uniform proportions across all possible sources. This indicates a high degree of uncertainty when attempting to use strontium to source timber samples. Only the San Mateo source emerges as a higher probability source for spruce (0.469) and fir (0.320). Original $^{87}Sr/^{86}Sr$ data were rounded to the 4th decimal place.

Table S9 Maize Source Strontium Isotope Data.

Table S10 Tree Source Strontium Isotope Data.

Table S11 Maize Strontium Isotope Data.

Table S12 Chaco Timber Strontium Isotope Data.

Acknowledgments

We wish to thank Thomas C. Windes for his advice and insights about Chaco archaeology and its tree-ring record, as well as his remarkable dedication and commitment to constructing an empirical dendrochronological dataset for the canyon. We also acknowledge with gratitude the support and interest of staff at Chaco Culture National Historic Park, especially Dabney Ford, Wendy Bustard, and Barbara West. Much of this analysis was supported by the University of New Mexico and we have benefited greatly from discussions with our UNM colleagues Patricia Crown, Hannah Mattson and Jennie Sturm. This material is based in part upon work supported by the National Science Foundation Graduate Research Fellowship Program under Grant No. DGE-0903444. We would like to thank two anonymous reviewers who gave important input that greatly improved the manuscript. We would also like to thank Larry Benson, Julio Betancourt, Jay Quade, and Nathan English who provided valuable comments and feedback as part of the PLoS One review process.

Author Contributions

Conceived and designed the experiments: BLD WHW MIH WD. Performed the experiments: BLD. Analyzed the data: BLD. Contributed reagents/materials/analysis tools: WHW MIH WD. Wrote the paper: BLD WHW MIH.

References

1. Vivian GR, Van West CR, Dean JS, Akins NJ, Toll MS, et al. (2006) Ecology and economy. In: The Archaeology of Chaco Canyon: An Eleventh-Century Pueblo Regional Center (ed. Lekson SH) School of American Press. Santa Fe, NM. pp. 45–66.
2. Benson L, Cordell L, Vincent K, Taylor H, Stein J, et al. (2003) Ancient maize from Chacoan Great Houses: Where was it grown? Proc Natl Acad Sci 100: 13111–13115.
3. Wills WH (2000) Political leadership and the construction of Chacoan Great Houses, A.D. 1020–1140. In: Alternative Leadership Strategies in the Prehistoric Southwest (ed. Mills B). University of Arizona Press, Tuscon. pp. 19–44.
4. Toll HW (2001) Making and breaking pots in the Chaco World. Am Antiq 66: 56–78.
5. Toll HW (2006) Organization of Production. In: The Archaeology of Chaco Canyon: An Eleventh-Century Pueblo Regional Center (ed. Lekson SH) School of American Press. Santa Fe, NM. pp. 117–151.
6. Crown PL, Hurst JL (2009) Evidence of cacao use in the prehispanic American Southwest. Proc Natl Acad Sci 106: 2110–2113.
7. Benson LV, Stein JR, Taylor HE (2009) Possible sources of archaeological maize found in Chaco Canyon and Aztec Ruin, New Mexico. J Archaeol Sci 36: 387–407.
8. Benson LV (2010) Who provided maize to Chaco Canyon after the mid-12th century drought? J of Arch Sci 37: 621–629.
9. Benson LV (2011) Factors controlling the pre-Columbian and early historic maize productivity in the American Southwest, Part 2: The Chaco Halo, Mesa Verde, Pajarito Plateau/Bandalier, and Zuni archaeological regions. J Arch Method and Theory 18(1): 61–109.
10. Benson LV (2012) Development and application of methods used to source prehistoric Southwestern maize: a review. J Arch Sci 39: 791–807.
11. Earle T (2001) Economic support of the Chaco Canyon society. Am Antiq 66: 26–35.
12. English NB, Betancourt JL, Dean JS, Quade J (2001) Strontium isotopes reveal distant sources of architectural timber in Chaco Canyon, New Mexico. Proc Natl Acad Sci 98: 11891–11896.
13. Grissino-Mayer HD (1996) A 2129 year annual reconstruction of precipitation for northwestern New Mexico, USA. In: Dean JS, Meko DM, Swetnam TW, editors. Tree Rings, Environment, and Humanity. Radiocarbon 1996, Department of Geosciences, The University of Arizona, Tucson: 191–204.
14. Windes TC, Ford D (1996) The Chaco Wood Project: The chronometric reappraisal of Pueblo Bonito. Am Antiq 61(2): 295–310.
15. Reynolds AC, Betancourt JL, Quade J, Patchett PJ, Dean JS, et al. (2005) 87Sr/86Sr sourcing of ponderosa pine used at Anasazi great house construction in Chaco Canyon, New Mexico. J of Arch Sci 32: 1061–1075.
16. Dorshow WB (2012) Modeling agricultural potential in Chaco Canyon during the Bonito phase: A predictive geospatial approach. J Arch Sci 39(7): 2098–2115.
17. Dorshow WB, Wills WH (2012) Agriculture and community in Chaco Canyon: Revisiting Pueblo Alto. J Arch Sci 31(2): 138–155.
18. Windes TC, McKenna PJ (2001) Going against the grain: wood production in Chacoan society. Am Antiq 66: 119–140.
19. Lekson SH, Gillespie WB, Windes TC (1984) Great Pueblo Architecture of Chaco Canyon. Publications in Archaeology 18B, National Park Service, US Department of the Interior. Albuquerque, NM.
20. Van Dyke RM (2007) The Chaco Experience: Landscape and Ideology at the Center Place. Santa Fe: School for Advanced Research Press. 344 p.
21. Lekson SH, Windes TC, Stein JR, Judge WJ (1988) The Chaco Canyon community. Scientific American 256: 100–109.
22. Neitzel JE (2003) The organization, function, and population of Pueblo Bonito. In: Neitzel JE, editor. Pueblo Bonito: Center of the Chacoan World. Washington, D.C: Smithsonian Institution Press. pp. 143–150.

23. Grebinger P (1973) Prehistoric social organization in Chaco Canyon, New Mexico: an alternative reconstruction. Kiva 39: 3–23.
24. Sebastian L (1992) The Chaco Anasazi: Sociopolitical Evolution in the Prehistoric Southwest. Cambridge: Cambridge University Press. 196 p.
25. Kantner J (1996) Political competition among the Chaco Anasazi of the American Southwest. J of Anth Arch 15: 41–105.
26. Kantner J, Kintigh KW (2007) The Chaco World. In: Lekson SH, editor. The Archaeology of Chaco Canyon: An Eleventh-Century Pueblo Regional Center. Santa Fe: School of American Research Press. pp. 153–188.
27. Dean JS, Warren RL (1983) Dendrochronology. In: Lekson SH, editor. The Architecture and Dendrochronology of Chetro Ketl. Albuquerque, New Mexico: National Park Service. pp. 105–240.
28. Windes TC, Ford D (1996) The Chaco wood project: the chronometric reappraisal of Pueblo Bonito. Am Antiq 61: 295–310.
29. Kirby S, Pohs K (2004) Southwest Regional GAP Analysis Project Final Report. USGS GAP Analysis Program. Available: http://earth.gis.usu.edu/swgap/geology.html. Accessed 2014 May 5.
30. Judd NM (1954) The Material Culture of Pueblo Bonito. In: Smithsonian Miscellaneous Collections 147(1). Washington, D.C: Smithsonian Institution.
31. Hall SA (1977) Late Quaternary sedimentation and paleoecological history of Chaco Canyon, New Mexico. Geological Society of America Bulletin 88: 1593–1618.
32. Hall SA (1988) Prehistoric vegetation and environment at Chaco Canyon. American Antiquity 53(3): 582–292.
33. Betancourt JL, Van Devender TR (1981) Holocene vegetation in Chaco Canyon, New Mexico. Science 214(4521): 656–658.
34. Betancourt JL, Dean JS, Hull HM (1986) Prehistoric long-distance transport of construction beams, Chaco Canyon, New Mexico. Am Antiq 51: 370–375.
35. Durand SR, Shelley PH, Antweiler RC, Taylor HE (1999) Trees, chemistry, and prehistory in the American Southwest. J of Archaeol Sci 26: 185–203.
36. Bentley RA (2006) Strontium isotopes from the earth to the archeological skeleton: a review. J Arch Method and Theory 13: 135–187.
37. Beard BL, Johnson CM (2000) Strontium isotope composition of skeletal material can determine the birth place and geographic mobility of humans and animals. Journal of Forensic Science 45(5): 1049–1061.
38. Hodel DA, Quinn RL, Brenner M, Kamenov G (2004) Spatial variation of strontium isotopes in the Maya region: A tool for tracking ancient human migration. J Archeol Sci 31: 585–601.
39. Poszwa A, Dambrine E, Ferry B, Pollier A, Loubet M (2002) Do deep tree roots provide nutrients to the tropical rainforest? Biogeochemistry 60: 97–118.
40. Sillen A, Hall G, Richardson S, Armstrong R (1998) 87Sr/86Sr ratios in modern and fossil food-webs of the Sterkfontein Valley: Implications for early hominid habitat preference. Geochim Cosmochim Acta 62: 2463–2478.
41. United States Geological Survey National Gap Analysis Program (2004) *Provisional Digital Land Cover Map for the Southwestern United States*. Version 1.0. RS/GIS Laboratory, College of Natural Resources, Utah State University.
42. Hughs RE (1998) On reliability, validity, and scale in obsidian sourcing research. In: Unit Issues in Archaeology: Measuring Time, Space, and Material (eds. Ramenofsky AF, Steffen A). University of Utah Press, Provo, UT pp. 103–114.
43. Wickham H (2009) ggplot2: Elegant graphics for data analysis. Springer: New York.
44. Auguie B (2012) gridExtra: Functions in Grid graphics. R version 0.9.1. Available: http://CRAN.R-project.org/package = gridExtra. Accessed 2014 May 5.
45. Erhardt EB, Bedrick EJ (2013) A Bayesian framework for stable isotope mixing models. Environ Ecol Stat 20: 377–397.
46. Phillips DL (2012) Converting isotope values to diet composition: the use of mixing models. Journal of Mammalogy 93(2): 342–352.
47. Inger R, Bearhop S (2008) Applications of stable isotope analyses to avian ecology. Ibis, 150(3): 447–461.
48. Phillips DL, Gregg JW (2001) Uncertainty in source partitioning using stable isotopes. Oecologia, 127(2): 171–179.
49. Parnell AC, Inger R, Bearhop S, Jackson AL (2010) Source partitioning using stable isotopes: coping with too much variation. PLoS One 5(3): e9672.
50. Newsome SD, Martinez del Rio C, Bearhop S, Phillips DL (2007) A niche for isotopic ecology. Frontiers in Ecology and the Environment 5(8): 429–436.
51. Moore JW, Semmens BX (2008) Incorporating uncertainty and prior information into stable isotope mixing models. Ecology Letters, 11(5): 470–480.
52. Parnell A, Jackson A (2013) SIAR: Stable isotope analysis in R. R package version 4.2. http://CRAN.R-project.org/package = siar. Accessed 2014 May 5.
53. Love DW (1983) Quaternary facies in Chaco Canyon and their implications for geomorphic-sedimentologic models. In: Wells SG, Love DW, Gardner TW, editors. Chaco Canyon Country. American Geomorphological Field Group. pp. 195–206.
54. Graustein WC (1989) $^{87}Sr/^{86}Sr$ Ratios Measure the Sources and Flow of Strontium in Terrestrial Ecosystems. Ecological Studies 68: 491–512.
55. Aubert D, Probst A, Stille P, Viville D (2002) Evidence of hydrological control of Sr behavior in stream water (Strengbach catchment, Vosges mountains, France). Appl Geochem 17: 285–300.
56. Montgomery J, Evans JA, Cooper RE (2005) Resolving archaeological populations with strontium mixing diagrams. Geophysical Research Abstracts, Vol. 7, 10020.
57. Cameron CM (2001) Pink chert, projectile points, and the Chacoan regional system. Am Antiq 66: 79–102.
58. Duff A, Moss JMM, Windes TC, Kantner J, Shackley MS (2012). Patterns in procurement of obsidian in Chaco Canyon and in Chaco-era communities in New Mexico as revealed by X-ray fluorescence. J of Archaeol Sci 39: 2995–3007.

2

Intensity of Territorial Marking Predicts Wolf Reproduction: Implications for Wolf Monitoring

Luis Llaneza[1]*, **Emilio J. García**[1], **José Vicente López-Bao**[2,3]*

1 A.RE.NA. Asesores en Recursos Naturales, S.L., Lugo, Spain, 2 Research Unit of Biodiversity (UO/CSIC/PA), Oviedo University, Mieres, Spain, 3 Grimsö Wildlife Research Station, Swedish University of Agricultural Sciences (SLU), Riddarhyttan, Sweden

Abstract

Background: The implementation of intensive and complex approaches to monitor large carnivores is resource demanding, restricted to endangered species, small populations, or small distribution ranges. Wolf monitoring over large spatial scales is difficult, but the management of such contentious species requires regular estimations of abundance to guide decision-makers. The integration of wolf marking behaviour with simple sign counts may offer a cost-effective alternative to monitor the status of wolf populations over large spatial scales.

Methodology/Principal Findings: We used a multi-sampling approach, based on the collection of visual and scent wolf marks (faeces and ground scratching) and the assessment of wolf reproduction using howling and observation points, to test whether the intensity of marking behaviour around the pup-rearing period (summer-autumn) could reflect wolf reproduction. Between 1994 and 2007 we collected 1,964 wolf marks in a total of 1,877 km surveyed and we searched for the pups' presence (1,497 howling and 307 observations points) in 42 sampling sites with a regular presence of wolves (120 sampling sites/year). The number of wolf marks was ca. 3 times higher in sites with a confirmed presence of pups (20.3 vs. 7.2 marks). We found a significant relationship between the number of wolf marks (mean and maximum relative abundance index) and the probability of wolf reproduction.

Conclusions/Significance: This research establishes a real-time relationship between the intensity of wolf marking behaviour and wolf reproduction. We suggest a conservative cutting point of 0.60 for the probability of wolf reproduction to monitor wolves on a regional scale combined with the use of the mean relative abundance index of wolf marks in a given area. We show how the integration of wolf behaviour with simple sampling procedures permit rapid, real-time, and cost-effective assessments of the breeding status of wolf packs with substantial implications to monitor wolves at large spatial scales.

Editor: Cédric Sueur, Institut Pluridisciplinaire Hubert Curien, France

Funding: Regional wolf surveys were funded by the Regional Government of Asturias, the Spanish Environmental Ministry and the Regional Government of Galicia. The funders had no role in study design, data collection and analysis, decision to publish, or preparation of the manuscript.

Competing Interests: The authors have declared that no competing interests exist.

* E-mail: jv.lopezbao@gmail.com

These authors contributed equally to this work.

Introduction

Large carnivores are actively managed in an ebb and flow between human interests and conservation goals. The relative weight of social, economic, conservation or ecological factors [1–6] tip the balance towards either side continuously, turning the management of large carnivores into one of the most complex tasks within wildlife management. To achieve an adaptive management approach [7], decision-makers require reliable and updated estimates of their status to establish, for instance, trends, conservation actions, sustainable hunting quotas or population controls [3,8,9]. Independent of the target, primary management step is the ability to obtain objective, reliable and economically affordable abundance estimates that are easily reproducible over time.

However, despite the fact that the wide range of methods available to survey these elusive species has [10,11], significantly increased due to promising advances in non-invasive molecular tools [10], camera trapping [12], and statistical procedures [13,14], the implementation of complex approaches over large spatial scales (i.e. regions or countries) or large populations of these species is resource demanding. Consequently, these methods are still commonly restricted to threatened species, small populations or small distribution ranges [14–16]. Because resources for wildlife management are limited [17] (especially deteriorated under the current global financial crisis) and budgets are generally dependent on the conservation status of species [18], the availability of handsome funds to survey large carnivores over large spatial scales is not always guaranteed. This uncertainty is especially important if species or populations are not threatened or are delisted from any protective conservation status [19]. But, the management of contentious species such as large canids, bears or felids must require regular estimations of their abundance regardless of their conservation status in order to guide decision-makers, especially for those populations living in human-dominated landscapes [20].

As a consequence of multiple conflicts surrounding wolves (*Canis lupus*), a development of simple and efficient tools to regularly assess the status of populations is a pressing need. Methods to survey wolf populations involve sign surveys, howling, radio tracking, questionnaires, or statistics of livestock damages and hunting bags [21–23]. More recently, the use of non-invasive samples became popular [24–26]. In Europe, wolf management relies on an estimated number of individuals or packs, with the former estimation being less accurate as population size or range increases, and the quality of both wolf estimates varies widely across countries and within populations (see http://www.kora.ch/sp-ois/wolf-ois) [27]. Thus, non-invasive samples or intensive radio-tracking are usually implemented over small spatial scales or small populations (<500 individuals) [16,24,26,28], whereas questionnaires or hunting statistics are commonly used over vast areas and large populations, increasing uncertainty (see http://www.kora.ch/sp-ois/wolf-ois) [27]. On the other hand, small populations are annually surveyed (e.g. Scandinavian or the Western Alps wolf populations) [16,28], whereas a lack of temporal continuity or partial assessment covering small population fractions (usually corresponding to different administrative limits such as counties or national parks) is normal in large populations. This is the case for the NW Iberian wolf population – the main wolf population in Western Europe [29] - where the last comprehensive view of the status of the whole population (as a result of several Spanish regional surveys and a Portuguese national census) was ten years ago (estimated population size between 1999 and 2003 of ca. 320 packs or ca. 2000–3000 individuals) [30], and only some regional surveys are regularly implemented (e.g. county of Asturias).

Wolf monitoring over large spatial scales is difficult. However, the integration of wolf marking behaviour with simple indexes resulting from sign counts (e.g. visual and scent marks such as faeces) may offer a good alternative to monitor wolves over large areas. Sign counts have been widely used as indexes of relative abundance of large carnivores. Government agencies have implemented them as a relative low-cost approach to gathering data at large scales using rangers, technical staff or volunteers [10,21]. Although there have been attempts to correlate these indexes with densities [31], few efforts have been carried out in order to link sign counts with demographic parameters such as breeding status. The most popular attempt is the use of snow-tracking to infer reproduction [28,32]; but it is a time-delayed method dependent on long periods of good snow conditions the next winter after the breeding period, absent, for instance, in most of the southern wolf's European range.

Territoriality in wolves is indicated by the means of visual and scent marks such as faeces, urine, ground scratching or anal gland segregations [32–36]. Pack members accumulate these marks, for instance, in the edges of their territories or the surroundings of the *rendezvous sites* [32–36]. As a general rule, only mature, dominant wolves, breeds every year within the pack [32], and they show an intense territorial marking behaviour compared to other members of the pack [27,37–39]. Thus, we hypothesised that as pack members, particularly the breeding pair, use visual and scent marks to indicate possession and defence of the pack's territory, high levels of marking behaviour around the breeding season in an area could reflect the existence of an organised pack and a successful breeding pair. Since a biological meaningful units to monitor wolf populations are the pack and the number of wolf reproductions, then high abundance of wolf marks around this period could be used as an indicator of successful reproduction (i.e. the presence of pups) [39] and may offer a cost-effective tool to monitor wolves (i.e. number of wolf reproductions). However, to date, no studies have attempted to establish a real-time link between wolf marking behaviour and breeding status in a pack [39]. We discuss the application of the method proposed here for wolf monitoring over large spatial scales (i.e. countries or entire wolf populations).

Methods

Study area

The study was carried out in the Cantabrian Mountains, N Iberia (ca. 9,000 km^2; Fig. 1a), covering the rugged region of Asturias and a small mountain area within Galicia (east of Lugo province, hereafter Lugo; Fig. 1a). Vegetation is mainly comprised by scrublands, woodlands, and grasslands (pastures). Scrublands are mainly composed by *Calluna vulgaris*, *Genista* spp., *Erica* spp. and *Vaccinium* spp..Woodlands are dominated by beech (*Fagus sylvatica*), oaks (mainly *Quercus petraea* and *Quercus pyrenaica*), birch (*Betula pubescens*) and anthropogenic chesnut trees (*Castanea sativa*). Isolated trees or small groups of holly (*Ilex aquifolium*) and rowan (*Sorbus aucuparia*) often occur scattered through mature or secondary forests. Scrublands are the predominant habitat type and the level of fragmentation of forests here is high [40]. Snow is present irregularly (both seasonally and annually) between December and March.

Collection of wolf marks

We used 42 sampling sites with a regular presence of wolves during the last two decades (39 in Asturias and 3 in Lugo). Each sampling site comprised 2–5 valleys and the surrounding mountains covering approximately 150–200 km^2 (see Fig. 1b). This area was of a similar size than the average home range size reported in wolves in N Iberia [41]. All sites were regularly sampled between 1994 and 2007 around the pup-rearing period (summer-autumn) within different regional wolf surveys (from 1994 to 2004 in Asturias and between 2006 and 2007 in Lugo). Information from these surveys was comparable since the same field procedures (see details below) were implemented in all surveys throughout the study period [42]. Each sampling site was surveyed on average 3 years (range 1–6), with a total of 120 sampling sites/year (hereafter sites, note that this is the sample size that will be used for subsequent analyses). Within each site, wolf marks were searched in transects along existing paths and trails, on foot or using a vehicle (<10 km h^{-1}) covering as much selected area as possible (see Fig. 1b).

Faeces and ground scratching marks serve as visual and scent marks in wolves [34–36]. Based on their visual function facilitating their detection by field observers with different levels of expertise compared with urine –which is also dependent on snow conditions-, we considered both visual wolf marks for this study (hereafter wolf marks). Shape, size, content, smell and spatial position (i.e. distance to villages) were, in combination, diagnostic attributes to determine wolf faeces. In those cases that there were clear doubts on the identity of the species, faeces were not collected. We considered the criteria used to identify wolf faeces and the experience of field-observers reliable since a posterior trial determining wolf faeces using these diagnostic attributes and a parallel DNA analyses confirmed that 90% of faeces were correctly assigned to wolves (n = 108, *unpublished data*). On the other hand, ground scratching marks were identified on the basis of size, length, intensity (i.e. excluding those marks compatible with dogs) [36]. The presence of wolf faeces accompanying this sign was an important factor determining wolf scratching.

As random sampling is not effective to locate wolf marks [42], surveys in this mountainous region were focused on landscape

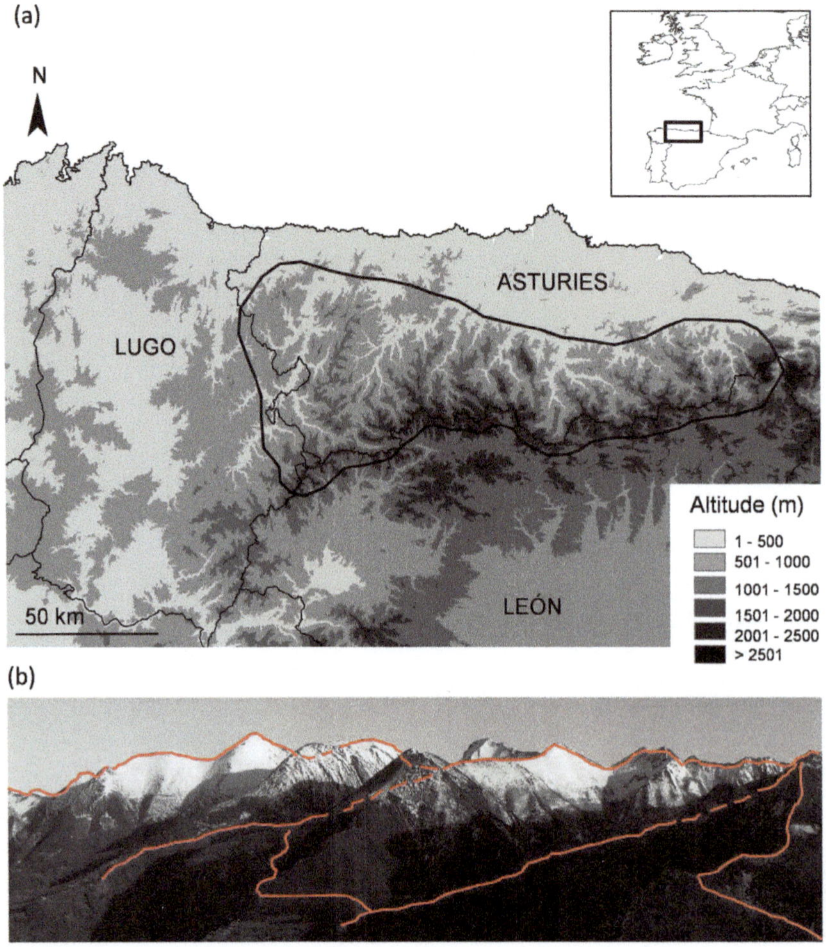

Figure 1. Study area and sampling design. (a) Map showing the study area located in the Cantabrian Mountains (N Spain). (b) Scheme showing the spatial distribution of transects within a typical sampling site.

elements often used by wolves as marking places (paths and trails, particularly focusing on junctions and mountain passes; see Fig. 1b) [34,42,43] where the probability of the detection of wolf marks by other intra and inter-pack individuals is maximized. Mountain passes may favour wolf movements between valleys in mountainous areas turning these places into important landmarks on a landscape level. One to several wildlife trails/paths usually cross mountain passes being an important place for territorial marking. For example, considering 10 transects crossing such passes in our study area, out of the 164 wolf marks found in a total of 83.5 km surveyed (an average of 8.5 km per transect), 82% of wolf marks were located in these particular conspicuous sites (ranging from 40% to 100% of wolf marks detected per transect).

The ability of wolves to deposit their marks in conspicuous sites (both on a large and fine spatial scale) [34,35,39,43] indicates a degree of control about where to locate their marks. For example, wolf faeces, whether deposited as scent marks or not, are recognized as important visual marks and a powerful sources of odour [34,35,39]. Although it is difficult to distinguish between faeces with and without (i.e. excretion) an intention of communication [35,38], pack members seems to use substrates and conspicuous places differently according to their position in the social hierarchy [38], and some faeces perhaps had other functions than territorial marking during the period of study (e.g. for marking empty food caches) [35,39], we considered that all faeces placed on the abovementioned landscape features represented some kind of marking behaviour [39]. This assumption also relies on the fact that wolf marks were deposited on conspicuous places on a landscape level that enhance them as marks [32,35,36,39,43]. This procedure decrease the probability to include faeces from young and lone wolves since they located more often their faeces off-trail [25,26,39]. Potential differences in marking behaviour among pack members [38] were not considered in this study. Furthermore, considering the posterior spatial distribution of rendezvous sites -based on the results from the howling and observation points - transects did not cross rendezvous sites, decreasing the probability to count pup faeces as wolf marks as well as avoiding the inclusion of notable aggregations of faeces in these sites as a consequence of its regular use by pack members (e.g. resting sites) [32].

Taking into account the landscape configuration and the availability of paths and trails, 1 to 9 transects were sampled within each site with an average number of 4 transects per site (median = 4), varying in length from 1 to 18 km (average length per transect = 4.1 km, $n_{transects} = 463$). All transects were conducted once around the pup rearing period. Seventy-seven percent of transects were surveyed between August and September, when all individuals of the pack are relatively stable around rendezvous sites [32]. The remaining surveys were carried out between

October and early November. The number of wolf marks by transect was noted.

Reproductive success

Two complementary methods, simulated howling and direct observation points, were used to assess wolf reproduction (i.e. presence of pups in the rendezvous sites). First, human-simulated howling elicited the response of adults and pups in the rendezvous site [44]. In each howling point, the observer emitted between 3 to 5 howls, each separated between 5 and 8 s long breaks [44]. If wolves did not reply within 2 minutes, then the observer repeated this process 2–3 times. The selection of the howling points was based on the landscape configuration (e.g. distribution of potential rendezvous sites based on the availability of refuge and areas with low human activity) [45], the environmental conditions (selection of the best places ensuring optimal conditions to simulate howling and to hear replies), and information gathered previously in the collection of wolf marks [42]. Howling sessions started at sunset and spanned the early night-time hours, avoiding rainy or windy nights, and were carried out between August and October when reply rate is remarkable [44] (occasionally early November). Second, observation points were used to contact pups in the rendezvous sites. The observer used 8× or 10× binoculars and telescopes with 20–60× zoom lenses to scan potential rendezvous sites and the surrounding areas during at least one hour. Observation points were carried out at sunrise and sunset in the same period of howling sessions.

We only considered reproduction when pups responded positively to howling or when they were observed at the rendezvous site or its surroundings. Sampling effort was nearly constant among sites with four days invested per site; which overall, accounted for 480 days along the study period.

Ethics statement

In Spain, wolves north of river Duero are in Annex V of the European Habitats Directive (92/43/EEC); being a game species in Galicia and a species with a special regime (no game species) in Asturias. Fieldwork procedures (collection of wolf marks and the methods used to confirm the presence of pups) were specifically approved by the Regional Governments of Asturias and Galicia as well as the Spanish Ministry of Environment (Picos de Europa National Park) as a part of regional and local wolf monitoring activities. We did not handle any individual.

Data analyses

For each site, we considered two measures of relative abundance of wolf marks. First, we calculated a mean relative index of abundance of wolf marks as the ratio between the total number of marks found and the sum of the length of all surveyed transects (in kilometres). Second, we identified the transect with the highest value of relative index of abundance (i.e. the ratio between the marks found and the length for each transect) and we considered this value as the maximum relative index of abundance per site. Moreover, we categorised all sites in a binary variable, according to the confirmed presence of pups.

Differences in sampling effort among sites could affect either the value of the relative indexes of abundance used or the probability to detect wolf reproduction, which could ultimately affect our results. Therefore, we first tested whether the sampling effort was different between sites with and without confirmed wolf reproduction by comparing transect length as well as the number of howling and observation points using Mann-Whitney U tests. While the former comparison gives us an idea about the homogeneity in the sampling effort to find wolf marks irrespective of the presence of pups, the two other comparisons inform how invested the effort in detecting pups was. We assumed higher values of howling points in areas without confirmed wolf reproduction would reduce the probability of false absences of pups.

We then evaluated the relationship between wolf reproduction (i.e. the presence of pups) and the mean and the maximum relative indexes of abundance of wolf marks (predictors). Since the same site was sampled several years, we built generalised linear mixed models (GLMMs) with binomial error distribution and logit link. We included site and year as random factors in the models. Although sampling effort per site (total transect length) did not differ between sites with and without a confirmed presence of pups or between months, its variance was considerable (72 km), which could potentially affect the studied relationships. Therefore, we included in the models a covariate, the total transect length, in order to statistically control their potential effect. Additionally, we included a second covariate in the models (month) to account for temporal differences in wolf mark detectability and abundance. Finally, as total transect length per site positively correlated with the number of transects (Spearman coefficient correlation, $r_s = 0.845$, $P<0.0001$) and the number of transects per site could influence the maximum relative index of abundance (the probability to find higher indexes increase with the number of transects), we included this factor as another covariate in the model with the maximum relative index of abundance. For each model, we estimated the marginal and the conditional R^2 following Nakagawa and Schielzeth (2013) [46]. Marginal R^2 represents the variance explained by fixed factors (relative indexes of abundance of wolf marks, total transect length, month, number of transects per site) whereas Conditional R^2 is interpreted as variance explained by both fixed and random factors. We predicted the probability of wolf reproduction based on the intensity of wolf marking behaviour and determined the cutting points for both relative abundance indexes when the probability of wolf reproduction was higher than 0.60, 0.80 and 0.99 (these values were selected according to their informative and conservative value from a management point of view). All GLMMs were fitted with SAS 9.2 (procedure GLIMMIX) [47].

Results

A total of 1,877 km were surveyed in 463 transects with an average length (± s.d.) of transects of 15.6±8.5 km per site. Overall, 1,964 wolf marks were found and 1,497 howling (mean number per site = 12.5, range 1–43) and 307 observations points (mean number per site = 2.5, range 1–18) were used to confirm the presence of pups. Out of the 120 sites surveyed, wolf reproduction was confirmed in 84 cases (70%).

Comparing sites with and without reproduction, average total transect length was similar, 14.8 km and 17.7 km respectively, and we did not detect significant differences between groups (Mann-Whitney U test, $P=0.151$, $n = 120$), indicating homogeneity in the sampling effort invested among sites in relation with the number of kilometers surveyed. However, the number of howling and observation points was different between groups. Whereas the mean number of howling points was significantly higher in sites without a confirmed reproduction (16 *vs.* 11; Mann-Whitney U test, $P=0.005$, $n = 120$); the number of observation points showed the opposite pattern (2 *vs.* 3; Mann-Whitney U test, $P=0.029$, $n = 120$). Although observation points were made in all sites regardless of the confirmation of reproduction, they particularly confirmed wolf reproduction in those sites where howling sessions were positive.

The mean number of wolf marks was ca. 3 times higher in sites with a confirmed presence of pups (20.3 *vs.* 7.2 marks in sites with and without confirmed reproduction, respectively) with the mean relative index of abundance lower in sites without confirmed reproduction (0.43; n = 36) compared to sites with confirmed reproduction (1.61; n = 84; Fig. 2). The same pattern was observed for the maximum relative index of abundance (1.19 *vs.* 3.16 respectively; Fig. 2). We found a significant relationship between the mean and the maximum relative abundance index and the probability of wolf reproduction ($\chi2^2 = 17.52$, d.f. = 1, $P<0.0001$ and $\chi2 = 14.55$, d.f. = 1, $P<0.001$, respectively; Table 1; Fig. 3) when month, total transect length and number of transects were controlled for (all covariates with a $P>0.285$; Table 1). The model using the mean relative abundance index of wolf marks showed a marginal R^2 of 0.79 and a conditional R^2 of 0.84, whereas in the case of the model considering the maximum relative abundance of wolf marks, marginal R^2 was 0.63 and conditional R^2 was 0.77. These results indicate that the explained variance by the random factors (pack and year) was small.

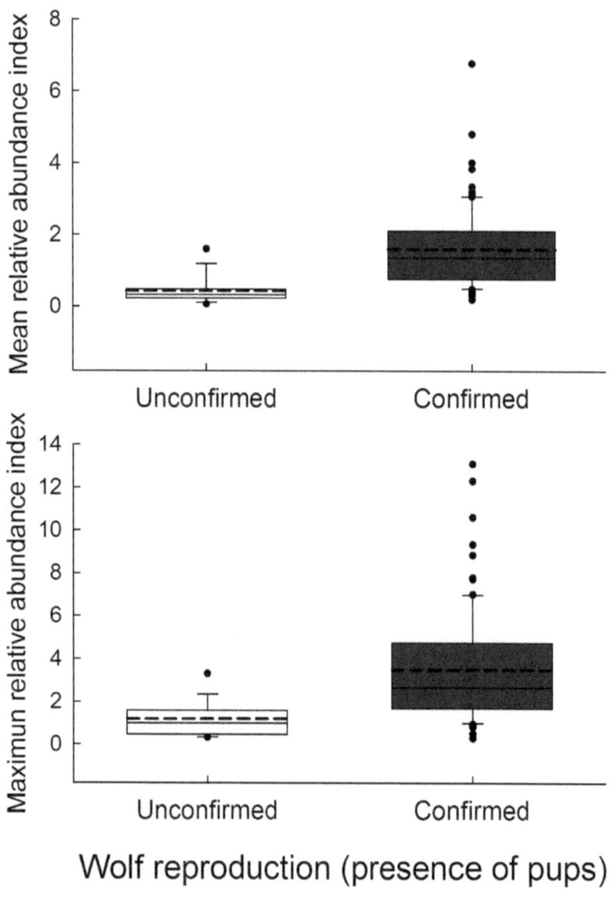

Figure 2. Box-plot showing mean and maximum relative abundance indexes of wolf marks in sites with and without confirmed wolf reproduction (presence of pups). Grey box indicates sites with confirmed wolf reproduction, whereas white box refers to sites without confirmed wolf reproduction. The continuous line within the box indicate the median; whereas the discontinuous line shows the mean value. The lower end of the box indicate the 25th percentile, the upper end of the box indicate the 75th percentile, and the error bars shows the 10th and 90th percentiles. Points denote outliers.

Our models predicted a probability of wolf reproduction higher than 0.60, 0.80 and 0.99 for a mean relative abundance index of 0.65, 0.93 and 1.84 wolf marks per km respectively; and a maximum relative abundance index of 0.71, 1.45 and 3.88 marks per km, respectively (Fig. 3). Predicted probabilities of pup presence by both indexes were positively correlated (Spearman coefficient correlation, rs = 0.843, $P<0.0001$). In fact, considering the three established cutting points and focusing on the mean relative abundance index, we observed 100% of agreement with the maximum relative abundance index in determining wolf reproduction for a probability of 0.60, 97% for a probability of 0.80, and 96% for a probability of 0.99. The proportion of false positives increased towards lower values of cutting points for the probability of pup presence (Table 2). Within each selected cutting point, this proportion was always higher for the maximum relative abundance index (Table 2), although differences were not significant (Z-test, all *P-values*>0.100; Table 2).

Discussion

Wolf marking behaviour and sign counts

Individual asymmetries in territorial marking behaviour can influence indices of abundance based on marks [25,26,48,49]. Thus, understanding the link between different marking behaviours and the abundance of marks is the only way to efficiently interpret monitoring procedures based on marks (i.e. sign counts). Territoriality in carnivorous mammals is commonly indicated by using scent or visual marking, such as urine, faeces or ground scratching, playing these marks an important role in territory maintenance and defence, indication of breeding status, demarcation of valuable places such as rich-food patches or home sites and intra- or inter-specific communication [34–36,50]. Reproductively active individuals - dominants in social species - assume a high cost in territorial marking behaviour compared to subordinates [38,51].

In the case of wolves, an assumption that only the mature pair breeds [32], translates into a higher territorial marking intensity by the breeding individuals [26,38,45], especially in conspicuous sites, to increase the probability of the mark's detection by conspecifics. On the contrary, lone wolves or young individuals tend to deposit less marks in these conspicuous sites compared to breeding wolves and other pack members [25,32,39]. In the case of lone wolves this behaviour may be particular important to decrease the probability of an encounter with territory holders. Higher levels of sex hormones detected in the wolf faeces located in particular conspicuous sites such as junctions supports these behavioural patterns [38]. As predicted, high levels of wolf marking behaviour around breeding season (pup rearing period) will reflect the existence of a successful breeding pair and the presence of pups. The observation that the number of wolf marks was three times higher in sites with a confirmed presence of pups compared with sites where wolves were present but pups were not confirmed is worth mentioning. Because pups are mainly stationary around rendezvous sites in the first months of life, they can not be responsible for a higher number of faeces in these areas. On the other hand, an absence of temporal differences in the intensity of territorial marking (we only detected a non-significant decrease in November, see Table 1) suggests constancy in this behaviour during this period.

Determining wolf reproduction

Simple mark/sign counts provide valuable information about presence, distribution or relative abundance of large carnivores [10]. However, gathering data on demographic parameters

Table 1. Parameter estimates (± s.e.) in the models testing the relationship between the mean and the maximum relative abundance of wolf marks (marks km^{-1}) and wolf reproduction.

		Mean relative abundance index	Maximum relative abundance index
Model-effect		**Parameter estimate (± s.e.)**	**Parameter estimate (± s.e.)**
Intercept		−1.88±1.14	−0.51±1.17
Relative abundance of wolf marks		3.53±0.84*	1.32±0.34*
Sampling effort (km)		−0.01±0.04	0.05±0.07
Number of transects		-	−0.21±0.30
Month	*September*	0.51±0.84	0.99±0.87
	October	0.39±1.07	0.79±1.05
	November	−1.48±1.27	−0.77±1.21

The level "month (August)" is included in the intercept.
** Significant at $P<0.001$.*
Note that the covariate "number of transects" was not included in the model with the mean relative abundance index (see text for details).

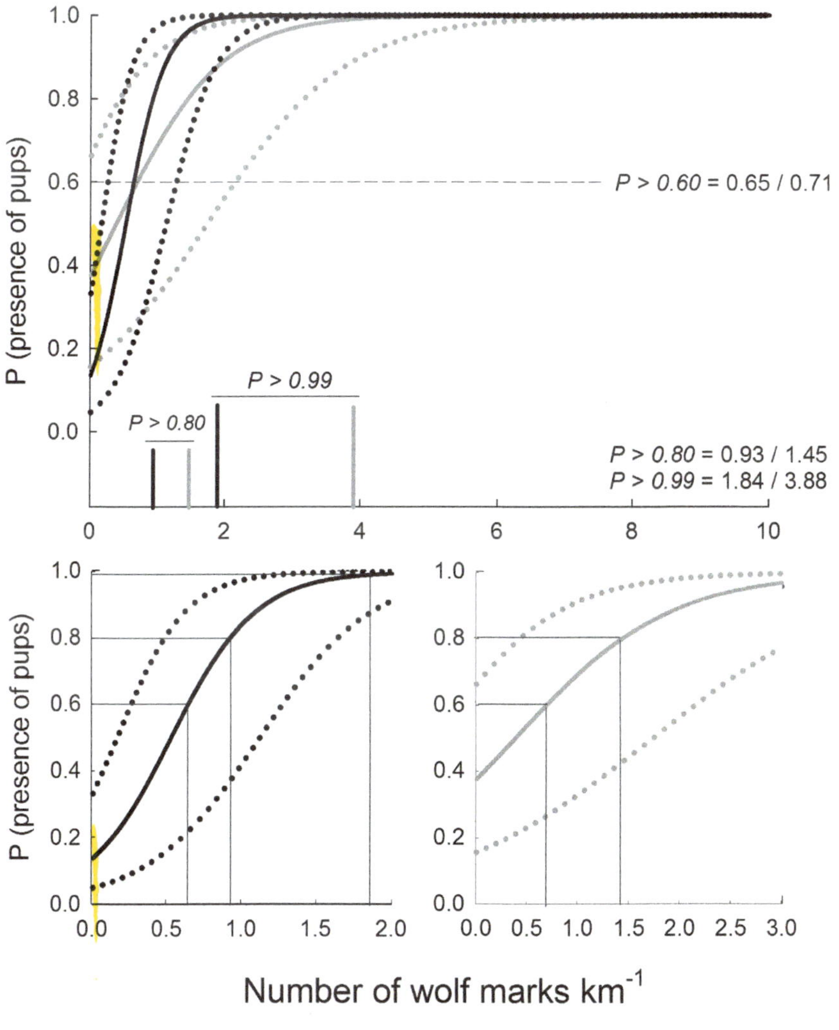

Figure 3. Predicted probability of presence of pups against the mean and the maximum relative abundance index of wolf marks. Continuous black and grey line refers to the mean and the maximum relative abundance index of wolf marks, respectively (± s.e., dotted lines). Striped line show the cutting point of 0.60; whereas the vertical black and grey bars show the cutting points of 0.80 and 0.99 for the mean and the maximum relative abundance index, respectively. Numbers for these cutting points refers to the number of wolf marks km^{-1} for the mean (left) and the maximum (right) relative abundance index. For clarity, we also show a zoom (bottom) on these predicted probabilities, with lines denoting the correspondence between the established cutting points and the values of the relative abundance indexes, for the first range of values of the number of wolf marks km^{-1}.

Table 2. Proportion of false positives (Type I error) using the 0.60, 0.80 and 0.99 cutting points for the predicted probability of wolf reproduction based on the mean and the maximum relative abundance index.

	Mean relative abundance index	Maximum relative abundance index
Cutting point	Proportion of false positives (number of total cases)	Proportion of false positives (number of total cases)
0.60	0.08 (75)	0.18 (101)
0.80	0.07 (61)	0.14 (80)
0.99	0.00 (26)	0.04 (22)

generally required other intensive, invasive, and expensive approaches (e.g. radiotracking or DNA analyses), often feasible for research-oriented efforts applied over small areas or small populations. Excluding snow-tracking – a time-delayed approach used to determine the occurrence of reproduction events based on the number of individuals detected (tracks) in the following winter - which shows several logistical constraints associated to snow conditions, spatial scale (high cost and significant field effort needed) and, sometimes, interpretation [28], to our knowledge, this is the first time that it is established a real-time relationship between intensity of territorial marking and the probability of wolf reproduction.

The level of confidence determining the probability of wolf reproduction in a given area could be adapted according to different management goals and population scenarios. For example, we recommend a reliable level in small populations (cutting points of 0.80 or 0.99) decreasing the probability of false positives (type I error, see Table 2), whereas this probability could be conservative (0.60) at large populations. The mean relative abundance index showed a low probability of false positive (below 0.10 for all cutting points); being zero for the cutting point of 0.99. Thus, as a general rule, we suggest a cutting point of 0.60 in the probability of wolf reproduction to monitor wolves on a regional scale and the use of the mean instead of the maximum relative abundance index of wolf marks. This option will reduce the influence of short transects with a high number of wolf marks or a small number of transects per site on the probability of wolf reproduction. By combining this method with different spatial criteria, such as the distance among packs in different scenarios (saturated, non-saturated or recovering areas) or simulated wolf territories, it would be possible to improve the estimates on the number of wolf packs. In addition, we highlight that a similar approach based on wolf marking behaviour could determine the probability of the occurrence of non-breeding packs. This application would require the understanding of differences in the intensity of territorial marking in areas with lonely or floaters individuals and areas with established non-breeding pairs or family groups.

Although our analyses were controlled for mark detectability, we acknowledge that prior to generalising the use of this method, it is necessary to develop similar trials in other landscape configurations and wolf populations evaluating quantitative changes in the values of the relative abundance indexes in established cutting points for the probability of wolf reproduction (0.60, 0.80 and 0.99). For example, several studies suggest that wolves use linear infrastructures to place a significant proportion of their territorial marks [31,32,34,42,43], but how landscape configuration influences this marking behaviour is unknown. Another important concern in the use of this method would be the feasibility in the identification of wolf marks. However, features described above may be efficiently used as diagnostic attributes, along with the training of observers, resulting in high correct assignment rates [52].

On the other hand, even although we used a multiple sampling approach to determine reproduction events, it is well-known that both methods are not completely efficient [53,54], and the existence of an important number of false absences could affect our results. However, in our case, sampling effort was higher in those sites where pups were not detected, decreasing the probability of false absences of wolf reproduction. In fact, for both relative abundance indexes, the 25th percentiles of sites with confirmed wolf reproduction did not overlap or showed a low overlapping with the 75th and 90th percentiles, respectively, of sites with unconfirmed wolf reproduction (Fig. 2), suggesting that the number of false absences was small. Finally, since the probability of wolf reproduction was independent of a sampling effort, indicating the importance of a stratified sampling to monitor wolves based on wolf marking behaviour, more studies are needed to optimize sampling design (e.g. minimum transect length per site). Moreover, because pack members seems to use substrates and conspicuous places differently for marking behaviour [38], the influence of different levels of territorial marking, according to individual attributes and the conspicuous sites considered, on the studied relationships deserves further investigation.

Our findings would objectively reduce the level of uncertainty in the estimation of the number of packs, particularly when they are not confirmed with other procedures [53]. For example, many monitoring programmes simulate howls to locate pups [22,23], but this method also has drawbacks in relation with the subjectivity of the observer determining the participation of pups in a chorus howl and the fact that wolves occasionally do not respond, as well as its logistical constraints when it is applied over large scales [54]. For wolf surveys covering large areas, it has been suggested that the effectiveness of simulated howling could be improved by selecting howling sites in those places with the largest concentrations of wolf marks during the breeding period [42]. In other words, selecting sites where the probability of wolf reproduction is high based on the intensity of territorial marking as this study shows. In wolf surveys based on howling, this method could add an objective approach to determine the probability of wolf reproduction when high relative abundance of wolf marks are found in a given area, but the presence of pups is not verified using howling.

Surveying wolves at large spatial scales

In the last years, the convenience of estimating the number of wolves or packs (or the number of reproductions) has been the focus of an intense and controversial debate. Finally, wolf biologists and managers seem to agree about the robust and biological meaning of packs as a unit to monitor wolves [16], which could be easily estimated with the method proposed here. However, monitoring and management are interactive processes, and therefore, management goals for each wolf population will determine the level of accuracy required in wolf estimates (individuals or packs). For example, at small populations such as the critically endangered population located in Sierra Morena (S Spain; <50 individuals) [27,29] the appropriate level of information would estimate both the number of individuals, using intensive approaches and the number of packs. But, the former level of accuracy may be unnecessary at large populations, such as the NW Iberian wolf population, assuming higher levels of uncertainty until the population would not be below a critical value, such as a given annual number of packs, number of reproductions or effective population size (minimum number of breeding pairs). In this regard, adopting a cost effective approach, the number of packs could serve as sentinel information to shift between intensive (individuals and packs) and simple and extensive (packs) monitoring procedures and vice versa.

We propose that this procedure could be easily exported to other wolf populations and countries. Across Europe, transboundary cooperation in wolf management is a pressing need [29]. This fact requires coordination among all authorities and actions including compatibility and/or convergence in monitoring methods and target information. Since this method does not rely upon complex procedures or equipment, it can be relatively straightforward and quick to conduct by rangers, technical staff or volunteers, and easily implemented in routine samplings. The use of the number of packs may turn estimates more homogeneous in a transboundary context.

Concluding Remarks

We show how the integration of wolf marking behaviour with simple sampling procedures (sign counts) permit rapid, real-time, and cost effective assessments of the breeding status of wolf packs with substantial implications to monitor wolves over large spatial scales. The assessment of the number of packs using this approach would provide a simple way to regularly assess the status of wolf populations providing acceptable estimates of demographic parameters such as reproduction success or annual effective population sizes. This information is essential for an efficient, adaptive management framework in wolf populations, even when populations are not threatened.

Acknowledgments

We are in debt to wolf biologists Victor Sazatornil, Vicente Palacios, Alberto Fernández, Javier Naves, Andrés Ordiz, Antonio Uzal and Miguel Rico who participated in the fieldwork. We thank to the staff, ranges and technicians of these institutions for their help. JVLB was supported by the Spanish Ministry of Economy and Competitiveness. This is scientific paper no. 1 from the Iberian Wolf Research Team (IWRT).

Author Contributions

Conceived and designed the experiments: JVLB LLL. Performed the experiments: LLL EJG JVLB. Analyzed the data: JVLB. Contributed reagents/materials/analysis tools: LLL JVLB. Wrote the paper: JVLB LLL.

References

1. Johansson M, Karlsson J, Pedersen E, Flykt A (2012) Factors Governing Human Fear of Brown Bear and Wolf. *Hum Dimens Wildl* 17: 58–74.
2. Thirgood S, Woodroffe R, Rabinowitz A (2005) The impact of human-wildlife conflict on human lives and livelihoods. In: Woodroffe R, Thirgood S, Rabinowitz A, editors. People and Wildlife, Conflict or Coexistence? Cambridge University Press, Cambridge, UK. pp. 13–26.
3. Treves A (2009) Hunting for large carnivore conservation. *J Appl Ecol* 46: 1350–1356.
4. Mech LD (1995) The challenge and opportunity of recovering wolf populations. *Conserv Biol* 9: 270–278.
5. Mech LD (2012) Is science in danger of sanctifying the wolf? *Biol Cons* 150: 143–149.
6. López-Bao JV, Sazatornil V, Llaneza L, Rodríguez A (2013) Indirect effects on heathland conservation and wolf persistence of contradictory policies that threaten traditional free-ranging horse husbandry. *Conserv Lett* 6: 448–455.
7. Holling CS (1978) Adaptive environmental assessment and management. John Wiley & Sons, New York.
8. Jhala Y, Qureshi Q, Gopal R (2011) Can the abundance of tigers be assessed from their signs? *J Appl Ecol* 48: 14–24.
9. Peek J, Dale B, Hristienko H, Kantar L, Loyd KA, et al. (2012) Management of large mammalian carnivores in North America. The Wildlife Society Technical Review 12–1. The Wildlife Society, Bethesda, Maryland, USA.
10. Long RA, MacKay P, Zielinski WJ, Ray JC (2008) Noninvasive survey methods for carnivores. Island Press, Washington D.C.
11. Boitani L, Powell RA (2012) Carnivore ecology and conservation: a handbook of techniques. Oxford University Press, London, UK.
12. O'Connell AF, Nichols JD, Karanth KU (2011) Camera traps in animal ecology. Methods and analyses. Springer, Tokyo.
13. Royle JA, Nichols JD, Karanth KU, Gopalaswamy AM (2009) A hierarchical model for estimating density in camera-trap studies. *J Appl Ecol* 46: 118–127.
14. Karanth KU, Gopalaswamy AM, Kumar NS, Vaidyanathan S, Nichols JD, et al. (2011) Monitoring carnivore populations at the landscape scale: occupancy modelling of tigers from sign surveys. *J Appl Ecol* 48: 1048–1056.
15. Kendall KC, States U, Survey G, Rocky N, Science M, et al. (2009) Demography and genetic structure of a recovering grizzly bear population. *J Wildl Manage* 73: 3–17.
16. Marucco F, Boitani L (2012) Wolf population monitoring and livestock depredation preventive methods in Europe. *Hystrix* 23: 1–4.
17. Leader-Williams N, Albon SD (1988) Allocation of resources for conservation. *Nature* 336: 533–535.
18. Palomares F, Rodríguez A, Revilla E, López-Bao JV, Calzada J (2011) Assessment of the conservation efforts to prevent extinction of the Iberian lynx. *Conserv Biol* 25: 4–8.
19. Gude JA, Mitchell MS, Russell RE, Sime CE, Bangs EE, et al. (2012) Wolf population dynamics in the U.S. Northern Rocky Mountains are affected by recruitment and human-caused mortality. *J Wildl Manage* 76: 108–118.
20. Llaneza L, López-Bao JV, Sazatornil V (2012) Insights into wolf presence in human-dominated landscapes: the relative role of food availability, humans and landscape attributes. *Divers Distrib* 18: 459–469.
21. Linnell JDC, Swenson JE, Landa A, Kvam T (1998) Methods for monitoring European large carnivores – A worldwide review of relevant experience. *NINA Oppdragsmelding* 549: 1–38.
22. Blanco JC, Cortés Y (2002) Ecología, censos, percepción y evolución del lobo en España: análisis de un conflicto. SECEM, Málaga. 176 pp.
23. Kunkel K, Mack CM, Melquist WE (2005) An Assessment of Current Methods for Surveying and Monitoring Wolves. Report: 1–79. 2005. Lapwai, Idaho, Nez Perce Tribe, Lapwai.
24. Caniglia R, Fabbri E, Cubaynes S, Randi E (2012) An improved procedure to estimate wolf abundance using non-invasive genetic sampling and capture – recapture mixture models. *Conserv Genet* 13: 53–64.
25. Marucco F, Pletscher DH, Boitani L, Schwartz MK, Pilgrim KL, et al. (2009) Wolf survival and population trend using non-invasive capture–recapture techniques in the Western Alps. *J Appl Ecol* 46: 1003–1010.
26. Galaverni M, Palumbo D, Fabbri E, Caniglia R, Greco C, et al. (2011) Monitoring wolves (*Canis lupus*) by non-invasive genetics and camera trapping: a small-scale pilot study. *Eur J Wildlife Res* 58: 47–58.
27. Kaczensky P, Chapron G, von Arx M, Huber D, Andrén H, et al. (2013) Status, management and distribution of large carnivores - bear, lynx, wolf and wolverine - in Europe. Report to the EU Commission, 272 p.
28. Liberg O, Aronson Å, Sand H, Wabakken P, Maartmann E, et al. (2012) Monitoring of wolves in Scandinavia. *Hystrix* 23: 29–34.
29. Linnell JDC, Boitani L (2012) Building biological realism into wolf management policy: the development of the population approach in Europe. *Hystrix* 23: 80–91.
30. Alvares F, Barroso I, Blanco JC, Correia J, Cortes Y, et al. (2005) Wolf status and conservation in theIberianPeninsula. Frontiers of wolf recovery: Southwestern U.S. and the world. Colorado Springs, Colorado.
31. Crête M, Messier F (1986) Evaluation of indices of gray wolf (*Canis lupus*), density in hardwood-conifer forests of Southwestern Quebec. *Can Field Nat* 101: 147–152.
32. Mech LD, Boitani L (2003) Wolves: behavior, ecology, conservation. University of Chicago Press.
33. Kleiman DG (1966) Scent marking in the Canidae. *Symp Zool Soc Lond* 18: 167–177.
34. Barja I, Miguel FJD, Bárcena F (2005) Faecal marking behaviour of Iberian wolf in different zones of their territory. *Folia Zool* 54: 21–29.
35. Peters RP, Mech LD (1975) Scent-marking in Wolves. *Am Sci* 63: 628–637.
36. Zub K, Theuerkauf J, Jedrzejewski W, Jedrzejewska B, Schmidt K, et al. (2003) Wolf pack territory marking in the Bialowieza Primeval Forest (Poland). *Behaviour* 140: 635–648.

37. Scandura M, Iacolina L, Capitani C, Gazzola A, Mattioli L, et al. (2011) Fine-scale genetic structure suggests low levels of short-range gene flow in a wolf population of the Italian Apennines. *Eur J Wildlife Res* 57: 949–958.
38. Barja I, Silván G, Rosellini S, Piñeiro A, Illera MJ, et al. (2008) Quantification of sexual steroid hormones in faeces of Iberian wolf (*Canis lupus signatus*): a non-invasive sex typing method. *Repro Domest Anim* 43: 701–707.
39. Rothman RJ, Mech LD (1979) Scent-marking in lone wolves and newly formed pairs. *Anim Behav* 27: 750–760.
40. García D, Quevedo M, Obeso JR, Abajo A (2005) Fragmentation patterns and protection of montane forest in the Cantabrian range (NW Spain). *Forest Ecol Manag* 208: 29–43.
41. García E, Llaneza L, Palacios V, López-Bao JV, Sazatornil V, et al. (2012) Primeros datos sobre la ecología espacial del lobo en Galicia. Abstract-book of the III Iberian Wolf Congress, pp. 44.
42. Llaneza L, Ordiz A, Palacios V, Uzal A (2005) Monitoring wolf populations using howling points combined with sign survey transects. *W B P* 1: 108–117.
43. Vilà C, Urios V, Castroviejo J (1994) Use of faeces for scent marking in Iberian wolves (*Canis lupus*). *Can J Zool* 72: 374–377.
44. Harrington FH, Mech LD (1982) An analysis of howling response parameters useful for wolf pack censusing. *J Wildl Manage* 46: 686–693.
45. Ausband DE, Mitchell MS, Doherty K, Zager P, Mack CM, et al. (2010) Surveying predicted rendezvous sites to monitor gray wolf populations. *J Wildl Manage* 74: 1043–1049.
46. Nakagawa S, Schielzeth H (2013) A general and simple method for obtaining R^2 from Generalized Linear Mixed-effects Models. *Methods Ecol Evol* 4: 133–142.
47. Littell RC, Milliken GA, Stroup WW, Wolfinger RD, Schabenberger O (2006) SAS for Mixed Models. 2nd ed. SAS Institute Inc., Cary, NC.
48. Palomares F, Roques S, Chávez C, Silveira L, Keller C, et al. (2012) High proportion of male faeces in Jaguar populations. *PLOS ONE* 7: e52923.
49. Ralls K, Sharma S, Smith DA, Bremner-Harrison S, Cypher BL, et al. (2010) Changes in Kit Fox Defecation Patterns During the Reproductive Season: Implications for Noninvasive Surveys. *J Wildl Manage* 74: 1457–1462.
50. Gorman ML (1990) Scent-marking strategies in mammals. *Rev Suisse Zool* 97: 3-29.
51. Sands J, Creel S (2004) Social dominance, aggression and faecal glucocorticoid levels in a wild population of wolves (*Canis lupus*). *Anim Behav* 67: 387–396.
52. Godinho R, Castro D, López-Bao JV, García E, Rio-Maior H, et al. (2011) Genetic assessment of error rates in identifying wolf (*Canis lupus*) faeces in four areas of the Iberian Peninsula. Abstract-book of the X Jornadas Españolas de Conservación y Estudio de Mamíferos: p. 64. Fuengirola, Spain.
53. Blanco JC, Cortés Y (2012) Surveying wolves without snow: a critical review of the methods used in Spain. *Hystrix* 23: 35–48.
54. Fuller TK, Sarnpson BA (1988) Evaluation of a simulated howling survey for wolves. *J Wildl Manage* 52: 60–63.

3

Spatial Heterogeneity in Ecologically Important Climate Variables at Coarse and Fine Scales in a High-Snow Mountain Landscape

Kevin R. Ford[1]*, Ailene K. Ettinger[1], Jessica D. Lundquist[2], Mark S. Raleigh[2], Janneke Hille Ris Lambers[1]

1 Department of Biology, University of Washington, Seattle, Washington, United States of America, **2** Department of Civil and Environmental Engineering, University of Washington, Seattle, Washington, United States of America

Abstract

Climate plays an important role in determining the geographic ranges of species. With rapid climate change expected in the coming decades, ecologists have predicted that species ranges will shift large distances in elevation and latitude. However, most range shift assessments are based on coarse-scale climate models that ignore fine-scale heterogeneity and could fail to capture important range shift dynamics. Moreover, if climate varies dramatically over short distances, some populations of certain species may only need to migrate tens of meters between microhabitats to track their climate as opposed to hundreds of meters upward or hundreds of kilometers poleward. To address these issues, we measured climate variables that are likely important determinants of plant species distributions and abundances (snow disappearance date and soil temperature) at coarse and fine scales at Mount Rainier National Park in Washington State, USA. Coarse-scale differences across the landscape such as large changes in elevation had expected effects on climatic variables, with later snow disappearance dates and lower temperatures at higher elevations. However, locations separated by small distances (~20 m), but differing by vegetation structure or topographic position, often experienced differences in snow disappearance date and soil temperature as great as locations separated by large distances (>1 km). Tree canopy gaps and topographic depressions experienced later snow disappearance dates than corresponding locations under intact canopy and on ridges. Additionally, locations under vegetation and on topographic ridges experienced lower maximum and higher minimum soil temperatures. The large differences in climate we observed over small distances will likely lead to complex range shift dynamics and could buffer species from the negative effects of climate change.

Editor: Francesco de Bello, Institute of Botany, Czech Academy of Sciences, Czech Republic

Funding: Research was supported by the University of Washington Department of Biology (KF, AE), the American Alpine Club (KF), the Seattle chapter of the ARCS Foundation (KF), an NSF Doctoral Dissertation Improvement Grant (DEB-1010787) (AE), the University of Washington Royalty Research Foundation (JH) and the US Department of Energy (DOE#DE-FC02 06ER64159) (JH). This material is based on work supported by the NSF Graduate Research Fellowship Program (DGE-0718124- KF, AE), NSF grant no. CBET-0931780 (JL) and by NASA Headquarters under the NASA Earth and Space Science Fellowship Program (NNX09AO22H) (MR). The funders had no role in study design, data collection and analysis, decision to publish, or preparation of the manuscript.

Competing Interests: The authors have declared that no competing interests exist.

* E-mail: kford10@gmail.com

Introduction

Biologists have long recognized the fundamental role climate plays in determining the geographic distributions of species and biomes [1–3]. As a result, climate change is expected to induce shifts in the geographic ranges of species. This prediction is supported by the many observations of upward or poleward range shifts over the last 100 years consistent with observed warming [4] as well as range shifts inferred from the fossil record [5,6]. Alarmingly, models of the impacts of future anthropogenic climate change on species ranges have forecasted widespread extinction risks as the climatic niche of many species disappears or shifts faster than species can likely migrate [7,8].

However, these projections of climate change-induced range shifts (and subsequent extinction risks) are sensitive to the spatial scale at which the analyses are conducted [9]. Most range shift assessments rely on gridded maps of climate variables with grid cell sizes ranging from 800×800 m (e.g. PRISM [10] and WorldClim [11]) up to 50×50 km (e.g. [12]). The finer scale maps (800×800 m grid cells) capture a wide variety of climatic patterns, but the scales of these maps are still far coarser than the scales at which organisms experience their environment. Thus, these climate maps may hide fine-scale differences in climate that are important for organism distributions [13]. For example, north and south facing slopes separated by tens of meters may receive different amounts of solar radiation and experience very different temperature regimes [14,15], which could lead to differences in species composition within these microhabitats.

The implication of significant fine-scale climatic heterogeneity that is not captured by coarse-scale climate maps is that projections based on these maps could fail to capture important range shift dynamics. For example, cool microhabitats (such as north-facing slopes in the Northern Hemisphere) near the contracting edge or core of a species' distribution may allow populations of that species to persist if individuals can disperse to them from warmer microhabitats (such as south-facing slopes in the Northern Hemisphere), even if most of the surrounding landscape becomes unsuitably warm (as long as these microhabitat types comprise a total area large enough to support a population [16]). At the same time, warm microhabitats beyond the advancing edge of a species' range may provide the first sites of colonization that allow that species to migrate to new locations. Thus, instead of needing to move hundreds of meters upward or hundreds of kilometers poleward to track suitable climate, many species may only need to move tens of meters from one microhabitat to another and could be buffered from the negative effects of climate change [17].

For such fine-scale climatic heterogeneity to strongly influence range dynamics, however, fine-scale differences in climate must be large relative to coarse-scale differences. We addressed this issue by examining the magnitude of fine-scale heterogeneity relative to coarse-scale heterogeneity in snow disappearance date and growing season soil temperature. Specifically, we deployed 284 microclimate sensors across a ~1500 m elevation gradient spanning forest, subalpine and alpine biomes at Mount Rainier National Park. Our objectives were to 1) quantify snow disappearance date and soil temperature as a function of coarse-scale differences in elevation and exposure to storm tracks (i.e. being on the windward or leeward side of the mountain) and fine-scale differences in vegetation structure or topography, 2) compare fine-scale differences in climatic variables (that would be missed by climate models) to coarse-scale differences (that would be captured by climate models), and 3) determine whether fine-scale patterns in climatic variables related to topography (but not vegetation structure) are correlated with fine-scale patterns in vegetation characteristics. We focus on snow disappearance date and growing season soil temperature because snow disappearance date influences the length of the growing season (especially important in this region where the growing season can be very short due to the persistence of large winter snowpacks) while soil temperature strongly influences plant growth rates and other physiological processes [18]. These variables have also been shown to be strongly associated with patterns of distribution, abundance, productivity and diversity of plant species, at our sites and others [19–21]. Additionally, both variables are likely to change in the coming decades as a result of anthropogenic climate change, with rising temperatures and declining snowpacks leading to warmer and longer growing seasons [22,23].

It has long been known that climate can vary dramatically at fine spatial scales (reviewed in [14] and [24]), but these patterns have only recently begun to be studied explicitly and systematically with respect to the impacts of climate change on species distributions. Studies have found that locations separated by only tens of meters experienced mean seasonal soil temperatures that differed by 3–7°C, equivalent to the average temperature difference experienced in locations separated by hundreds of meters in elevation or hundreds of kilometers in latitude [15,21,25–28]. Moreover, such large differences in temperature are known to strongly influence organismal performance [18,29] and are greater than the expected increase in temperature due to climate change in many parts of the globe [30]. Similarly large differences were also observed in air temperature, snow cover duration or snow disappearance date over fine spatial scales in these studies. Our paper builds on these case studies and is notable for its large sample size of 284 sensors (important for assessing microclimate patterns in a statistically rigorous way), its explicit comparison of coarse- and fine-scale climatic heterogeneity (important for assessing the biases of coarse-scale models) and the broad environmental gradients covered (important for assessing how widespread these biases may be).

Methods

Ethics Statement

We obtained the necessary scientific research permits from Mount Rainier National Park, where all data collection occurred, before conducting this study. We did not sample any protected species.

Study Area

Mount Rainier National Park encompasses 95,354 ha of land in the western Cascade Mountains in Washington State, USA (Figure 1). The region experiences a temperate, maritime climate with mild, dry summers and cool, wet winters that produce large snowpacks. Elevation ranges from 518 m in the deep valley floors to 4392 m at the peak of Mount Rainier, the volcano located in the middle of the Park. The mountainous terrain produces steep climatic gradients: temperature decreases and precipitation increases with elevation, while the rainshadow effect produced by the volcano leads to lower precipitation on the east side of the Park. There are two primary climate stations in the Park. At the station located at 842 m elevation, mean annual temperature is 6.6°C and mean annual precipitation is 2030 mm; at the 1654 m station, mean annual temperature is 3.7°C and mean annual precipitation is 3005 mm (1981–2010 normals, NOAA National Climate Data Center – www.wrcc.dri.edu/Climsum.html).

The large climatic gradients create three major biomes in the Park. The forest biome extends from the lowest elevations of the Park up to about 1600–2000 m and is dominated by coniferous trees. The subalpine biome typically extends about 300 m above the upper limit of the forest and is a mosaic of conifer tree patches and subalpine meadows. The alpine biome occupies the highest elevations, stretching from 1900–2300 m to the summit of Mount Rainier, and consists of large patches of mostly continuous alpine meadows (dominated by forbs, grasses and dwarf shrubs) near the lower limit of the biome, with exposed rock, glaciers, bare soil, and cryptogams (mostly mosses, lichens, algae and cryptobiotic soil crusts) predominating above.

Study Design

From September 2009 through October 2010, we deployed 284 soil temperature sensors (HOBO Pendants made by the Onset Computer Corporation and iButtons made by Maxim Integrated Products) across Mount Rainier at elevations ranging from 638 m to 2164 m as part of two different plant ecology studies where microclimate was measured as an explanatory variable. The first study took place in the forest biome and spanned the elevational range of forests on the south side of Mount Rainier. The second study took place in the subalpine and alpine biomes, with study sites set up at the lower limit of the subalpine biome and the upper limit of alpine meadows on three sides of the mountain (Figure 1). Microclimate data from these studies were ideally suited for our questions as they covered large elevational gradients with sensors at each location stratified by vegetation or topographic features expected to influence microclimate. Due to differences in study design (described below), we analyzed the microclimate data from the two studies separately. The sensors remained in place and logged data until we collected them in September/October 2010.

For each sensor, we calculated the values of four climatic variables: snow disappearance date, and average daily mean, maximum and minimum soil temperature. We could assess snow cover from soil temperature measurements because snow insulates soil from fluctuations in air temperature so that temperatures beneath the snowpack in this region remain constant around 0°C. Thus, the soil temperature data allowed us to determine whether

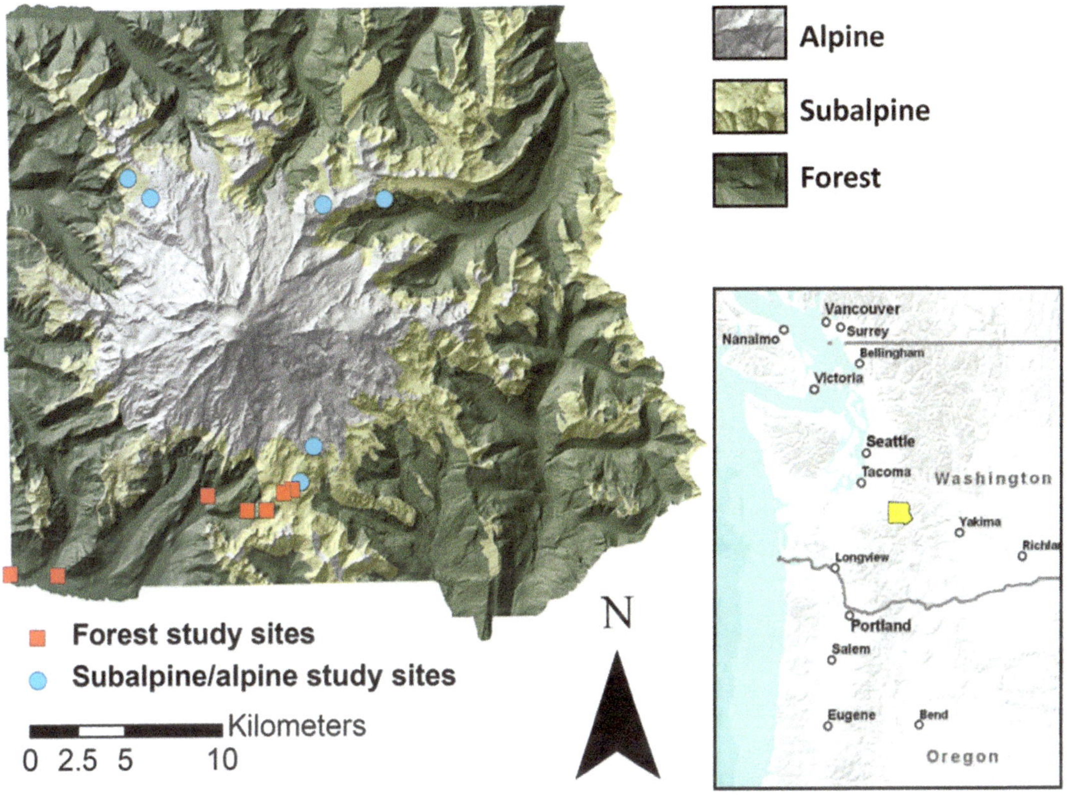

Figure 1. Study area. Mount Rainier National Park and its three major biomes, along with study site locations. Shading depicts topographic relief.

snow was covering the sensor for each day the sensor was deployed using an algorithm that considers daily temperature ranges and maxima [31,32]. We calculated average daily mean, maximum and minimum soil temperature for periods in Summer/Fall 2010 when all sensors in a study experienced snow-free conditions and reported data. This period was August 14 through October 3, 2010 for the sensors in the forest study and August 11 through August 18, 2010 for the sensors in the subalpine/alpine study. This meant we only used a portion of the snow-free temperature data for some sensors, even though temperatures outside this period are likely to also be ecologically relevant. However, it was necessary to use the same time period for all sensors in a study so that temporal differences in the snow-free period between locations did not confound our analysis of spatial differences in temperature.

Arrays of sensors were deployed at 13 sites throughout the Park. For the forest study, we established three study sites along an elevation gradient in Summer 2009, allowing us to calculate snow disappearance date in 2010. We quantified growing season soil temperature in Summer/Fall 2010 at these three sites plus an additional four sites along the same elevation gradient. Within each site, we deployed sensors under gaps in the forest canopy caused by tree falls ("gaps") and in locations under intact canopy within 20 m of one of the gaps ("non-gaps"). Each study site contained five of these gap/non-gap pairs. Gaps were ~130 m^2 on average. Within each gap or non-gap location, we placed one sensor in a 5.5×1.5 m area where all understory vegetation up to 2 m tall had been experimentally removed since early Summer 2009 (the "removed" plot) and one sensor at an adjacent location 2 m away where the vegetation had been left undisturbed (the "control" plot) (Figure 2A, Table 1).

For the subalpine/alpine study, we quantified microclimate at study sites on three sides of the mountain (south, northwest and northeast) which have different exposures to storm tracks and experience different precipitation regimes. On each side, we established two study sites, one close to the lower limit of the subalpine biome and one close to the upper limit of continuous alpine meadows, (below this limit, the ground is mostly vegetated while above it is almost entirely rock, glaciers and bare soil). These sites were about 200–300 m apart in elevation on each side of the mountain. Within each site, we established six linear transects that ran from a depression in the landscape up to a ridge (transects parallel to the slope) and were about 20 m in length. Within each transect, two sensors were located in the depression and two sensors were located on the ridge (Figure 2B, Table 1).

At each of the sensors in the subalpine/alpine biomes (where fine-scale sensor placement was stratified by topographic position and not vegetation structure), we measured vegetation characteristics in order to compare patterns in microclimate to ecological patterns. At the study sites near the lower limit of the subalpine biome, where closed-canopy forests transition to open meadows with increasing elevation, we measured percent cover by tree canopy above each sensor using a spherical densiometer (a gridded, hemisphere-shaped mirror used to estimate percent cover by foliage above a point on the ground), allowing us to assess the density of trees (higher values of tree canopy cover implies more or bigger trees). At the study sites near the upper limit of alpine meadows, where meadows transition to bare ground with increasing elevation, we estimated the percent of ground covered by vegetation at each sensor using a square-shaped PVC frame (1×1 m) placed on the ground adjacent to the sensor. String tied to the PVC frame created 100 evenly spaced grid points, allowing us

Figure 2. Temperature sensor deployment. Sensor deployment in (A) forest and (B) subalpine/alpine biomes. At each elevation in the forest biome (A), sensors were placed in gaps in the forest canopy (top left) and non-gaps with intact forest canopy (top right). Within each of these canopy types, sensors were located in plots where understory vegetation was removed (bottom left) and control plots where it was left undisturbed (bottom right). In the subalpine/alpine biomes (B), temperature sensors were located along transects running from depressions in the landscape to ridges.

to count the number of grid points overlaying vegetation in the area within the frame.

Data Analysis

We used linear mixed effects models (LMMs) to characterize the relationships between the potential drivers of climate and each of the four climatic response variables [33]. The LMMs allowed us to estimate the effects of explanatory variables and their two-way interactions on the response variable ("fixed effects") while statistically controlling for the effects of randomly selected experimental units on the response variable ("random effects"). At the forest sites, the drivers of climate were elevation, canopy structure (gap vs. non-gap) and understory structure (removed vs. control treatments), with gap/non-gap pair designated as a random effect. At the subalpine/alpine sites, the drivers of climate were side of the mountain (south vs. northwest vs. northeast), elevation (upper limit of alpine meadows vs. lower limit of the

Table 1. Details of temperature sensor deployment.

Biome(s)	Time span of deployment	# sites	# sensors per site	Total # sensors	Type of sensor	Sensor accuracy	Data logging interval	Sensor location
Forest	Summer 2009- Fall 2010	3	8	24	HOBO Pendants made by the Onset Computer Corporation	±0.53°C from 0° to 50°C	2 hours	Soil surface
Forest	Summer-Fall 2010	7*	20	140	HOBO Pendants made by the Onset Computer Corporation	±0.53°C from 0° to 50°C	1 hour	Soil surface
Subalpine/alpine	Summer 2009- Summer 2010	6	24	144	iButtons made by Maxim Integrated Products	±0.5°C from −10° to 65°C, or ±1°C from −30° to 70°C†	1, 2 or 4 hours†	3 cm below soil surface

*These sites include the three sites with sensors deployed in Summer 2009.
†Differences in accuracy and logging intervals were due to differences in the specific model of iButton sensor used.

subalpine biome, treated as a categorical variable since there were only two values of elevation on each side of the mountain) and topographic position (depression vs. ridge), with sensor transect designated as a random effect. We verified that the residuals of these models were normally distributed, to validate our use of linear mixed effects models (rather than generalized linear mixed effects models).

For each model, we used Akaike's information criterion (AIC) to select the most parsimonious combination of fixed and random effects to derive the "best-fit" model. Specifically, we used a three-step process following [33] where we (1) used AIC to determine the optimal random effects structure, selecting amongst several LMMs (fit with restricted maximum likelihood) that had different random effect terms (no random effects, random intercepts, random slopes or both), but the same fixed effect terms (which included all main effect and two-way interaction terms for each explanatory variable); (2) determined the optimal combination of fixed effect terms by using AIC to select amongst models (fit with maximum likelihood) with all possible combinations of fixed effect terms (but sharing the optimal random effects structure selected in the first step); and (3) fit a model with the random effects structure selected in the first step (which could be no random effects) and the fixed effects structure selected in the second step and considered this model to be our final "best-fit" model. This final model was fit with restricted maximum likelihood if it included random effects or maximum likelihood if it did not. All models were fit in R version 2.12.0 using the lme4 package for the LMMs [34,35]. See Appendix S1 for more details of the model fitting and selection procedure.

We assessed the significance of the model coefficients using Markov chain Monte Carlo sampling implemented with the MCMCglmm package in R [36] or *t*-tests (when no random effects were included in the best-fit model). We then used the explanatory variable coefficients of the best-fit models to calculate the magnitude of differences in microclimate response variables relative to differences in the explanatory variables. For example, if the coefficients related to topographic position in the model of snow disappearance date at the subalpine/alpine sites indicated that the difference between ridges and depressions was 20 days, controlling for differences in other variables, then the effect of topographic position on snow disappearance date would be equal to 20 days. In order to compare the effects of elevation (which we consider a coarse-scale driver of climate) to the effects of other explanatory variables, we calculated the difference in snow disappearance date or temperature between two points 100 m apart in elevation for each model (controlling for differences in other variables). Like differences in climate amongst different sides of the mountain, differences in climate caused by a 100 m difference in elevation can typically be captured by coarse-scale climate models (e.g. PRISM), while differences caused by vegetation structure and fine-scale topography cannot. If one of the explanatory variables was not included in the best-fit model, we included the main effect of that variable in the final model for comparative purposes. This happened for one explanatory variable in one model (the understory structure variable in the snow disappearance date model in the forest study).

For sites in the subalpine/alpine biomes, we also fit linear models (LMs) to characterize the relationships between each of the four microclimate variables and percent cover by tree canopy at the lower elevation sites, and the relationships between each of the four microclimate variables and percent cover by ground vegetation at the higher elevation sites, for a total of eight LMs. In these models, the response variable was the vegetation characteristic while the explanatory variables were the microclimate variable, side of the mountain (included as a covariate) and their interaction. Using LMMs with sensor transect designated as a random effect did not improve model fit for any of the relationships, so we used the simpler LMs for all of these analyses. Next, we used AIC to select the best-fit LM. With one exception, this best-fit model included both the microclimate variable and side of the mountain, but not their interaction, as explanatory variables. In these cases, we used *t*-tests to assess the significance of the microclimate variable coefficient in the best-fit model in order to assess the significance of that microclimate variable. When modeling percent tree canopy cover at the lower elevation sites as a function of average daily minimum temperature and side of the mountain, neither explanatory variable nor their interaction was included in the best-fit model (i.e. the best-fit model was the null model with only an intercept). For this situation, we performed a *t*-test on the minimum temperature coefficient in a model that included both minimum temperature and side of the mountain, but not their interaction, as explanatory variables (i.e. a model with the same structure as the best-fit model for all the other vegetation-climate analyses) in order to assess the significance of minimum temperature.

We calculated the proportion of variance in the response variable explained by variance in the fixed effect explanatory variables (r^2) for all models, following [37].

Results

Variations in climate were explained by both coarse- and fine-scale drivers, with best-fit models having r^2 values ranging between 0.20 and 0.94 (Table 2). As expected, higher elevations experienced later snow disappearance dates and lower average daily mean and minimum temperatures (Figure 3A, B, D). However, the relationship between elevation and average daily maximum temperature was weak, and variability in this parameter was dominated by variability amongst locations at similar elevations (Figure 3C). At the subalpine/alpine sites, snow disappearance date and temperature varied depending on what side of the mountain sensors were on – e.g. the south side experienced later snow disappearance dates on average, probably because this side of the mountain is the most exposed to winter storms and receives the largest amount of winter precipitation. However, there were also substantial differences amongst locations at similar elevations for each of these variables that could be attributed to vegetation structure or topographic position (Figures 3, 4). We also assessed heterogeneity in growing degree days (GDD – calculated as the sum of daily mean soil temperatures for all days where the daily mean soil temperature was over 5°C), which showed patterns very similar to those of snow disappearance date (results not shown due to limitations of the data – sensors were not deployed long enough to estimate GDD for the full year or growing season, which could bias comparisons of GDD amongst locations).

Forest Biome: Stratification by Vegetation Structure at Fine Scales

Snow disappearance date was later at higher elevations and in canopy gaps, while understory vegetation structure had little effect (Figure 3A). We also found that the effect of fine-scale differences in canopy structure (gaps vs. non-gaps) on snow disappearance date was similar to the effects of coarse-scale differences in elevation (100 m elevation differences) (Figure 4A). Thus, snow disappearance date differed as much at fine scales (where locations differed in forest canopy structure) as it did over coarse spatial scales.

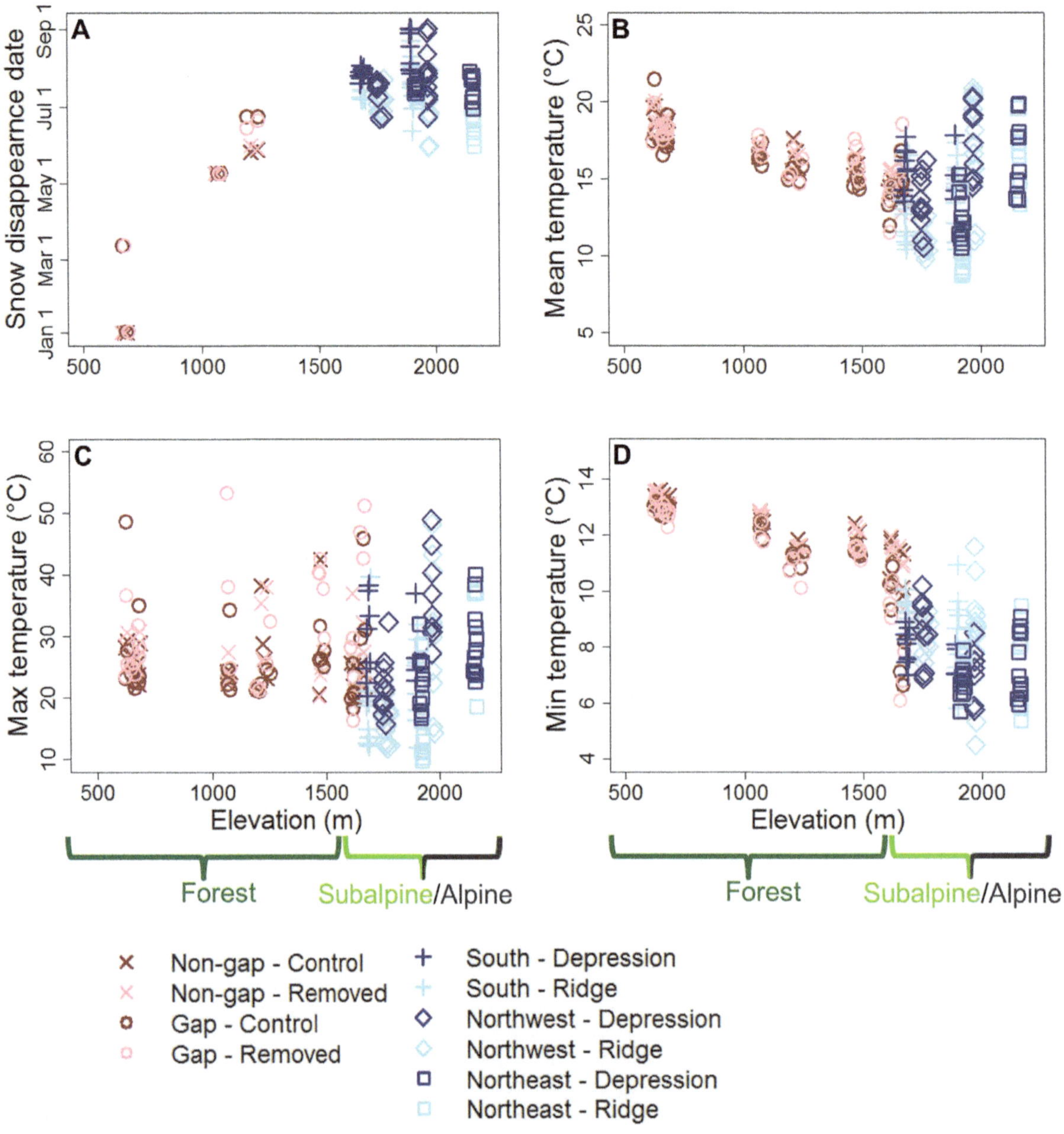

Figure 3. Patterns in climate. (A) Snow disappearance date in 2010 and average daily (B) mean, (C) maximum and (D) minimum soil temperature for a representative week during the growing season (August 11–18, 2010) plotted against elevation. Note the differences in scale on the axes showing temperature values. Points represent individual sensors with symbol type and color designating sampling stratification for forest (dark and light red) and subalpine/alpine sites (dark and light blue). "Non-gap"/"gap" refer to canopy structure categories while "control"/"removed" refer to understory structure categories (forest sites). "South"/"northwest"/"northeast" refer to sides of the mountain while "ridge"/"depression" refer to topographic positions (subalpine/alpine sites). Approximate biome ranges are shown below the elevation axes.

As expected, growing season soil temperature generally declined with increasing elevation. Canopy gaps had higher maximum temperatures, but lower minimum and mean temperatures relative to non-gaps (Figure 3B–D). Canopy structure had a similar or greater effect on temperature than a 100 m change in elevation for average daily maximum and minimum temperature (Figure 4C, D). Locations where understory vegetation was removed experienced higher maximum and mean temperatures, but lower minimum temperatures, relative to control plots where vegetation was undisturbed (Figure 3B–D). Understory structure had a greater effect on average daily maximum temperature than a 100 m change in elevation, but had weaker effects on average daily mean and minimum temperature (Figure 4B–D). Overall, there was about as much heterogeneity in temperature at fine scales (differing vegetation structure) as there was at coarse scales (100 m differences in elevation).

Subalpine/alpine Biomes: Stratification by Topographic Position at Fine Scales

Snow disappearance date was later on the south side of the mountain, at higher elevations and in topographic depressions (Figure 3A). Furthermore, the effect of fine-scale topographic differences (depressions vs. ridges) on snow disappearance date was

Table 2. Best-fit models for the climatic response variables.

<table>
<tr><th>Study</th><th>Climatic response variable</th><th>Model formula*</th><th>r^2</th></tr>
<tr><td>Forest: stratification by vegetation structure</td><td>Snow disappearance date</td><td>SDD = f { elev+canopy+understory[†]+ (1|pair) }</td><td>0.94</td></tr>
<tr><td>Forest: stratification by vegetation structure</td><td>Average daily mean temperature</td><td>Tmean = f { elev+canopy+understory+elev:canopy+elev:understory+canopy:understory+(0+canopy|pair) }</td><td>0.35</td></tr>
<tr><td>Forest: stratification by vegetation structure</td><td>Average daily maximum temperature</td><td>Tmax = f { elev+canopy+understory+elev:canopy+elev:understory+canopy:understory+(0+canopy|pair) }</td><td>0.40</td></tr>
<tr><td>Forest: stratification by vegetation structure</td><td>Average daily minimum temperature</td><td>Tmin = f { elev+canopy+understory+elev:canopy+canopy:understory+(0+canopy|pair) }</td><td>0.20</td></tr>
<tr><td>Subalpine/alpine: stratification by topographic position</td><td>Snow disappearance date</td><td>SDD = f { side+elev+topo+elev:topo+side:elev+(1|tran) }</td><td>0.60</td></tr>
<tr><td>Subalpine/alpine: stratification by topographic position</td><td>Average daily mean temperature</td><td>Tmean = f { side+elev+topo+elev:topo+side:elev }</td><td>0.60</td></tr>
<tr><td>Subalpine/alpine: stratification by topographic position</td><td>Average daily maximum temperature</td><td>Tmax = f { side+elev+topo+elev:topo+side:elev }</td><td>0.52</td></tr>
<tr><td>Subalpine/alpine: stratification by topographic position</td><td>Average daily minimum temperature</td><td>Tmin = f { side+elev+topo+side:elev }</td><td>0.30</td></tr>
</table>

*Parameters in bold have significant coefficients ($p<0.05$). For the forest biome study, elev = elevation; canopy = forest canopy. structure, gap or non-gap; understory = understory vegetation structure, removed or control; pair = gap/non-gap pairings. For the subalpine/alpine biomes study, side = side of the mountain, south or northwest or northeast; elev = elevation; topo = topographic position, ridge or depression; tran = sensor deployment transect. The colon indicates an interaction effect between two explanatory variables. The parentheses indicate the term is a random effect – all other terms are fixed effects. If *fe* is a particular fixed effect and *re* is a particular random effect, then (1|*re*) indicates the intercept was allowed to vary randomly with respect to *re* while (0+*fe*|*re*) indicates the interaction of *fe* and *re* was allowed to vary randomly. The case of both the intercept and interaction being allowed to vary randomly was not included in any of the best-fit models.

[†]Understory was not included in the best-fit model, based on AIC, but was retained for comparative purposes.

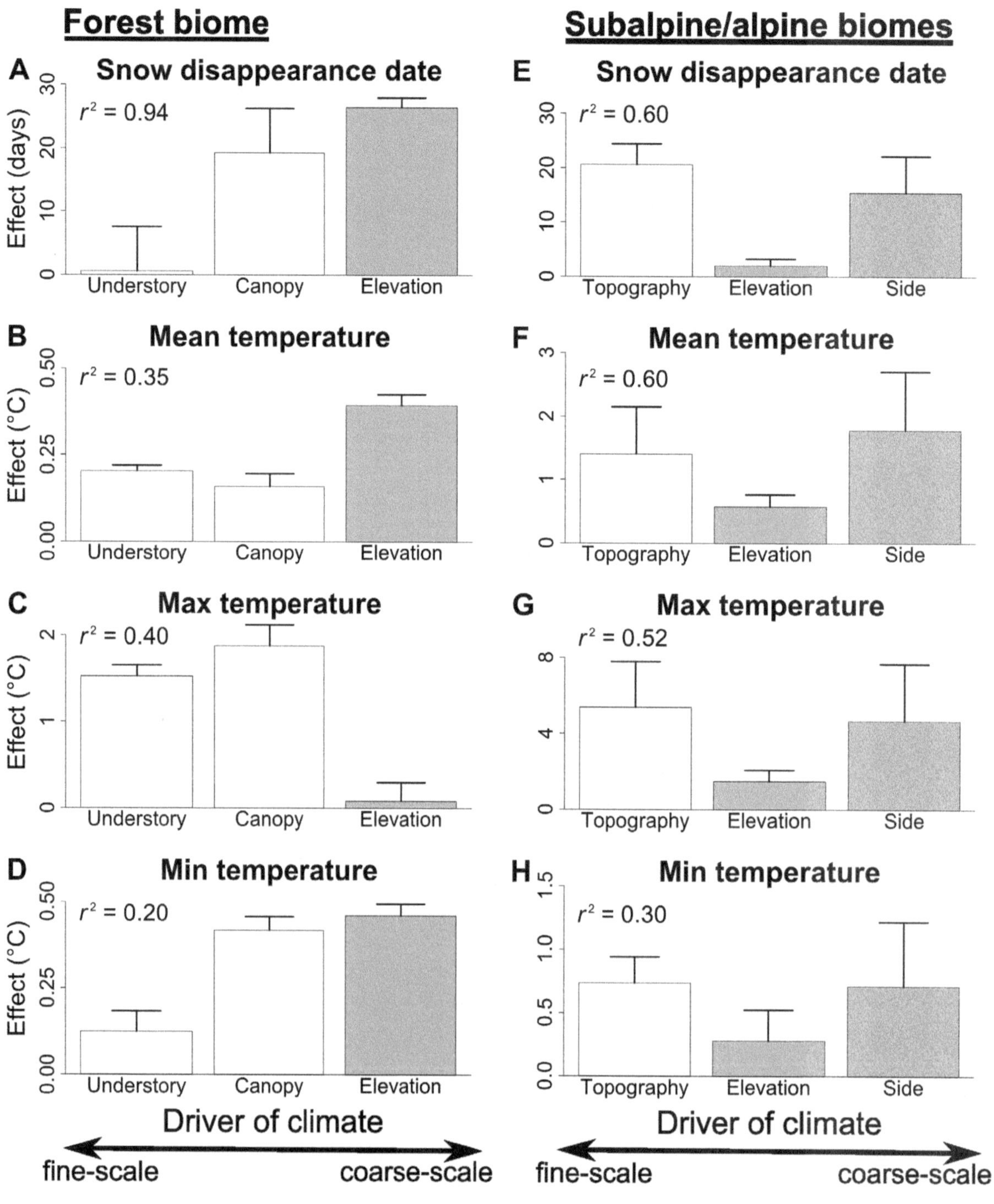

Figure 4. Effects of fine- and coarse-scale drivers of climate. The effects of fine- and coarse-scale drivers of climate on snow disappearance date and the average daily values of mean, maximum and minimum growing season soil temperature. Bars show differences in snow disappearance date or temperature attributed to the effect of different drivers of climate by the best-fit model, with standard error. The effect of elevation was standardized to the effect of a 100 m difference in elevation. Bars filled with gray represent drivers that are coarse enough in scale to be captured by typical climate models (>1 km) while unfilled bars represent drivers too fine in scale to be captured by these models (≤20 m). Fine-scale drivers of climate often had a greater effect on snow or soil temperature than coarse-scale drivers.

similar to the effects of coarse-scale differences in elevation (100 m difference in elevation) and side of the mountain (Figure 4E). In other words, snow disappearance date differed as much over fine spatial scales as it did over coarse spatial scales.

Growing season soil temperatures during our sampling period were lower on the northeast side of the mountain than on the northwest and south sides, potentially because the meadows are at higher elevations on this side of the mountain. On a given side of the mountain, higher elevations (the upper limit of alpine meadows) had higher mean and maximum temperatures, but lower minimum temperatures, than lower elevations (the lower limit of the subalpine biome). Compared to ridges, depressions had

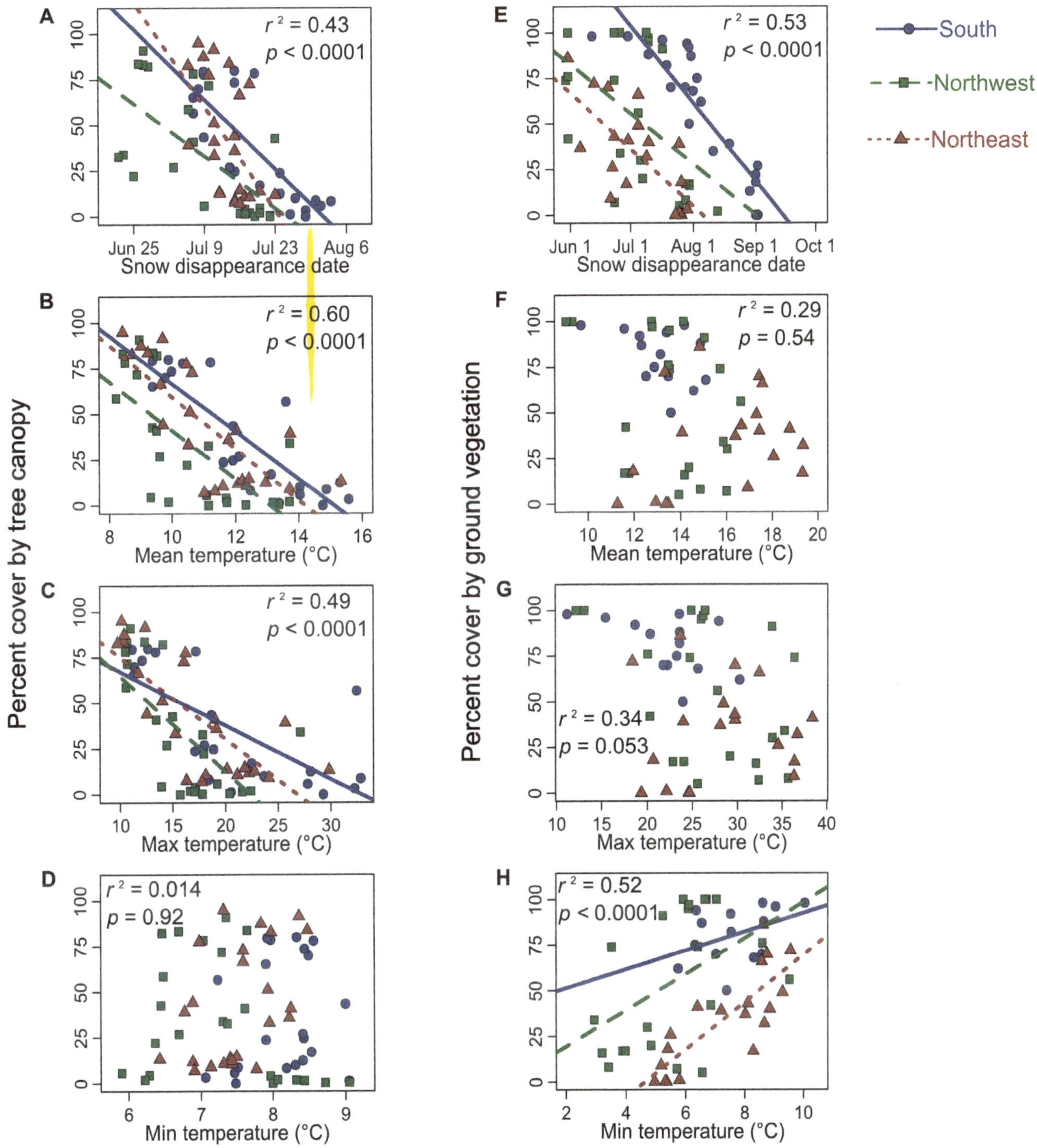

Figure 5. Relationships between vegetation characteristics and microclimate. (A–D) Percent cover by tree canopy at sites near the lower limit of the subalpine biome and (E–H) percent cover by ground vegetation at sites near the upper limit of alpine meadows plotted against the four microclimate variables (snow disappearance date and average daily mean, maximum and minimum soil temperature) on each of the three sides of the mountain. The r^2 values are for models that included the microclimate variable and side of the mountain as explanatory variables, while the p values indicate the significance of the microclimate variable in these models. Regression lines are shown for significant p values (<0.05).

higher mean and maximum temperatures but lower minimum temperatures (Figure 3B–D). We found that the effects of fine-scale topographic differences (depressions vs. ridges) were similar to the effects of coarse-scale differences in elevation and side of the mountain for average daily mean, maximum and minimum temperature (Figure 4F–H). Overall, there was as much heterogeneity in temperature at fine scales as there was at coarse scales.

Several microclimate variables were significantly correlated with vegetation characteristics (Figure 5). At study sites near the lower limit of the subalpine biome, percent cover by tree canopy was lower where snow disappearance date was later and average daily mean and maximum soil temperatures were higher ($p<0.0001$). At study sites near the upper limit of alpine meadows, percent cover by ground vegetation was lower where snow disappearance date was later and average daily minimum soil temperature was lower ($p<0.0001$).

Discussion

Our study suggests that climatic heterogeneity at the fine spatial scales most organisms experience their environment is substantial, implying that projections based on coarse-scale climate models will not capture the full complexity of range shifts in response to climate change. Specifically, we found large differences in snow disappearance date and growing season soil temperatures over small distances (Figures 3, 4), differences that were sometimes as large as those experienced when travelling hundreds of meters upward in elevation or several kilometers to a different side of the mountain. These microclimate variables have been shown to be strongly associated with plant species distributions and abundances [20,21], suggesting that the microclimate heterogeneity we observed is important for plant communities. We also found that vegetation characteristics (canopy and ground vegetation cover) can be strongly correlated with the microclimate variables influenced by fine-scale topographic features (Figure 5), further suggesting that the fine-scale climatic heterogeneity we observed is ecologically important. However, because we did not measure species distributions or abundances in this study, we cannot conclusively state that the microclimate heterogeneity we observed is linked to species distributions or abundances at Mount Rainier. Nonetheless, understanding fine-scale climatic heterogeneity will likely be critical for management, as cool or snowy microhabitats could provide an important buffer against the negative effects of climate change on biodiversity. Thus, when assessing potential species range shifts in response to climate change, it is critical for ecologists to consider fine-scale patterns in climate in addition to other important factors such as broad-scale climate patterns, dispersal constraints, biotic interactions and evolutionary dynamics.

Explanations of Fine-scale Climatic Heterogeneity

In the forest biome, a complex interplay between elevation and vegetation structure is likely responsible for the heterogeneous patterns in snow disappearance date and soil temperature we observed. For example, locations under tree canopy gaps likely experienced later snow disappearance dates than locations under an intact canopy (Figure 3A) because tree canopies intercept snowfall where it can rapidly sublimate or melt instead of being incorporated into the snowpack on the ground [38]. Tree canopies also increase incoming longwave radiation (which increases ablation rates) and this effect can sometimes be greater than the effect of canopies decreasing incoming shortwave radiation by shading the snowpack (which reduces ablation rates), leading to a net effect of canopies increasing ablation rates [39]. Although the presence of trees has also been shown to lead to longer snow persistence by shading the snowpack and decreasing wind speeds (reducing incoming sensible and latent heat fluxes) [38], these effects appear to be relatively weak at our study sites. Increased shading from tree canopies and understory vegetation in forest locations probably led to substantially lower maximum soil temperatures (Figure 3C). But these low sky exposure locations also experienced higher minimum soil temperatures (Figure 3D), probably due to vegetation emitting more longwave radiation (which warms the surface) than the night sky [14]. Differences in mean soil temperatures appeared to be the net effect of these two counteracting influences of sky exposure, with mean soil temperatures being higher in the shadier non-gap locations, but lower in the shadier undisturbed understory vegetation locations (Figure 3B).

Similarly, in the subalpine/alpine biomes we found that both coarse- and fine-scale features had large effects on climate. For example, snow disappeared substantially later from depressions in the landscape than from ridges only ~20 m away, likely because snow typically collects in these depressions while it is blown off of ridges and because shading from surrounding slopes can reduce ablation rates [24]. Feedbacks between vegetation and climate are also likely to influence fine-scale climatic variability. At the lower elevation sites, for example, patches of trees with trunks sticking out above the snowpack emit substantial amounts of longwave radiation which quickens the ablation of snow next to the tree patch and can lead to earlier snow disappearance dates. Trees can also intercept snowfall, reducing snowpack accumulation under canopy and resulting in earlier snow disappearance [38]. These effects can lead to a positive feedback, where trees establish in microsites with earlier snow disappearance dates (e.g. ridges), and the established trees lead to even earlier snow disappearance dates and more tree establishment. This result is consistent with previous studies from subalpine meadows in the region that have documented increased tree establishment on ridges that tend to have earlier snow disappearance dates [40,41].

A striking pattern to emerge from our data was that mean and maximum soil temperatures were greater at higher elevations within the subalpine/alpine biomes (though minimum soil temperatures were lower). Feedbacks between climate and vegetation likely play important roles in producing this temperature pattern. First, tree cover declines with increasing elevation, leading to less shading and potentially higher daytime soil temperatures, especially during the sunny growing season when our data were collected. This explanation is supported by the negative correlation we observed between percent canopy cover (a measure of tree density) and mean/maximum soil temperature in the subalpine/alpine biomes (Figure 5B, C). Second, ground vegetation density declines with increasing elevation, which can lead to lower organic matter content in the soil and lower soil moisture levels. The lower moisture levels probably cause the soil to have a lower heat capacity, leading to greater temperature change per unit of energy input and hence higher maximum temperatures and lower minimum temperatures. This second explanation is supported by the pattern of low soil organic matter content and soil water holding capacity at high elevations in Mount Rainier's subalpine/alpine biomes (Appendix S2). Soil characteristics also have important effects on vegetation in subalpine/alpine environments [42], creating the possibility for complex feedbacks amongst soil, vegetation and climate. These two hypotheses are not mutually exclusive, and further study is needed to assess the importance of each. Regardless, our results suggest that even if patterns in climate are ultimately responsible for patterns in vegetation, the feedback effect of vegetation on soil temperature (either directly, or indirectly through the effects of vegetation on soil characteristics which then affect temperature) appears to at times be stronger than the original forcing of physiographic effects on soil temperature.

Implications of Fine-scale Climatic Heterogeneity for Species Distributions in a Warming World

Since snow disappearance date and growing season temperature vary dramatically over short distances, species whose distributions are primarily constrained by these climate variables may not need to migrate long distances to remain in suitable habitat even when there are large changes in climate. For example, in the subalpine/alpine biomes we found that the average difference in snow disappearance date between depressions and ridges separated by only ~20 m was often one month or more. This is an especially large difference considering the ground is typically only free of snow for 3–5 months out of the year at these elevations. Snow manipulation experiments have shown that differences of this magnitude have large impacts on the phenology, species composition, diversity and productivity of plant communities [20]. Thus, these snowy microhabitats have the potential to serve as refugia for species in a warmer world and provide linkages to new areas of suitable climate, implying that fine-scale climatic heterogeneity could buffer species from climate change [27,43,44], as it may have done during past periods of rapid climate change [45]. Given that we did not stratify our sensors along all gradients likely to produce fine-scale differences in climate (e.g. wind direction, aspect – [14,24]), our results may even be an underestimate of the magnitude of fine-scale heterogeneity.

The importance of topographic heterogeneity for creating climatic heterogeneity shown in this study and others [15,21,25–28] also suggests that mountainous regions will be important for providing climatic refugia in a warming world. However, mountain biotas will still likely face unique challenges. For example, organisms currently living on or near summits will not be able to shift upwards to track suitable climate, and deep valleys between mountains will likely pose serious obstacles to poleward shifts [7]. Broad-scale modeling will continue to be important for addressing these problems. Furthermore, fine-scale environmental heterogeneity does not guarantee that biodiversity will be buffered from climate change. It is possible for heterogeneity to produce small, isolated patches of habitat that cannot support many species, producing a negative effect on diversity [16]. Thus, whether the net effect of heterogeneity on maintaining diversity will be positive during a period of rapid climate change remains an open question.

Different kinds of cool or snowy microhabitats will likely differ in their effectiveness as refugia in a warming world. First, the abundance of microhabitat types will strongly influence how effective they can be as refugia. For example, depressions in the landscape in the subalpine/alpine biomes may have a high likelihood of serving as refugia because they are a common topographic feature. Second, the longevity of microhabitat types will affect their ability to act as refugia. For example, canopy gaps may disappear relatively quickly as trees establish in them, forcing species that might use gaps as refugia to migrate amongst gaps, which may not be possible for some species (though others may be adapted to this migration). In contrast, depressions in the landscape could provide more long-term refugia. Third, the temporal climatic heterogeneity experienced in microhabitat types may affect how well they can serve as refugia. For example, gaps had lower daily minimum and higher daily maximum temperatures compared to non-gaps, showing that these microhabitats experience a wide variety of temperatures. This heterogeneity may favor some species but not others. A final complicating factor influencing how and whether microhabitat types can serve as climatic refugia are the non-climatic conditions associated with them. For example, depressions may differ from other topographic positions in soil characteristics, which could prevent some species from using them as snowy microrefugia.

An important caveat to these findings is that they are based on one year of data. Spatial patterns in climate can change from year to year due to differences in prevailing synoptic weather patterns [46–48], so the patterns we observed in the year we conducted this study may not represent typical spatial patterns. However, the year of our study was a fairly typical year in terms of snow disappearance date and in terms of growing season air temperature for the past few decades (Appendix S3). And although spatial patterns in climate can vary year to year, the patterns are generally constant from one year to the next, especially in terms of snow [49–52]. For example, locations with later snow disappearance dates in one year tend to have later snow disappearance dates in other years, even though the spatially averaged snow disappearance date varies from year to year.

Challenges and Opportunities for Management

To best protect biodiversity in a period of rapid climate change, conservation biologists and resource managers will require realistic assessments of future species distributions [53]. Thus, incorporating fine-scale climatic heterogeneity is essential for improving projections of species range shifts and extinction risks. Current coarse-scale models of the relationships between climate and species distributions ignore fine-scale heterogeneity and may therefore overestimate the distance species must migrate to track suitable climate (because forecasted range shifts are necessarily at the resolution of the model), and overpredict habitat loss and extinction risks ([54–57], but see [58]). Ecologists have previously criticized these bioclimate envelope models for only predicting where the climate that is currently associated with a species distribution will shift to and failing to account for biotic factors that could affect a species' ability to track these climate shifts (dispersal limitations, biotic interactions and evolutionary changes) [59]. However, the limitation of model spatial resolution could undermine predictions not only of species' abilities to track shifts in climate but also of the climate shifts themselves.

In addition to more realistically forecasting range shifts, knowledge of fine-scale climatic heterogeneity may also allow managers to increase species and ecosystem resilience to climate change. For example, protecting microhabitats with cooler temperatures or later snow disappearance dates could become increasingly important as climate change occurs because these microhabitats may provide critical refugia for species. Additionally, our results suggest that planting seedlings and sowing seeds at a variety of microhabitats when restoring degraded sites is an important bet-hedging strategy because it could increase the probability that species establish in microsites that remain suitable as climate change progresses, even if those microsites are only marginally suitable now. Thus, information on fine-scale climate heterogeneity has the potential to be useful for natural area protection and restoration when taken together with other important factors such as edaphic constraints, biotic interactions, genetic diversity and financial costs [60]. More detailed and longer term studies are needed to assess whether microclimate heterogeneity can contribute substantially to plant establishment and restoration efforts in a warming world.

Conclusions

We have shown that snow disappearance date and growing season soil temperature vary dramatically over small distances due to differences in vegetation structure and topography. In fact, fine-scale features such as gaps in the forest canopy or small depressions in the landscape can produce differences in snow disappearance date and soil temperature as large as those produced by shifting hundreds of meters up a mountain slope. This large degree of fine-scale spatial heterogeneity may provide an important buffer against the negative effects of rapid climate change, as many species may only need to migrate tens of meters from one microhabitat to another in order to track suitable climate, as opposed to shifting hundreds of meters upward in elevation or hundreds of kilometers poleward. Climate change will undoubtedly pose serious threats to biodiversity, but knowledge of fine-scale climatic heterogeneity may allow managers to better assess and potentially alleviate some of these threats.

Acknowledgments

Thanks to Erin Curtis, Rachel Konrady, Tony Krueger, Jack Lee, Gerald Lisi, Anna O'Brien, Mitch Piper, Elli Jenkins Theobald and Courtney Wenneborg for field assistance. We also want to thank Lou Whiteaker and Bret Christoe at Mount Rainier National Park for advice and help with permitting as well as Josh Lawler, Jeremy Littell, Regina Rochefort, Catharine Copass Thompson, Pete Wyckoff and two anonymous reviewers for providing helpful feedback on the manuscript.

Author Contributions

Conceived and designed the experiments: KF AE JHRL. Performed the experiments: KF AE. Analyzed the data: KF. Contributed reagents/materials/analysis tools: KF AE MR JHRL. Wrote the paper: KF AE JL MR JHRL.

References

1. Humboldt Av, Bonpland A (1807) Essay on the geography of plants. Paris: Levrault Schoell. 155 p.
2. Merriam CH, Steineger L (1890) Results of a biological survey of the San Francisco Mountain region and the desert of the Little Colorado, Arizona. North American Fauna Report 3. Washington, DC: US Department of Agriculture, Division of Ornithology and Mammalogy. 136 p.
3. Whittaker RH (1975) Communities and ecosystems. New York: Macmillan. 385 p.
4. Chen I, Hill JK, Ohlemueller R, Roy DB, Thomas CD (2011) Rapid range shifts of species associated with high levels of climate warming. Science 333: 1024–1026. doi: 10.1126/science.1206432.
5. Davis M, Shaw R (2001) Range shifts and adaptive responses to Quaternary climate change. Science 292: 673–679. doi: 10.1126/science.292.5517.673.
6. Huntley B (2005) North temperate responses. In: Lovejoy TE, Hannah L, editors. Climate change and biodiversity. New Haven, Connecticut: Yale University Press. pp. 109–124.
7. Fischlin A, Midgley GF, Price JT, Leemans R, Gopal B, et al. (2007) Ecosystems, their properties, goods, and services. In: Parry ML, Canziani OF, Palutikof JP, Linden PJvd, Hanson CE, editors. Climate change 2007: Impacts, adaptation and vulnerability. Contribution of Working Group II to the Fourth Assessment Report of the Intergovernmental Panel on Climate Change. Cambridge, UK and New York, NY, USA: Cambridge University Press. pp. 211–272.
8. Loarie SR, Duffy PB, Hamilton H, Asner GP, Field CB, et al. (2009) The velocity of climate change. Nature 462: 1052–1057. doi: 10.1038/nature08649.
9. Levin SA (1992) The problem of pattern and scale in ecology. Ecology 73: 1943–1967.
10. Daly C, Halbleib M, Smith JI, Gibson WP, Doggett MK, et al. (2008) Physiographically sensitive mapping of climatological temperature and precipitation across the conterminous United States. Int J Climatol 28: 2031–2064. doi: 10.1002/joc.1688.
11. Hijmans RJ, Cameron SE, Parra JL, Jones PG, Jarvis A (2005) Very high resolution interpolated climate surfaces for global land areas. Int J Climatol 25: 1965–1978. doi: 10.1002/joc.1276.
12. Thuiller W, Lavorel S, Araujo MB, Sykes MT, Prentice IC (2005) Climate change threats to plant diversity in Europe. Proc Natl Acad Sci U S A 102: 8245–8250. doi: 10.1073/pnas.0409902102.
13. Guisan A, Zimmermann NE (2000) Predictive habitat distribution models in ecology. Ecol Model 135: 147–186.
14. Geiger R, Aron RH, Todhunter P (2009) The climate near the ground. Lanham, MD: Rowman & Littlefield Publishers, Inc. 623 p.
15. Ackerly DD, Loarie SR, Cornwell WK, Weiss SB, Hamilton H, et al. (2010) The geography of climate change: Implications for conservation biogeography. Divers Distrib 16: 476–487. doi: 10.1111/j.1472-4642.2010.00654.x.
16. Tamme R, Hiiesalu I, Laanisto L, Szava-Kovats R, Partel M (2010) Environmental heterogeneity, species diversity and co-existence at different spatial scales. Journal of Vegetation Science 21: 796–801. doi: 10.1111/j.1654-1103.2010.01185.x.
17. Willis KJ, Bhagwat SA (2009) Biodiversity and climate change. Science 326: 806–807. doi: 10.1126/science.1178838.
18. Larcher W (2003) Physiological plant ecology: Ecophysiology and stress physiology of functional groups. Berlin: Springer-Verlag. 513 p.
19. Franklin JF, Moir WH, Hemstrom MA, Greene SE, Smith BG (1988) The forest communities of Mount Rainier National Park. Scientific Monograph Series No. 19. Washington, DC: US Department of the Interior, National Park Service. 194 p.
20. Wipf S, Rixen C (2010) A review of snow manipulation experiments in Arctic and alpine tundra ecosystems. Polar Res 29: 95–109. doi: 10.1111/j.1751-8369.2010.00153.x.
21. Scherrer D, Körner C (2011) Topographically controlled thermal-habitat differentiation buffers alpine plant diversity against climate warming. J Biogeogr 38: 406–416. doi: 10.1111/j.1365-2699.2010.02407.x.
22. Mote P, Salathé EP (2009) Future climate in the Pacific Northwest. In: Elsner MM, Littell J, Binder LW, editors. The Washington climate change impacts assessment. Seattle, WA: Climate Impacts Group, Center for Science in the Earth System, Joint Institute for the Study of the Atmosphere and Oceans, University of Washington. pp. 21–43.
23. Elsner MM, Cuo L, Voisin N, Deems JS, Hamlet AF, et al. (2009) Implications of 21st century climate change for the hydrology of Washington State. In: Elsner MM, Littell J, Binder LW, editors. The Washington climate change impacts assessment. Seattle, WA: Climate Impacts Group, Center for Science in the Earth System, Joint Institute for the Study of the Atmosphere and Oceans, University of Washington. pp. 69–106.
24. Clark MP, Hendrikx J, Slater AG, Kavetski D, Anderson B, et al. (2011) Representing spatial variability of snow water equivalent in hydrologic and land-surface models: A review. Water Resources Research 47: doi:10.1029/2011WR010745.
25. Fridley JD (2009) Downscaling climate over complex terrain: High finescale (<1000 m) spatial variation of near-ground temperatures in a montane forested landscape (Great Smoky Mountains). Journal of Applied Meteorology and Climatology 48: 1033–1049. doi: 10.1175/2008JAMC2084.1.
26. Millar CI, Westfall RD (2010) Distribution and climatic relationships of the American pika (*Ochotona princeps*) in the Sierra Nevada and western Great Basin, USA; periglacial landforms as refugia in warming climates. Arctic Antarctic and Alpine Research 42: 76–88. doi: 10.1657/1938-4246-42.1.76.
27. Scherrer D, Körner C (2010) Infra-red thermometry of alpine landscapes challenges climatic warming projections. Global Change Biol 16: 2602–2613. doi: 10.1111/j.1365-2486.2009.02122.x.
28. Wundram D, Pape R, Loeffler J (2010) Alpine soil temperature variability at multiple scales. Arctic Antarctic and Alpine Research 42: 117–128. doi: 10.1657/1938-4246-42.1.117.
29. Deutsch CA, Tewksbury JJ, Huey RB, Sheldon KS, Ghalambor CK, et al. (2008) Impacts of climate warming on terrestrial ectotherms across latitude. Proc Natl Acad Sci U S A 105: 6668–6672. doi: 10.1073/pnas.0709472105.
30. Meehl GA, Stocker TF, Collins WD, Friedlingstein P, Gaye AT, et al. (2007) Global climate projections. In: Solomon S, Qin D, Manning M, Chen Z, Marquis M, et al, editors. Climate change 2007: The physical science basis. Contribution of Working Group I to the Fourth Assessment Report of the Intergovernmental Panel on Climate Change. Cambridge, UK and New York, NY, USA: Cambridge University Press. pp. 747–845.
31. Lundquist JD, Lott F (2008) Using inexpensive temperature sensors to monitor the duration and heterogeneity of snow-covered areas. Water Resour Res 44: W00D16. doi: 10.1029/2008WR007035.
32. Raleigh MS, Rittger K, Moore C, E., Henn B, Lutz J, A., etal. (2013) Ground-based testing of MODIS fractional snow cover in subalpine meadows and forests of the Sierra Nevada. Remote Sensing of Environment 128: 44–57.
33. Zuur AF, Ieno EN, Walker NJ, Saveliev AA, Smith GM (2009) Mixed effects models and extensions in ecology with R. New York: Springer. 574 p.
34. R Development Core Team (2010) R: A language and environment for statistical computing.
35. Bates D, Maechler M, Bolker B (2011) lme4: Linear mixed-effects models using S4 classes. R package version 0.999375-39.
36. Hadfield JD (2010) MCMC methods for multi-response generalized linear mixed models: The MCMCglmm R package. Journal of Statistical Software 33: 1–22.
37. Xu RH (2003) Measuring explained variation in linear mixed effects models. Stat Med 22: 3527–3541. doi: 10.1002/sim.1572.
38. Varhola A, Coops NC, Weiler M, Moore RD (2010) Forest canopy effects on snow accumulation and ablation: An integrative review of empirical results. Journal of Hydrology 392: 219–233. doi: 10.1016/j.jhydrol.2010.08.009.
39. Sicart JE, Pomeroy JW, Essery RLH, Hardy J, Link T, et al. (2004) A sensitivity study of daytime net radiation during snowmelt to forest canopy and

atmospheric conditions. J Hydrometeorol 5: 774–784. doi: 10.1175/1525-7541(2004)005<0774:ASSODN>2.0.CO;2.

40. Rochefort RM, Peterson DL (1996) Temporal and spatial distribution of trees in subalpine meadows of Mount Rainier National Park, Washington, USA. Arct Alp Res 28: 52–59.
41. Zald HSJ, Spies TA, Huso M, Gatziolis D (2012) Climatic, landform, microtopographic, and overstory canopy controls of tree invasion in a subalpine meadow landscape, Oregon Cascades, USA. Landscape Ecol 27: 1197–1212. doi: 10.1007/s10980-012-9774-8.
42. Körner C (2003) Alpine plant life: Functional plant ecology of high mountain ecosystems, 2nd edition. Berlin: Springer. 344 p.
43. Dobrowski SZ (2011) A climatic basis for microrefugia: The influence of terrain on climate. Global Change Biol 17: 1022–1035. doi: 10.1111/j.1365-2486.2010.02263.x.
44. Keppel G, Van Niel KP, Wardell-Johnson GW, Yates CJ, Byrne M, et al. (2012) Refugia: Identifying and understanding safe havens for biodiversity under climate change. Global Ecol Biogeogr 21: 393–404. doi: 10.1111/j.1466-8238.2011.00686.x.
45. Hof C, Levinsky I, Araujo MB, Rahbek C (2011) Rethinking species' ability to cope with rapid climate change. Global Change Biol 17: 2987–2990. doi: 10.1111/j.1365-2486.2011.02418.x.
46. Lundquist JD, Cayan DR (2007) Surface temperature patterns in complex terrain: Daily variations and long-term change in the central Sierra Nevada, California. Journal of Geophysical Research-Atmospheres 112: D11124. doi: 10.1029/2006JD007561.
47. Bennie JJ, Wiltshire AJ, Joyce AN, Clark D, Lloyd AR, et al. (2010) Characterising inter-annual variation in the spatial pattern of thermal microclimate in a UK upland using a combined empirical-physical model. Agric For Meteorol 150: 12–19. doi: 10.1016/j.agrformet.2009.07.014.
48. Lundquist JD, Minder JR, Neiman PJ, Sukovich E (2010) Relationships between barrier jet heights, orographic precipitation gradients, and streamflow in the northern Sierra Nevada. J Hydrometeorol 11: 1141–1156. doi: 10.1175/2010JHM1264.1.
49. Erickson T, Williams M, Winstral A (2005) Persistence of topographic controls on the spatial distribution of snow in rugged mountain terrain, Colorado, United States. Water Resour Res 41: W04014. doi: 10.1029/2003WR002973.
50. Deems JS, Fassnacht SR, Elder KJ (2008) Interannual consistency in fractal snow depth patterns at two Colorado mountain sites. J Hydrometeorol 9: 977–988. doi: 10.1175/2008JHM901.1.
51. Sturm M, Wagner AM (2010) Using repeated patterns in snow distribution modeling: An Arctic example. Water Resour Res 46: W12549. doi: 10.1029/2010WR009434.
52. Egli L, Jonas T, Gruenewald T, Schirmer M, Burlando P (2012) Dynamics of snow ablation in a small alpine catchment observed by repeated terrestrial laser scans. Hydrol Process 26: 1574–1585. doi: 10.1002/hyp.8244.
53. Hannah L, Midgley G, Andelman S, Araujo M, Hughes G, et al. (2007) Protected area needs in a changing climate. Frontiers in Ecology and the Environment 5: 131–138.
54. Austin MP, Van Niel KP (2011) Impact of landscape predictors on climate change modelling of species distributions: A case study with *Eucalyptus fastigata* in southern New South Wales, Australia. J Biogeogr 38: 9–19. doi: 10.1111/j.1365-2699.2010.02415.x.
55. Barrows CW, Murphy-Mariscal ML (2012) Modeling impacts of climate change on Joshua trees at their southern boundary: How scale impacts predictions. Biol Conserv 152: 29–36. doi: 10.1016/j.biocom.2012.03.028.
56. Luoto M, Heikkinen RK (2008) Disregarding topographical heterogeneity biases species turnover assessments based on bioclimatic models. Global Change Biol 14: 483–494. doi: 10.1111/j.1365-2486.2007.01527.x.
57. Randin CF, Engler R, Normand S, Zappa M, Zimmermann NE, et al. (2009) Climate change and plant distribution: Local models predict high-elevation persistence. Global Change Biol 15: 1557–1569. doi: 10.1111/j.1365-2486.2008.01766.x.
58. Trivedi MR, Berry PM, Morecroft MD, Dawson TP (2008) Spatial scale affects bioclimate model projections of climate change impacts on mountain plants. Global Change Biol 14: 1089–1103. doi: 10.1111/j.1365-2486.2008.01553.x.
59. Pearson RG, Dawson TP (2003) Predicting the impacts of climate change on the distribution of species: Are bioclimate envelope models useful? Global Ecol Biogeogr 12: 361–371.
60. Rochefort RM, Kurth LL, Carolin TW, Riedel JL, Mierendorf RR, et al. (2006) Mountains. In: Apostol D, Sinclair M, editors. Restoring the Pacific Northwest: The art and science of ecological restoration in Cascadia. Washington, DC: Island Press. pp. 241–275.

Narrow-Front Loop Migration in a Population of the Common Cuckoo *Cuculus canorus*, as Revealed by Satellite Telemetry

Mikkel Willemoes[1]*, Roine Strandberg[2], Raymond H. G. Klaassen[2,3], Anders P. Tøttrup[1], Yannis Vardanis[2], Paul W. Howey[4], Kasper Thorup[1], Martin Wikelski[5,6], Thomas Alerstam[2]

1 Center for Macroecology, Evolution and Climate, Natural History Museum of Denmark, University of Copenhagen, Copenhagen, Denmark, **2** Department of Biology, Lund University, Lund, Sweden, **3** Dutch Montagu's Harrier foundation and Animal Ecology Group, Centre for Ecological and Evolutionary Studies, Groningen University, Groningen, the Netherlands, **4** Microwave Telemetry Inc, Columbia Maryland, United States of America, **5** Max Planck Institute for Ornithology, Department of Migration & Immuno-ecology, Radolfzell am Bodensee, Germany, **6** Department of Biology, University of Konstanz, Konstanz, Germany

Abstract

Narrow migration corridors known in diurnal, social migrants such as raptors, storks and geese are thought to be caused by topographical leading line effects in combination with learning detailed routes across generations. Here, we document narrow-front migration in a nocturnal, solitary migrant, the common cuckoo *Cuculus canorus*, using satellite telemetry. We tracked the migration of adult cuckoos from the breeding grounds in southern Scandinavia (n = 8), to wintering sites in south-western Central Africa (n = 6) and back to the breeding grounds (n = 3). Migration patterns were very complex; in addition to the breeding and wintering sites, six different stopover sites were identified during the 16,000 km annual route that formed a large-scale clockwise loop. Despite this complexity, individuals showed surprisingly similar migration patterns, with very little variation between routes. We compared observed tracks with simulated routes based on vector orientation (with and without effects of barriers on orientation and survival). Observed distances between routes were often significantly smaller than expected if the routes were established on the basis of an innate vector orientation programme. Average distance between individuals in eastern Sahel after having migrated more than 5,000 km for example, was merely 164 km. This implies that more sophisticated inherent guiding mechanisms, possibly involving elements of intermediate goal area navigation or more elaborate external cues, are necessary to explain the complex narrow-front migration pattern observed for the cuckoos in this study.

Editor: Csaba Moskát, Hungarian Academy of Sciences, Hungary

Funding: The work was supported by the Danish National Research Foundation http://dg.dk/en/and Swedish Research Council http://www.vr.se/inenglish. The funders had no role in study design, data collection and analysis, decision to publish, or preparation of the manuscript.

Competing Interests: The authors have the following interests. Paul Howey is the director of Microwave Telemetry, the company that makes the satellite tags used for tracking the birds. There are no patents, products in development or marketed products to declare. This does not alter the authors' adherence to all the PLOS ONE policies on sharing data and materials, as detailed online in the guide for authors.

* E-mail: mwkristensen@snm.ku.dk

Introduction

Every year billions of birds migrate between Europe and sub-Saharan Africa [1,2]. Especially for the smaller, nocturnal migrants, little is known about population specific migration schedules, migration routes, stopover sites and non-breeding ranges [3]. For some diurnally migrating birds, and particularly for soaring migrants utilising thermal updrafts, it is well known that populations follow specific and narrow migration corridors. Typically these birds avoid large water bodies and high mountains, resulting in funnel-shaped flyways converging at specific points [4]. Well known migration hot spots, like the Strait of Gibraltar, Bosporus and the Isthmus of Panama, are good examples of such convergence points, although it is not clear how many birds do not use these specific concentration points. For many nocturnal, solitary migrants, for example most songbirds, migration corridors are generally much less condensed (often several hundred kilometres broad as seen from e.g. bird ringing atlases of Scandinavian birds [5,6]). In social migrants, young birds on their first migration receive guidance from experienced birds as in day-migrating geese, storks, cranes [7,8,9] and possibly raptors [10]. In contrast, young solitary migrants are supposedly only led by their innate migration programme. It is generally believed that the innate migration programme is based on a simple vector orientation principle (a clock-and-compass strategy) [4,11], perhaps with some external influences. An additional component of true navigation is often considered unlikely among naïve birds on their first migratory journey (but see [12,13,14]). As the scatter between individuals is always increasing with distance travelled in migration routes resulting from a clock-and-compass strategy [15], some external factors (topography, weather, habitat etc) or the possibility of goal area navigation are necessary to explain convergence of migration tracks [16,17].

Currently, full migration patterns of many species of birds are being revealed (e.g. [18,19,20]), though large diurnally migrating species are still over-represented in this field [21]. Nocturnal, solitary migrants are typically small and it has so far not been possible to track these migrants using accurate methods such as satellite telemetry or GPS. Here, we present the first migration tracks of the common cuckoo *Cuculus canorus*, which is a primarily nocturnal, solitary migrant [2,22,23]. The common cuckoo is an obligate nest parasite and young cuckoos are raised by a host species. Because the adults leave the breeding grounds well before the young, it can be ruled out that young, naïve cuckoos learn from experienced adults [24]. Hence, cuckoo migration is the ultimate example of a genetic basis of migration patterns. However, little is known about its migration. Cuckoos are rarely seen in high concentrations on the typical migration hotspots along the Mediterranean coast and they are thought to migrate over a broad front and possibly overfly northern Africa without stopping [24,25]. Ringing has only provided sparse information of the migration route. From southern Scandinavia, two birds ringed in Denmark were recovered in Italy in spring [5] and four birds ringed in Sweden were recovered around the Mediterranean in autumn; two in eastern Italy, one in Greece and one in north-western Egypt [26]. No ringing recoveries from Scandinavian cuckoos are available from sub-Saharan Africa.

We describe the entire annual migration cycle of adult common cuckoos from their breeding sites in southern Scandinavia to the winter grounds in south-western Central Africa and back. The tracks revealed that birds from this population were migrating in a large clockwise loop during their annual cycle, with the different individuals following closely similar routes using several stopover areas in both temperate and tropical zones without diverging from each other to any large degree. This similarity between individual routes in such a complex geographic migration pattern seems to be difficult to reconcile with the assumption that individual migration patterns have been established as a result of an innate vector orientation programme during the first annual migratory journey.

For adult birds, it is assumed that they have learnt about suitable stopover and wintering areas during their first journey as young birds, allowing them to navigate back to these familiar areas during subsequent journeys of their lives [4,27]. This means that the spatial variation between adult individuals to a large degree is expected to reflect the scatter caused by the vector orientation during the individuals' first journeys, while consistency within individuals in their successive journeys, showing movements towards the same stopover/wintering/breeding areas, reflects the learned ability of navigation/homing.

According to the principles of vector orientation alone, we should always expect the scatter between the routes of different individuals to increase with distance along the migration route. However, there are three main factors that may limit or reduce this increase in spatial scatter between individuals. (*i*) The scatter between individuals will increase with distance at a slower rate with increasing number of orientation steps/vectors (or with shorter step/vector lengths) used by migrants to cover a certain distance (with the successive steps/vectors drawn from a circular distribution with a given mean and variation). Hence, the frontal geographic scatter is expected to be approximately proportional to $(n)^{-1/2}$, where n is the number of orientation steps to reach a certain total distance (cf. [13,14,15,28,29]). (*ii*) If vector orientation is combined with a response to topographical features (large water bodies, mountains or other habitats acting as ecological barriers) this will have the effect of adjusting individual routes towards more favourable habitats and away from large barriers, which will decrease the scatter between routes. (*iii*) Differential survival of juveniles on their first journey could have the effect that only individuals following specific favourable routes by their vector orientation programme will survive. This may lead to a reduced spatial scatter among adult individuals as a result of mortality of juveniles travelling outside the most favourable migration corridor. Of course, the mortality loss of juveniles must be substantial to explain a very narrow corridor for the surviving population of adult individuals.

In order to investigate if vector orientation, alone or in combination with the effects of topographical response and differential survival, can explain the observed pattern of complex loop migration of the cuckoos, we compare observed tracks with tracks simulated according to a clock-and-compass migration programme based on the migratory behaviour shown by the tracked birds. Particularly we investigate if the concentration of tracks is similar between simulated and observed routes.

We demonstrate that it is highly unlikely that the complex loop migration of southern Scandinavian cuckoos is generated by a vector orientation programme, not even if such a programme is combined with inherent responses to topographical features and with strong mortality of juveniles on the flanks of the primary loop migration corridor. We infer that these migrants are likely to rely on more sophisticated inherent instructions along their migration route, to allow the evolution of complex and narrowly defined population-specific migration routes.

Methods

Ethics statement

Animal work in Denmark was approved by the Danish Nature Agency by permission to the Copenhagen Bird Ringing Centre (J.nr. SN 302-009). In Sweden, the study was approved by the Ethical Committee for Animal Experiments in Malmö/Lund (M112-09). The Danish Nature Agency gave permission to work on their land in Denmark.

Tracking

Eight adult common cuckoos, five males and three females, were tagged on the breeding grounds in eastern Denmark and southern Sweden (55.68–56.05°N, 12.30–14.53°E) and followed from the breeding season in 2010 to the breeding season in 2011 or until transmission stopped. Three birds transmitted the entire annual cycle returning to the breeding grounds. One stopped transmitting in north-central Europe, one in the southern Sahara, one in the winter quarters, one in West Africa and one in northern Africa north of the Sahara.

The birds were caught using mist nets and tape lure in wet land areas with plenty of bushes (primarily *Salix spp*) and heavy reed *Phragmites australis* growth. Cuckoos from these habitats and in this region are typically using Eurasian reed warbler *acrocephalus scirpaceus* as a host species. The tagged birds were sexed using plumage, biometric and voice characters. The transmitters, Solar PTT-100s (Microwave Telemetry Inc.), were fitted to the birds as a back-pack using a body harness made from a 2 mm braided nylon string. The weight of the transmitters was 5 g corresponding on average to 4.3% (3.7–5.0%) of the body weight of the cuckoos. Transmission was scheduled on a 10 hours on 48 hours off duty cycle.

Geographical positions of the transmitters were obtained from ARGOS/CLS Service Argos [30]. Position estimates from ARGOS are assigned to a location quality class (3-0, A–B and Z; 3 has highest and Z lowest accuracy). We excluded all positions of class Z. During transmission periods, we determined whether a bird was travelling or stationary. A bird was considered stationary, if distances moved were small (<10 km; high quality class positions) or in random directions (low quality positions). For stationary birds, only the highest quality position from each 10 h transmission period was included. For travelling birds, all positions during the 10 h transmission periods were used for analyses. For several transmission cycles no positions were obtained, especially in Europe during autumn where the birds passed through an area with well known poor satellite reception, probably due to a high level of background noise in this area [31].

Staging periods were defined as those lasting a minimum of 5 days. To give an overview of the migration pattern, we defined major migration steps and staging areas. Migration steps were defined as steps between staging areas shown by at least half of the individuals (ESM table S1). Individual staging areas were considered the same if they were in the same overall region and were in the same stage of the migration (e.g. staging areas in Hungary and Bulgaria were considered the same staging area since they were all in south-eastern Europe and following a stopover in northern Europe). In some cases, we lacked the positions necessary to determine whether a bird used a specific stopover area. For the same reason, departure and arrival dates at stopover areas are in some cases uncertain.

We defined the mean stopover position as the average of longitude and latitude for all birds using that stopover area. The distance between the birds at each stopover was calculated as the average of the distances from each tracked individual to the mean position in that stopover area.

Simulations

We simulated several possible migration scenarios based on a clock-and-compass migration system (Table 1). The clock-and-compass system was modelled using vector summation [12,13,14,15] according to each specific scenario based on the directional concentration and step lengths estimated from the set of migration vectors identified above (the steps between stopovers). In other studies using vector summation, each migration step is normally modelled as a vector with a specified length and a direction taken randomly from a directional distribution, typically the von Mises distribution [14,32,33]. We used this approach when simulating migration with short steps, and used random recombination of the actual observed vectors when simulating long steps (see below).

We investigated if the observed narrow loop migration pattern could be explained by a clock-and-compass system, using the following migration scenarios (Table 1): (*i*) Adults using single vector steps between each stopover and winter site, not considering topography. (*ii*) As (*i*), except that the vector step length was 100 km (i.e. many steps between each stopover). (*iii*) Migration of adult birds, but with mortality resulting from barrier crossing taken into account. Since mortality also occurred among the adults we tracked, we did not remove the entire route, but only the part that comes after a barrier. Step length was 100 km. (*iv*) Adults encountering impassable barriers (mountains or ocean) choose another random direction to circumvent it if possible. Step length was 100 km. (*v*) Adult birds following the routes arisen from juvenile vector orientation, but with juvenile mortality resulting from barrier crossing implemented. As the mortality occurs prior to our tracking (during the birds' first migration cycle as juveniles), entire maladaptive routes were removed. This was modelled with single steps between stopovers as in (*i*). (*vi*) As (*v*), except that step length was 100 km.

Simulations excluding effects of barriers: models (i) and (ii). Our basic simulations assumed that migration routes were established by the use of a clock-and-compass strategy between the successive stopover and winter sites. Under this assumption, we fitted two models. In the first, the distance between stopovers was covered in one single step. For this simulation, the vector summation was done by simply using the vectors that the birds used for travelling between stopovers, i.e. random combinations of the individual vectors recorded for each of the eight main steps of the annual migration loop. In the second model, we used a more typical step length of 100 km [12,14]. Shorter steps result in less scattered routes, but as it is unlikely that daily migration steps (flight steps during a night) were shorter than 100 km (this data) or that orientation is re-established more frequently than at each daily flight step [34,35], we consider this a conservative approach. The concentration of the directional distribution to be used with the smaller step length was found as the one in which vector addition resulted in a mean vector with concentration and length equal to the mean step vector observed for tracked birds at each of the eight different main steps between stopover areas. For each model, we simulated 1000 tracks.

Simulation with removal of part of the track after reaching barriers: model (iii). The topographical structures considered as barriers were high mountains and large water bodies. Mountains higher than 1700 meters and water surfaces

Table 1. Description of the different migration scenarios simulated. See Methods for more details.

Scenario	Name	Topography effects	Description
(i)	LONG	Not included	Migration simulated as a clock-and-compass strategy with single steps between each stopover and winter sites (8 steps in total). Each step is drawn randomly between the vectors observed in the tracked birds. This scenario is valid both for adults following the same route as they did as juveniles and for adults using a clock-and-compass strategy directly
(ii)	SHORT	Not included	Same as (i) LONG above, except that the step length is 100 km. Directions are drawn randomly from a circular distribution that fits the concentrations at each stopover.
(iii)	SHORT_Mort	Direct mortality	Migration of adult birds with birds crossing mountains or ocean dying, i.e. tracks are discontinued after crossing. Otherwise as (ii) SHORT.
(iv)	SHORT_Dir_Change	Change of migration direction	Migration simulated as a clock-and-compass strategy with birds encountering impassable mountains or ocean choosing another random direction to circumvent it if possible. This scenario is valid both for adults following the same route as they did as juveniles and for adults using a clock-and-compass strategy directly
(v)	LONG_Juv_Mort	Juvenile mortality	Migration of adult birds following the routes followed as juveniles but with juvenile mortality taken into account. Thus, juvenile birds crossing mountains or ocean are assumed to die and thus, these tracks are removed. Step length as in (i) LONG.
(vi)	SHORT_Juv_Mort	Juvenile mortality	As above (v) LONG-Juv-Mort, except that steps are modelled as in (ii) SHORT.

more than 200 kilometres from the nearest coast were considered as barriers. If adult birds migrate by a vector orientation program mortality from barrier crossing could potentially cause route convergence. We modelled the migration of such adults by assuming that barrier crossing would result in mortality and we terminated all simulated tracks from the point where a track reached a barrier. To be comparable with our tracked adults, we included tracks up until the point of termination, since we also have incomplete routes in the tracked birds.

Simulation including directional changes as response to barriers: model (iv). Here, we assumed that a bird encountering a barrier would choose a new direction for a 100 km step (from the same von Mises distribution). After 100 unsuccessful attempts of finding a random direction that would lead it around the barrier, it was allowed to pass. This scenario imitates a situation where the birds follow a clock-and-compass system, but have an innate response to try to find a new direction when they face a barrier (except if going around the barrier would infer getting too far off course, in which case they cross it).

Simulation with removal of entire tracks reaching barriers: models (v) and (vi). For models incorporating barriers, the assumptions differed between models of juvenile and adult migration. In scenarios (v) and (vi) we assume that the birds migrate by a vector orientation program only as juveniles, while survivors after the first journey (adults) stick to the same or a very similar route as they used as juveniles based on navigation cues that they have learned during the first journey. Assuming mortality of migrating juveniles when crossing barriers could cause convergence in adults. We can therefore directly compare the simulated routes avoiding barriers (i.e. we see only those not dying from barrier crossing) with the observed routes for adults. Under this scenario, we removed the complete tracks of those birds that cross a barrier both for the long and the short step lengths.

Evaluation of model fit to observed data. After all simulations, we calculated the distance between birds at each stopover site. We ran 10,000 bootstraps of n randomly drawn simulated routes and calculated the distance to the mean stopover position (n is the number of tracked birds we have positions for at the respective stopover). All simulations and bootstraps were done in R2.14.0 [36] using packages: "CircStats", "geosphere", "maptools" and "sp" [37,38,39,40]. All route representations were done in ArcMap10 [41].

Results

Tracking

The cuckoos followed very similar routes (Fig. 1) with average total distances travelled in autumn of 7118 km and in spring 9136 km (see Table 2 for a summary of migration timing, distance and speed). They stopped over for prolonged periods of time in north-central Europe (n = 8 out of 8 alive and transmitting individuals), south-eastern Europe (n = 4/7, in three birds it was not possible to determine if they used this stopover due to missing transmissions though), eastern Sahel (n = 6/6), south-western Central Africa in winter (n = 6/6), north-western Central Africa (n = 5/5), West Africa (n = 5/5) and southern Europe (n = 3/3). The migration corridor was remarkably narrow with average distances to the mean position on each stopover ranging from 123–456 km, all smaller than the 480 km in the over-wintering area. In the breeding area the average distance was 59 km before migration and 15 km after returning from migration (because only Danish birds returned). Birds crossed the Sahara in autumn in eastern Libya or western Egypt and Sudan (n = 4). In spring they moved north from West Africa into Mali and Niger and further through Algeria (n = 4), and presumably crossed the Mediterranean Sea between Tunisia and Italy (n = 3; see ESM Figure S1 and Table S1 for details on individual route, timing and stopover duration). The stopovers in north-central Europe lasted on average 26±20 days (mean±sd), in south-eastern Europe 26±19 days, in eastern Sahel 39±19 days, in the over-wintering site in south-western Central Africa 100±15 days, in north-western Central Africa 35±18 days, in West Africa 31±15 days, and in southern Europe 10±1 days before returning to the breeding sites after more than 10 months leaving only 49±3 days for the annual residence time in the breeding area. Two of the three individuals that were tracked for a complete migration cycle returned to the same breeding site, while the third bird moved to a site 8 km from where it was caught. The median return date (17/5) is rather late compared to literature data, but fits well with peak arrival this year (Fig S2).

Clock-and-compass simulations

Tracks simulating a clock-and-compass migration system (Fig. 2) showed a significantly larger scatter than the observed tracks (Fig. 3; for statistical test values see Table S2). Independently of step length, the simulated routes diverged significantly more than the observed tracks after only three steps (Fig. 3 and ESM Table S2). When simulating smaller steps, the scatter of the simulated tracks was similar to the observed tracks in the winter site in southwest Central Africa (Fig. 3) but still larger at stopovers. Taking topography into account improved the model fit, though neither direct mortality nor avoidance of barriers caused the same degree of route convergence as observed in the tracked birds (Fig. 3). Similar patterns were found when assuming that the routes taken by juveniles using vector orientation with barrier-related mortality is reflected in the routes of adults (Fig. 3). Pure vector orientation is normally expected to be associated with a pattern of increasing scatter with increasing distance. However, this was not observed for the simulations after the birds' departure from West Africa in spring, which was due to the fact that (a) the simulations during the final part of spring migration were based on a reduced sample of only three very similar tracks and (b) the birds moved northwards and the absolute distances in frontal spread with a given angular spread in courses will be smaller for a movement from the equator and northwards compared to a corresponding southward movement towards the equator.

Discussion

We found a very narrow migration corridor in a nocturnal, solitary migrant, and the tracks often converged on stopovers along the route. Since this contradicts the pattern expected from a pure clock-and-compass system, such a narrow corridor is highly surprising for a nocturnal solitary migrant relying solely on genetic information. Even though the overwintering range appeared relatively small, the distance between birds was largest in winter. Stopover habitat was primarily open farmland or wetlands (NC Europe, SE Europe, E Sahel, W Africa, and S Europe). The wintering sites and the north-western Central African spring stopover sites were situated in more forested habitats. We have tracked relatively few birds, and with more individuals tracked we would likely see some individuals with divergent routes, this will however most likely be few individuals and not alter the extent of the corridor to any large degree.

Observed routes did not fit the expectations from simple clock-and-compass migration scenarios, even if taking topography such as mountain and ocean barriers into account. Avoidance of geomorphological structures during vector summation reduced the

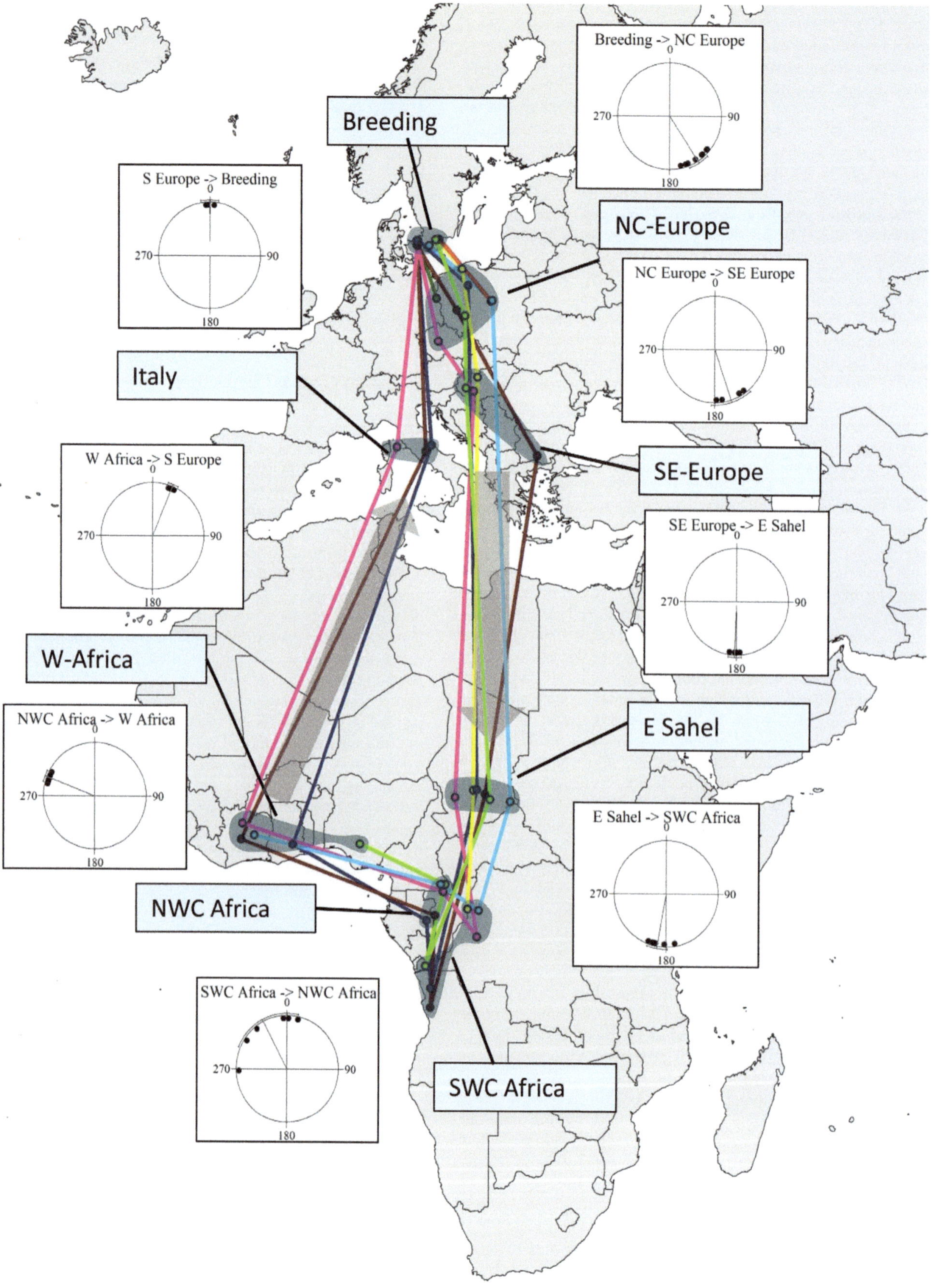

Figure 1. Staging areas of eight satellite-tracked adult common cuckoos with vector directions between stopovers indicated by inserted orientation diagrams. Lines are connecting staging sites and do not necessarily represent the paths followed (see ESM Fig. S1). Mercator projection.

spread in migration routes, but still observed tracks were more concentrated than simulated tracks, at least for some stopovers. Even allowing directional changes induced by barriers, as could be possible both for juvenile and adult birds, resulted in more scatter than was actually observed. That adults do not use vector orientation alone is according to theory, since they should be experienced and able to navigate [4,9]. However, that a population of adult birds would have such a narrow migration corridor is unexpected and hard to explain by vector orientation among juvenile birds, as it implies that only the juveniles that followed the correct route during their first migration survived. Our simulation scenarios of juveniles migrating by vector summation followed by adult route replication were characterized by unreasonably high mortality costs (98%), as also found by Thorup and Rabøl [14] for cases of long-distance songbird migration (up to 90% in autumn migration alone), and still the scatter between locations of surviving individuals (simulations) were larger than that observed (Fig. 3).

A possible alternative explanation for the simulations to be insufficient in explaining route convergence is that the birds are influenced by additional factors such as winds, odour guidance or habitat leading lines, where specific habitat types could be attracting and possibly guiding birds (e.g. rivers or valleys). However, a more likely explanation is that the cuckoos could rely on true navigation. The fact that the tracks were extremely close to each other in for example southern Chad in autumn or Italy in spring suggests that the birds are navigating towards specific *en route* goal areas as proposed by Rabøl [13]. Finally, it cannot be completely excluded that some kind of social interactions could influence the routes chosen by the birds even though this appears unlikely given the mostly solitary behaviour observed in cuckoos during migration [23], and none of the birds tracked in this study were travelling together.

Whether vector summation is sufficient to explain observed migration patterns is a controversial question [33]. As the first two steps (within Europe) in this study's simulations were not significantly different from the observed, it points to the possible sufficiency of vector summation to explain observed patterns in some studies [12,29] of intra-European migration (but see [33]). All three studies which concluded that vector summation is insufficient to explain the migration patterns observed [14,42,43] simulated long-distance migration into Africa, as in this study.

The route and stopover areas of the southern Scandinavian population of cuckoos are likely an adaptation to fit the spatio-temporally optimal route in relation to habitat, available resources, safe barrier crossing and dominating wind systems, the details of which are subject to future research. If so, even small changes in the breeding range could result in different routes being optimal and different populations even within Europe would have different migration routes and stopovers, also even if they share the winter range. The arrival to southern Chad for instance is very well timed with peak primary productivity following the rainy season. Comparable habitats with similar seasonal timing occur over most of the southern Sahel though, why other stopover areas might be more cost-effective for other populations. Indeed, most British cuckoos seem to migrate along different routes, indicating population specific migration patterns [44], although they follow a similar clockwise loop and winter in the same region. More detailed comparative studies of different populations are in progress. The existence of narrow migratory corridors and specific stopover areas point to the possible importance of conservation measures *en route* as e.g. stopover habitat degradation, high hunting pressures or collision risks will affect the entire population using a flyway [45,46,47,48]. This is well-known within conservation of e.g. shorebirds, geese and raptors, but our results suggest that this should not be neglected in the smaller nocturnal migrants.

The documentation of a highly unexpected narrow migration corridor in a nocturnal migrant calls for further studies. A more experimental approach, involving displacements in space and time, is necessary to test whether young and adult cuckoos really

Table 2. Summary table of the autumn and spring migration of eight adult common cuckoos tracked from southern Scandinavia.

Season	Bird	Migration period	Total distance (km)	Total time (days)	Average speed (km/d)	Total travelling days	Av. travelling speed (km/d)
Autumn	Male 19150†	12 Jul–2 Sep†	4158†	52†	80†	?	?
	Male 49466	8 Aug–14 Oct	6261	67	93	<16	>391
	Female 57372	4 Jul–18 Nov	7013	137	51	<27	>260
	Male 57374	1 Jul–2 Nov	6712	124	54	?	?
	Female 36328	18 Jul–22 Oct	6786	96	71	?	?
	Male 36331	6 Jul–19 Nov	8378	136	62	<36	>233
	Female 36332†	30 Jun–6 Sep†	398†	68†	6†	1†	398†
	Male 36487	30 Jun–5 Nov	7559	128	59	<23	>329
	Average‡	10 Jul–3 Nov	7118	115	65	26	303
Spring	Female 57372†	16 Feb–15 May†	5604†	88†	64†	<27†	>208†
	Male 57374†	5 Mar–1 May†	3752†	57†	66†	9†	417†
	Female 36328	30 Jan–2 Jun	9100	123	74	22	414
	Male 36331	26 Feb–17 May	9729	80	122	28	347
	Male 36487	28 Jan–9 May	8578	101	85	25	343
	Average‡	14 Feb–19 May	9136	101	94	25	368

†Transmitters stopped prior to migration completion.
‡Only complete migrations are included in the averages.
Distances are travelled distances along the routes followed, but not including movements within a staging area. Due to missing transmissions, the exact number of travel days was difficult to determine in several cases and the listed numbers are maxima (consequently travelling speeds are minima).

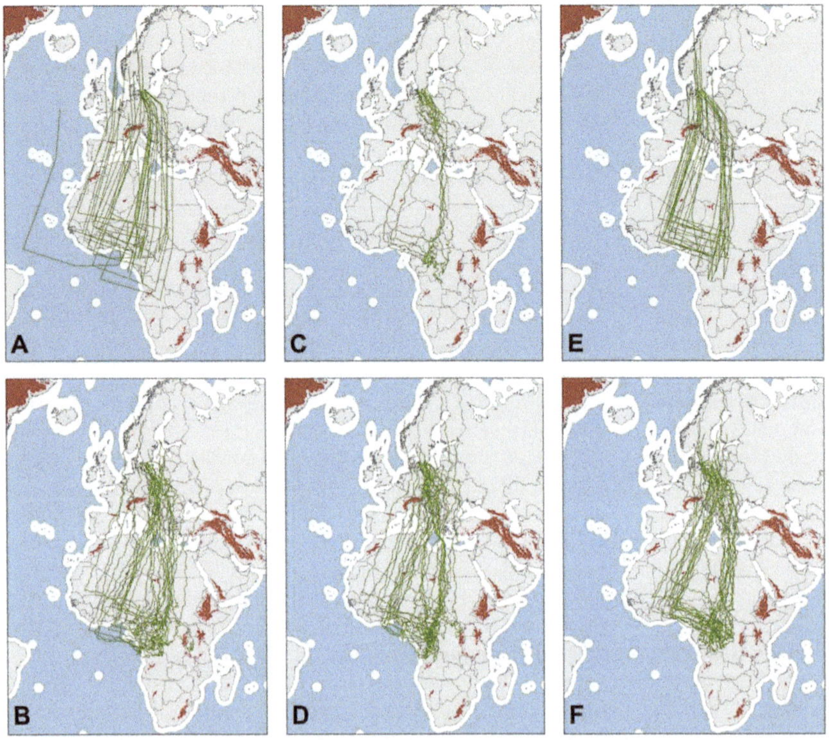

Figure 2. Maps showing examples of 15 random migration routes simulated by vector summation in six different ways. A: simulations using the LONG model, B: SHORT, C: SHORT_Mort, D: SHORT_Dir_Change, E: LONG_Juv_Mort, F: SHORT_Juv_Mort (see Table 1). Blue parts of maps indicate water bodies considered a barrier (>200 km from the coast) and red parts indicate mountain barriers (more than 1700 m above sea level). Mercator projection.

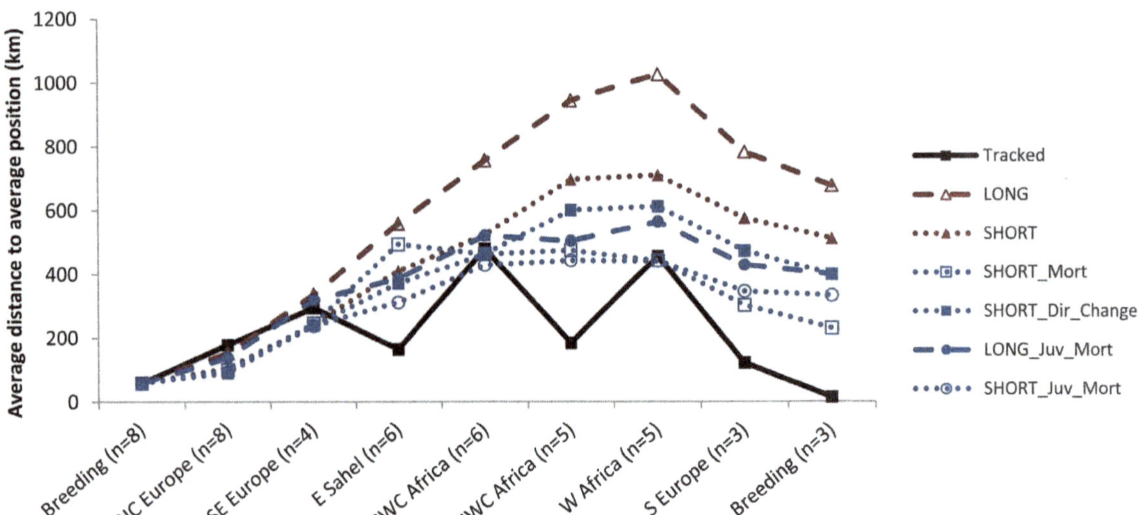

Figure 3. Distances among individual common cuckoos on tracked (solid line) and simulated stopovers during the whole annual migration (see Table 1 for explanation of the types of simulations). In the tracked birds, the values represent average distance from the birds to the average position at that stopover. For the simulated birds, the values are averages of 10,000 bootstraps of average distance from the birds to the average position in that specific stopover. In NC Europe and SE Europe no simulations were significantly different from the tracked birds. In E Sahel, NWC Africa, S Europe and Breeding all simulations were significantly more scattered than the tracked birds. In SWC Africa and W Africa only LONG was significantly more scattered than the tracked individuals (statistical comparisons based on bootstraps in Table S2).

navigate towards specific stopover sites (perhaps based on "sign-posts" of unknown nature) or whether they are guided by other cues related to the landscape.

Supporting Information

Figure S1 Tracks of eight individual common cuckoos as recorded by satellite telemetry. For stationary birds, only the highest quality position from each 10 h transmission period was included. For travelling birds, all positions during the 10 h transmission periods were used for analyses (cf. Methods). Mercator projection.

Figure S2 Timing of arrival of Common cuckoos in Denmark in 2011. The phenology of new unique locations with observations of cuckoos in Denmark in 2011. Note the main peak of new locations (indicating massive arrival) is in mid May. Source: DOFbasen, www.dofbasen.dk, Dansk Ornitologisk Forening (BirdLife Denmark). Accessed 2013 Nov 27.

Table S1 Timing of migration stages in eight adult common cuckoos tracked by satellite telemetry. When departure or arrival was in a period of missing transmissions, the dates are not known exactly as indicated with "<" or ">". The symbol "-" indicates that a bird stopped transmitting before that stopover, whereas "?" indicates that we have no positions for the bird at that stopover, but it could have used it during a period with missing transmissions. Step distance and step direction are the basis for the actual vectors in the first simulations, and are the distances and directions travelled from the preceding stopover to the current. Median/mean are medians of departure/arrival dates for the specific stopover, and means of durations, distances and directions for the specific stopover.

Table S2 Simulating distances between individuals. Simulation results based on vector summation (10,000 bootstraps) of distances between individuals from the four different types of simulation compared with observed data for the cuckoos recorded by satellite tracking. Unshaded rows show the average distances between individuals ±standard deviation, and shaded rows show p-values for simulated routes to be lower than observed data.

Acknowledgments

We thank Troels Eske Ortvad, Jens Mikkel Lausten and Jesper Johannes Madsen for field assistance. We are grateful to the Danish Nature Agency for permission to work at their sites. We acknowledge BirdLife Denmark (Dansk Ornitologisk Forening) for use of data for Fig S2. R.S., R.H.G.K., Y.V. and T.A. are associated with the Centre for Animal Movement Research at Lund University.

Author Contributions

Conceived and designed the experiments: MiW RS RHGK APT KT MaW TA. Performed the experiments: MiW RS RHGK APT KT TA. Analyzed the data: MiW RS RHGK APT YV KT TA. Contributed reagents/materials/analysis tools: PWH MaW TA. Wrote the paper: MiW RS RHGK APT YV PWH KT MaW TA.

References

1. Hahn S, Bauer S, Liechti F (2009) The natural link between Europe and Africa-2.1 billion birds on migration. Oikos 118: 624–626.
2. Moreau RE (1972) The Palearctic-African Bird Migration Systems. London: Academic Press Inc.
3. Bowlin MS, Bisson IA, Shamoun-Baranes J, Reichard JD, Sapir N, et al. (2010) Grand Challenges in Migration Biology. Integr Comp Biol 50: 261–279.
4. Berthold P (2001) Bird Migration. Oxford: Oxford Univ Press.
5. Bønløkke J, Madsen JJ, Thorup K, Pedersen KT, Bjerrum M, et al. (2006) Dansk Trækfugleatlas (Danish Bird Migration Atlas). Humlebæk: Rhodos.
6. Fransson T, Hall-Karlsson S (2008) Svensk ringmärkningsatlas (Swedish bird ringing atlas), Vol. 3. Stockholm.
7. Sterbetz I, Szijj J (1968) Zugverhalten der Rothalsganz (*Branta ruficollis*) in Europa. Vogelwarte 24: 266–277.
8. Schüz E. Überblick über die Orientierungsversuche der Vogelwarte Rossitten (jetzt: Vogelwarte Radolfzell). In: Hörstadius Seditor; 1951. Almquist & Wiksell, Uppsala. pp. 249–268.
9. Essen Lv (1982) An effort to reintroduce the lesser white-fronted goose (*Anser erythropus*) into the Scandinavian mountains. Aquila 89: 103–105.
10. Mellone U, Liminana R, Mallia E, Urios V (2011) Extremely detoured migration in an inexperienced bird: interplay of transport costs and social interactions. Journal of Avian Biology 42: 468–472.
11. Berthold P (1996) Control of Bird Migration. London: Chapman & Hall.
12. Mouritsen H (1998) Modelling migration: the clock-and-compass model can explain the distribution of ringing recoveries. Anim Behav 56: 899–907.
13. Rabøl J (1978) One-direction orientation versus goal area navigation in migratory birds. Oikos 30: 216–223.
14. Thorup K, Rabøl J (2001) The orientation system and migration pattern of long-distance migrants: conflict between model predictions and observed patterns. J Avian Biol 32: 111–119.
15. Alerstam T (2000) Bird migration performance on the basis of flight mechanics and trigonometry. In: Blake RW, Domenici P, editors. Biomechanics in Animal Behaviour. Oxford: BIOS Scientific Publishers Limited. pp. 105–124.
16. Fransson T, Jakobsson S, Kullberg C (2005) Non-random distribution of ring recoveries from trans-Saharan migrants indicates species-specific stopover areas. J Avian Biol 36: 6–11.
17. Alerstam T (2006) Conflicting evidence about long-distance animal navigation. Science 313: 791–794.
18. Egevang C, Stenhouse IJ, Phillips RA, Petersen A, Fox JW, et al. (2010) Tracking of Arctic terns *Sterna paradisaea* reveals longest animal migration. Proc Natl Acad Sci USA 107: 2078–2081.
19. Hawkes LA, Balachandran S, Batbayar N, Butler PJ, Frappell PB, et al. (2011) The trans-Himalayan flights of bar-headed geese (*Anser indicus*). Proc Natl Acad Sci USA 108: 9516–9519.
20. Tøttrup AP, Klaassen RHG, Strandberg R, Thorup K, Kristensen MW, et al. (2012) The annual cycle of a trans-equatorial Eurasian-African passerine migrant: different spatio-temporal strategies for autumn and spring migration. Proceedings of the Royal Society B-Biological Sciences 279: 1008–1016.
21. Bridge ES, Thorup K, Bowlin MS, Chilson PB, Diehl RH, et al. (2011) Technology on the Move: Recent and Forthcoming Innovations for Tracking Migratory Birds. Bioscience 61: 689–698.
22. Cramp S, editor (1985) Handbook of the birds of Europe, the Middle East, and North Africa. New York: Oxford University Press.
23. Rappe A (1965) Notes sur le passage diurne de la tourterelle des bois *Streptopelia turtur* et du coucou *Cuculus canorus*. Aves 2: 27–31.
24. Seel DC (1977) Migration of the northwestern European population of the cuckoo (*Cuculus canorus*), as shown by ringing. Ibis 119: 309–322.
25. Moreau RE (1961) Problems of Mediterranean-Saharan migration. Ibis 103a: 373–427.
26. Fransson T, Österblom H, Hall-Karlsson S (2008) Svensk ringmärkningsatlas. Vol. 2. Stockholm.
27. Alerstam T, Hake M, Kjellen N (2006) Temporal and spatial patterns of repeated migratory journeys by ospreys. Anim Behav 71: 555–566.
28. Wallraff HG (1978) Social interrelations involved in migratory orientation of birds - possible contribution of field studies. Oikos 30: 401–404.
29. Mouritsen H, Mouritsen O (2000) A mathematical expectation model for bird navigation based on the clock-and-compass strategy. J Theor Biol 207: 283–291.
30. Argos (2011) Argos User's Manual. Available: http://www.argos-system.org/web/en/76-user-s-manual.php. Accessed 2013 Nov 27.
31. Microwave Telemetry Inc.(2005) Argos Performance in Europe. Microwave Telemetry Inc Newsletter, Winter 2005. Available: http://www.microwavetelemetry.com/uploads/newsletters/winter_05page8.pdf. Accessed 2013 Nov 27.
32. Fisher NI (1995) Statistical analysis of circular data: Cambridge University Press.
33. Thorup K, Rabøl J, Madsen JJ (2000) Can clock-and-compass explain the distribution of ringing recoveries of pied flycatchers? Anim Behav 60: F3–F8.
34. Cochran WW, Mouritsen H, Wikelski M (2004) Migrating songbirds recalibrate their magnetic compass daily from twilight cues. Science 304: 405–408.
35. Wiltschko R, Munro U, Ford H, Wiltschko W (2001) Orientation in migratory birds: time-associated relearning of celestial cues. Anim Behav 62: 245–250.
36. R Development Core Team (2011) R: A language and environment for statistical computing. Vienna, Austria: R Foundation for Statistical Computing. Available: http://www.R-project.org. Accessed 2013 Nov 27.

37. Agostinelli C (2009) CircStats: Circular Statistics, S-plus original by U. Lund "Topics in circular Statistics" (2001). R package version 0.2–4. http://CRAN.R-project.org/package = CircStats.
38. Lewin-Koh NJ, Bivand R, and contributions by: Pebesma EJ, Archer E, Baddeley A, et al. (2012) maptools: Tools for reading and handling spatial objects. R package version 0.8–16. http://CRAN.R-project.org/package = maptools.
39. Hijmans RJ, Williams E, Vennes C (2012) geosphere: Spherical Trigonometry. R package version 1.2–28. http://CRAN.R-project.org/package = geosphere.
40. Pebesma EJ, Bivand RS (2005) Classes and methods for spatial data in R. Available: http://cran.r-project.org/doc/Rnews/.
41. ESRI (2011) ArcGIS. In: ESRI, editor. 10.0 ed. Redlands, CA Environmental Systems Research Institute.
42. Thorup K, Alerstam T, Hake M, Kjellen N (2003) Can vector summation describe the orientation system of juvenile ospreys and honey buzzards? - An analysis of ring recoveries and satellite tracking. Oikos 103: 350–359.
43. Strandberg R, Klaassen RHG, Hake M, Olofsson P, Alerstam T (2009) Converging migration routes of Eurasian hobbies *Falco subbuteo* crossing the African equatorial rain forest. Proc R Soc B 276: 727–733.
44. BTO (2012) Tracking cuckoos to Africa... and back again. British Trust for Ornithology. Available: http://www.bto.org/science/migration/tracking-studies/cuckoo-tracking. Accessed 2013 Nov 27.
45. Julliard R, Bargain B, Dubos A, Jiguet F (2006) Identifying autumn migration routes for the globally threatened Aquatic Warbler *Acrocephalus paludicola*. Ibis 148: 735–743.
46. Royal Society for the Protection of Birds (2009) Hunting danger to one of the world's most threatened birds. Available: http://www.rspb.org.uk/news/details.aspx?id = tcm:9-213689. Accessed 2013 Nov 27.
47. Tellería JL (2009) Potential impacts of wind farms on migratory birds crossing Spain. Bird Conservation International 19: 131–136.
48. Tøttrup AP, Klaassen RHG, Kristensen MW, Strandberg R, Vardanis Y, et al. (2012) Drought in Africa Caused Delayed Arrival of European Songbirds. Science 338: 1307.

5

Prioritizing Populations for Conservation Using Phylogenetic Networks

Logan Volkmann[1], Iain Martyn[2], Vincent Moulton[3], Andreas Spillner[4], Arne O. Mooers[1,2]*

1 Department of Biological Sciences, Simon Fraser University, Burnaby, British Columbia, Canada, **2** Interdisciplinary Research in the Mathematical and Computational Sciences (IRMACS) Centre, Simon Fraser University, Burnaby, British Columbia, Canada, **3** School of Computing Sciences, University of East Anglia, Norwich, United Kingdom, **4** Department of Mathematics and Computer Science, University of Greifswald, Greifswald, Germany

Abstract

In the face of inevitable future losses to biodiversity, ranking species by conservation priority seems more than prudent. Setting conservation priorities within species (i.e., at the population level) may be critical as species ranges become fragmented and connectivity declines. However, existing approaches to prioritization (e.g., scoring organisms by their expected genetic contribution) are based on phylogenetic trees, which may be poor representations of differentiation below the species level. In this paper we extend evolutionary isolation indices used in conservation planning from phylogenetic trees to phylogenetic networks. Such networks better represent population differentiation, and our extension allows populations to be ranked in order of their expected contribution to the set. We illustrate the approach using data from two imperiled species: the spotted owl *Strix occidentalis* in North America and the mountain pygmy-possum *Burramys parvus* in Australia. Using previously published mitochondrial and microsatellite data, we construct phylogenetic networks and score each population by its relative genetic distinctiveness. In both cases, our phylogenetic networks capture the geographic structure of each species: geographically peripheral populations harbor less-redundant genetic information, increasing their conservation rankings. We note that our approach can be used with all conservation-relevant distances (e.g., those based on whole-genome, ecological, or adaptive variation) and suggest it be added to the assortment of tools available to wildlife managers for allocating effort among threatened populations.

Editor: Tom Gilbert, Natural History Museum of Denmark, Denmark

Funding: Funded by the Natural Sciences and Engineering Research Council of Canada (Discovery grant to AOM), http://www.nserc-crsng.gc.ca/, Simon Fraser University (fellowship to LAV and IM), and Yale University (YIBS fellowship, AOM). The funders had no role in study design, data collection and analysis, decision to publish or preparation of the manuscript.

Competing Interests: The authors have declared that no competing interests exist.

* E-mail: amooers@sfu.ca

Introduction

Extinctions due to human impacts are now unavoidable: even optimistic scenarios predict significant changes in biodiversity by the year 2100 [1,2], with most extinction starting with the loss of isolated populations [3,4].

One prime conservation goal is to preserve genetic variation [5,6], both as a representation of past evolution and raw material for future evolution [7] and, potentially, as a surrogate for improved ecosystem function [8]. However, not all genetic lineages are equally important, with more isolated lineages warranting additional interest because of their expected contribution to total variation [5,8,9]. Indices of evolutionary isolation have been developed to rank species on a phylogenetic tree based on unique and shared evolutionary history (*e.g.*, [10–13]). These metrics use rooted phylogenetic trees with edge lengths as input (Figure 1), and rank tips with less shared history as requiring more urgent conservation attention. For example, the Zoological Society of London has made this approach operational in their "Edge of Existence" programme (www.edgeofexistence.org). In the United States, taxonomic distinctiveness is one of several explicit criteria for prioritizing conservation attention [14]. The extension to populations within species would seem to be straightforward.

Below the species level, Ryder [15] advocated the use of *evolutionarily significant units* (ESUs) to identify populations with genetic variation in need of long-term conservation; this was expanded by Moritz [16] and Waples [17] to the concepts of *management units* (MUs) and *distinct population segments* (DPS), respectively, for species that had undergone more recent range fragmentation. All these population-based approaches have enjoyed wide usage in population genetic studies (*e.g.*, [18–21]), and are the basis for identifying populations worthy of protection in law. Importantly, ESUs assume that the relationships among populations can be represented by a bifurcating tree. However, bifurcating trees often fail to capture the relationships among populations [22]. DPSs and MUs can deal with populations that have more complex interrelationships (Figure 1), but neither of these designations is designed to prioritize among populations. This would seem a major shortcoming if populations do need to be prioritized for conservation attention.

Previous authors have shown that the logic of measuring and maximizing phylogenetic diversity [6], which forms the basis for tree-based prioritization schemes, can be generalized to phylogenetic networks [23–25]. Here we show that the prioritization approaches for trees can also be adapted for populations within species by extending evolutionary isolation indices from trees to networks. We develop efficient algorithms to compute these

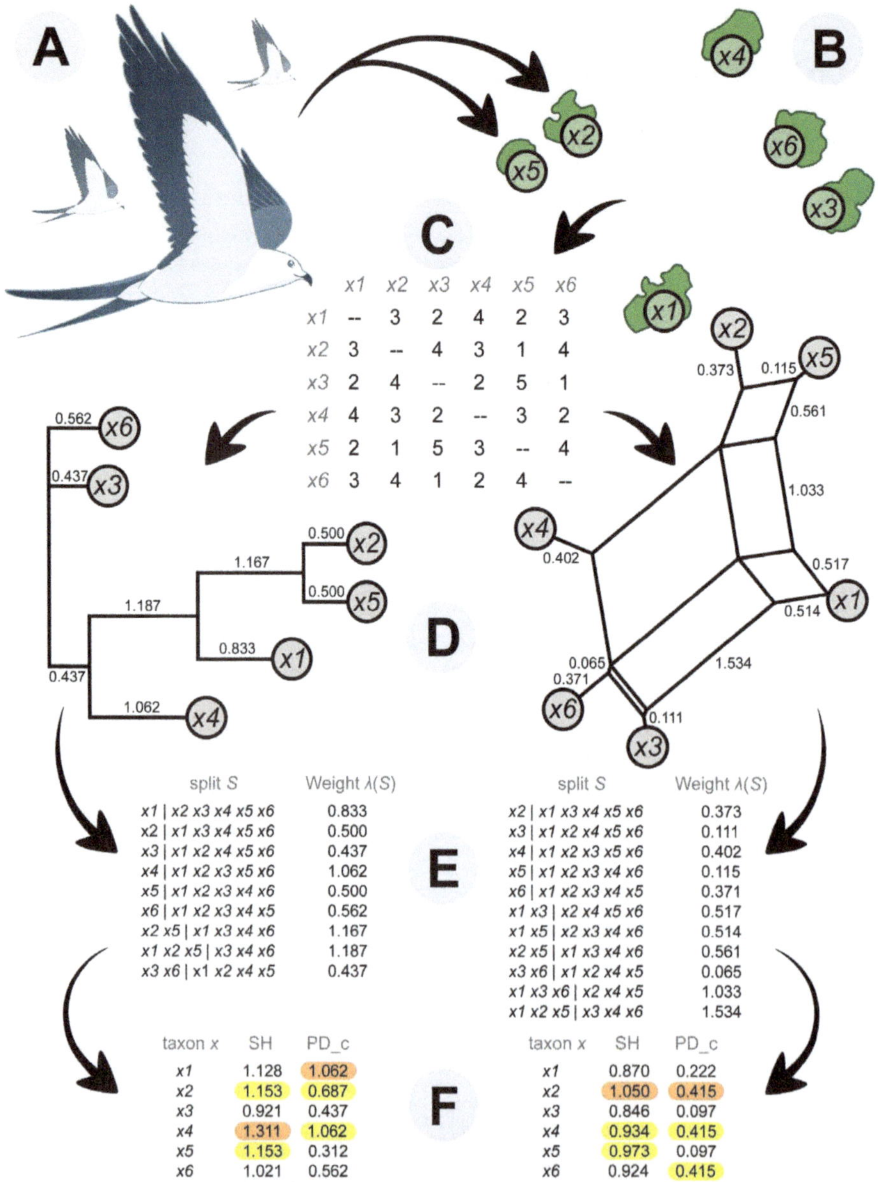

Figure 1. Using pairwise distances to rank species or populations. Consider a hypothetical group of taxa (A)—a set of closely-related species or populations of a single species—that is distributed across several islands in an archipelago (B). Differences among the taxa, labeled *x1* through *x6*, can be organized into a pairwise distance matrix (C). We can represent this matrix either as a phylogenetic tree or as a phylogenetic network (D), where a set of weighted *splits* describes the relationships among the taxa (E). Altogether, these splits represent the group's phylogenetic diversity (PD). By selecting subsets of splits that exclude a given taxon, we can calculate each taxon's contribution to the total PD of the tree or network (F). The *Shapley metric* (SH) and *expected PD complementarity* (PD_c) are different approaches for ranking taxa based on split data. Note that the highest-scoring taxa (highlighted values) can differ considerably depending on the type of metric used and whether the splits come from a tree or network. We discuss the reasons for these differences and methods for ranking taxa in Section (*ii*) of the main text.

indices for NeighborNet networks [26,27], and illustrate their use with heuristic data from two imperiled species, the spotted owl (*Strix occidentalis* Xantus de Vesey 1860) and the mountain pygmy-possum (*Burramys parvus* Broom 1896). The new approach to assessing population differentiation might be of immediate practical use to those tasked with managing discrete populations of a threatened species, and may allow for new policy associated with conservation triage [28].

Methods

We present our approach for prioritizing populations in three steps. First, we briefly review the various approaches for measuring diversity and evolutionary isolation on bifurcating trees of taxa. We then review the properties of NeighborNet networks as a representation of pairwise evolutionary distances and describe how to prioritize taxa by their expected contribution to biodiversity. In File S1, we outline efficient algorithms for estimating evolutionary isolation on NeighborNet networks. Finally, we illustrate the new method of population prioritization using two small published datasets.

(i) Diversity measures on trees and networks

The concept of evolutionary isolation can be understood in terms of a species' biological distinctiveness, which we might measure by comparing its adaptive or non-adaptive traits to those of related species. More generally, our goal is to measure a taxon's contribution to the current and/or future "diversity" in a set of taxa. Several different approaches for quantifying such diversity have been proposed. One of the earliest, described by Weitzman [23], is *expected diversity*. Rather than score taxa individually, this approach seeks to identify the set of taxa that will retain the most diversity on a future tree, given some measure of diversity and a probability of persistence for each potential combination of taxa. Although Weitzman's original diversity metric was rather general, he did consider an example of biological character-state differences that could be represented on a phylogenetic tree.

On such a tree, every taxon contributes an amount of unique evolutionary information denoted by the length of the branch (or *edge*) linking it to all other taxa (Figure 1) [6,23]. This length may be calibrated in units of time (*e.g.*, millions of years) or in raw or inferred genetic distances. Looking specifically at biological systems, Witting and Loeschcke [29] and Faith and Walker [30] combined Weitzman's [23] expected diversity framework with Faith's [6] concept of *phylogenetic diversity* (PD), the latter which specifically calculates the sum of all branch lengths on a tree (see next section). Like Weitzman [23], this *expected PD* approach can be used to identify a set of taxa that maximizes the amount of total tree length retained, given a set of extinction probabilities for the tips.

The related *k of n problem* [6] seeks to identify the most diverse subset of k taxa (*i.e.*, the one that maximizes PD) on a tree of size n. Faith [31] and Weitzman [32] explored the special case where $k=1$, which Faith [33] refers to as the *PD complementarity* of a given taxon.

An independently-derived approach based on Game Theory ([10], first published 2005) explicitly considers the individual contribution of each taxon to future diversity. Like Weitzman's [23] expected diversity framework, all possible subsets of taxa on a tree may persist. By calculating the amount of unique information each taxon contributes to future subsets (*i.e.*, the average length of the edge linking the taxon to all possible future trees), one can rank taxa in order of their relative impact on future diversity. This *Shapley metric* (SH) is almost identical to the ad-hoc *evolutionary distinctness* (ED) metric used by the Zoological Society of London in their Edge of Existence programme (www.edgeofexistence.org). The major difference between the two is that the ED metric is explicitly measured on a rooted tree, as opposed to the more general undirected graph that SH takes as input [34].

The Shapley metric was further refined by Steel *et al.* [35] and named HED (for *heightened evolutionary distinctiveness*). HED is the expected contribution of a given taxon to future subsets of taxa where the subsets are weighted by their probability of persistence. In this case, the focal taxon is assumed to persist (*i.e.*, its probability of extinction does not affect its HED score). On trees, HED is formally equivalent to a form of PD complementarity where the contribution of a taxon is measured with respect to all possible subsets, each weighted by their probability of persistence [33]. Weitzman [32] also arrived at this formulation ten years earlier, which he termed the "distinctiveness" of a taxon, in the context of his "Noah's Ark Problem" of biodiversity preservation. Using Faith's [33] terminology, HED, which combines the concepts of expected PD with PD complementarity, might be considered *expected PD complementarity*.

As a final antecedent, Minh *et al.* [24,25,36] extended PD to phylogenetic networks and presented algorithms for solving the k of n problem to maximize diversity for a given subset size. They referred to this metric as *split diversity* (SD).

In this context it should be possible to measure the PD contribution of individual taxa on a phylogenetic network. Critically for our purposes, the two metrics we use here (SH and HED) do not require a rooted phylogenetic tree, and so can be adapted to networks in the same way that PD indices can [24,25,27,36]. SH and HED are formally defined in File S1 and discussed further below. In short, if we do not have probabilities of extinction for taxa, we assume all future subsets of taxa are equally likely, and calculate SH. If we can estimate (even broadly) the probabilities of persistence of all taxa, we can weight future subsets by their probability, and use HED.

(ii) Interpreting phylogenetic networks, Shapley values, and HED

NeighborNet [26] is a method that permits the representation of pairwise distances between taxa in the form of a network. An important property of NeighborNet networks is that they permit the representation of relationships among the underlying taxa that cannot be depicted on any phylogenetic tree. For example, to the extent that populations exchange migrants, the between-population genetic distance data (F_{ST}) may yield many alternative trees, none of which accurately reflect the actual relationships among these populations (*e.g.*, [37]). The NeighborNet framework, by contrast, accommodates for such phylogenetic uncertainty and will always yield a single network with positive edge lengths, permitting calculation of SH and HED. If a pairwise distance matrix is tree-like (*i.e.*, yields only one possible phylogeny) the resulting NeighborNet output will resemble a phylogenetic tree. Where there is no tree-like history, a network representation should be more informative. Indeed, for many distance matrices (including Example A below, results not shown), the assumptions necessary to produce a tree are not met, and a neighbor-joining tree, for example, produces negative edge lengths. Here, a network representation would definitely be preferred [26].

An example of a very simple matrix of pairwise distances and the resulting network is depicted in Figure 1. Each edge or set of parallel edges in the network corresponds to a partition of the underlying set of taxa into two non-overlapping subsets, called a *split* (S). The *edge length* reflects the *weight* of the split ($\lambda(S)$)—in other words, a component of the pairwise distance (F_{ST}, for example) separating any two taxa. Thus, just as a phylogenetic tree represents a collection of weighted splits (Σ) [38], where each branch of the tree denotes a split, a NeighborNet network represents a weighted collection of splits of the underlying set of taxa. As Figure 1 illustrates, the distance between two tips on a network (*i.e.*, the shortest path between two taxa) represents the observed distance in the distance matrix.

Whether represented on a tree or a network, every split system contains information on the overall diversity of its constituent taxa [5,23]. The conservation planning metric *phylogenetic diversity* (PD) [6] can be calculated for split systems as

$$PD_{(\Sigma,\lambda)}(Y) = \sum_{\substack{A|B\in\Sigma \\ A\cap Y\neq 0, B\cap Y\neq 0}} \lambda(A|B)$$

where Y is a subset of taxa on the tree or network and $\lambda(A|B)$ is the weight of the split between two non-overlapping groups A and

B of taxa. Note that the overall PD for both trees [6,39] and networks [24,25,36] is simply the sum of all split weights (Figure 1).

A very simple approach for measuring an individual taxon's PD contribution, illustrated in Figure 1, is to consider the change in PD when this taxon is removed from the tree or network [40]. This *PD complementarity* (PD_c) metric can be expressed as

$$PD\_c(x) = PD_{(\Sigma,\lambda)}(X) - PD_{(\Sigma,\lambda)}(\bar{X})$$

where X is the set of all taxa in the tree or network and $\bar{X}$ is the subset where a given taxon x has been removed from the underlying distance matrix.

We can also extend the metrics SH and HED from trees to NeighborNet networks using similar ideas for extending PD calculations from trees (*e.g.*, [6,27,29]) to networks (*e.g.*, [24,25,36,41]). On a tree, the Shapley value (ψ_x^{sh}) for taxon x can be defined as the mean split weight of the set of splits defining $x|\bar{X}$, where $\bar{X}$ represents all unique possible subsets of the taxon set X that do not contain x. Importantly, Haake *et al.* [10] present a formal proof that the Shapley value for x can also be calculated as a weighted sum of all the edge lengths on a tree, with the weights determined by the sizes of the sets containing x. This can be presented compactly using split notation as

$$\psi_x^{sh}(\Sigma,\lambda) = \sum_{S\in\Sigma} \frac{|\bar{S}(x)|}{|X||S(x)|}\lambda(S)$$

where (Σ,λ) is the set of splits defined by the network and their weights, $|X|$ is the total number of taxa, $|S(x)|$ is the size of a split set containing the taxon x, $|\bar{S}(x)|$ is the size of the complementary set that does not contain x, and $\lambda(S)$ (following the notation from Minh *et al.* [24,25]) is the split weight, equal to the edge length separating $S(x)$ from $\bar{S}(x)$. To calculate the Shapley value for taxon $x1$ in the network in Figure 1, we take the first split $x2|x1,x3,x4,x5,x6$ to be composed of $|S(x1)|=5$ and $|\bar{S}(x1)|=1$ and $\lambda(S)=0.373$, the second split $x3|x1,x2,x4,x5,x6$ to be composed of $|S(x1)|=5$ and $|\bar{S}(x1)|=1$ and $\lambda(S)=0.111$ and so on. With a taxon set containing six elements, $|X|=6$ and the Shapley value for taxon $x1$ is 0.870 (Figure 1).

As with a phylogenetic tree, the sum of Shapley values will always equal the sum of all parallel split weights in the network. Because the shape of a network reflects the relative distances among its taxa, we should expect outlying taxa (*i.e.*, those connected to the rest of the network by long edges, like taxon $x4$) to show higher values for ψ_x^{sh}. Thus, the Shapley values calculated for a network can reflect the relative degree of isolation of each taxon based on molecular, morphological, or any other relevant distance measure.

Though conceptually similar, the calculation of HED (ψ_x^{hed}) is somewhat more complex, as it accounts for differences in the probability of extinction $p(y)$ for each taxon:

$$\psi_x^{head}(\Sigma,\lambda) = $$

$$\sum_{S\in\Sigma}\left(\prod_{y\in(S(s)-\{x\})} p(y)\right)\cdot\left(1-\left(\prod_{y\in(\bar{S}(x))} p(y)\right)\right)\cdot\lambda(S)$$

Here, the first product operator considers $p(y)$ for every taxon in $S(x)$ but excludes $p(y)$ for taxon x itself [27,35]. The second product operator considers $p(y)$ for every taxon in $\bar{S}(x)$. Unlike SH, the sum of HED scores will not equal the sum of split weights in the split system. We also note that $\lambda(S)$ will influence HED more strongly than $p(y)$ for outlying taxa. Thus, the ranking order for highly isolated populations should be similar for SH and HED, regardless of which populations have a higher extinction probability.

A more detailed mathematical treatment of the SH and HED metrics and efficient algorithms for their computation are given in File S1. For the datasets in this paper, we used the implementation of NeighborNet in the SplitsTree software package [41] to compute networks. For a given matrix of pairwise distances, this yields the network together with the corresponding collection of weighted splits. We also developed custom R scripts (available in File S1) [42] to compute SH and HED on the outputs from SplitsTree.

(iii) Application

We present SH and HED ranking for two datasets based on putatively neutral genetic markers. In the first example (*A*), the size of each population (and hence the probability of extinction for each population) is not known. In the second example (*B*), population sizes are known, allowing us to estimate separate probabilities of extinction for each population.

We selected our two examples based on the following criteria: (1) The species as a whole is of conservation interest (*i.e.*, vulnerable, endangered, or critically endangered), (2) its distribution is fragmented (*i.e.*, we can define multiple populations), (3) sampling efforts have covered its entire range, and (4) genetic analyses have been published or the raw sequence data made publicly available.

Readers should note that the primary goals of this article are to introduce and illustrate our network ranking approach, not to advocate new management decisions for the taxa described below.

Example A.. Spotted owls (*Strix occidentalis*) are distributed throughout late-succession conifer forests in western North America [43]. Four subspecies are currently recognized (Figure 2a): *S. o. caurina* from southern British Columbia to northwest California, *S. o. occidentalis* in California and Nevada, *S. o. lucida* in Utah, Colorado, Arizona, New Mexico, and northern Mexico, and *S. o. juanaphillipsae* in central Mexico [44,45]. Populations in the United States continue to decline due largely to poor timber harvesting practices, but also as a result of climate change and the westward expansion of barred owls (*S. varia* Barton 1799) [46]. *S. o. caurina* (the northern spotted owl) and *S. o. lucida* (the Mexican spotted owl) are threatened subspecies under the United States' Endangered Species Act, and *S. o. occidentalis* (the California spotted owl) is a subspecies of special concern in the state of California [47]. Spotted owls in the American Southwest "sky islands" (mostly *S. o. lucida*) are particularly fragmented and perhaps most suitable for population-level conservation [48]. Although genetic data for the Mexican subspecies remain poor, we can construct a reasonably complete representative phylogenetic network for subspecies in the United States.

Spotted owl mitochondrial sequences were obtained from Genbank (accession numbers AY833608–AY833644, AY836774–AY836776, DQ230843–DQ230888) and aligned in Mega v. 5 [49] using MUSCLE [50]. These sequences comprise about 1105 bp of the control (D-loop) region and represent 86 haplotypes from 32 populations in the United States and Mexico (Figure 2b; Table 1) [48,51]. We ran a standard analysis of molecular variance (AMOVA) [52] on all 298 aligned sequences in Arlequin v. 3.5 [53] using the Kimura 2-Parameter model [54] to compute distances among haplotypes (Φ_{ST}). This procedure

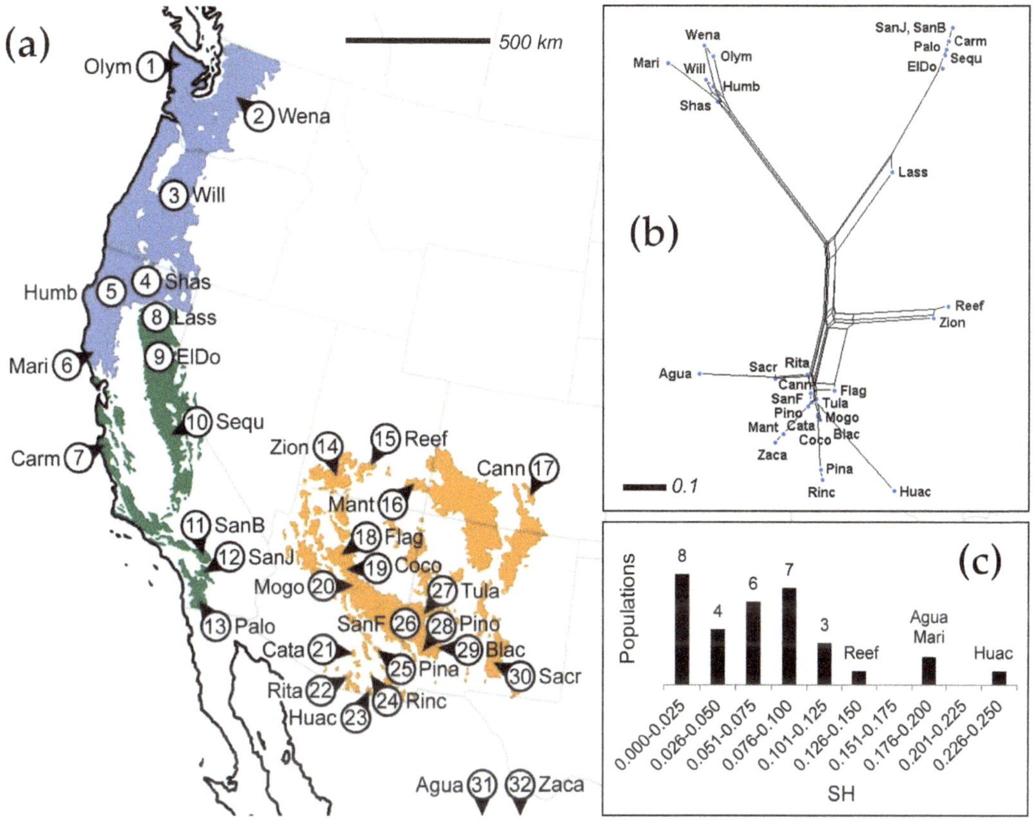

Figure 2. Conservation prioritization of spotted owl (*Strix occidentalis*) populations. (a) Distribution of spotted owls in the United States and the populations sampled by Barrowclough *et al.* [48,51]. Shaded areas denote suitable habitat based on forest cover data [73]. Colors denote the subspecies *S. o. caurina* (blue), *S. o. occidentalis* (green), and *S. o. lucida* (orange). Populations 31 and 32 represent the *S. o. juanaphillipsae* subspecies in Mexico (range not shown). (b) NeighborNet of sampled populations based on mtDNA differentiation (pairwise Φ_{ST} values). (c) Histogram of SH values, highlighting the populations with the highest scores. See Table 1 for an explanation of abbreviations used.

generated a pairwise differentiation matrix for the 32 populations (Table S1). A NeighborNet based on this matrix (Figure 2b) [26] was then constructed in SplitsTree v. 4.11 [41] under default assumptions. Negative Φ_{ST} values were treated as being equal to zero. Because the size of each population is not known, for the purposes of illustration, we gave each population an extinction probability $p_i = 0.5$ when calculating HED—an approach similar to the "PD50" metric used by FISHBASE (www.fishbase.org) [55].

Example B. Mountain pygmy-possums (*Burramys parvus*) are alpine specialists restricted to three small regions of the Australian Alps (Figure 3a). The species depends on block streams and block fields found above 1,400 meters—habitats less than 10 km^2 in total extent [56]. The areas where mountain pygmy-possums still occur are particularly sensitive to destruction and fragmentation. Surveys conducted in the 1990s estimated the adult population size to be 2,600 [57]. A decade later this number had decreased to below 2,000 [56], with signs of continued decline [58]. At present, the IUCN lists mountain pygmy-possums as critically endangered [59].

Because of its restricted distribution and high extinction risk, the species has been subject to extensive population genetic research [58,60–62]. Unlike our example with spotted owls, direct estimates of population sizes are available, within-population sample sizes are uniformly large, and genetic data are available across the mountain pygmy-possums' entire range. This provides us with an opportunity to compare SH to HED and assess the effect of variable population sizes on conservation ranking.

We used a published matrix of genetic differentiation (F_{ST}) based on data from 8 microsatellite loci [58] to construct a phylogenetic network for 13 mountain pygmy-possum populations (Figure 3b). Our methods for generating NeighborNet outputs, and for computing SH and HED, were the same as above.

We modeled the probabilities of extinction for individual populations (p_i) of a given size (n_i) as a negative exponential

$$p_i \propto e^{-cn_i}$$

where the constant of proportionality c is $-\ln(P)/N$, with P being the probability that the entire species goes extinct and N being the total census size of the species (the sum of n_i). We used a conservative 100-year extinction probability for the entire species, $P = 0.4$, to derive HED (see [63]).

Results

Example A

As expected for a set of lineages with a recent history of gene flow, the network for spotted owls is quite non tree-like (Figure 2b). However, populations with the greatest degrees of genetic differentiation, relative to all other populations, occupy nodes subtending the longest edges. Populations at relatively isolated

Table 1. Spotted owl populations sampled by Barrowclough *et al.* [48,51] and ranked by Shapley value (SH) and heightened evolutionary distinctiveness (HED).

Pop.	Code	Subspecies	State	Sampling Locality	*n* ind.	*n* hap.	SH	HED
1	Huac	*lucida*	AZ	Huachuca Mountains	5	2	0.242	7.431E-03
2	Agua	*juanaphillipsae*	—	Aguascalientes, Sierra Fria, Mexico	1	1	0.191	3.983E-03
3	Mari	*caurina*	CA	Marin County	8	3	0.177	4.067E-03
4	Reef	*lucida*	UT	Capitol Reef National Park	9	4	0.133	3.683E-03
5	Wena	*caurina*	WA	Cascade Range	10	8	0.111	1.777E-03
6	Olym	*caurina*	WA	Olympic Peninsula	10	4	0.106	1.729E-03
7	Zion	*lucida*	UT	Zion National Park	7	4	0.105	2.851E-03
8	SanB	*occidentalis*	CA	San Bernardino Mountains	15	1	0.091	9.418E-04
9	SanJ	*occidentalis*	CA	Mount San Jacinto	15	1	0.091	9.418E-04
10	Rinc	*lucida*	AZ	Rincon Mountains	8	4	0.089	2.542E-03
11	Will	*caurina*	OR	Willamette National Forest	15	8	0.081	7.621E-04
12	Carm	*occidentalis*	CA	Carmel Valley	10	1	0.079	5.187E-04
13	ElDo	*occidentalis*	CA	El Dorado National Forest	15	4	0.077	6.214E-04
14	Zaca	*juanaphillipsae*	—	Zacatecas, Sierra de Urica, Mexico	1	1	0.076	2.079E-03
15	Palo	*occidentalis*	CA	Mount Palomar	8	1	0.074	3.994E-04
17	Humb	*caurina*	CA	Humboldt and Siskiyou Counties	30	11	0.072	4.528E-04
16	Sequ	*occidentalis*	CA	Sierra National Forest	15	6	0.071	3.422E-04
18	Pina	*lucida*	AZ	Pinaleno Mountains Graham County	4	2	0.071	1.605E-03
19	Shas	*caurina*	CA	Klamath and Shasta National Forests	16	8	0.067	3.638E-04
20	Mant	*lucida*	UT	Manti-La Sal National Forest	2	2	0.054	1.276E-03
21	Sacr	*lucida*	NM	Sacramento Mountains	8	6	0.047	1.284E-03
22	Lass	*occidentalis*	CA	Lassen National Forest	11	6	0.041	1.006E-04
23	Flag	*lucida*	AZ	San Fransisco Peaks	4	4	0.027	2.144E-04
24	Blac	*lucida*	NM	Black Range	8	6	0.026	3.461E-04
25	Coco	*lucida*	AZ	Coconino Plateau	15	9	0.022	1.039E-04
26	Cata	*lucida*	AZ	Santa Catalina Mountains	5	3	0.021	6.359E-05
27	Pino	*lucida*	NM	Pinos Altos Mountains	5	4	0.020	2.832E-05
28	SanF	*lucida*	NM	San Fransisco Mountains	7	4	0.017	1.020E-06
29	Cann	*lucida*	CO	Near Canon City	4	4	0.017	8.505E-07
30	Rita	*lucida*	AZ	Santa Rita Mountains	4	4	0.017	1.374E-05
31	Mogo	*lucida*	AZ	Mogollon Mesa	8	6	0.017	8.578E-07
32	Tula	*lucida*	NM	Tularosa Mountains	15	12	0.017	2.034E-06

Number of individuals (*n* ind.), number of haplotypes (*n* hap.), SH, and HED scores from the present study are reported.

nodes, such as those from Mount San Jacinto and the Huachuca Mountains, share few mutations with neighboring populations and subsequently exhibit higher pairwise Φ_{ST} values (Table 2; Table S1). Conversely, the (uncorrected) pairwise Φ_{ST} values for closely-related populations are either negative (as great as -1) or close to zero, indicating higher levels of genetic differentiation within these populations than among them [52].

We observe strong geographic structure across the United States consistent with current subspecific designations (Figure 2b). Populations of *S. o. lucida* exhibit a more star-like phylogenetic network that may reflect historical isolation in the "sky islands" of the American Southwest [48]. The intermediate position of the Lassen National Forest population, in contrast, may be due to its location near the point of contact between southern *S. o. caurina* and northern *S. o. occidentalis* [51].

The results of our SH and HED ranking are shown in Table 1. As expected, populations at relatively isolated tips score higher than those closer to the interior of the network (Figure 2b, c). The rankings are highly consistent between the two metrics (Spearman rank correlation =0.91), and the same populations receive top ranking for both SH and HED.

Example B

As with spotted owls, the most genetically differentiated populations of mountain pygmy-possums, namely those in the northern and southern areas of their range, occupy nodes that are separated from most other populations by long edges (Figure 3b). Overall the structure of our network is in good agreement with the species' present distribution (Figure 3a). Given the habitat requirements and limited dispersal ability of mountain pygmy-possums, it is not likely that Mount Buller and Kosciusko National Park still exchange migrants with the Bogong High Plains [58]. In contrast, the close grouping of central populations in our

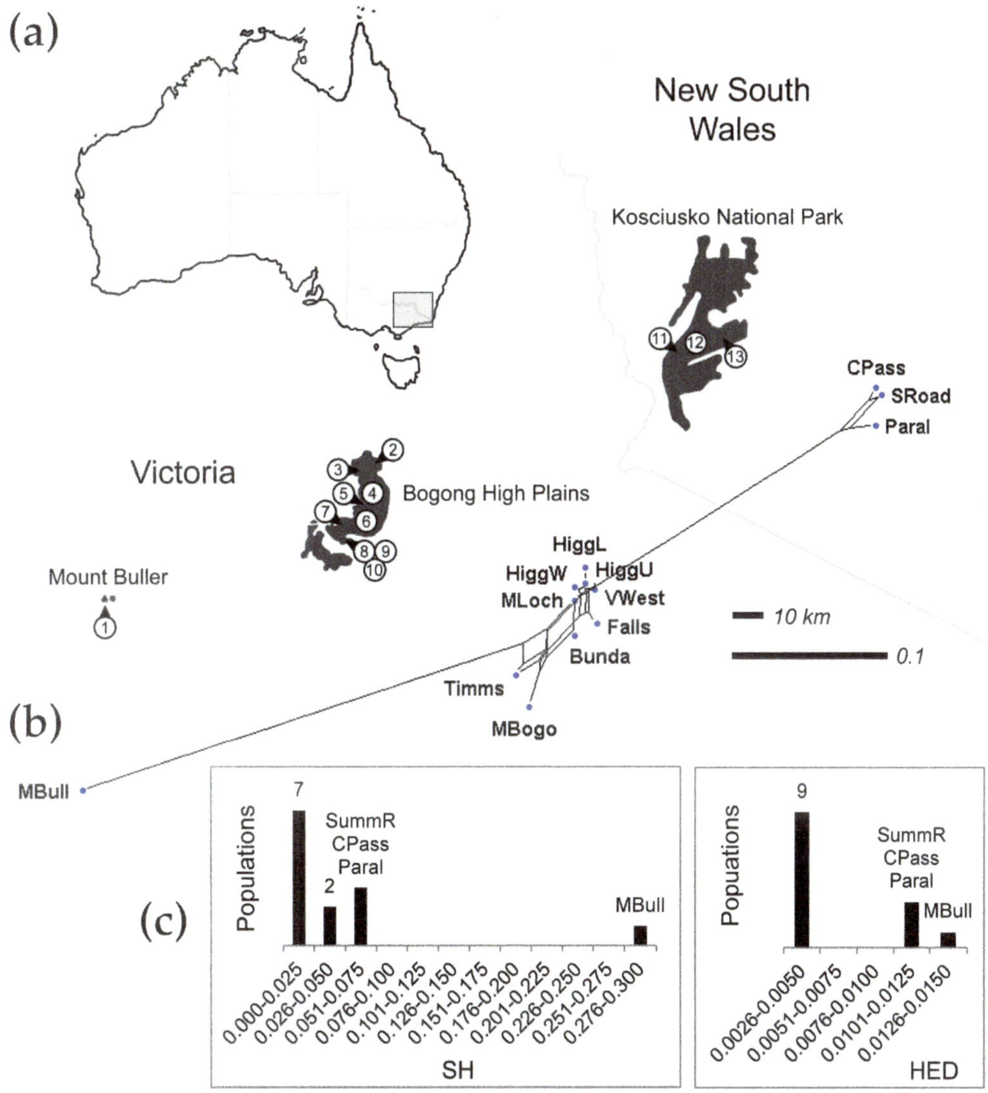

Figure 3. Conservation prioritization of mountain pygmy-possum (*Burramys parvus*) populations. (a) Distribution of mountain pygmy-possums in Australia (gray inset), showing populations sampled by Mitrovski *et al.* [58]. Shaded areas denote suitable habitat above 1,400 m. (b) NeighborNet of sampled populations based on microsatellite differentiation (pairwise F_{ST} values). (c) Histograms of SH and HED values, highlighting the populations with the highest scores. See Table 2 for an explanation of abbreviations used.

phylogenetic network, and subsequently their low SH and HED, is consistent with a shared history and/or recent gene flow.

The ranking results are shown in Table 2. Again, the phylogenetic network for mountain pygmy-possums reflects geographic distribution. Although we did not make *a priori* group assignments based on sampling location, the 13 populations still partition into northern, central, and southern regions. Again, outlying populations on the network tend to receive higher SH and HED scores. Unsurprisingly, the small and isolated Mount Buller population consistently ranks highest. For HED, no bias towards small or large populations is apparent; populations with high extinction probabilities do not necessarily receive high scores [35]. Again, although ranking order changes slightly between SH and HED, the two methods provide roughly equivalent rankings (Spearman rank correlation = 0.97, Figure 3c). High-ranking populations are similar in both cases.

We note that SH and HED calculations on a network consider a taxon's distance from *all other* taxa. Thus, although the three northern populations are closely related to each other, they still receive high SH and HED scores because of the long branches separating them from the central and southern populations (Table 2; Figure 3c).

Discussion

The premise of conservation below the species level is not novel. Faith's original [6] discussion of prioritizing taxa also considered populations on a tree. Several economically-important taxa have received population-level management since the late 1980s, *e.g.*, Atlantic salmon (*Salmo salar* Linnaeus 1758) [64], brown trout (*Salmo trutta* Linnaeus 1758) [65] and yellowfin tuna (*Thunnus albacares* Bonnaterre 1788) [66]. Managing species at the population level implies at least an informal ranking scheme, one which would rely, for example, on estimates of habitat patch size or effective population size [67]. Habitat degradation, climate change, and the demands of a growing human population have

Table 2. Mountain pygmy-possum populations sampled by Mitrovski *et al.* [58] and ranked by Shapley value (SH) and heightened evolutionary distinctiveness (HED).

Pop.	Code	Region	Sampling Locality	*n* ind.	*n* all.	*r*	*N*	*pi*	SH	HED
1	MBull	Southern	Mount Buller	66	3.38	2.29	150	0.9072	0.292	1.497E-02
2	CPass	Northern	Charlottes Pass	44	6.00	5.21	45	0.9712	0.072	1.153E-02
3	Paral	Northern	Paralyser	40	6.63	5.77	22	0.9858	0.071	1.089E-02
4	SummR	Northern	Summit Road	43	6.13	5.16	25	0.9839	0.068	1.138E-02
5	MBogo	Central	Mount Bogong	42	6.50	5.77	100	0.9372	0.045	4.497E-03
6	Timms	Central	Timm Spur	120	6.88	5.32	120	0.9251	0.028	3.859E-03
7	Falls	Central	Falls Creek	35	6.63	5.73	30	0.9807	0.020	3.051E-03
8	HiggL	Central	Mount Higginbotham L	17	5.25	5.25	50	0.9681	0.019	2.850E-03
9	Bunda	Central	Bundara	78	7.25	5.45	120	0.9251	0.018	2.858E-03
10	HiggW	Central	Mount Higginbotham W	59	7.63	6.01	250	0.8502	0.013	2.659E-03
11	VWest	Central	Pretty Valley West	69	7.00	5.95	50	0.9681	0.010	2.541E-03
12	HiggU	Central	Mount Higginbotham U	56	7.25	5.76	50	0.9681	0.010	2.420E-03
13	Mloch	Central	Mount Loch	93	7.00	5.80	400	0.7714	0.010	2.646E-03

Number of individuals (*n* ind.), number of alleles (*n* all.), allelic richness (*r*), and adult population sizes (*N*) are reported from previously-published data. Probabilities of extinction (*pi*, with $P = 0.4$), SH, and HED scores from the present study are also shown.

ensured the continued fragmentation of species' ranges over the next century (see, *e.g.*, pikas (*Ochotona princeps* Richardson 1828) [68]). In the midst of such rapid change, managing an imperiled species over its entire range may no longer be feasible, such that population rankings may be necessary.

Phylogenetic diversity measures have previously been adapted for non-treelike population genetic data (*e.g.*, [24,25,36]). However, the PD complementarity scores that can be obtained from these methods are contingent, *i.e.*, subject to change if extinction alters the shape of the network. Ours is the first ranking scheme to consider a taxon's contribution to all possible future networks (*sensu* Weitzman [23]), a potentially relevant framework for preserving future biodiversity. Given the stochastic nature of extinction, the general ranking systems offered by SH and HED may be more useful to wildlife managers than those that only consider the present structure of a phylogenetic network. Unlike previous approaches based on PD (*e.g.*, [24,25,36]), SH and HED rankings allow one to lengthen or shorten the list of taxa to conserve in the event that resources become more or less available.

Molecular techniques are now inexpensive and robust enough to make population genetic sampling a standard component of conservation planning, and we argue that a phylogenetic network approach offers insight into a species' population structure complementary to the current statistical assessments of differentiation employed by MUs and DPSs [16,17]. We encourage researchers to employ such networks in future population genetic studies to provide conservation agencies with more informative analysis of datasets. Genotyping at multiple loci will provide more accurate estimates of population differentiation and allow for more sophisticated analyses of conservation-relevant processes such as recent demographic history and gene flow [69].

We acknowledge that the mathematical shortcomings of Φ_{ST} and F_{ST} estimators [70] may influence the magnitude and ranking of SH and HED scores, depending on the number of loci measured and the distribution of genetic diversity in a set of taxa. Our intention here is not to solve these theoretical problems but to demonstrate our network-based prioritization method with existing data. Newly-developed metrics such as Jost's D can be used to calculate SH and HED just as readily as traditional Φ_{ST} and F_{ST} distances, and we encourage the use of such unbiased estimators whenever such data are available. Indeed, any conservation metric of difference (*e.g.*, ecological, genomic, adaptability) can be used.

Several properties of the networks described here invite further investigation. In both our heuristic datasets, geographically peripheral populations are more genetically isolated, meaning they would rank highly on SH and HED. However, this was based on only very few putatively neutral markers. Two related questions concern how processes such as demographic history and current patterns of gene flow map onto genetic isolation as we measure it here, and also how phylogenetic networks map onto networks produced from ecological data (*e.g.*, niche use differences among populations).

We do not advocate relying solely on genetic isolation when deciding where and how resources should be allocated at the population level. Total population genetic diversity (*i.e.*, number of haplotypes) might also be considered. We note that in our examples, low-ranking populations tend to be geographically close to one another, meaning that their haplotypes are often shared. Important differences in ecology and adaptability [71] and current and future connectivity [72], must also be considered. However, our network approach and ranking system based on genetic differentiation can supplement existing systems of MUs and DPSs to improve the conservation of evolutionarily distinct populations in a world of increasing pressures and limited resources.

Supporting Information

File S1 Mathematical treatment of SH and HED and annotated R code for calculating both metrics.

Table S1 Pairwise genetic distances (Φ_{ST}) for spotted owl (*Strix occidentalis*) populations based on data from Barrowclough *et al.* [48,51], with negative values set to zero.

Acknowledgments

We thank the Isaac Newton Center (Cambridge University) for facilitating our collaboration, and M Hart, JB Joy, M Steel, three anonymous reviewers and the FAB Lab at Simon Fraser University for discussion and comments on the manuscript. All R scripts used in this paper are available in File S1.

Author Contributions

Conceived and designed the experiments: AOM VM AS. Performed the experiments: LV IM AS. Analyzed the data: LV AOM. Contributed reagents/materials/analysis tools: IM AS. Wrote the paper: LV AOM.

References

1. Sala OE, Chapin FS, Armesto JJ, Berlow E, Bloomfield J, et al. (2000) Global biodiversity scenarios for the year 2100. Science 287: 1770–1774.
2. Jenkins M (2003) Prospects for biodiversity. Science 302: 1175–1177.
3. Hughes JB, Daily GC, Ehrlich PR (1997) Population diversity: its extent and extinction. Science 278: 689–692.
4. Ceballos G, Ehrlich PR (2002) Mammal population losses and the extinction crisis. Science 296: 904–907.
5. Vane-Wright RI, Humphries CJ, Williams PH (1991) What to protect?—Systematics and the agony of choice. Biol Conserv 55: 235–254.
6. Faith DP (1992) Conservation evaluation and phylogenetic diversity. Biol Conserv 61: 1–10.
7. Bowen BW, Roman J (2005) Gaia's handmaidens: the Orlog model for conservation biology. Conserv Biol 19: 1037–1043.
8. Redding DW, Mooers AØ (2010) Can systematists help decide the relative worth of bits of biodiversity? Systematist 32: 4–8.
9. May RM (1990) Taxonomy as destiny. Nature 347: 129–130.
10. Haake CJ, Kashiwada A, Su FE (2008) The Shapley value of phylogenetic trees. J Math Biol 56: 479–497.
11. Pavoine S, Ollier S, Dufour AB (2005) Is the originality of a species measurable? Ecol Lett 8: 579–586.
12. Redding DR, Mooers AØ (2006) Incorporating evolutionary measures into conservation prioritization. Conserv Biol 20: 1670–1678.
13. Isaac NBJ, Turvey ST, Collen B, Waterman C, Baillie JEM (2007) Mammals on the EDGE: conservation priorities based on threat and phylogeny. PLoS One 2: e296.
14. Fay JJ, Thomas WL (1983) Endangered species listing and recovery priority guidelines. Fed Regist 48: 43098–43105.
15. Ryder OA (1986) Species conservation and systematics: the dilemma of subspecies. Trends Ecol Evol 1: 9–10.
16. Moritz C (1994) Defining 'evolutionarily significant units' for conservation. Trends Ecol Evol 9: 373–375.
17. Waples RS (1991) Pacific salmon, *Oncorhynchus* spp., and the definition of 'species' under the endangered species act. Mar Fish Rev 53: 11–22.
18. Firestone KB, Elphinstone MS, Sherwin WB, Houlden BA (1999) Phylogeographical population structure of tiger quolls *Dasyurus maculatus* (Dasyuridae: Marsupialia), an endangered carnivorous marsupial. Mol Ecol 8: 1613–1625.
19. Gorbics CS, Bodkin JL (2001) Stock structure of sea otters (*Enhydra lutris kenyoni*) in Alaska. Mar Mammal Sci 17: 632–647.
20. Haig SM, Wagner RS, Forsman ED, Mullins TD (2001) Geographic variation and genetic structure in spotted owls. Conserv Genet 2: 25–40.
21. Jensen EL, Govindarajula P, Russello MA (2013) When the shoe doesn't fit: applying conservation unit concepts to western painted turtles at their northern periphery. Conserv Genet. doi: 10.1007/s10592-013-0535-2.
22. Fitch WM (1997) Networks and viral evolution. J Mol Evol 44: S65–S75.
23. Weitzman ML (1992) On diversity. Q J Econ 107: 363–405.
24. Minh BQ, Pardi F, Klaere S, von Haeseler A (2009) Budgeted phylogenetic diversity on circular split systems. IEEE/ACM Trans Comput Biol Bioinform 6: 22–29.
25. Minh BQ, Klaere S, von Haeseler A (2009) Taxon selection under split diversity. Syst Biol 57: 586–594.
26. Bryant D, Moulton V (2004) Neighbor-Net: an agglomerative method for the construction of phylogenetic networks. Mol Biol Evol 21: 255–265.
27. Martyn I, Kuhn TS, Mooers AØ, Moulton V, Spillner A (2012) Computing evolutionary distinctiveness indices in large scale analysis. Algorithms Mol Biol 7: 6.
28. Bottrill MC, Joseph LN, Carwardine J, Bode M, Cook C, et al. (2008) Is conservation triage just smart decision making? Trends Ecol Evol 23: 649–654.
29. Witting L, Loeschcke V (1995) The optimization of biodiversity conservation. Biol Conserv 71: 205–207.
30. Faith DP, Walker PA (1996) DIVERSITY – TD. In: Faith DP, Nicholls AO, editors. BioRap: rapid assessment of biodiversity. Volume 3: tools for assessing biodiversity priority areas. Canberra: The Australian BioRap Consortium. pp. 63–74.
31. Faith DP (1996) Conservation priorities and phylogenetic pattern. Conserv Biol 10: 1286–1289.
32. Weitzman ML (1998) The Noah's Ark Problem. Econometrica 66: 1279–1298.
33. Faith DP (2008) Threatened species and the potential loss of phylogenetic diversity: conservation scenarios based on estimated extinction probabilities and phylogenetic risk analysis. Conserv Biol 22: 1461–1470.
34. Hartmann K (2013) The equivalence of two phylogenetic biodiversity measures: the Shapley value and the Fair Proportion index. J Math Biol 67: 1163–1170.
35. Steel M, Mimoto A, Mooers AØ (2007) Hedging our bets: the expected contribution of species to future phylogenetic diversity. Evol Bioinform Online 3: 237–244.
36. Minh BQ, Klaere S, von Haeseler A (2007) Phylogenetic diversity on split networks. Technical Report NI07090-PLG. Cambridge: Isaac Newton Institute. 19 p.
37. Poczai P, Mátyás KK, Szabó I, Varga I, Hyvönen J, et al. (2011) Genetic variability of thermal *Nymphaea* (Nymphaeaceae) populations based on ISSR markers: implications on relationships, hybridization, and conservation. Plant Mol Biol Rep 29: 906–918.
38. Buneman P (1971) The recovery of trees from measures of dissimilarity. In: Hodson FR, Kendall DG, Tautu P, editors. Mathematics in the archaeological and historical sciences. Edinburgh: Edinburgh University Press. pp. 387–395.
39. Steel M (2005) Phylogenetic diversity and the greedy algorithm. Syst Biol 54: 527–529.
40. Faith DP (1994) Phylogenetic pattern and the quantification of organismal biodiversity. Phil Trans R Soc Lond B 345: 45–58.
41. Huson DH, Bryant D (2006) Application of phylogenetic networks in evolutionary studies. Molec Biol Evol 23: 254–267.
42. R-Development Core Team (2009) R: a language and environment for statistical computing. Vienna: R Foundation for Statistical Computing. 1706 p.
43. Gutiérrez RJ, Franklin AB, Lahaye WS (1995) Spotted owl (*Strix occidentalis*). In: Poole A, editor. The birds of North America online. Ithaca: Cornell Lab of Ornithology. Available: http://bna.birds.cornell.edu/bna/species/179. Accessed 28 November 2011.
44. Dickerman RW (1997) Geographic variation in southwestern US and Mexican spotted owls, with the description of a new subspecies. In: Dickerman RW, editor. The era of Allen R. Phillips: a festschrift. Albuquerque: Horizon Communications. pp. 45–48.
45. Barrowclough GF, Gutiérrez RJ, Groth JG (1999) Phylogeography of spotted owl (*Strix occidentalis*) populations based on mitochondrial DNA sequences: gene flow, genetic structure, and a novel biogeographic pattern. Evolution 53: 919–931.
46. US Fish and Wildlife Service (2011) Revised recovery plan for the northern spotted owl (*Strix occidentalis caurina*). Portland, OR: US Fish and Wildlife Service. 277 p.
47. Gould GI (1985) Management of spotted owls by the California Department of Fish and Game. In: Gutiérrez RJ, Carey AB, editors. Ecology and management of the spotted owl in the Pacific Northwest. Portland, OR: US Department of Agriculture, Forest Service. pp. 21–26.
48. Barrowclough GF, Groth JG, Mertz LA, Gutiérrez RJ (2006) Genetic structure of Mexican spotted owl (*Strix occidentalis lucida*) populations in a fragmented landscape. Auk 123: 1090–1102.
49. Tamura K, Peterson D, Peterson N, Stecher G, Nei M, et al. (2011) MEGA5: molecular evolutionary genetics analysis using maximum likelihood, evolutionary distance, and maximum parsimony methods. Mol Biol Evol 28: 2731–2739.
50. Edgar RC (2004) MUSCLE: multiple sequence alignment with high accuracy and high throughput. Nucl Acids Res 32: 1792–1797.
51. Barrowclough GF, Groth JG, Mertz LA, Gutiérrez RJ (2005) Genetic structure, introgression, and a narrow hybrid zone between northern and California spotted owls (*Strix occidentalis*). Mol Ecol 14: 1109–1120.
52. Excoffier L, Smouse PE, Quattro JM (1992) Analysis of molecular variance inferred from metric distances among DNA haplotypes: application to human mitochondrial DNA restriction data. Genetics 131: 479–491.
53. Excoffier L, Laval G, Schneider S (2005) Arlequin (version 3.0): An integrated software package for population genetics data analysis. Evol Bioinform Online 1: 47–50.
54. Kimura M (1980) A simple method for estimating evolutionary rates of base substitutions through comparative studies of nucleotide sequences. J Mol Evol 16: 111–120.
55. Faith DP, Reid CAM, Hunter J (2004) Integrating phylogenetic diversity, complementarity, and endemism for conservation assessment. Conserv Biol 18: 255–261.
56. Heinze D, Broome LS, Mansergh I (2004) A review of the ecology and conservation of the mountain pygmy-possum *Burramys parvus* (Broom 1896). In: Hillis DM, Moritz C, Mable BK, editors. The biology of the Australian possums and gliding possums. Sunderland, MA: Sinauer Associates. pp. 254–267.
57. Mansergh I, Broome L (1994) The mountain pygmy-possum of the Australian Alps, Australia. Sydney: New South Wales University Press. 114 p.
58. Mitrovski P, Heinze DA, Broome L, Hoffmann AA, Weeks AR (2007) High levels of variation despite genetic fragmentation in populations of the endangered mountain pygmy-possum, *Burramys parvus*, in alpine Australia. Mol Ecol 16: 75–87.

59. Menkhorst P, Broome L, Driessen M (2008) Burramys parvus. IUCN Red List of Threatened Species. Available: http://www.iucnredlist.org/details/full/3339/0. Accessed 28 November 2011.
60. Osborne MJ, Norman JA, Christidis L, Murray ND (2000) Genetic distinctness of isolated populations of an endangered marsupial, the mountain pygmy-possum, *Burramys parvus*. Mol Ecol 9: 609–613.
61. Mitrovski P, Heinze DA, Guthridge K, Weeks AR (2005) Isolation and characterization of microsatellite loci from the Australian endemic mountain pygmy-possum, *Burramys parvus* Broom. Mol Ecol Notes 5: 395–397.
62. Mitrovski P, Hoffmann AA, Heinze DA, Weeks AR (2008) Rapid loss of genetic variation in an endangered possum. Biol Lett 4: 134–138.
63. Mooers AØ, Faith DP, Maddison WP (2008) Converting endangered species categories to probabilities of extinction for phylogenetic conservation prioritization. PLoS One 3: e3700.
64. Fontaine PM, Dodson JJ, Bernatchez L, Slettan A (1997) A genetic test of metapopulation structure in Atlantic salmon (*Salmo salar*) using microsatellites. Can J Fish Aquat Sci 54: 2434–2442.
65. Ferguson A (1989) Genetic differences among brown trout, *Salmo trutta*, stocks and their importance for the conservation and management of the species. Freshw Biol 21: 35–46.
66. Ward RD (1995) Population genetics of tunas. J Fish Biol 47: 259–280.
67. Lande R, Barrowclough GF (1987) Effective population size, genetic variation, and their use in population management. In: Soulé ME, editor. Viable populations for conservation. Cambridge: Cambridge University Press. pp. 87–123.
68. Galbreath KE, Hafner DJ, Zamudio KR (2009) When cold is better: climate-driven elevation shifts yield complex patterns of diversification and demography in an alpine specialist (American pika, *Ochotona princeps*). Evolution 63: 2848–2863.
69. Marko PB, Hart MW (2011) The complex analytical landscape of gene flow inference. Trends Ecol and Evol 26: 448–456.
70. Jost L (2008) GST and its relatives do not measure genetic differentiation. Mol Ecol 17: 4015–4026.
71. Bonin A, Nicole F, Pompanon F, Miaud C, Taberlet P (2007) Population adaptive index: a new method to help measure intraspecific genetic diversity and prioritize populations for conservation. Conserv Biol 21: 697–708.
72. Geffen E, Waidyaratne S, Dalén L, Angerbjörn A, Vila C, et al. (2007) Sea ice occurrence predicts genetic isolation in the artic fox. Mol Ecol 16: 4241–4255.
73. US Geological Service (2013) The national atlas of the United States of America. Washington, DC: US Department of Interior, Geological Service. Available: http://www.nationalatlas.gov/biology.html. Accessed 28 November 2011.

Spatio-Temporal Characteristics of Global Warming in the Tibetan Plateau during the Last 50 Years Based on a Generalised Temperature Zone - Elevation Model

Yanqiang Wei[1,2]*, Yiping Fang[1]

1 Institute of Mountain Hazards and Environment, Chinese Academy of Sciences, Chengdu, Sichuan, P.R. China, **2** University of Chinese Academy of Sciences, Beijing, P.R. China

Abstract

Temperature is one of the primary factors influencing the climate and ecosystem, and examining its change and fluctuation could elucidate the formation of novel climate patterns and trends. In this study, we constructed a generalised temperature zone elevation model (GTEM) to assess the trends of climate change and temporal-spatial differences in the Tibetan Plateau (TP) using the annual and monthly mean temperatures from 1961–2010 at 144 meteorological stations in and near the TP. The results showed the following: (1) The TP has undergone robust warming over the study period, and the warming rate was 0.318°C/decade. The warming has accelerated during recent decades, especially in the last 20 years, and the warming has been most significant in the winter months, followed by the spring, autumn and summer seasons. (2) Spatially, the zones that became significantly smaller were the temperature zones of −6°C and −4°C, and these have decreased 499.44 and 454.26 thousand sq km from 1961 to 2010 at average rates of 25.1% and 11.7%, respectively, over every 5-year interval. These quickly shrinking zones were located in the northwestern and central TP. (3) The elevation dependency of climate warming existed in the TP during 1961–2010, but this tendency has gradually been weakening due to more rapid warming at lower elevations than in the middle and upper elevations of the TP during 1991–2010. The higher regions and some low altitude valleys of the TP were the most significantly warming regions under the same categorizing criteria. Experimental evidence shows that the GTEM is an effective method to analyse climate changes in high altitude mountainous regions.

Editor: Vanesa Magar, Plymouth University, United Kingdom

Funding: This work was financially supported by the research grant from the National Basic Research Program of China (973 Program, No. 2010CB951704). The funders had no role in study design, data collection and analysis, decision to publish, or preparation of the manuscript.

Competing Interests: The authors have declared that no competing interests exist.

* E-mail: weiyq@imde.ac.cn

Introduction

Contemporary global warming has received worldwide attention as increasing numbers of climate fluctuations and catastrophes have been reported in recent years. The Intergovernmental Panel on Climate Change (IPCC) Fourth Assessment Report (AR4) indicated that warming has increased since the 1910s, that the global annual mean temperature increased by 0.74°C from 1906 to 2005, and that this temperature is likely to increase 1.1–6.4°C by 2100 [1]. In China, the temperature has increased 0.4–0.5°C from 1860 to 2005, and the increase in winter temperatures has been apparent since 1951; nineteen 'green winters' have been experienced since 1986/1987 [2]. Regarding global warming, the projected climate trends are emphasised by the existing literature. The accumulating evidence suggests a strong and sustained increase in the global average temperatures since the 1970s, which is superimposed on the more gradual warming of Earth over the entire 20th century, and increasing levels of uncertainties in future weather patterns.

Other representative climate indicators, e.g., snow, glaciers and permafrost of the cryosphere, particularly in high altitude mountains, have proven more sensitive and vulnerable to climate change than other land surface components. The warming climate has led to the melting and shrinking of glaciers and permafrost. As Oerlemans concluded, the worldwide extent of glacier retreat between 1884 and 1978 was proportional to the extent of global warming (0.66±0.1°C) over the same period, and the total area of the world's mountain glaciers are predicted to decrease by 1/3–2/3 in the 21st century [3,4,5]. Furthermore, overall warming could lead to accelerated ice mass loss, and accelerated iceberg melting is contributing significantly to the increase in sea levels.

Unfortunately, climate changes at the three poles of Earth, especially in high altitude regions, are more severe than in middle and low altitude regions but have not received adequate attention [5,6,7,8,9]. High altitude regions do not face the direct influences and hazards that global warming presents to coastal and relatively low altitude regions (from rising sea levels); however, other subtle changes in heterogeneous systems, particularly in hydrological cycles and ecosystems, are much more significant at these elevations because of the impact on atmospheric circulation, species redistribution and ecosystem processes at different altitudes and locations and in different climate regimes. In addition to these concerns, high altitude regions often have their own unique characteristics that relate to global warming. Several studies have reported that the surface climate changes at high elevations show more pronounced warming trends than those of low altitude areas, i.e., an elevation dependency of climatic warming [8,10,11].

Currently, the relevant studies of temperature rise and isotherm uplift as a result of global warming in high altitude regions are mainly focused on the following areas: (1) Snowline or equilibrium line altitude (ELA) elevation with glacier shrinkage [12,13,14,15]; (2) Treeline and ecotone rises toward the poles or higher altitudes in response to warming [16,17,18,19,20]; (3) Climate belt or species range shifts poleward or towards higher elevations [21,22,23,24]; and (4) Land use cover change (LUCC) [9,25,26]. These aspects have provided a general framework for climate change in high altitude regions, but climate changes in high altitude mountains are quite different from those of flatlands because of their complicated topography conditions and vertical zonality [21,22]. However, because of the lack of adequate observational data at the spatial and temporal resolution levels required for climate research, it has only been in the last few decades that investigations into contemporary climate change in high altitude mountains have been conducted [7,9]. Analysis of the climate change differences in both the spatial and temporal scales of high altitude mountains, as well as detailed studies of alpine climate change, are necessary.

The Tibetan Plateau (TP) is called 'the Antenna of the Earth,' as it is highly sensitive to climate change, and climate indicators in this area are much more representative than those of lower altitude regions [27,28]. Warming in the TP has been mainly characterised by an increase in temperature and stable precipitation accompanied by a decrease in the potential evaporation capacity in recent years, according to previous studies of contemporary environmental change on the plateau [29,30,31]. Observational data from recent Chinese investigations on glacier and permafrost degradation and its environmental effects in the TP indicate that a large portion of the TP has experienced significant warming since the mid-1950s [7,31,32]. The permafrost were undergoing accelerated degradation, while mountain glaciers had extensive melting period and intensifying ablation, and accelerated shrunk in the last half of the 20th century [32,33,34,35]. A great deal of evidence has been obtained from the elevated snowline/ELA and treeline resulting from glacier retreat caused by temperature increases in the TP [15,20]. The rapid changes in climate conditions, especially in the permafrost and glaciers, have already caused deterioration of the environment and ecosystem of the plateau [36,37]. Human-induced warming and LUCC in the TP have had a more significant impact on the local and regional scales than non-anthropogenic influences [38,39,40]. In addition, numerical simulations indicate that the air temperature of the TP will continue to increase in the 21st century, which is projected to improve the frigid and arid environment on the plateau [7,10,41]. Regarding the warming trend differences, Yang et al. [42] considered the air temperature increases to be most significant in the central, eastern, and northwestern parts of the plateau. Liu et al. [43] found that the warming trends in winter night-time temperatures were among the highest when compared with the surrounding regions of the TP. After examining the spatial and temporal variations of the monthly mean temperature at 71 stations with elevations above 2000 m asl. in the eastern and central TP during 1961–2004, You et al. [29] concluded that the northeastern subregion of the TP had the most significant warming trends, especially in winter and autumn. Vertically, Liu and Chen [44] revealed a more pronounced warming pattern at high elevations based on examination of the temporal trends in monthly surface temperatures from 97 weather stations above 2000 m asl. in the eastern TP and its vicinity during 1955–1996, a conclusion that has since been confirmed by numerical experiments [10]. This tendency may continue in future climate change scenarios [8]. However, You et al. [45] found that there was no relationship between elevation and temperature extremes in the eastern and central TP. There were also no simple linear relationships between the magnitudes of elevation and temperature trends based on the 71 homogenised surface stations with elevations above 2000 m asl. in the eastern and central TP [46].

Temperature is one of the key indicators of climate and usually manifests the properties of the climate, and changes in temperature trends and spatial distribution patterns usually suggest similar changes in climate. However, analysis of temperature zone or climate zone shifts in high altitude mountains is scarce in the existing literature. The objectives of this paper are the following: (i) to establish a generalised temperature zone model for high altitude regions and use it to evaluate the trends of contemporary climate change in the TP and (ii) to examine the differences of the warming trends in the temporal and spatial dimensions. The organisation of the paper is as follows: (1) Materials and methods, in which we establish a Generalised Temperature zone Elevation Model (GTEM); (2) Results, in which we use the TP as a case study and calculate the areas of each temperature zone every 5 years and analyse the shift of the GTEM and temperature changes in both the horizontal and vertical dimensions under climate change; (3) Discussion, in which we compare our results with previous relevant studies of the TP and discuss the warming trend differences in these temporal-spatial analyses; and (4) Conclusions, in which the outcomes of this study are summarized.

Materials and Methods

Basic Models

(i) Mean troposphere lapse rate. The lapse rate is defined as the decrease rate of temperature with height for an atmospheric variable. Because the energy of the aerosphere results from long-wave radiation emitted by the Earth, temperature decreases as height increases. Thus, the adiabatic lapse rate is the rate of decrease with height and not simply the rate of change. In general, an adiabatic lapse rate is the negative of the rate of temperature change over the altitude change:

$$\gamma = -\frac{dT}{dz} \tag{1}$$

Where γ is the lapse rate given in units of temperature divided by units of altitude, T is temperature, and z is altitude [47]. If the air is dry and the lift process is adiabatic, the rate of temperature γ fall is 1°C per 100 meters of lift or 5 1/2°F per 1000 feet of lift. Because air always contains some moisture, the Average Adiabatic Lapse Rate (AALR) is the mean lapse rate of the temperature and is between the dry adiabatic and the unsaturated moist adiabatic rates below the condensation level at approximately 0.65°C per 100 meters or 3.3°F per 1000 feet. Therefore in the troposphere, as the elevation increases, the temperature decreases with the AALR and in the vertical dimension and behaves as an isotherm.

(ii) Temperature zone Elevation Model (TEM). The relationship between the temperature zone and elevation can be informed by the AALR, and in the vertical dimension, the isotherms can be projected into the projection layer. As an alpine region usually contains a series of mountains with different shapes and sizes, the shapes of the mountains must be simplified to create a representative but manageable model. A mountain can be simplified to a circular cone shape, as shown in Fig. 1, and an alpine region can be considered to contain a series of such circular cones of different sizes.

In Fig. 1, C_1, C_2, … C_n are a series of isotherms on the circular cone, and their areas that are projected on the projection layer are

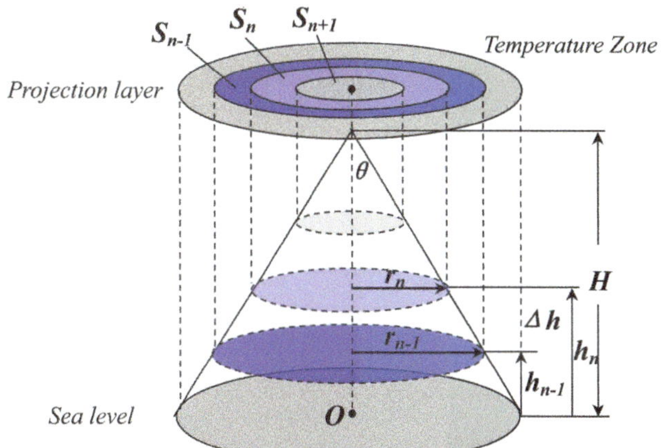

Figure 1. A cone-like mountain model. The elevation/temperature contours are projected on the projection layer and get a series of Temperature Zones (e.g., S_{n-1}, S_n, S_{n+1}, ...).

the temperature belts that need to be measured. The temperature zone can be defined as the loop between two adjacent isotherms on the projection layer. In the model, these zones are the loops *1*, *2*, ... *n*. The elevation interval Δh can be expressed as the absolute elevation between location *n* and location *n-1*:

$$\Delta h = h_n - h_{n-1} = \frac{T_n - T_{n-1}}{\Delta t} \tag{2}$$

Because the vertical temperature lapse rate Δt is an invariant, it remains stable in a certain location (certain longitude and latitude); here, Δt is equal to the AALR. The elevation interval Δh in the vertical direction is also an invariant because the isotherm interval $T_n - T_{n-1}$ is unchanged.

To measure the areas of the temperature loops, the radii of these circles are also required. The radii of these circles are defined by $r_1, r_2, \cdots, r_{n-1}, r_n$. The areas of these temperature loops S_n with elevation h_n can be expressed as

$$S_{n-1} = S_{n-1}' - S_n' \tag{3}$$

$$S_{n-1}' = \pi r_n^2, S_n' = \pi r_n^2, \tag{4}$$

In which S_n' is the area of the contour's horizontal section, and θ is the angle between the cone generatrix and vertical axis. Therefore,

$$r_n = (H - h_n)\tan\theta$$

$$r_{n-1} = (H - h_{n-1})\tan\theta \tag{5}$$

Let $\Delta h = h_n - h_{n-1}$ so $h_{n-1} = h_n - \Delta h$. Substituting h_n into (2)–(5) yields S_{n-1},

$$S_{n-1} = S_{n-1}' - S_n'$$

$$= \pi[(H - h_{n-1})\tan\theta]^2 - \pi[(H - h_n)\tan\theta]^2$$

$$= \pi[(H - h_{n-1})\tan\theta]^2 - \pi[(H - h_{n-1} - \Delta h)\tan\theta]^2$$

$$= 2\pi(H - h_{n-1} - \Delta h)\Delta h \tan^2\theta + \pi\Delta h^2 \tan^2\theta$$

Consequently, $S_n = 2\pi(H - h_n - \Delta h)\Delta h \tan^2\theta + \pi\Delta h^2\tan^2\theta$

$$= -2\pi\Delta h \tan^2\theta * h_n + \pi(2H - \Delta h)\Delta h \tan^2\theta$$

$$\textit{Simplified}, \; S_n = -k * h_n + b \tag{6}$$

In which $k = 2\pi\Delta h \tan^2\theta > 0$, and $b = \pi(2H - \Delta h)\Delta h \tan^2\theta > 0$; obviously, both of these are invariants. In equation (6), if $S_n = 0$, then $h_n = b/k$; if $h_n = 0$, then $S_n = b$. There is an obvious linear relationship between the temperature zone (S_n) and elevation (h_n). As the elevation h_n increases, the temperature decreases, and the area of the temperature zone (S_n) decreases correspondingly. The area between two adjacent isotherms on the projection layer (the temperature zone) is only directly associated with the elevation. These variables have a simple negative linear correlation in the TEM (Fig. 2).

(iii) Generalised Temperature zone Elevation Model (GTEM). It is clear that the temperature zone is only dependent on the elevation and that the variables have a negative linear correlation in equation (6). In the TEM, the mountain is simplified to a circular cone and has a specified surface in which the temperature zone can be calculated by a simple linear function. However, in actuality, the elevations of mountains with relatively young ages nearly follow the normal distribution [48,49]. This distribution may be defined as

$$f(x) = \frac{1}{\sqrt{2\pi}\sigma}\exp\left(-\frac{(x-\mu)^2}{2\sigma^2}\right) \tag{7}$$

The distribution is always described as $f(x)\tilde{N}(\mu,\sigma^2)$, $(\sigma > 0, -\infty < x < +\infty)$. Based on (6), the function (6) may be

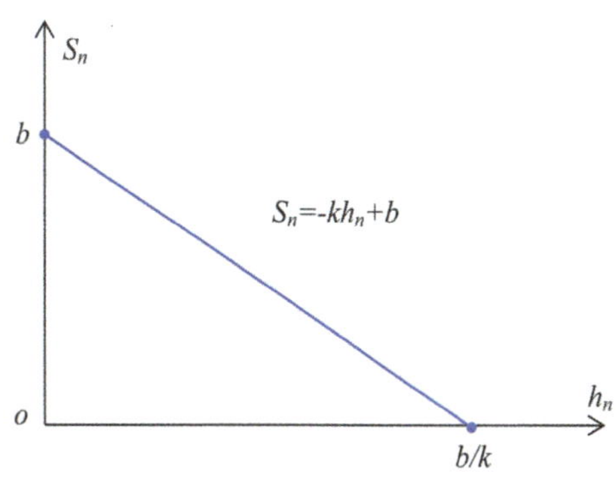

Figure 2. The relationship between the temperature zone (S_n) and the elevation (h_n) in Temperature zone - Elevation Model (TEM). Clearly, they are the linear relationship as h_n increases, the area of temperature zone S_n decreases correspondingly.

simplified to $S=kh+b$, and when the independent variable x is substituted in (7), the distribution density function can be represented as

$$f(h)=\frac{1}{\sqrt{2\pi}\sigma}\exp\left(-\frac{(kh+b-\mu)^2}{2\sigma^2}\right)$$

This may be transformed as

$$f(h)=\frac{1}{k^2}*\frac{1}{\sqrt{2\pi}(\frac{\sigma}{k})^2}\exp\left(-\frac{(h-\frac{\mu-b}{k})^2}{2(\frac{\sigma}{k})^2}\right) \quad (8)$$

Comparing the forms of (8) and (7), $f(h)$ is observed to obey the normal distribution, with the mean $\frac{\mu-b}{k}$, the standard deviation $\frac{\sigma}{k}$ and the coefficient $\frac{1}{k^2}$, namely,

$$f(h)\sim N\left(\frac{\mu-b}{k},\left(\frac{\sigma}{k}\right)^2\right), (\sigma>0, b>0, -\infty<h<+\infty) \quad (9)$$

This expression (9) demonstrates that the distribution of temperature zones also obeys a normal distribution with respect to elevation. The statistic map of temperature zones with elevation is shown in Fig. 3.

(iv) The shift of the GTEM under climate change. Generally, the distribution of temperature zones can be plotted, as shown in Fig. 3, and remains invariable for a period of time. However, if the climate regime changes, as occurs with rising temperatures, then the curve will move accordingly and become deformed. The isotherms become elevated due to warming over a wide spectrum of elevations in mountainous areas, especially in the plateaus. Accordingly, the area of the temperature zones with the same values will change. As a result, the total area $S_{t>0°C}$ of the temperature zones accumulating upwards to a particular value (e.g., 0°C towards the bottom of the mountain) is increased, and the total area $S_{t<0°C}$ accumulating downwards from this value (towards the top of the mountain) shrinks, or vice versa (Fig. 4). If the area of every temperature zone is plotted in the X-Y coordinate system according to expression (9), the normal distribution curve of the total area for every temperature zone will move towards the positive X-axis (the right direction) under conditions of global warming as the temperature rises. If the climate is cooling, then the curve will move in the opposite direction towards the negative X-axis (the left direction, Fig. 5).

Therefore, the GTEM model can be used to judge the presence of climate warming. A simple method to detect climate change is to calculate the areas of all of the temperature zones, plot them in the coordinate system and fit them with the normal distribution curve; if the curve moved towards the right, the climate is warming, while if it is moving in the opposite direction (towards the left of the X-axis), the climate is cooling. The gap between the two adjacent curves shows the rate of change and reveals the magnitude of climate change. However, the GTEM curve will be stable in lower altitude, flat regions due to their narrow range of elevations and monotonic topographies; thus, the model is unsuitable for such areas.

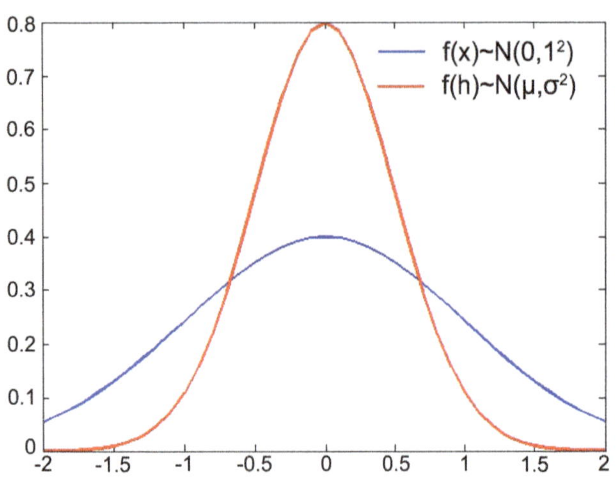

Figure 3. The curves of standard normal distribution $f(x)\sim N(0,1^2)$ (the blue) and $f(h)\sim N(\mu,\sigma^2)$ (the red). It is much peaked with smaller σ under the same μ.

Study Area

The GTEM model should be applied in mountains or high elevation regions. If the climate is warming quickly, then a flat plain at low altitude would not show characteristics of climate change in the GTEM. If the altitude of isotherm exceeds the maximum elevation of the plain, then the temperature zone's area, e.g., $S_{t>0°C}$, will remain unchanged. Considering these requirements of the model, we chose the Tibetan Plateau (TP) as the study area. The TP is distinguished by its high altitude and cold environment and has been called the Roof of the World. After the Hungarian geologist Lóczy [50] first studied the region and defined its boundaries, the TP has become an area of international scientific observation and focused research in recent decades. However, there is not a definitive consensus on the boundaries and extent of the TP. Zhang et al. [51] have suggested that the range of the TP include the Himalayan Mountains in the south and the Kunlun Mountains, the northern margin of Altun and Qilian Mountains in the north, along with western reaches to the Pamirs and the Karakorum Mountains and eastern reaches to the Chengdu plain. In this paper, we chose the Chinese region of the TP along with the surrounding altitudes over 3500 m above sea level (asl.) as our study area. The southern boundary of the TP

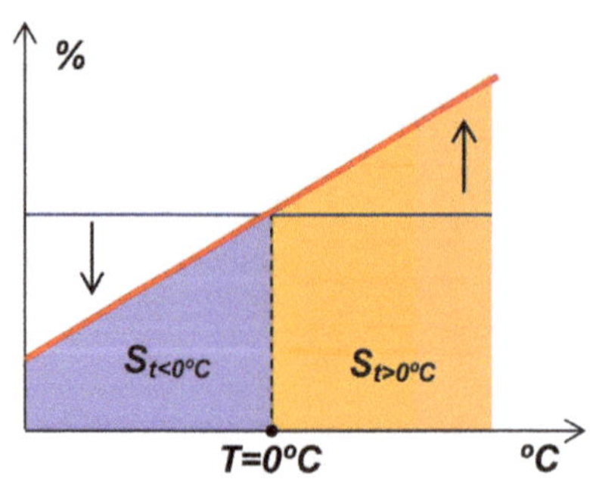

Figure 4. The total area of temperature zones below the Zero Isotherm ($S_{t<0°C}$) shrinks and the total area of temperature zones above the Zero Isotherm ($S_{t>0°C}$) increases under global warming.

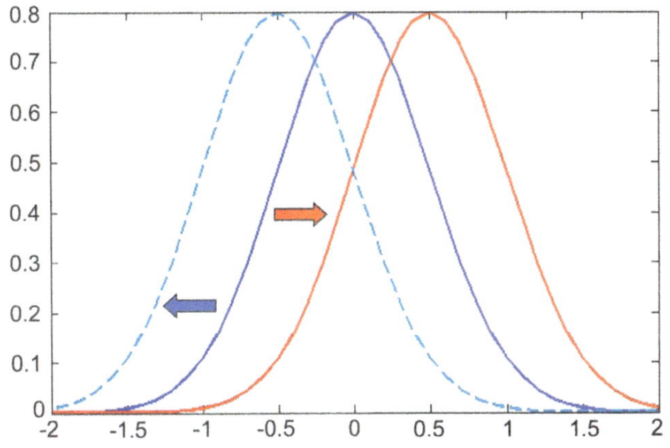

Figure 5. The moved curve of the normal distribution of temperature zones under climate change. If the climate is warming, then the $S_{t<0°C}$ getting shrunk and the $S_{t>0°C}$ getting raised and the curve will move towards the positive X-axis (the red), and when the climate is getting cold it will move towards the negative X-axis (the blue).

is the Chinese National Boundary (Fig. 6). The area of the TP is approximately 2.57 million square kilometres (sq km).

Data Acquisition and Methods

The DEM (Digital Elevation Model) data with cell size 90 m * 90 m were downloaded from the WIST geo-database of NASA (http://reverb.echo.nasa.gov/reverb/redirect/wist, or http://srtm.csi.cgiar.org/SELECTION/inputCoord.asp) and were re-sampled to the resolution of 1 km * 1 km for temperature rasterization. The annual mean surface air temperature data and the monthly mean surface air temperatures were provided by the National Climate Center of China Meteorological Administration (http://cdc.cma.gov.cn/home.do). The annual temperature data were derived from the annual mean temperature records of 686 weather stations in China dating from 1961 to 2010. To calculate the temperature zone's area, the raster temperature data with the GRID format is needed. Currently, there are several methods to get the rasterized temperature data according to the literature (e.g., [52,53,54,55]). Ninyerola et al. [54] and Liao et al. [55] conclude that the Multiple Linear Regression combined with Spatial Interpolation is the minimum error approach to get the rasterized temperature data with the GRID format especially in the complex topography areas. And comparing with other interpolation methods, the Inverse Distance Weighting (IDW) is optimum for temperature interpolation [53,54,55]. Considering the complex topography and the uneven and very sparse distribution of the observation stations in the western part of the TP, in the paper, the Multiple Linear Regression combined with the Spatial Interpolation is used to rasterize the temperatures for reducing the rasterizing errors as much as possible. Based on this method, the temperature can be considered as the quantitative regular part A of environment (e.g., longitude, latitude and altitude) and the non-regular part B (e.g., topography, atmospheric circulation) that can not be quantified. The regular part A can be calculated by the regression function, and the non-regular part B is

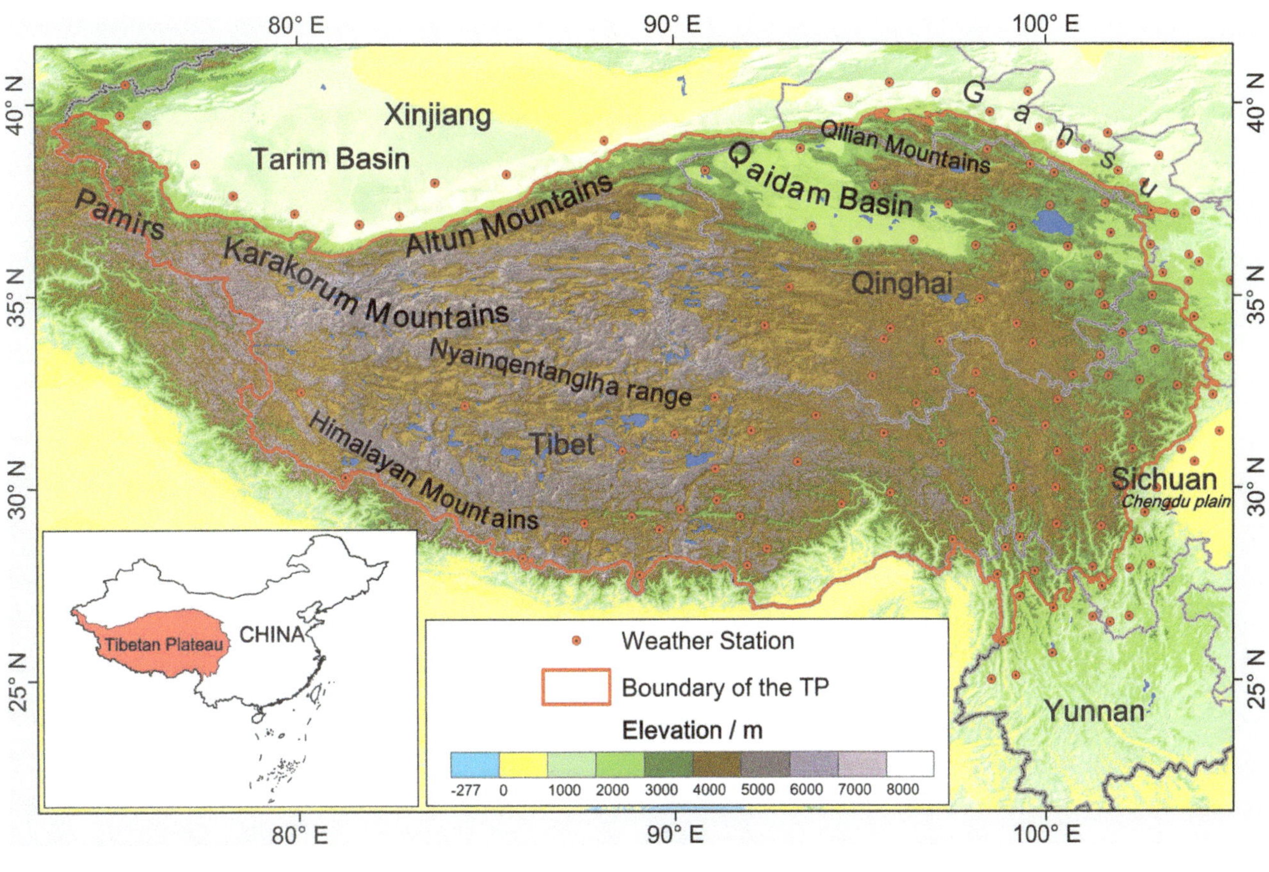

Figure 6. The location of the study area of the Tibetan Plateau (TP) in China.

the *residual error*. The actual temperature is the sum of the calculated temperature A from the regression function and the *residual error* B. The expression is

$$T = A + B \tag{10}$$

As A is the actual temperature regression function of latitude, longitude and altitude, the expression is changed to.

$$T = D + a * x + b * y + c * z + B \tag{11}$$

T (°C) is the calculated actual temperature in the meteorological station. D is the regression constant; a, b, c are the regression coefficients of the multiple linear regression function; x is the *Latitude* (°), y is the *Longitude* (°) and z is the *Elevation* (m asl.). B is the *residual error* between the calculated temperature by the regression function and the observation temperature. The regression functions of the annual mean temperatures in the 144 meteorological stations across the TP and its vicinity (160 km enlarged region, almost 1.2 degree in latitude/longitude; There are 9 incomplete records stations on the TP.) in the 5-year intervals are calculated. The x is the *Latitude* (°), y is the *Longitude* (°) and z is the *Elevation* (m asl.), $N = 135$, $p<0.0001$. The regression functions are listed as follows:

$$T_{1961} = 57.8480 - 0.8071x - 0.1215y - 0.0048z, \ (R^2 = 0.9294)$$

$$T_{1965} = 57.2511 - 0.7496x - 0.1316y - 0.0050z, \ (R^2 = 0.9393)$$

$$T_{1970} = 61.9674 - 0.8333x - 0.1564y - 0.0048z, \ (R^2 = 0.9393)$$

$$T_{1975} = 57.3541 - 0.8213x - 0.1129y - 0.0047z, \ (R^2 = 0.9275)$$

$$T_{1980} = 59.5747 - 0.7609x - 0.1521y - 0.0048z, \ (R^2 = 0.9381)$$

$$T_{1985} = 61.2535 - 0.7969x - 0.1602y - 0.0048z, \ (R^2 = 0.9257)$$

$$T_{1990} = 58.6094 - 0.7271x - 0.1518y - 0.0048z, \ (R^2 = 0.9389)$$

$$T_{1995} = 59.0470 - 0.8251x - 0.1271y - 0.0047z, \ (R^2 = 0.9230)$$

$$T_{2000} = 58.8704 - 0.7416x - 0.1478y - 0.0048z, \ (R^2 = 0.9263)$$

$$T_{2005} = 57.4895 - 0.7754x - 0.1238y - 0.0046z, \ (R^2 = 0.9213)$$

$$T_{2010} = 58.9779 - 0.7616x - 0.1416y - 0.0046z, \ (R^2 = 0.9223)$$

From the regressions, all of the regression functions have high correlation coefficients R^2 (all of the R^2 are above 0.92) and $P<0.0001$. The regression functions are reliable to calculate the temperatures in each station. So, the regular part A can be calculated by the regression functions per pixel by the DEM (Digital Elevation Model, contains the latitude, longitude and elevation in each pixel) on the whole region year by year. The *residual error* B is the difference that the calculated temperature by the regression function subtract the observation temperature in each station. After the calculation, the *residual errors* are interpolated into the whole region by the IDW with the same resolution of 1 km * 1 km as part A. The sum of the two parts is the rasterized temperature on the whole TP in the current year. The other years can be repeated using the same method.

Additionally, to calculate the area of each temperature zone, the reliability and accuracy test of the rasterized temperature data is needed. We compared the rasterized temperatures (Rasterized T) with the observation temperatures (Recorded T) in the 82 stations on the TP in 2000, 2005 and 2010 (Fig. 7, The other years are similar to them and have not been listed). The correlation coefficients R^2 are 0.9912, 0.9872 and 0.9842, respectively. The mean relative errors are 9.6%, 12.1% and 9.7%, respectively. And the mean relative errors below 5% account for 72%, 67% and 68% of the total tested stations, respectively. The tests demonstrate that the rasterized temperature data have the high accuracies and they are reliable for the area calculation.

The temperature zones are divided for every 2°C. The 1 km*1 km raster temperature data were converted into shape files to calculate the area of each zone with the same temperature. After geographic projection and clipping by the range of the TP, the area of each temperature zone (S_n) was calculated. To simplify the calculations while maintaining the trends of the climate change in recent decades, the data were filtered over 5-year intervals. All of these analyses were calculated using ESRI ArcGIS 10. In the discussion section, the monthly surface mean temperature data are the original records of each meteorological station, but they have not been converted to the GRID format.

Results

The Dynamic of the GTEM Curve

Model test. The statistical results of the elevation in the study area are plotted based on the GTEM (Fig. 8). The horizontal-axis of the figure represents the elevations of the TP, and the vertical-axis of the figure represents the statistical numbers of cells with the same elevation. Approximately 80 per cent of the elevations are between 3000 to 5500 m asl., and the peak value is located at approximately 5000 m asl. This result demonstrates that most parts of the TP have high altitudes and that alpine characteristics are prominent in the region.

The elevations of the TP manifest an approximately skew-normal distribution, as shown in Fig. 8. To determine the distribution regime of the elevation, the Kolmogorov-Smirnov test (KS-test) was used in this study, as the KS-test has the advantage of making non-parametric assumptions about the distribution of data [56]. The total number of grids is 162686, the mean elevation is 4384.85 m asl. and the std. deviation of the elevation is 868.045 m. Under the KS-test, the coefficient of skewness (Sk) and the coefficient of kurtosis (Ku) were calculated. The results show that the coefficient of skewness $Sk = -1.177<0$, the coefficient of kurtosis $Ku = 1.840>0$, and the 2-tailed $P = 0<0.05$. Therefore, the distribution pattern of elevation in the TP shows a negative skew-normal distribution with a steep kurtosis of 1.840. Azzalini [48] and Mudholkar et al. [49] have

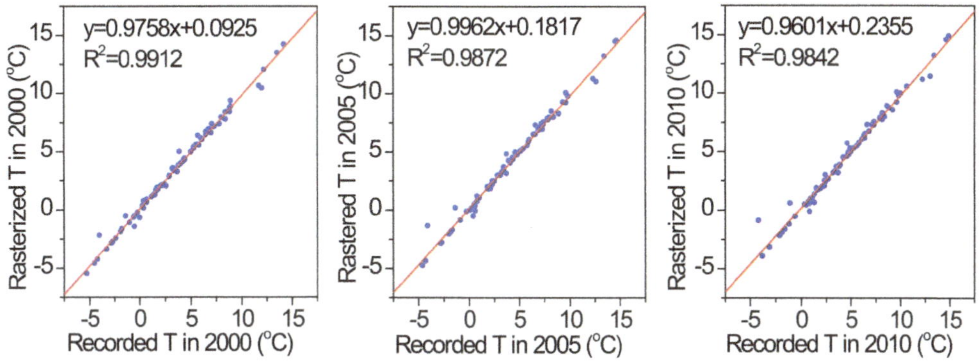

Figure 7. The relationship between the rasterized temperature (Rasterized T) and the observation temperature (Recorded T) in the 82 meteorological stations on the TP in 2000, 2005 and 2010. (The other years are similar to them and have not been listed).

proven that the skew-normal distribution has an equal density function after it is reparameterised using a skewness parameter ε. Accordingly, the function (7) is proper for this study, and the GTEM is correct for modelling the distribution of temperature zones.

The forward moving of the GTEM curve. The shift scenarios of the GTEM under climate change are discussed in Fig. 5. To examine the trends of climate change in the TP over the past 50 years, the 2°C interval annual mean temperature zones for every 5 years were calculated and fitted by Gauss Curve. The statistical results are shown in Fig. 9.

In the illustration of the distribution of 2°C interval annual mean temperature zones, the curves of the earlier years are coloured in shades of navy blue, while those of recent years are coloured in shades of red. From these distribution curves, several obvious conclusions can be drawn:

(1) Representative alpine characteristics. The distribution characteristics of the temperature zones show that the TP is frigid and cold, and the surface annual mean temperatures in most parts of the TP are below 0°C. When compared with low altitude areas at the same latitudes, the TP has distinct alpine cold climate characteristics. Areas with an average temperature below 0°C comprised up to 68.7 per cent of the study region from 1961 to 1980, although this ratio has decreased gradually in recent years. Because the snow lines in the TP are generally between 4200 to 5200 m asl. [57,58,59], most parts of the TP were in permafrost zones, and above a certain altitude, they were also covered by glaciers and permanent snow.

(2) The distributions of the temperature zones are concentrated in the central peaks of the curves. In the bell curves of the distribution series, the areas of temperature zones between −10 to 8°C accounted for 92.4 per cent of the total. The summits were located in the range from −6 to 2°C. These high concentrations of low temperature zones revealed that most parts of the TP were considered to be in cold conditions

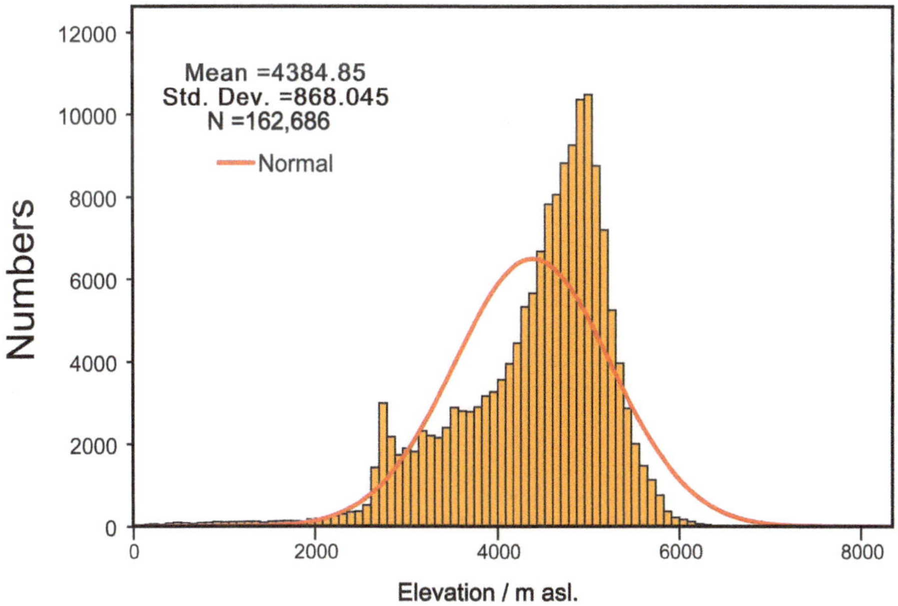

Figure 8. The distribution pattern of the elevations in the TP and their statistical numbers. It can be proved that it manifests a skew-normal distribution with a steep kurtosis.

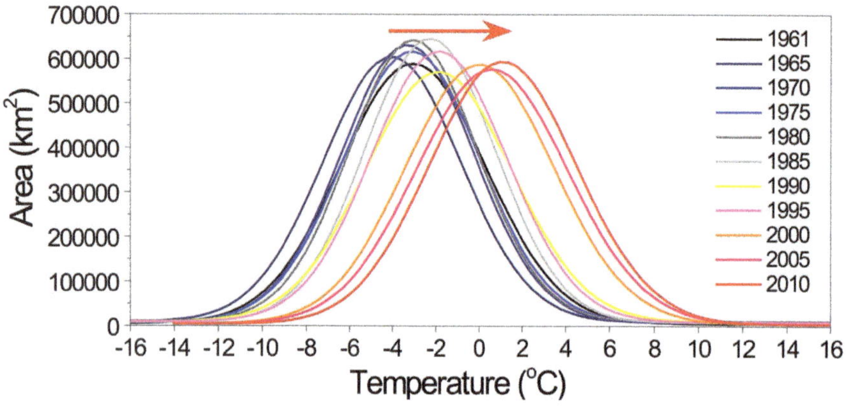

Figure 9. The distributions and changes of the 2°C interval annual mean temperature zones in the characteristic years.

and that the peaks and extremely low altitude valley regions accounted for a small proportion of the overall area.

(3) The curve is moving forward. Although there exist small fluctuations in the GTEM curves, the trend of the curve has moved forward gradually over the last 50 years based on the distributions of the temperature zones in Fig. 9. The direction of the red arrow in the diagram indicates the moving tendency of the temperature zone for this period. The forward moving trend demonstrates that the climate in the TP has gradually warmed from 1961 to 2010, as suggested by the hypothesis and scenario analyses in Fig. 5. This robust warming was manifested by elevation of the isotherms in the vertical dimension, and the isotherm projections on the projection layer were changed accordingly. The changes are manifested by the enlargement of the cumulative temperature zones above a certain level (e.g., 0°C) and the shrinking of the cumulative temperature zones below this level (Fig. 10). With these changes in temperature zones, their distributions also changed.

(4) Warming has accelerated, especially in the last 20 years. From the moving trend of the curves, the increasing gaps between adjacent curves illustrate that the curve is accelerating and that warming has become more severe, particularly after 1990. Although in 1995 the rate of warming retreated to levels comparable with the overall trend of the GTEM curves, the whole tendency of the curve is moving in the warming direction, and the gradually enlarged gaps between adjacent curves demonstrate that this movement is accelerating. These results show the global warming present in the TP and demonstrate that the GTEM is a sensitive and effective tool to detect climate changes in alpine regions.

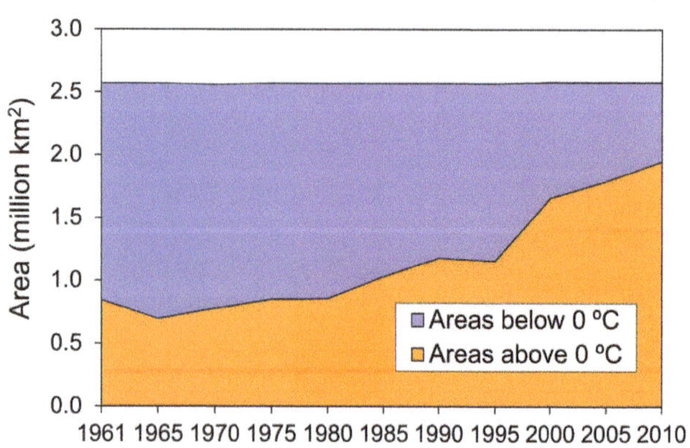

Figure 10. The changes of temperature zones in the TP in the past 50 years classified by 0°C annual mean temperature zone. The areas below 0°C (on the top of the TP) were getting shrunk, whereas the areas above 0°C (on the bottom of the TP) were getting extended over the study period.

The Spatial Changes of Temperature Zones

Horizontal changes. Because spatial warming trends differ depending on the location, especially in regions with complicated topographies, spatial analysis of these changes is indispensable to examining the warming differences. A direct and effective method for investigating changes in the spatial dimension is to examine the changes in the areas of temperature zones.

The total areas of each 2°C annual mean temperature zone over the 5-year intervals were calculated (Fig. 11). The three most reduced temperature zones were the −6°C, −4°C, and −8°C zones. The −6°C temperature zone was the fastest zone and had decreased 499.4 thousand sq km at an average rate of 25.1% every five years from 1961 to 2010. The second most quickly shrinking temperature zone was the −4°C, which shrank 454.3 thousand sq km from 1961 to 2010 at an average rate of 11.7% every five years. During the period from 1961 to 1990, however, the −4°C zone was spreading gradually; after 1990, it began to shrink significantly. Simultaneously, however, several warmer temperature zones have been increasing in size, especially over the last 20 years. The top three most enlarged temperature zones were the 0°C, 2°C, and 4°C zones. These zones comprised a large proportion of the TP at the start of the study period, accounting for 14.8%, 9.3% and 3.4%, respectively, of the total area in 1961 but had gradually increased to 20.8%, 21.4% and 15.2%, respectively, in 2010. The average increasing rates of these zones were 4.5%, 9.9% and 17.8%, respectively, every five years.

It is obvious from the distribution curves in Fig. 11 that the most prominent changes were concentrated in certain temperature zones, especially the −6°C and −4°C zones. Low temperature zones with extremely high elevations at the roof of the TP and warmer temperature zones with low altitudes in the valley regions of the TP experienced lesser changes in comparison with other zones in the middle of the region when the climate was warming. Further analyses demonstrated that these zones only accounted for a small proportion of the area of the TP. Consequently, these small proportions meant that the regions were represented at the bottom

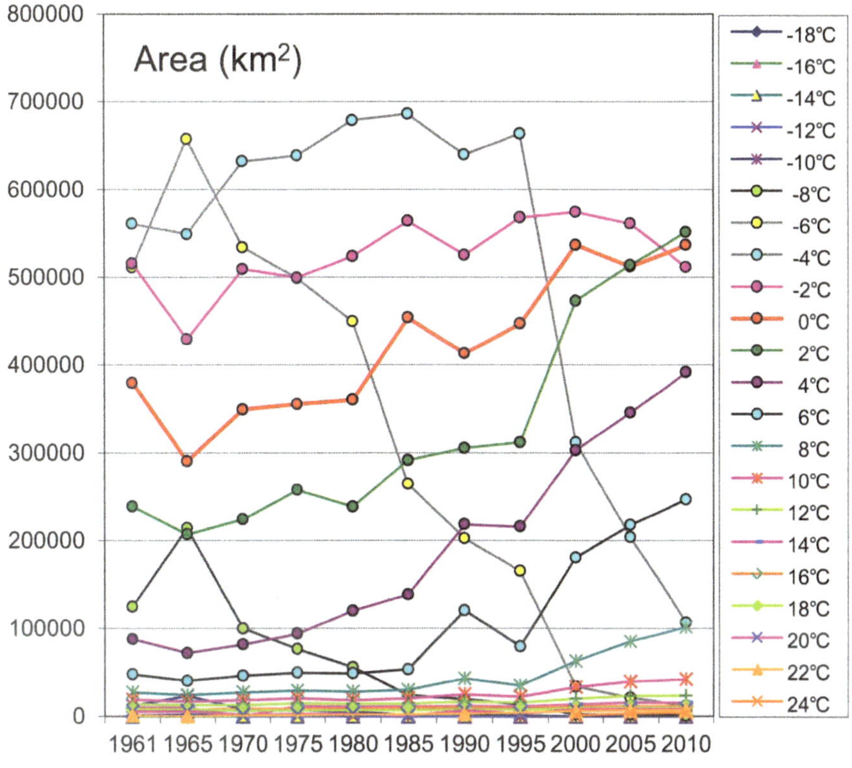

Figure 11. The areas of each 2°C annual mean temperature zones in every 5-year intervals.

of the diagram. In the TP, approximately 80 per cent of elevations are between 3000 to 5500 m asl., and the annual average temperatures of these regions range from −8°C to −4°C (Fig. 11). These areas account for the vast majority of the TP, with slight temperature variations, and the changes to such large areas are prominent. Therefore, the statistics indicate that these regions are more sensitive and responsive to climate warming. This pattern also explains the result that under the same absolute temperature increment, the mountainsides with gentle slopes in the GTEM are much more prone to change in the projected areas than the flatlands.

In addition to the changes of each individual temperature zone, the cumulative temperature zones within a certain spectrum also exhibited the same statistical warming tendencies. The trends of the cumulative temperature zones within certain scopes, i.e., the zones below −6°C, −4°C, −2°C etc., were calculated for further analyses (Fig. 12).

The calculated results revealed that four cumulative temperature zones below −6°C, −4°C, −2°C and 0°C had shrunk, while the remaining zones were enlarging simultaneously. The top three most reduced zones were the cumulative temperature zones below −6°C, −4°C and −2°C. The absolute decreases for each zone were 636.22 thousand sq km, 1.09 million sq km and 1.1 million sq km, respectively, from 1961 to 2010, which equated to decreases of 98.05%, 90.14% and 63.45%, respectively. On the whole, the cumulative temperature zones below 0°C had declined 937.64 thousand sq km from 1961 to 2010, a decrease of 44.54%. In terms of the rate, the fastest shrinking zones were the cumulative temperature zones below −6°C, followed by the −4°C, −2°C and 0°C. These statistical results demonstrate that the fastest shrinking zones were the higher elevations of the TP and that these regions were more sensitive and responsive to climate warming.

Although the total area of the zone between −4 to 0°C had expanded at the beginning of the study period, it had also begun to shrink after 1990. We doubt that the abnormality observed in 1995 is the manifestation of the complicated topography and the vibration of the 5-year interval temperature. However, each of the cumulative temperature zones above 0°C had expanded simultaneously over the past 50 years. These zones occupied almost one third of the total area of the TP, and the absolute increase in area of these cumulative temperature zones was 714.51 thousand sq km, an increase of 89.7% from 1961 to 2010 (Table S1).

The shrinking and expanding magnitudes of each cumulative temperature zone in the horizontal dimension exhibited the trends and amplitudes of climate changes in the spatial dimension. The magnitudes and rates of change in different temperature zones varied from place to place, as have the differently accelerating rates of global warming in recent decades. On the whole, the cumulative temperature zones below −6°C in the northwestern and central TP were the most prominently reduced zones during the last 50 years. To clearly display the changes in the spatial dimension, the spatial distributions of the temperature zones were plotted in decade intervals from 1961 to 2010 (Fig. 13).

The spatial distributions of the temperature zones were nearly consistent with the elevations of the TP. The colder zones were located in the Kunlun-Hoh Xil-Tanggula Mountains, northwest of the TP, and the Himalayas-Gangdise Mountains, southwest of the TP. The warmer zones (shown in warm red colours) were located in the Brahmaputra valleys and the regions near India and Myanmar and Yunnan and the Chengdu plain, southeast of the plateau margin. In the northern TP, the vicinity of the Qaidam Basin and Qinghai Lake had relatively low altitudes of approximately 2800–3200 m asl. and was another warmer temperature zone.

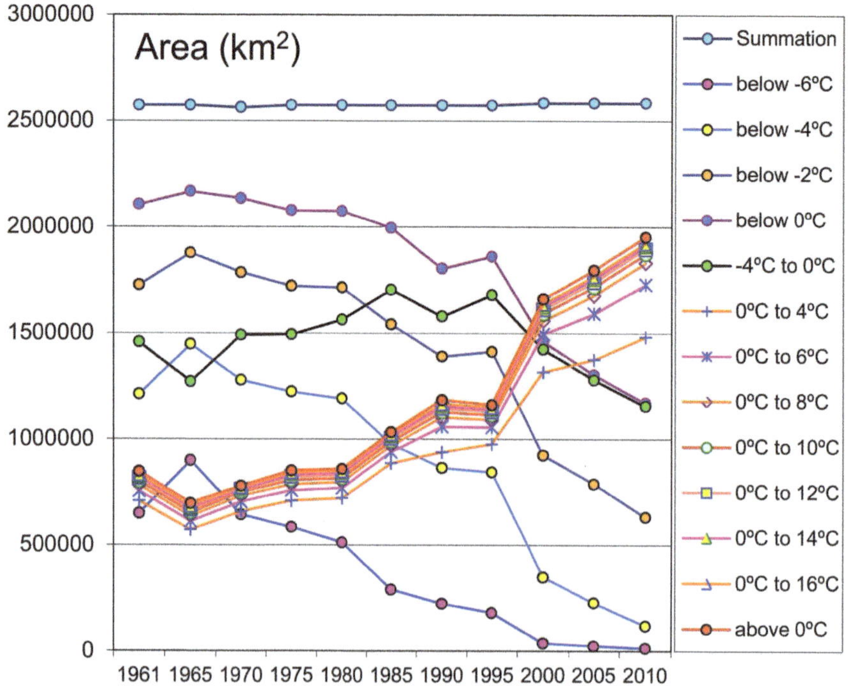

Figure 12. The total areas of each cumulative temperature zones within different ranges in every 5-year interval.

To distinguish the colder temperature zones from the warmer zones, the colder annual mean temperature zones were shown in navy blue colours (Fig. 13). The cumulative temperature zones below −6°C were 648.86 thousand sq km in 1961, accounting for more than 1/4 of the total area of the TP. From 1961 to 2010, however, these areas had shrunk 636.22 thousand sq km and accounted for less than one per cent (0.5%) of the final total, a similar trend as that shown for the −6°C temperature zone (Fig. 14). The warmer temperature zones, in the Qaidam Basin and the southeast of the TP in the Brahmaputra valleys, under the same categorizing criteria, had expanded correspondingly. Obvious changes could also be found in the vicinity of Qinghai Lake, and the trends of these expansions were remarkable. These regions have flat terrain (average altitudes are 3000 m asl.), and with climate warming, the absolute surface mean temperature of the regions had increased almost 2°C during the last 50 years. The shifts of the enlarged temperature zones (warm red colours) and the shrunken zones (deep blue colours) demonstrated that the TP has experienced a significant warming period during the last 50 years. The changes of these temperature zones in the horizontal dimension were the results of vertical elevation of the isotherms. Spatial analyses of the changes can provide further understanding of the detailed warming differences in the spatial dimension.

Vertical changes. The shifts of isotherms and temperature zones are the most direct manifestations of climate change. With the increase in surface air temperatures, isotherms will be vertically elevated, and the temperature zones will change simultaneously. To further examine the trends of climate changes in recent years, especially in the vertical dimension, the mean elevations of each temperature zone in the characteristic years were calculated (Fig. 15).

The trends of the average elevations of the temperature zones were much smoother in the middle elevations of the TP (approximately 2000 to 4000 m asl.) and exhibited only small fluctuations before 1990, as seen in the statistical results in Fig. 15. The small uplift zones in the middle of the TP, e.g., the 6°C zones, had ascended only slightly, an average of 3287 to 3529 m asl. from 1961–2010. Nevertheless, the colder temperature zones on the top of the TP (shown in deep blue colours) and the warmer temperature zones on the bottom of the TP (in red colours) had increased prominently, especially over the past 20 years. For the elevated zones, the −12 to −6°C (above 5000 m asl.) zones were remarkably elevated. The elevations of the virtual isotherms are convincing evidence for global warming, and they displayed a gradual vertical elevation along with the warming. A representative ascending temperature zone is the 0°C zone, which ascended from 4014 to 4728 m asl. at an average rate of 14.3 m per year during from 1961–2010. On the whole, there was an abrupt significant increase in the elevation amplitudes since the beginning of 1990s. We hypothesise that an accelerated period of warming has occurred over the past two decades, but further proof and discussion are needed.

Discussion

Temperature is a critical factor of climate change. Because the TP is a large and unique alpine climate zone, its climate characteristics are very different from the surrounding low altitude areas at the same latitudes. Warming trends in different parts of the TP are also quite different. The search for early signals of temperature variation in the TP may be applicable to assessments of the environmental impacts of global warming in other high altitude regions. Although there is some consensus regarding the warming trends in the TP, the short observation periods of the meteorological stations (most of the stations in the TP were established in 1961 or later), their altitude limitations (there are very few stations above 5000 m asl.) and their uneven distribution, especially in the sparse northwestern TP, have limited comprehensive research of the temperature variation in the TP. In this paper, we used the temperature zone as our analysis perspective

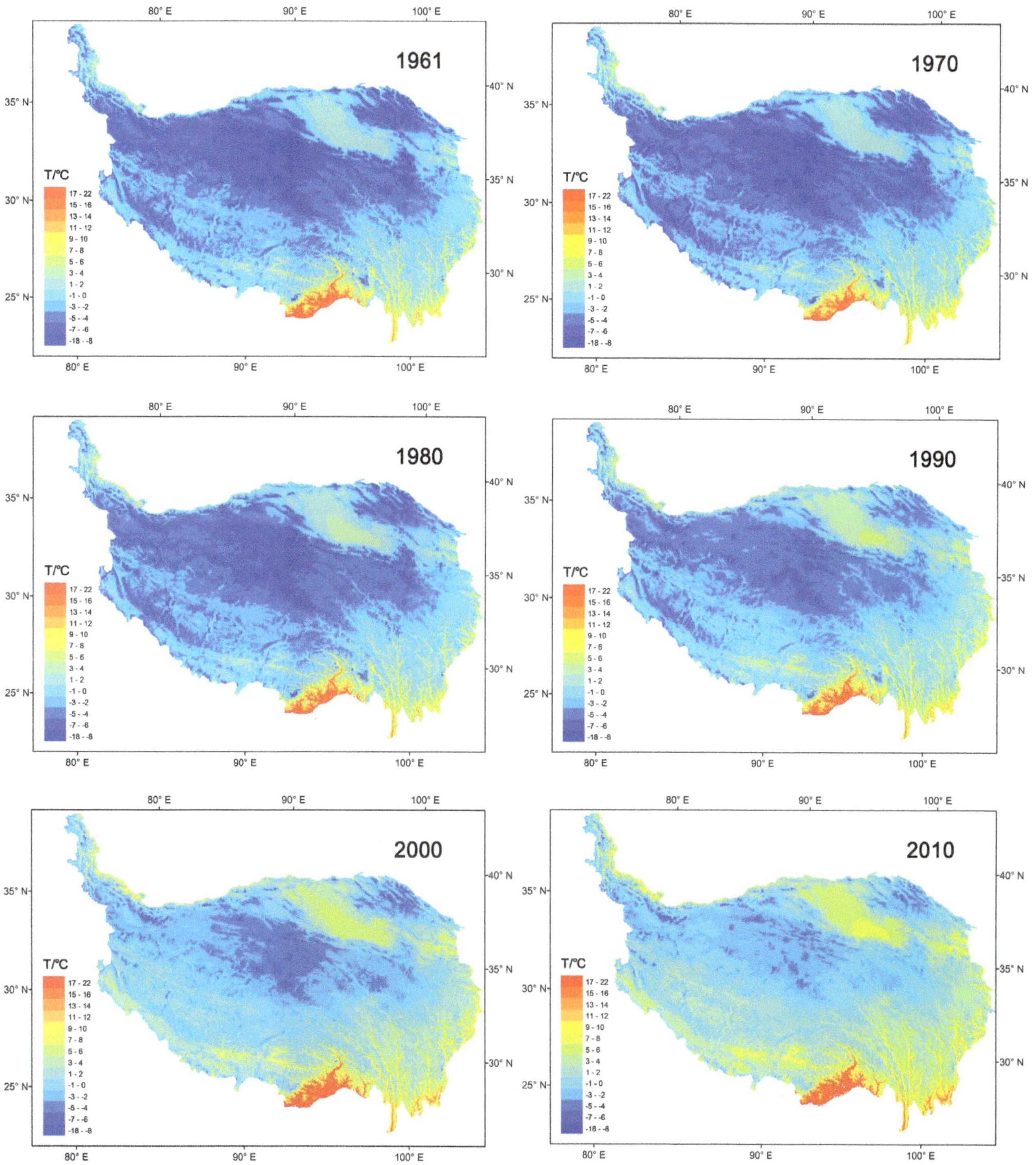

Figure 13. The distribution of the temperature zones in the TP in the characteristic years from 1961 to 2010 under the same categorizing criteria.

and examined the changes in these zones. The calculated results of the GTEM demonstrated that the TP has undergone robust warming from 1961–2010. The temperature changes in the spatial dimension can be clearly examined by the GTEM. In the following section, several relevant issues are briefly discussed. We hope that our perspectives, which are supported by our analyses, will generate further discussion on this subject.

Warming Trends

While the primary manifestations of climate change are physical in nature, such as alterations in temperature and precipitation, rising sea levels, and the increased frequency of extreme weather events, global change is a persistent and fluctuating phenomenon with significant uncertainty and shows the retreat, convolution, acceleration and deterioration of its effects [9,41,60]. During a

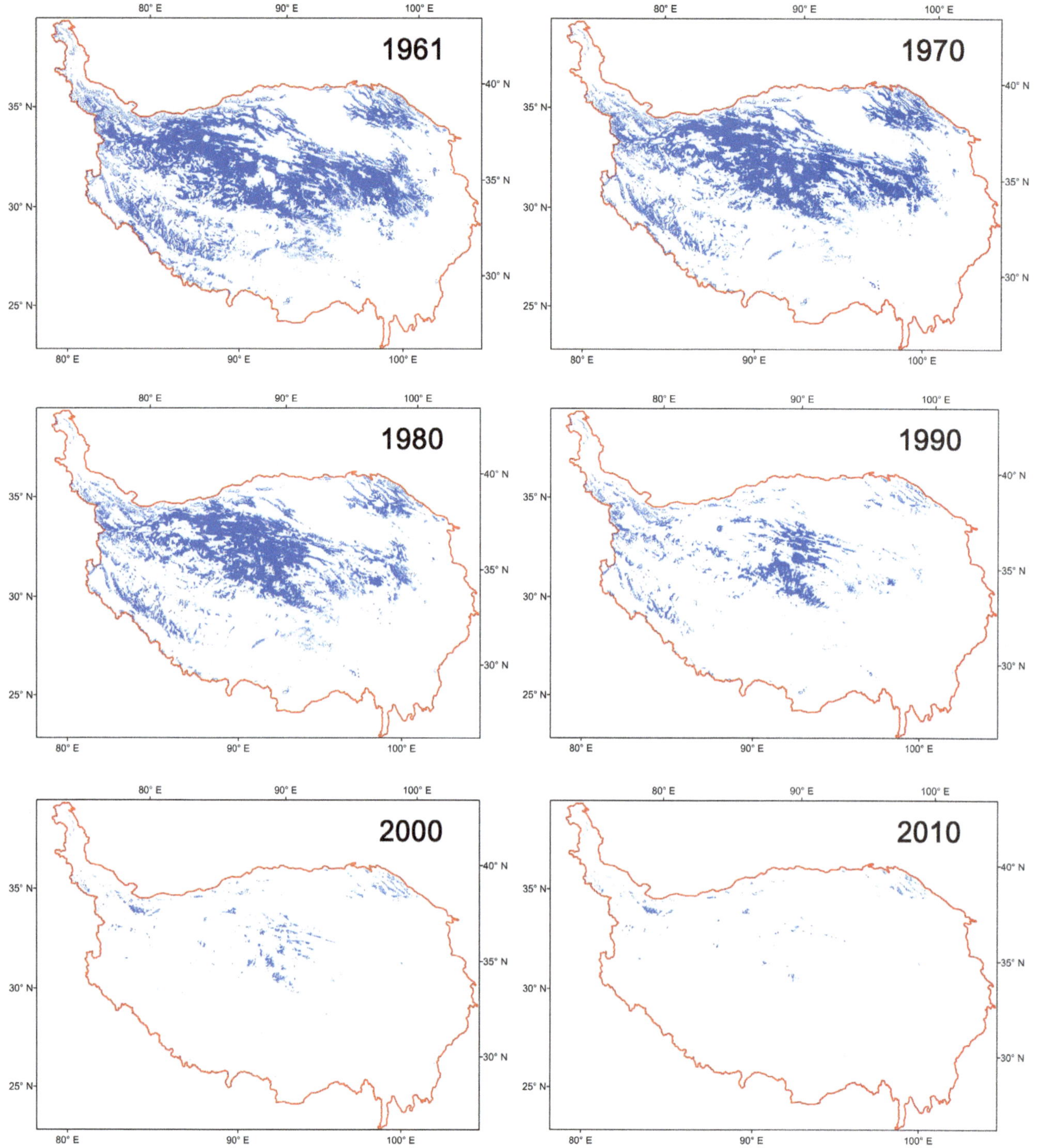

Figure 14. The change of the −6°C temperature zone in the TP in the characteristic years from 1961 to 2010.

short climatic period, however, change is always dominated by one type of climate regime, e.g., warming or cooling [3,61]. The consensus is that global warming has been accelerating since the 1910s, and the trend of this warming is still an uncertainty under different development patterns and greenhouse gas emission scenarios [1,60,62]. As temperature variation is a primary manifestation of climate change, this paper has just discussed the temperature trends from the perspective of the temperature zone in the high elevation TP over the past 50 years and explicitly analysed the warming trend differences in the temporal-spatial dimensions with the GTEM using meteorological station records. The statistical results show that the curve of the GTEM has moved forward and that the gaps between adjacent curves have become larger in the TP, especially in the last 20 years of our monitoring period (Fig. 9). This phenomenon shows that the TP has undergone a significant warming period in recent decades and

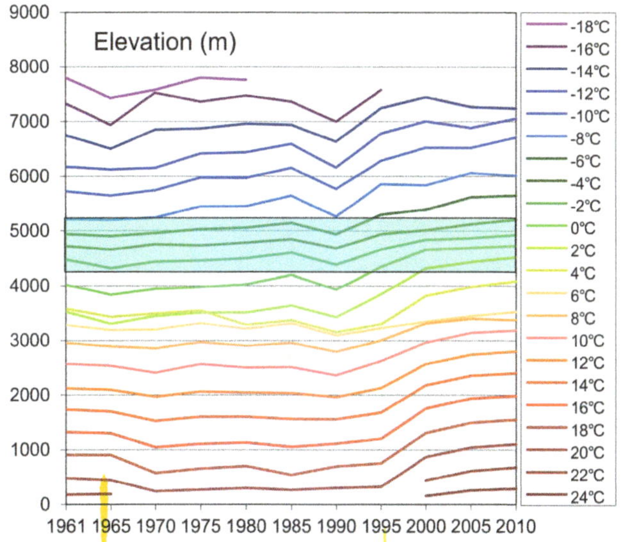

Figure 15. The mean elevations of each temperature zones in the characteristic years. The blue belt is the mean range of snowline/ELA of glaciers in the TP. The range is about 4200–5200 m asl. on average.

that this trend has gradually accelerated. This conclusion is consistent with previous studies of the TP based on surface temperature analyses, e.g., [27,43,44,63]. Further analyses by the GTEM showed that the −6°C and −4°C temperature zones were the most significantly decreased zones and had fallen to 499.44 and 454.26 thousand sq km from 1961 to 2010 at average rates of 25.1% and 11.7% every 5 years, respectively. These regions were located in the northwestern and central TP and at high elevations. The analyses in the vertical dimension showed that, with the increase in the isotherms under global warming, the mean elevations of each temperature zone had increased at different rates. The significantly elevated zones were on the higher and lower elevations of the TP, while trends of the zones in the middle elevations of the TP were much more moderate (Fig. 15). Considering the distribution map of the temperature zone, the most significant altered regions were located in the northwestern and central TP and in several low altitude areas, e.g., the Brahmaputra valley, the Qaidam Basin and the vicinity of Qinghai Lake.

In this paper, our analysis was inspired by the isotherms and based on the perspective of temperature zone changes. The isotherms were elevated, and the temperature zones changed simultaneously due to global warming. Similar changes were also observed in other climate indicators, e.g., the snowline and treeline. Global warming has also caused shifts in the snowline, treeline, species composition and land cover in alpine regions because of altered climate regimes.

As the cryosphere is an amplifier of global warming, the equilibrium line altitude (ELA) or snowline has been elevated more significantly in response to rising temperature and the elevation of isotherms, with glacier shrinking being particularly evident in the Alps and Central Asia [32,64]. The average altitude of the snowline/ELA in the TP reported in the literature is approximately 4200 to 5200 m asl. [9,15,32]. Based on the GTEM, this altitude spectrum was in the range of the −6 to −2°C temperature zones, and more importantly, the mean elevations of these zones were increasing with climate warming, especially over the last 20 years (Fig. 15). The outcomes are consistent with previous studies based on ELA monitoring. One such study, conducted by Vincent et al. [12], compared two mountainous regions of the world and concluded that alpine glaciers are strongly influenced by air temperature increases behind the elevated ELA in the Alps. In the TP and its surroundings, there are approximately 46300 glaciers with a total glacial area of approximately 59400 sq km and total volume of approximately 5600 cu km [32]. Because the TP contains considerable proportions of the glaciers in Asia, its changes are more representative as a way to monitor the effects of climate change. Yao et al. [32] found that glacial retreat in the TP and its surroundings has been a characteristic of the area since the 1960s and has intensified in the past 10 years. With the retreat, most of the snowlines of the glaciers had been elevated, and the retreat had caused an increase of more than 5.5% in river runoff from the plateau; simultaneously, glacial retreat had also caused rising lake levels in areas with large glacier coverage. Tobias [13] and Kutuzov [14] used remote sensing data to investigate the changes to glaciers and found that the temperature increases are two times higher than the global average in northern Tien Shan and very small increases in average precipitation were the main reasons for the snowline rise. Ding et al. [65] showed that more than 80% of glaciers in western China have retreated, losing 4.5% of their combined area, and that the mean ELA has become prominently elevated. Similar results were obtained by Li et al. [15] in an examination of the ELA of the Hailuogou glacier on Mt. Gongga (7556 m asl.), east of the TP; the data show that the absolute retreat ELA of the Hailuogou glacier since the 1930s has amounted to 300 m. The ELA is generally considered to respond positively to temperature and negatively to precipitation [66,67,68]. As a critical indicator of climate change, the ELA/snowline provides sufficiently robust proof of the process and further supports the conclusions from the GTEM. However, these observations are nearly sparsely located in the gigantic TP, and the observation periods are very short; it is difficult to detect climate change differences in the spatial dimension, even with remote sensing data, as they just cover a short observation period.

From the perspective of the treeline and ecotone, shifts of species towards the poles or to higher altitudes have been reported in response to recent global warming trends [16,17,18]. In general, the warmer current climate translates into better conditions for the recruitment and growth of trees at the upper limits of their ranges and facilitates the upward expansion of their population, while at lower limits it causes declines in growth and regeneration, resulting in the replacement of species by trees more suited to warmer climates [19,23]. These trends are quite significant and are relatively easy to observe in alpine regions by comparisons with historic photographs. Baker and Moseley [20] used historic climate data and repeated photographs to assess alpine treeline and glacial recession in the southeastern TP (Yunnan province). These authors found that the warming rate in the most recent two decades was 0.06°C/year and that the warming was causing the retreat of glaciers and the vertical advance of the alpine treeline. Chen et al. [69] tracked the five travels of Ernest Henry Wilson (1876–1930) in western China from 1899 to 1918, then repeated the photos he had taken in his travels and compared them with the originals. The authors found that there was an obvious warming trend over the past 100 years, not only in specific areas but throughout all of Western China. Of the photos, 39 indicated that the treeline had elevated and that the vegetation had changed to suit the altered conditions. These comparative studies further support the results of our analyses based on the GTEM in the paper. As photographic technology is relatively new and long-term document photos are so rare, however, there is a limit to the power of comparative studies in this field.

The existing evidence suggests that global warming is driving the shrinkage of mountain glaciers and melting of permafrost, along with shifts in climate belts or species ranges poleward and towards higher elevations [21,22,23]. To confirm these shifts of climate belts and species redistributions under global warming, Hughes [24] studied the Australian continent, a remarkably flat region: 99% of the land area is less than 1000 m asl., and few summits exceed 2000 m asl., with the highest peak, Mt. Kosciuszko, at only 2228 m asl. As there is a lack of topographic relief, there is a limit on the ability of many species to shift to higher elevations as temperatures increase. Therefore, keeping pace with shifting climate zones requires long overland journeys to new geographic areas for the majority of species. This situation can be considered to be an exceptional example in the GTEM because when a region has a narrow spectrum of altitude, the GTEM curve will stay invariable with warming. Therefore, high elevation mountainous areas are always refuges for species sensitive to global warming [70,71]. In the TP, with its wide spectrum of elevation and vertical zones, relevant research has demonstrated that species redistribution is highly associated with the magnitudes of the warming trend and precipitation conditions [20,69]. However, the shifts in species ranges are due to the comprehensive effects of climate change, and it is difficult to attribute species migration to temperature increase alone. As a manifestation of global change, however, species shifts should be integrated with temperature data in comparative studies. Similar evidence has been reported in studies of climate change-induced land use cover change (LUCC) in high altitude mountain regions [25,41]. Furthermore, these changed land use patterns could lead to changes in livelihoods [72,73]. The same trends have been reported in different mountainous areas, e.g., the Andes [9,26]. In the TP, Byg and Salick [41] considered morality and lifestyle to be the other main impact factors of LUCC besides climate warming. In addition to the climate factors, other anthropogenic factors, e.g., fire, should not be neglected [20]. Overgrazing and intensive human activity can also aggravate the deteriorating trends of LUCC [38,40,69].

In conclusion, there has been a robust warming period in the TP, especially in recent decades, and our GTEM results are consistent with previous relevant research related to temperature zone changes in the region. Most of these observational studies are short-term case studies, scattered widely over the large area of the TP, making it difficult to examine the warming trend differences in the spatial dimension; however, these studies cover different indicators of global warming to provide concrete proof of the temperature zone changes in the TP and further support our conclusions based on the GTEM in this paper. The integration of experimental and observational data with statistical meteorological records is vital for a synthesised analysis.

Warming Trends with Temporal Scales

It has been demonstrated, based on the GTEM using the surface annual mean temperature, that the TP experienced a significant warming process over the monitoring period from 1961–2010 and that the warming trend has accelerated, especially in the recent 20 years. To compare this accelerating process on different temporal scales and further analyse the differences in seasonal warming trends, we calculated the warming trends of each meteorological station during different periods. To examine the differences in the spatial dimension, we divided the 144 stations into 3 elevation zones at 2000 m intervals, i.e., 0–2000 m asl., 2000–4000 m asl. and above 4000 m asl. Warming trends over the past 20, 30 and 50 years for the different elevation zones were calculated (Fig. 16).

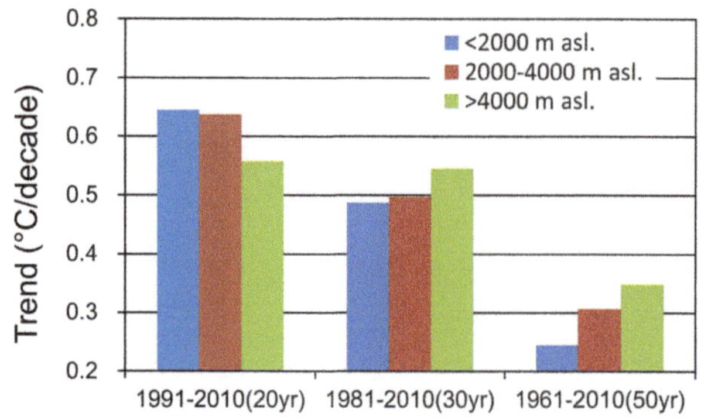

Figure 16. The average warming trends over the past 20, 30 and 50 years in different elevation zones in the TP and its surroundings.

The statistical results showed that the warming trends in the TP and its surroundings have accelerated from 1961–2010. The mean warming trends for these 50 years were 0.24, 0.31 and 0.35°C/decade at lower, middle and higher elevations, respectively. However, in the past 30 years, the warming rates have increased to 0.49, 0.5 and 0.55 °C/decade, respectively. In the last 20 years, the TP has experienced an even more accelerated warming period, and the warming trends have increased to 0.644, 0.637 and 0.557°C/decade, respectively. These statistical results are consistent with the conclusions in the GTEM that the TP had experienced statistically significant warming over the past 50 years and that the warming trend was more prominent in the past 20 years (Fig. 15). In Fig. 15, the smoothly fluctuating elevation lines have significant increased since 1990, and their amplitudes have been increasing greatly year by year. Within the statistical ranges of the paper, the 22°C and 24°C temperature zones had appeared by the end of the study period (the dark red lines at the bottom of the figure). Similar conclusions can be found in several papers, e.g., Liu and Chen [44], Liu et al. [43] and Kang et al. [7]. Using 97 stations located above 2000 m asl. on the TP, Liu and Chen [44] found that the TP had experienced a warming period and that this warming had accelerated from the late 1990s. After analysing the trends of daily and monthly maximum and minimum surface air temperatures, the accelerated warming trends, especially since the 1990s, were confirmed [43]. The accelerated warming trend demonstrated that the global warming in the TP was significant over the past two decades. The changes of in the temperature rates demonstrate the warming trends in the TP and simultaneously support the results of the GTEM.

Because it used data from annual mean temperature records, the GTEM neglected seasonal warming differences. We used the monthly surface mean temperature records as seasonal analysis data and calculated the warming trends for each meteorological station. To compare the trends between different seasons, we defined spring as March to May, summer as June to August, autumn as September to November and winter as December to February of the following year. The elevation was divided into 10 subzones at 500-m intervals to monitor the differences in the vertical dimension. In every zone, the mean linear warming trend was calculated for each season (Fig. 17).

Comparing the trends in different seasons, the seasonal differences are quite clear, as seen in Fig. 17. The winter was the fastest warming season during the period from 1961–2010, with a mean warming trend of 0.473°C/decade. The maximum increase rate in winter was 0.668°C/decade at the 3.5–4 km asl.

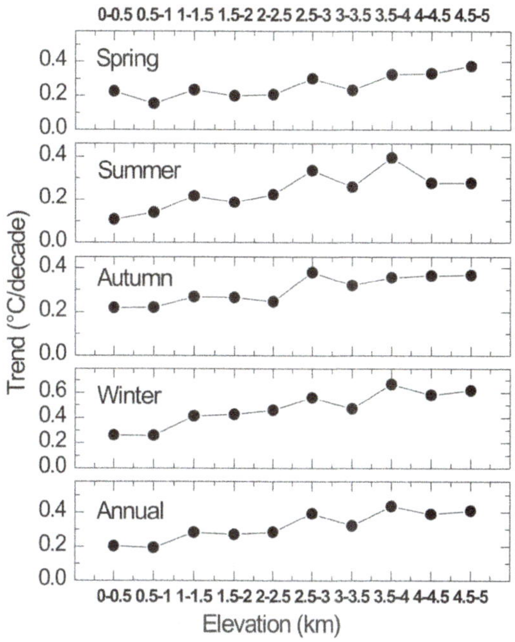

Figure 17. The mean linear warming trends of different seasons in each of the 500-m interval elevation zones during the period 1961–2010 in the TP and its surroundings.

level. The season with the minimum increase was summer, with a warming trend of 0.242°C/decade. The annual temperature increase rate in the TP was 0.318°C/decade during the period 1961–2010, which closely approached the autumn rate of 0.301°C/decade. Further analysis of the monthly mean temperature increase rates classified into 2000-m intervals demonstrated that the warming trends were more prominent in January, February, November and December (Fig. 18). Climate warming was most robust in the winter months, followed by the spring and autumn months. The maximum increase rates were in January and December during winter, while the minimum rates were in May and August in the warmer seasons. These statistically updated analyses on the observational data further confirmed the conclusions from previous seasonal analyses [10,27,29,44]. For example, Liu and Chen [44] examined monthly mean temperature trends at 197 stations in and surrounding the TP and detected more pronounced warming in winter. Wang et al. [63] analysed the annual mean surface temperatures increase rates averaged over 90 stations in the TP during 1960–2007 and obtained a value of 0.36°C/decade, higher than the 0.318°C/decade rate of this paper and double the previous estimate of Liu and Chen [44]. However, these authors reached the same conclusion that the warming trends were more significant in winter than in summer. Generally, these differences in rate were highly related to the choices of stations and research periods [7,28].

While there is a consensus on the seasonal warming trends in recent years in the TP and its surroundings, the main factors and mechanisms underlying this phenomenon remain unclear. Duan and Wu [74] believed that the changes in cloud amount most likely contribute to the recent climate warming in the TP. As the TP is one of the most sensitive areas of snow feedback on Earth, snow/ice-albedo feedback is also considered to be one of the factors contributing to the recent warming in the TP [44]. Based on overall warming, other factors caused by human beings, e.g., anthropogenic greenhouse gas emissions and overgrazing induced

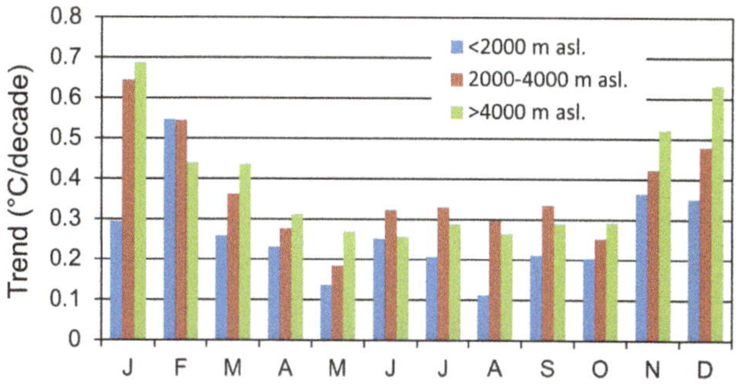

Figure 18. The monthly mean linear warming trends of the 2000-m intervals during the period 1961–2010 in the TP and its surroundings.

land use changes, are generally considered to be the main causes of climate warming in the TP [1,38,39,74]. Nevertheless, the exact reasons underlying the seasonal warming trend differences will require further investigation.

Warming Trends with Elevations

Interestingly, an analysis of the warming trend differences in the vertical dimension suggests that the warming rates are increasing with the elevation in high altitude regions, a hypothesis known as the elevation dependency of climate warming [8,9,43,75]. In our analyses, the GTEM results show that the −4°C, −6°C and −8°C temperature zones (zone mean elevations are 4860, 5176 and 5568 m asl., respectively, Fig. 15) have become significantly smaller during the past 50 years and especially in the past 20 years (Fig. 11). Similar trends can be found in the cumulative temperature zones below −6°C, −4°C, −2°C and 0°C (the mean elevations are above 4214 m asl., Fig. 12). As these zones are located in the higher elevations of the TP, their rapid reductions demonstrate that the top of the TP has undergone a robust warming period during the past decades. We tracked the changes in the −6°C temperature zone during the period from 1961–2010, and the zone nearly disappeared by the end of the monitoring period, as seen in the last three figures of Fig. 14. During this process, the mean elevation of the zone had increased by 715 m, from 4940 to 5655 m asl. Although the warmer temperature zones were concentrated in the narrow valleys and lower altitude regions of the TP, they still accounted for a large proportion of its total area (the areas above 0°C accounted for nearly 4/5 of the TP in 2010, Fig. 10). Accordingly, slight temperature increases would bring about a considerable enlargement of these temperature zones. Further analysis shows that the Qaidam Basin, the vicinity of Qinghai Lake, the southeast TP and the Brahmaputra valleys were the most remarkable warming regions under the same categorizing criteria in the figures (Fig. 13). Based on the GTEM, the warming trends were most significant at higher elevation of the TP and in several low altitude valley regions over the study period from 1961–2010. These results make the existence of the elevation dependency of climate warming in the TP questionable.

An operational method of analysing the problem is to divide the meteorological stations into different classes according to the elevation scale and calculate the statistical linear mean trend of each segment. To determine if the phenomenon is a universal or regional trend, however, the results must be compared over different temporal scales and regions.

First, we divided the 144 stations into three elevation zones at 2000-m intervals. From the statistical results (Fig. 16), it is clear that the highest warming rate was at the top of the TP (above

4000 m asl.), the second was in the middle (2000–4000 m asl.) and the minimum was at the bottom (below 2000 m asl.) during the period from 1961–2010. Similar trends are displayed in Fig. 18 based on the monthly dataset analysis. In the majority of the months, the highest warming rates were found in the elevation zones above 4000 m asl. at the top of the TP, followed by the middle and the bottom. To confirm these results, further analyses using a higher elevation resolution of 500-m intervals were conducted and demonstrated that the statistical warming trends increased with elevation for each of the seasons (Fig. 17). Although there were some slight fluctuations between different elevations, the statistical upward trends were still remarkable. Warming trends in Fig. 17 also show that the elevation dependency of climate warming was much more obvious in winter than in summer. These outcomes are consistent with those of previous studies [8,44]. To diminish the statistical errors resulting from low resolutions and monitor the subtle changes of the warming rates at different altitudes, we also calculated the mean linear warming rates for each of the elevation regions in 250-m intervals (Fig. 19).

These results showed that there was a statistically significant warming trend with the altitude increase in the TP during 1961–2010. Although there were great fluctuations across different elevation zones, warming rates still demonstrated an increasing trend with elevation. When compared with the trends of the 2000-m interval and the 500-m interval, the 250-m interval displayed greater variation. There are 13 stations with elevations below 1000 m asl. and only 1 station (Gerze station, No: 55248, 4763 m asl.) near 5000 m asl.; unfortunately, there were no stations located above 5000 m asl. The lack of a sufficient number of stations for the statistical calculations, especially in the western and central TP, led to large fluctuations in the calculated results. Despite these fluctuations, the statistical warming trends did increase with elevation, suggesting that an elevation dependency for climate warming in the TP was present during the period 1961–2010. These conclusions are consistent with previous studies by Liu and Chen [44], who examined the 97 stations located above 2000 m asl. in the TP from 1955 to 1996, and by Liu et al. [8], who monitored the 116 weather stations in and surrounding the eastern TP during 1961–2006.

However, further analysis showed that the tendency of the warming trend to increase with elevation in the TP has gradually declined in recent decades. As discussed above, an elevation dependency of climate warming did exist in the TP during the period 1961–2010, and the statistical results showed that the same trends were maintained during the period 1981–2010. During the accelerating warming process of the TP in recent decades, however, the gaps of the warming trends between the top, middle and bottom regions have narrowed (Fig. 16). Warming trends at the bottom of the TP accelerated more quickly in comparison with the other two areas until in the last 20 years (1991–2010), warming rates at the bottom had exceeded those of the middle and the top. This result demonstrates that, in the past 20 years, the bottom of the TP has experienced an accelerating warming process and that its warming rates have exceeded those of the middle and the top of the TP. In conclusion, the statistical warming rates were higher at the top of the TP from the perspective of entire study period, but quickly accelerated warming at the bottom of the region has altered this situation in the past 20 years. These results explain why the GTEM displays the Qaidam Basin, the vicinity of Qinghai Lake, the southeast of the TP and the Brahmaputra valleys as the most remarkable warming regions under the same categorizing criteria (Fig. 13).

The literature analysis also shows that there have been many previous studies doubting the existence of the elevation dependency of climate warming as a universal phenomenon but still considering it a regional trend due to the uncertainty behind its reasons and mechanisms [45,46,76,77]. Vuille and Bradley [76] found that warming trends generally decreased with elevation increases in the tropical Andes but that the reason was unknown. Pepin and Seidel [77] found no statistical relationship between the magnitude of temperature trends and elevation in a global analysis of temperature trends for high elevation regions. You et al. [45] examined 11 indices of temperature extremes at 71 stations with elevations above 2000 m asl. in the eastern and central TP during 1961–2005 and found no significant correlations between the elevation and the trend magnitude of temperature extremes; they emphasised that the topographic type and degree of urbanisation were both factors with a strong influence on the analysis. The authors have presented similar conclusions in their subsequent research [46]. Even in approval studies, e.g., [8,44,75], there was no consensus explaining the phenomenon, and most of the conclusions were based on the statistics of meteorological records.

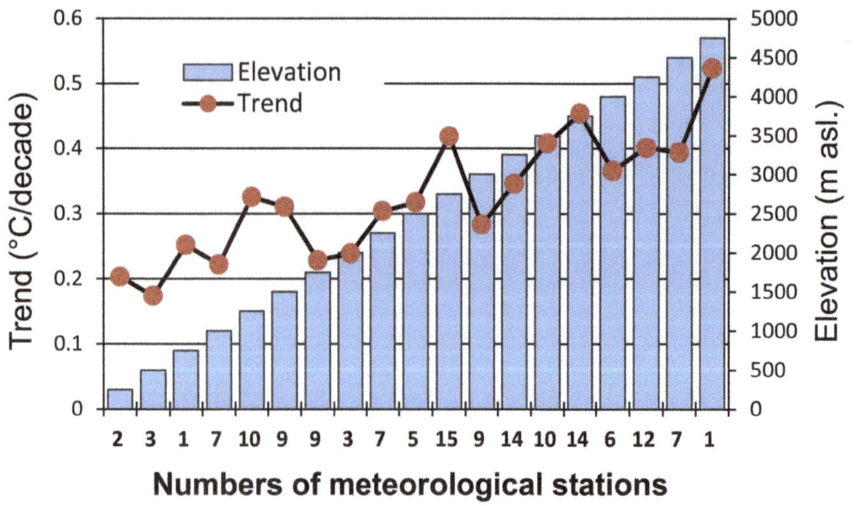

Figure 19. The mean warming trends of every elevation zones. The zones were divided by a 250-m interval based on the 144 meteorological stations in the TP and its surroundings during the period 1961–2010. The labels of the horizontal axis are the numbers of meteorological stations.

Although an elevation dependency for climate warming existed in the TP and its surroundings during 1961–2010, this tendency has been gradually weakening, and both the GTEM and the warming trend-elevation statistics showed that the warming trends were more prominent at the bottom of the TP than at the top in the last 20 years. The elevation dependency of climate warming appears to be an exception or regional trend than a universal law, as surmised from meteorological records and case studies, and the reasons underlying the phenomenon are still unclear. However, the model remains a useful way to analyse warming trend differences in the vertical dimension in high altitude regions.

Data Precision

The uncertainty and complexity of climate change are significant obstacles to its study, not only because it is difficult to simulate the trends of a large climate system under different topographical and regional conditions in high altitude mountainous regions but also because the extent of the impact, the magnitudes and adaptations on different levels vary depending on location [1,78,79]. However, climate change is a lengthy process that is constantly changing and adapting. To simulate the trends on different temporal scales and examine the warming trend differences in the spatial dimension, a Generalised Temperature zone Elevation Model (GTEM) has been established in this paper. Compared to the statistical results of the monthly datasets and the changes in relevant climate indicators observed in the literature, the GTEM was shown to be an effective method for analysing climate changes in the spatial dimension, especially in high altitude regions. As the annual mean temperature raster dataset was used in the model with 5 year-intervals and not with every year considered individually, there may be issues with the precision and generality of the data. Generally, the scale and precision of the data are the two basic sources of error [80]. In large-scale geo-data calculations, high data resolution may produce data redundancy and errors in the analysis process and may not always yield more believable results. Liao and Bai [81] discussed the errors of vector data changed to raster and emphasised that the data scale and land surface conditions (e.g., landscape or topography) were the key components of errors. Therefore, the choice of a suitable data scale and proper data criteria are crucial to the analysis. In this paper, when considering the amount and availability of the data, the annual mean temperature data were filtered by 5-year intervals, and the temperature zones were divided by 2°C intervals. Further analysis showed that the 5-year interval temperature data and the 2°C interval temperature zones could properly detect the trends of climate changes, and although we adopted these conventions, some details of more subtle temperature changes were inevitably ignored. The annual temperature data, when filtered by the 5-year intervals, may have included years with extremely high or low values. The outlier in 1995 demonstrates this situation. Additionally, unnecessarily high resolution (e.g., 1°C intervals) would have divided the temperature zones into very narrow belts with corresponding elevation gaps smaller than those of the 2°C intervals, obscuring the changes caused by the warming process. Further analysis showed that several 2°C intervals displayed inconclusive patterns (in Fig. 11, the area of the −4°C temperature zone increased before 1995 but declined afterwards, and in Fig. 12, the areas of the −4°C–0°C zones increased before 1995 but declined afterwards). These 2°C intervals, in effect, summarised two 1°C interval temperature zones with independent increase/decrease rates. However, the overall trends of warming manifested in the GTEM and the accompanying analyses demonstrated that the model was an effective way to analyse warming at different temporal scales and spatial dimensions. Further studies may utilise different data resolutions and compare the precision of different datasets to improve the analyses.

Conclusions

According to analyses discussed above, we draw the following conclusions:

(1) The GTEM model is an effective method for illustrating the distribution of temperature and can depict changes both in the temporal and spatial dimensions. The forward moving distribution curve of the temperature zones demonstrated that the TP has undergone a robust warming period from 1961–2010, especially in the past 20 years, and the mean warming rate over this period was 0.318°C/decade. Monthly data analyses showed that climate warming was most prominent in the winter months, followed by the spring, autumn, and Summer.

(2) In terms of the spatial changes, the most reduced temperature zones were the −6°C and −4°C zones, which decreased 499.44 and 454.26 thousand sq km from 1961 to 2010 at average rates of 25.1% and 11.7% every 5 years, respectively. These quickly shrinking zones were located in the northwestern and central TP. Based on the GTEM, the warming trends were more prominent at the top of the TP and in several low altitude valley regions during 1961–2010, especially in the last 20 years.

(3) An elevation dependency of climate warming in the TP did exist during the period 1961–2010, but the tendency has gradually weakened over recent decades due to the more rapid warming rate at the bottom than in the middle or at the top of the TP during 1991–2010. This result caused the GTEM to identify the Qaidam Basin, the vicinity of Qinghai Lake, the southeast of the TP and the Brahmaputra valleys as the most remarkable warming regions under the same categorizing criteria.

(4) Comparison with relevant studies on other indicators of climate change in the literature, i.e., the snowline/ELA, treeline, species and LUCC, provided concrete support for the results of this paper.

Acknowledgments

We are grateful for very extensive and constructive comments by two anonymous reviewers, who have greatly contributed to the value of this paper. We also thank the National Climate Center of China Meteorological Administration for providing the meteorological data for this research.

Author Contributions

Conceived and designed the experiments: YW YF. Performed the experiments: YW. Analyzed the data: YW. Contributed reagents/materials/analysis tools: YF YW. Wrote the paper: YW YF.

References

1. IPCC (2007) Summary for Policymakers. Climate Change 2007: The Physical Science Basis. Contribution of Working Group I to the Fourth Assessment Report of the Intergovernmental Panel on Climate Change.
2. China Meteorological Administration (2006) Climate and Environment in China. Beijing: Science Press.
3. Oerlemans J (1994) Quantifying Global Warming from the Retreat of Glaciers. Science 264: 243–245.
4. Oerlemans J, Nick FM (2006) Modelling the advance–retreat cycle of a tidewater glacier with simple sediment dynamics. Global and Planetary Change 50: 148–160.
5. Oerlemans J (2009) Freezes, floes and the future. Nature 462: 572–573.
6. Yao T, Thompson L, Yang W, Yu W, Gao Y, et al. (2012) Different glacier status with atmospheric circulations in Tibetan Plateau and surroundings. Nature Clim Change 2: 663–667.
7. Kang SC, Xu YW, You QL (2010) Review of climate and cryospheric change in the Tibetan Plateau. Environmental Research Letters: 8 pp.
8. Liu XD, Cheng ZG, Yan LB, Yin ZY (2009) Elevation dependency of recent and future minimum surface air temperature trends in the Tibetan Plateau and its surroundings. Global and Planetary Change 68: 164–174.
9. Beniston M, Diaz HF, Bradley RS (1997) Climate change at high elevation sites: an overview. Climatic Change 36: 233–251.
10. Chen B, Chao WC, Liu X (2003) Enhanced climatic warming in the Tibetan Plateau due to doubling CO2: a model study. Climate Dynamics 20: 401–413.
11. Berthier E, Toutin T (2008) SPOT5-HRS digital elevation models and the monitoring of glacier elevation changes in North-West Canada and South-East Alaska. Remote Sensing of Environment 112: 2443–2454.
12. Vincent C, Ribstein P, Favier V, Wagnon P, Francou B, et al. (2004) Glacier fluctuations in the Alps and in the tropical Andes. Comptes Rendus Geosciences 337: 97–106.
13. Tobias B (2007) Climate change and glacier retreat in northern Tien Shan (Kazakhstan/Kyrgyzstan) using remote sensing data. Global and Planetary Change 56: 1–12.
14. Kutuzov S, Shahgedanova M (2009) Glacier retreat and climatic variability in the eastern Terskey–Alatoo, inner Tien Shan between the middle of the 19th century and beginning of the 21st century. Global and Planetary Change 69: 59–70.
15. Li ZX, He YQ, Yang XM, Theakstone WH, Jia WX, et al. (2010) Changes of the Hailuogou glacier, Mt. Gongga, China, against the background of climate change during the Holocene. Quaternary International 218: 166–175.
16. Parolo G, Rossi G (2008) Upward migration of vascular plants following a climate warming trend in the Alps. Basic and Applied Ecology 9: 100–107.
17. Lee TD, Barrett JP, Hartman B (2005) Elevation, substrate, and the potential for climate-induced tree migration in the White Mountains, New Hampshire, USA. Forest Ecology and Management 212: 75–91.
18. Díaz-Varela RA, Colombo R, Meroni M, Calvo-Iglesias MS, Buffoni A, et al. (2010) Spatio-temporal analysis of alpine ecotones: A spatial explicit model targeting altitudinal vegetation shifts. Ecological Modelling 221: 621–633.
19. Peñuelas J, Ogaya R, Boada MS, Jump A (2007) Migration, invasion and decline: changes in recruitment and forest structure in a warming-linked shift of European beech forest in Catalonia (NE Spain). Ecography 30: 829–837.
20. Baker B, Moseley R (2007) Advancing Treeline and Retreating Glaciers: Implications for Conservation in Yunnan, P.R. China. Arctic, Antarctic, and Alpine Research 39: 200–209.
21. Colwell RK, Brehm G, Cardelús CL, Gilman AC, Longino JT (2008) Global Warming, Elevational Range Shifts, and Lowland Biotic Attrition in the Wet Tropics. Science 322: 258–261.
22. Bonnaventure PP, Lewkowicz AG (2011) Modelling climate change effects on the spatial distribution of mountain permafrost at three sites in northwest Canada. Climatic Change 105: 293–312.
23. Jump AS, Mátyás C, Peñuelas J (2009) The altitude-for-latitude disparity in the range retractions of woody species. Trends in Ecology and Evolution 24: 694–701.
24. Hughes L (2010) Climate change and Australia: key vulnerable regions. Reg Environ Change 11: S189–S195.
25. Soini E (2005) Land use change patterns and livelihood dynamics on the slopes of Mt. Kilimanjaro, Tanzania. Agricultural Systems 85: 306–323.
26. Vuille M, Bradley RS, Werner M, Keimig F (2003) 20th Century Climate Change in the Tropical Andes: Observations and Model Results. Climatic Change 59: 75–99.
27. Schwalb A, Schütt B, Fang X (2010) Climate evolution and environmental response on the Tibetan Plateau. Quaternary International 218: 1–2.
28. Sivakumar MVK, Stefanski R (2011) Climate Change in South Asia. In: Lal R, Sivakumar MVKVK, Faiz SMAMA, Mustafizur Rahman AHMHMM, Islam KRR, editors. Climate Change and Food Security in South Asia: Springer Netherlands. pp. 13–30.
29. You QL, Kang SC, Pepin N, Flügel WA, Sanchez Lorenzo A, et al. (2010) Climate warming and associated changes in atmospheric circulation in the eastern and central Tibetan Plateau from a homogenized dataset. Global and Planetary Change 72: 11–24.
30. Yang T, Hao XB, Shao QX, Xu CY, Zhao CY, et al. (2011) Multi-model ensemble projections in temperature and precipitation extremes of the Tibetan Plateau in the 21st century. Global and Planetary Change 80–81: 1–13.
31. Zhao L, Ping CL, Yang DQ, Cheng GD, Ding YJ, et al. (2004) Changes of climate and seasonally frozen ground over the past 30 years in Qinghai-Xizang (Tibetan) Plateau, China. Global and Planetary Change 43: 19–31.
32. Yao TD, Pu JC, Lu AX, Wang YQ, Yu WS (2007) Recent Glacial Retreat and Its Impact on Hydrological Processes on the Tibetan Plateau, China, and Surrounding Regions. Arctic, Antarctic, and Alpine Research 39: 642–650.
33. Yang MX, Nelson FE, Shiklomanov NI, Guo DL, Wan GN (2010) Permafrost degradation and its environmental effects on the Tibetan Plateau: A review of recent research. Earth-Science Reviews 103: 31–44.
34. Jin HJ, Yu QH, Wang SL, Lü LZ (2008) Changes in permafrost environments along the Qinghai-Tibet engineering corridor induced by anthropogenic activities and climate warming. Cold Regions Science and Technology 53: 317–333.
35. Xue X, Guo J, Han BS, Sun QW, Liu LC (2009) The effect of climate warming and permafrost thaw on desertification in the Qinghai-Tibetan Plateau. Geomorphology 108: 182–190.
36. Lin XW, Zhang ZH, Wang SP, Hu YG, Xu GP, et al. (2011) Response of ecosystem respiration to warming and grazing during the growing seasons in the alpine meadow on the Tibetan plateau. Agric Forest Meteorol 151: 792–802.
37. Leemans R, Eickhout B (2004) Another reason for concern: regional and global impacts on ecosystems for different levels of climate change. Global Environmental Change Part A 14: 219–228.
38. Cui XF, Graf HF, Langmann B, Chen W, Huang RH (2006) Climate impacts of anthropogenic land use changes on the Tibetan Plateau. Global and Planetary Change 54: 33–56.
39. Du MY, Kawashima S, Yonemura S, Zhang XZ, Chen SB (2004) Mutual influence between human activities and climate change in the Tibetan Plateau during recent years. Global and Planetary Change 41: 241–249.
40. Fang YP, Qin DH, Ding YJ, Yang JP, Xu KY (2011) The impacts of permafrost change on NPP and implications: A case of the source regions of Yangtze and Yellow Rivers. Journal of Mountain Science 8: 437–447.
41. Byg A, Salick J (2009) Local perspectives on a global phenomenon–Climate change in Eastern Tibetan villages. Global Environmental Change 19: 156–166.
42. Yang B, Bräuning A, Liu JJ, Davis ME, Shao YJ (2009) Temperature changes on the Tibetan Plateau during the past 600 years inferred from ice cores and tree rings. Global and Planetary Change 69: 71–78.
43. Liu XD, Yin ZY, Shao XM, Qin NS (2006) Temporal trends and variability of daily maximum and minimum, extreme temperature events, and growing season length over the eastern and central Tibetan Plateau during 1961–2003. Journal of Geophysical Research: D19109.
44. Liu XD, Chen BD (2000) climatic warming in the Tibetan Plateau during recent decades. Int J Climatol: 1729–1742.
45. You QL, Kang SC, Pepin N, Yan YP (2008) Relationship between trends in temperature extremes and elevation in the eastern and central Tibetan Plateau, 1961–2005. Geophys Res Lett 35: L04704.
46. You QL, Kang SC, Pepin N, Flügel W-A, Yan YP, et al. (2010) Relationship between temperature trend magnitude, elevation and mean temperature in the Tibetan Plateau from homogenized surface stations and reanalysis data. Global and Planetary Change 71: 124–133.
47. Jacobson MZ (2005) Fundamentals of Atmospheric Modeling (2nd ed.).
48. Azzalini A (1985) A class of distributions which includes the normal ones. Scandinavian Journal of Statistics 12: 171–178.
49. Mudholkar GS, Hutson AD (2000) The epsilon–skew–normal distribution for analyzing near-normal data. Journal of Statistical Planning and Inference 83: 291–309.
50. Lóczy (1899) The Scientific Results of the Expedition Through Eastern Asia with B. Szechenyi. In Encyclopaedia Britannica The New Encyclopaedia Britannica, CD2000: 2000.
51. Zhang YL, Li BY, Zheng D (2002) A discussion on the boundary and area of the Tibetan Plateau in China (in chinese). Geographical Research 21: 1–8.
52. DeGaetano AT, Belcher BN (2007) Spatial Interpolation of Daily Maximum and Minimum Air Temperature Based on Meteorological Model Analyses and Independent Observations. Journal of Applied Meteorology and Climatology 46: 1981–1992.
53. Kurtzman D, Kadmon R (1999) Mapping of temperature variables in Israel: sa comparison of different interpolation methods. Climate Research 13: 33–43.
54. Ninyerola M, Pons X, Roure JM (2007) Objective air temperature mapping for the Iberian Peninsula using spatial interpolation and GIS. International Journal of Climatology 27: 1231–1242.
55. Liao SB, Li ZH (2004) Study on methodology for rasterizing accumulated temperature data (in chinese). Geographical Research 23: 633–640.
56. Simos GM (2007) A Kolmogorov–Smirnov type test for skew normal distributions based on the empirical moment generating function. Journal of Statistical Planning and Inference 137: 2681–2688.
57. Aizen VB, Aizen EM, Nikitin SA (2002) Glacier regime on the northern slope of the Himalaya (Xixibangma glaciers). Quaternary International 97–98: 27–39.

58. Casal TGD, Kutzbach JE, Thompson LG (2004) Present And Past Ice-Sheet Mass Balance Simulations For Greenland And The Tibetan Plateau. Climate Dynamics 23: 407–425.
59. Li ZX, He YQ, Pu T, Jia WX, He XZ, et al. (2010) Changes of climate, glaciers and runoff in China's monsoonal temperate glacier region during the last several decades. Quaternary International 218: 13–28.
60. Kerr RA (2007) Global Warming Is Changing the World. Science 316: 188–190.
61. Sundal AV, Shepherd A, Nienow P, Hanna E, Palmer S, et al. (2011) Melt-induced speed-up of Greenland ice sheet offset by efficient subglacial drainage. Nature 469: 521–524.
62. Kintisch E, Kerr RA (2007) Global Warming, Hotter Than Ever. Science 318: 1846–1847.
63. Wang B, Bao Q, Hoskins B, Wu GX, Liu YM (2008) Tibetan Plateau warming and precipitation changes in East Asia. Geophys Res Lett 35: L14702.
64. Lowell TV (2000) As climate changes, so do glaciers. Proceedings of the National Academy of Sciences 97: 1351–1354.
65. Ding YJ, Liu SY, Li J, Shangguan DH (2006) The retreat of glaciers in response to recent climate warming in western China. Annals of Glaciology 43: 97–105.
66. Zemp M, Hoelzle M, Haeberli W (2007) Distributed modelling of the regional climatic equilibrium line altitude of glaciers in the European Alps. Global and Planetary Change 56: 83–100.
67. Yang B, Bräuning A, Dong ZB, Zhang ZY, Jiao KQ (2008) Late Holocene monsoonal temperate glacier fluctuations on the Tibetan Plateau. Global and Planetary Change 60: 126–140.
68. Benn DI, Lehmkuhl F (2000) Mass balance and equilibrium-line altitudes of glaciers in high-mountain environments. Quaternary International 65–66: 15–29.
69. Chen H, Yin KP, Wang HY, Zhong SX, Wu N, et al. (2011) Detecting One-Hundred-Year Environmental Changes in Western China Using Seven-Year Repeat Photography. PLoS ONE 6: e25008.
70. Beardall J, Stojkovic S, Larsen S (2009) Living in a high CO_2 world: impacts of global climate change on marine phytoplankton. Plant Ecology & Diversity 2: 191–205.
71. Harley CDG, Paine RT (2009) Contingencies and compounded rare perturbations dictate sudden distributional shifts during periods of gradual climate change. Proceedings of the National Academy of Sciences 106: 11172–11176.
72. Bradley RS, Vuille M, Diaz HF, Vergara W (2006) Threats to water supplies in the tropical Andes. Science 312: 1755–1756.
73. Thornton PK, van de Steeg J, Notenbaert A, Herrero M (2009) The impacts of climate change on livestock and livestock systems in developing countries: A review of what we know and what we need to know. Agricultural Systems 101: 113–127.
74. Duan A, Wu G (2005) Role of the Tibetan Plateau thermal forcing in the summer climate patterns over subtropical Asia. Climate Dynamics 24: 793–807.
75. Beniston M, Rebetez M (1996) Regional behavior of minimum temperatures in Switzerland for the period 1979–1993. Theoretical and Applied Climatology 53: 231–243.
76. Vuille M, Bradley RS (2000) Mean annual temperature trends and their vertical structure in the topical Andes. Geophys Res Lett: 3885–3888.
77. Pepin NC, Seidel DJ (2005) A global comparison of surface and free-air temperatures at high elevations. J Geophys Res 110: D03104.
78. Adger WN (2006) Vulnerability. Global Environmental Change 16: 268–281.
79. Lahsen M, Sanchez-Rodriguez R, Lankao PR, Dube P, Leemans R, et al. (2010) Impacts, adaptation and vulnerability to global environmental change: challenges and pathways for an action-oriented research agenda for middle-income and low-income countries. Current Opinion in Environmental Sustainability 2: 364–374.
80. Shortridge AM (2004) Geometric variability of raster cell class assignment. International Journal of Geographical Information Science 18: 539–558.
81. Liao SB, Bai Y (2010) A new grid-cell-based method for error evaluation of vector-to-raster conversion. Computational Geosciences 14: 539–549.

7

Strong Phylogeographic Structure in a Millipede Indicates Pleistocene Vicariance between Populations on Banded Iron Formations in Semi-Arid Australia

Heidi Nistelberger[1]*, Margaret Byrne[2], David Coates[2], J. Dale Roberts[1]

1 School of Animal Biology, The University of Western Australia, Crawley, Western Australia, Australia, 2 Science Division, Department of Parks and Wildlife, Bentley Delivery Centre, Bentley, Western Australia, Australia

Abstract

The Yilgarn Banded Iron Formations of Western Australia are topographical features that behave as terrestrial islands within the otherwise flat, semi-arid landscape. The formations are characterised by a high number of endemic species, some of which are distributed across multiple formations without inhabiting the intervening landscape. These species provide an ideal context for phylogeographic analysis, to investigate patterns of genetic variation at both spatial and temporal scales. We examined genetic variation in the spirostreptid millipede, *Atelomastix bamfordi*, found on five of these Banded Iron Formations at two mitochondrial loci and 11 microsatellite loci. Strong phylogeographic structuring indicated the five populations became isolated during the Pleistocene, a period of intensifying aridity in this landscape, when it appears populations have been restricted to pockets of moist habitat provided by the formations. The pattern of reciprocal monophyly identified within the mtDNA and strong differentiation within the nuclear microsatellite data highlight the evolutionary significance of these divergent populations and we suggest the degree of differentiation warrants designation of each as a conservation unit.

Editor: Ulrich Joger, State Natural History Museum, Germany

Funding: 1. Cliffs Natural Resources (http://www.cliffsnaturalresources.com/EN/Pages/default.aspx), 2. Holsworth Wildlife Research Endowment (http://www.jason.edu.au/scholarship/4394). The funders had no role in study design, data collection and analysis, decision to publish, or preparation of the manuscript.

Competing Interests: The authors would like to confirm that Cliffs Natural Resources, a commercial company contributed funding to this study. This does not alter the authors' adherence to all the PLOS ONE policies on sharing data and materials.

* E-mail: heidi.nistelberger@grs.uwa.edu.au

Introduction

Islands provide a unique context for studying the evolutionary processes that drive species distributions. They are natural laboratories, allowing examination of processes of speciation, and patterns of colonisation, persistence, dispersal and extinction [1]. The genetic consequences of island distributions are well documented by phylogeographic and population genetic studies that infer historical patterns of natural and sexual selection, drift and their interactions with historical and ongoing dispersal [2–5]. The study of terrestrial or 'habitat' islands appears to offer a different challenge to oceanic islands, as the matrix between islands can be more permeable than water barriers, resulting in potentially complex spatial and temporal patterns of genetic structure [6,7]. The most conspicuous and well-studied terrestrial islands include mountain Ranges that form 'sky islands', such as the tepui summits of South America, the Western Ghats of southern India and the Madrean and Rocky Mountain sky island systems of central and northern America [6,8–12]. Here, 'island species' occupy disjunct populations on either mountain tops or the lowland valleys separating mountains. Other, less conspicuous topographic habitat islands are found throughout the world and include granite outcrops, inselbergs and Banded Iron Formations (BIFs), many of which display high endemism. Although topographically less dramatic, these provide equally compelling bases for phylogeographic study [13–17].

Inferring the evolutionary history of terrestrial island endemics can be difficult given that isolation is not always absolute, either spatially and temporally. Dispersal events can occur across valleys and lowlands for mountain-top endemics or over mountains for valley floor endemics, depending on the vagility of the organism and their environmental requirements [6,7]. Despite this, phylogeographic patterns, although potentially complex, may reveal the degree of connectivity or isolation experienced by island populations.

Restricted gene flow between island populations over long periods of time, will result in substantial genetic divergence due to genetic drift and/or, local adaptation [18]. Given sufficient time, populations may show patterns of reciprocal monophyly, where all individuals can be traced to a common ancestor within that population [19,20]. Recurrent migration and gene flow between islands on the other hand, has a homogenising effect on genetic structure unless there is strong selection favouring local genotypes on each island. For example, populations of yellow bellied marmots, restricted to the Great Basin mountain-tops of North America, were believed to have been marooned on the rocky outcrops of mountain tops following habitat contraction during the Pleistocene [21]. A pattern of isolation by distance and high genetic diversity showed that occasional dispersal events across the arid scrublands of the Basin have been an important process connecting mountain-top populations [21]. A contrasting pattern was found in the tree *Eucalyptus caesia*, endemic to granite outcrop 'islands' of Western Australia. This species has high levels of genetic divergence with little geographic pattern to the distribution of chloroplast haplotypes, reflecting long-term isolation and persistence without connectivity and emphasising the role of genetic drift in this landscape [22].

The Yilgarn Banded Iron Formations (BIFs) of south-western Australia are ancient, island-like rock outcrops that provide an ideal system in which to investigate island population dynamics. They border the transitional rainfall zone, a biogeographic region separating the high rainfall forest region from the arid interior and are characterised by high species endemism and richness [14,23]. BIFs vary widely in height (highest point approx. 700 m above sea level), shape and distance from one another and are comprised of banded iron talus slopes and areas of weathered duricrust [24]. These topographically complex formations occur in an otherwise flat landscape that is dominated by clay, silt and sand plains. The region has been geologically stable, un-glaciated and above sea level since the Permian (250 mya) [25–27]. Nonetheless there have been substantial changes in climate and vegetation communities over this time. At the beginning of the Cenozoic (65 mya) the region was warm, humid and dominated by temperate rainforest [28]. Over time, the landscape became increasingly arid, beginning approximately 11 mya in the mid-Miocene, shown by the absence of regular flows in paleo-drainage systems [28]. Another aridification event occurred 7 mya in the late Miocene with several more throughout the Pliocene, shown by changes in vegetation structure in palynological deposits to more arid-adapted forms [28,29]. By the early Pleistocene (approx. 2.6 mya), the modern climate had more or less become established but fluctuations throughout this period continued, associated with successive glacial and interglacial periods, the latter with intensifying arid conditions [28].

Historically, the altitudinal relief of BIFs and their complex structure, including deep valleys and shaded gullies as well as exposed outcrops and talus slopes [24], would have provided an abundance of niche habitats that allowed for the persistence of both mesic and xeric-adapted species during changing climates [30]. Whilst this may account for the high species diversity found on the BIFs today, the high endemism [14,23,31] also suggests long periods of isolation, allowing for speciation and the development and persistence of a unique flora and fauna. The Yilgarn BIF endemics consist of species that are localised (found on one BIF), or regional (located on several BIFs but not in the intervening landscape) [23]. Regional endemics provide an opportunity to study phylogeographic patterns, to determine what processes have caused current distributions, and historically, how diversity has evolved and accumulated. The floral assemblages of the Yilgarn BIFs are well known, with many examples of regional endemics [23]. Although less is known of the faunal assemblages, taxa that exhibit short-range endemism consistent with limited dispersal capacity, such as millipedes, mygalomorph spiders, land snails and troglofauna [32–34], may also exhibit patterns of regional endemism, persisting in micro-habitats provided by the BIF.

We investigated phylogeographic pattern in the spirostreptid millipede *Atelomastix bamfordi*, known to occur on five BIFs on the Yilgarn Plateau in Western Australia. Millipedes lack the waxy cuticle that protects other arthropods from desiccation and are reliant on moist environments for survival [32,35]. *Atelomastix* from the family Iulomorphidae, is a predominantly south-western Australian genus with many species distributed across the wetter parts of the southern and south-western coastline and mesic inland Ranges, including the Stirling and Porongorup Ranges [36,37]. Almost all species have restricted distributions and have been characterised as short-range endemics (geographic range <10000 km^2) [32]. *Atelomastix bamfordi* is the most easterly occurring member of this genus. It inhabits moist habitats provided by BIFs, such as the base of rocky slopes, under large rocks and under dense leaf litter. The life cycle is probably annual with adult millipedes dying prior to, or during the harsh summer months (pers. comm. M Harvey). The species has been reported as abundant on the large Koolyanobbing Range [33] but appears scarcer on the drier BIFs to the north and east. Increased exploration and mining of iron ore within the region [23] has resulted in the need for a delicate balance between mining activities and conservation of BIF endemic species. Understanding the patterns and evolution of diversity in these sites may guide two key conservation issues: i) the viability and conservation of highly endemic species ii) the conservation of landscape features which maximise the chances of maintaining patterns of dispersal, gene flow and local adaptation that have shaped the history of extant diversity and continue to be major factors in the persistence of this diversity.

Although unconnected today, BIF populations of *A. bamfordi* may have repeatedly come into contact through population expansion associated with historical climate change in this landscape. If this were the case, we would expect periods of gene exchange between neighbouring BIF populations to result in a pattern of isolation-by-distance, where proximate populations are more genetically related, or, if gene exchange has been extensive, to genetic homogeneity of populations [38]. Alternately, the current distribution may be due to contraction of a formerly widespread ancestor followed by long-term isolation and persistence on BIF islands. If this were the case, we would expect to see signatures of deep genetic divergence amongst BIF populations but no clear pattern of isolation by distance. Haplotypes would be conserved and there may be evidence of reciprocal monophyly, if sufficient time for lineage sorting has occurred [39]. This is the first phylogeographic study to examine patterns of population level divergence in a regional BIF endemic species from this important, biodiverse region.

We evaluated the genetic variation present in *A. bamfordi* to investigate evolutionary relationships and patterns of genetic diversity and structure across BIF populations. Our goals were:

1. To determine patterns of genetic differentiation among the five known BIF populations and to use those data to infer the impact of historical processes on the distribution of millipedes across the landscape.
2. To assess genetic diversity and structure to determine how limited dispersal capabilities affects structure within BIF millipede populations.
3. To use patterns of genetic differentiation, both within and among BIFs to inform conservation and management of this species.

Materials and Methods

Sampling and DNA extraction

A. bamfordi had been recorded from four of the Yilgarn BIFs: Koolyanobbing (KOO), Windarling (WIN), Mt Jackson (MTJ) and the Die Hardy Range (DH) (pers. comm. R. Teale) [33]. This study extended the known distribution to include the Helena Aurora (HA) Range. Four more BIFs in the region (Mt Dimer, Mt Correll, Hunt Range and Mt Manning) were also searched but no specimens were found. Specimens from the Windarling and

Koolyanobbing Ranges, preserved in 100% ethanol were available from the Western Australian Museum's Department of Terrestrial Zoology collection (Table S1). For the remaining BIFs (DH, MTJ and HA) sampling was spread across the geographic extent of each BIF as far as was practical to avoid sampling closely related individuals. For some BIFs this was not possible as millipedes were either not found and/or there was no suitable habitat. The same search procedure was applied to all BIFs, so we believe it is reasonable to assume that the specimens collected reflect the genetic diversity contained within each site. The five BIFs vary in size, topographic complexity and orientation. Koolyanobbing is large, approximately 12 km in length and reaching heights of 500 m. This BIF is topographically complex and contains many sites with habitat suitable for millipedes. Helena Aurora is also large (~12 km) and elevated (600–700 m height), but is comprised of more rounded hills with gentle slopes. Millipedes here were associated with drainage lines coming off the peaks. The Die Hardy Range is large and X-shaped, a topology providing many shaded slopes that supported habitat suitable for millipedes. The two smaller Ranges Windarling and Mt Jackson, have a narrow, east to west orientation and are therefore subject to an abundance of northern sunlight. These smaller Ranges did not possess as many water-gaining sites and millipedes were scarcer. Invertebrate surveys conducted on the intervening 'flats' between the Yilgarn BIF have not recorded *A. bamfordi* (pers. comm. R. Teale). We also searched any water-gaining or densely vegetated sites between BIFs and although these recovered other millipedes from the genus *Antichiropus*, *A. bamfordi* was not found. Thus, to the best of our knowledge, *A. bamfordi* populations are restricted to BIF habitat and are not presently connected by populations in intervening habitat. The field study did not involve any endangered or protected species and none of the field sites from which we sampled (i.e. excluding KOO and WIN) required access permission, with the exception of Mt Jackson (permission required from Cliffs Natural Resources) and Mt Manning (Regulation 4 permit required from the Western Australian Department of Parks and Wildlife (DPaW)). A license to take fauna for scientific purposes was obtained prior to sampling from DPaW. GPS coordinates of the sampling locations are provided in Table S1.

In total, 77 individuals from five BIFs were sequenced at two mitochondrial regions (77 at 16 S, 75 at CO1) and genotyped at 11 nuclear microsatellite loci. DNA was extracted from the legs or from the crushed heads of juvenile specimens stored in 100% ethanol using a Qiagen DNeasy Kit, or a salt-based DNA extraction protocol [40]. Elutions were performed in 100 µl or 50 µl of TE buffer respectively. Legs and head tissue were homogenized in the lysis buffer prior to addition of proteinase K using mini micropestles (Geneworks).

Mitochondrial DNA amplification and sequencing

We sequenced two partial regions of the mitochondria, the Cytochrome Oxidase Subunit 1 (CO1) and the large Ribosomal Subunit 16 S. For amplification of CO1 we used the Folmer primers LCO1490 (5′GGTCAACAAATCATAAAGATAT-TGG3′) and HCO2198 (5′TAAACTTCAGGGTGACCAAA-AAATCA3′) [41] and for 16 S, the 'universal' primers 16Sar (5′CGCCTGTTTTTCAAAAACAT3′) and 16Sbr (5′CCGGTT-TGAACTCAGATCATGT3′) [42]. We also amplified two nuclear gene regions, the Internal Transcribed Spacer 1 (ITS-1) [43] and the 18S rRNA gene [44]. ITS-1 sequenced products showed evidence of site heterogeneity and were excluded from use, and 18 S sequences were monomorphic. Amplifications were performed in 25 µl reactions containing 1× PCR buffer, 3 mM $MgCl_2$, 200 µM dNTP's, 0.4 µM of each forward and reverse primer, 0.25 µg of Bovine Serum Albumin (Fisher Biotech), 0.3 U of Platinum Taq DNA Polymerase (Invitrogen) and approximately 20 ng/µl DNA. The thermal cycler profile for both CO1 and 16 S involved an initial denaturation at 94°C for 3 min, followed by 35 cycles of 94°C for 30 sec, 46°C for 30 sec and 72°C for 30 sec followed by a final extension at 72°C for 2 min. PCR products were purified using an UltraClean PCR Cleanup kit (Mo-Bio technologies, Geneworks) and the cleaned product was quantified using a Nanodrop (Nanodrop Technologies). Sequencing reactions (1/8) were carried out in 10 µl volumes containing 1 µl of Big Dye Terminator, 1.5 µl of 5× sequencing buffer, 1 µl of 3.2 pmol primer and approximately 40 ng of purified PCR product. Thermal cycling conditions for sequencing reactions involved an initial denaturation at 96°C for 2 min, followed by 25 cycles of 96°C for 10 sec, 46°C for 5 sec and 60°C for 4 min. Sequenced products were cleaned prior to electrophoresis using a standard ethanol precipitation method (Applied Biosystems). Products were sequenced on an Applied Biosystems 3730 capillary sequencer (Murdoch University).

Mitochondrial DNA analysis

Sequence data were aligned using Clustal W in the program BioEdit [26] and finalised by eye with all mutations checked against raw sequence data. We used AMOVA, implemented in Arlequin v. 3.5 [45] to determine the structuring of mitochondrial DNA diversity amongst BIF populations using the number of pairwise base-pair differences and tested for significance with 10000 permutations. Bayesian analysis was used to visualise genetic relationships and to estimate coalescence times of the populations in BEAST v. 1.7.5 [46]. Data were partitioned into the CO1 and 16 S datasets, enforcing monophyly for all *A. bamfordi* samples, with a SRD06 codon-partitioned model of sequence evolution applied to the CO1 dataset and an HKY+G model to the 16 S partition. Due to an absence of fossil data and geological calibration points, estimates of population coalescence times were generated by applying the average arthropod mutation rate of 2.3% per million years to the partitioned data set [47]. We applied a strict clock to both partitions and used a Coalescent-Constant Size tree prior for four independent runs of 40 million generations, sampling every 1000 generations. Log files were combined in Log Combiner and the first 25% discarded as burnin. Visual inspection of the traces and high ESS values (>200) confirmed stationarity had been reached in Tracer (BEAST package). Tree files were combined in LogCombiner and the final file annotated in TreeAnnotater (BEAST package).

To distinguish between incomplete lineage sorting and migration for the two populations that were paraphyletic (see results), we performed nested model likelihood ratio tests of various isolation versus migration models in 'L mode' in the program IMa2 [48]. IMa2 uses Markov chain Monte Carlo simulations of gene genealogies to estimate divergence times of populations and effective population sizes of extant and ancestral populations [49]. To assess whether migration or drift was responsible for the paraphyly observed in the two populations we performed a run on the two populations in 'M mode' using the following command line priors: q200, t10, m1, b100000, l1000000, hfg, hn6, ha0.96, hb0.9. Convergence was assessed by monitoring ESS values (all >20 000), chain swap rates (>80%) and running the same command line four times independently using different seed values to ensure consistency of results. The output was then used to test the likelihood of 24 different migration/isolation models outlined in Hey and Nielsen [50] using the following command line: q200, t10, m1, r0, c2 b100000 l1000000. Likelihood ratio tests were used to reject models that were significantly worse than the full model, and the remaining models were compared using the AIC criterion (AIC = −2(log(P)+2K)) to identify the model of best fit, where K is the number of parameters in the statistical model and L is the maximised value of the likelihood function [51].

To assess the historical stability of BIF population sizes we performed tests of neutrality that can also serve as a measure of demographic expansion or contraction. Estimates of Tajima's D were calculated in Arlequin. Diversity statistics including haplotype diversity, nucleotide diversity (Pi), and nucleotide divergence (Jukes and Cantor) between populations were calculated in DnaSP v. 5.10 [52].

To test for isolation by distance within BIF populations using the combined sequence data, we conducted mantel tests in R using the Vegan package [53]. Permutations were performed 10000 times on matrices of the number of pairwise sequence differences and geographic distance. The R statistic is based on Pearson's product-moment correlation coefficient. We plotted the number of pairwise differences on geographic distance and established a linear relationship meaning no transformations of the data were required.

Nuclear microsatellite genotyping and data analysis

Genotyping of the 11 microsatellite markers was carried out according to Nistelberger et al. [54]. We checked all pairwise combinations of loci for linkage disequilibrium by performing exact tests in Genepop v.1.2 [55] and applying a sequential Bonferroni correction to determine significance [56]. There was no significant association among loci following sequential Bonferroni correction. Traditional F statistics were of limited use in our study due to the likelihood that our BIF 'populations' do not represent true panmictic populations given the distances over which individuals were sampled and the limited dispersal capabilities of *A. bamfordi*. This was confirmed with all populations showing significant deviation from Hardy-Weinberg equilibrium, with the exception of Windarling, where individuals had been sampled in close proximity to one another. Rarefied allelic richness provided a better estimator of genetic diversity, given the sampling, and was calculated in HP-Rare v 1.0 [57]. The partitioning of genetic diversity amongst BIF populations was assessed using an Analysis of Molecular Variance (AMOVA) implemented in Arlequin with significance determined with 10000 permutations. An analysis of genetic structure across the five BIF populations was assessed using a Bayesian clustering algorithm implemented in TESS ver. 2.3 [58]. This program assumes proximate interactions between individuals with spatial information prescribed at the individual level. We ran TESS with no admixture, owing to the limited dispersal capabilities of millipedes and performed 100 iterations of K values between 1 and 7, with a burnin length of 20000, 100000 sweeps and a spatial interaction parameter of 0.8. Lowest DIC values for each K value were plotted in order to determine the optimal K value. The results were then used in the downstream program CLUMPP [59] which aligns cluster assignment across the replicate analyses and ensures the correct assignment of K. Structure output graphics were created using DISTRUCT [60].

To further reconstruct the phylogenetic relationships of populations using microsatellite data we generated Cavalli-Sforza chord distances (D_C) [61] between populations and determined support using 100 bootstrap replications in the program Populations 1.2.30 [62]. This geometric distance measure does not make any biological assumptions about the data and can be more useful for phylogenetic inference than estimates that use stepwise mutation models when small population sizes are involved [61,63]. Distances were imported into TreeView for visualisation [64]. To assess isolation by distance within BIF populations using microsatellite data we correlated pairwise population matrices of genetic distance and geographic distance and tested for significance using a Mantel test with 9999 permutations in the program IBDWS v 3.23 [65].

Results

Genetic variation

Mitochondrial DNA data. Of the two mitochondrial regions amplified, CO1 was the most variable, with 45 haplotypes identified from the 75 individuals sequenced (two WIN samples did not amplify at CO1). The variation was composed of 54 transitions, eight transversions and two multistate substitutions. The 16 S region yielded 24 haplotypes from the 77 individuals sequenced and included 21 transitions, one transversion and one single base pair indel. Combining the two regions resulted in a total, aligned sequence length of 1114 bp and 48 haplotypes. The most prevalent haplotype (Hap 47 WIN) had a frequency of 23%, followed by Hap 35 (MTJ) and Hap 39 (HA) with frequencies of 8%, and Hap 33 (MTJ) and Hap 40 (HA) with 6%. The remaining haplotypes were only found in one or two individuals. Measures of genetic diversity were consistently lowest in the Windarling population and highest in Koolyanobbing. Haplotype diversity ranged from 0.295 (WIN) to 0.986 (KOO) with an average of 0.794. Nucleotide diversity ranged from 0.06% (WIN) to 8.5% (KOO) with an average of 4%. No populations departed from the neutral model of evolution except for Windarling (D = −1.775) (Table 1). Measures of diversity in Windarling are likely to be biased due to restricted sampling. Millipedes had been previously collected from two, closely located sites on this BIF (<260 m apart), and although noted to be rare along the length of the Range (pers. comm. R. Teale), further sampling was not possible due to restricted access associated with mining activity.

There was a large and highly significant degree of genetic structuring of the mitochondrial genetic diversity, with most variation (AMOVA) occurring among BIF populations (71.56%) rather than within (28.4%). The greatest genetic divergence occurred between the Mt Jackson and Helena Aurora populations (approx. 40 km apart) and the least between the two most spatially isolated populations, Koolyanobbing and Die Hardy (approx. 100 km apart) (Table 2, Fig. 1).

Nuclear microsatellite data. Microsatellite genetic diversity was low, both within and amongst BIF populations. Rarefied allelic richness ranged from 1.89 (HA) to 3.28 (KOO) with an average of 2.47. There was a highly significant degree of structuring of the microsatellite diversity with the majority of variation (68.11%) maintained within populations, compared to (31.89%) among populations.

Phylogeographic structure and population genetic structure

There was evidence of strong phylogeographic structure within *A. bamfordi*, with minimal sharing of CO1 and 16 S haplotypes amongst individuals within BIF populations and no sharing of haplotypes between populations. Bayesian analysis identified five genetic clades in which populations were reciprocally monophyletic, with the exception of Koolyanobbing and Helena Aurora for which some samples were paraphyletic (Fig. 2). This was due to a subset of six individuals from the north of the Koolyanobbing Range being grouped with samples from Helena Aurora (Fig. 1). Support for the five genetic clades was high (posterior probabilities ≥0.85) although the relationships between the clades was poorly supported. Dating estimates showed coalescence during the

Table 1. Diversity statistics for the five *Atelomastix bamfordi* BIF populations for mitochondrial DNA data and nuclear microsatellite data (bold), standard error in parentheses: N, number of individuals; # haps, number of haplotypes; HD, haplotype diversity; Pi, nucleotide diversity; D, Tajima's D; n, average number of individuals genotyped; AR^, rarefied allelic richness; * P value<0.05.

	Population		#				Haplotype	Genbank Accession #			
Population	**abbrev.**	**N**	**haps**	**HD**	**Pi**	**D**	**numbers**	**CO1**	**16 S**	**n**	**AR^**
Die Hardy Range	DH	16	14	0.983 (0.002)	0.0050 (0.000)	−0.569	19–32	KC689871–KC689883	KC689837–KC689841	**13.7**	**2.62 (0.34)**
Helena Aurora Range	HA	14	8	0.890 (0.004)	0.0042 (0.000)	−0.733	38–45	KC689889–KC689894	KC689845–KC689851	**13.9**	**1.89 (0.28)**
Koolyanobbing Range	KOO	21	18	0.986 (0.001)	0.0085 (0.000)	−0.316	1–18	KC689853–KC689870	KC689829–KC689836	**21.7**	**3.28 (0.40)**
Mt Jackson	MTJ	11	5	0.818 (0.008)	0.0017 (0.000)	−0.234	33–37	KC689884–KC689888	KC689842–KC689844	**9.6**	**2.57 (0.36)**
Windarling Range	WIN	13	3	0.295 (0.012)	0.0006 (0.000)	−1.775*	46–48	KC689895–KC689897	KC689852	**14**	**1.97 (0.27)**

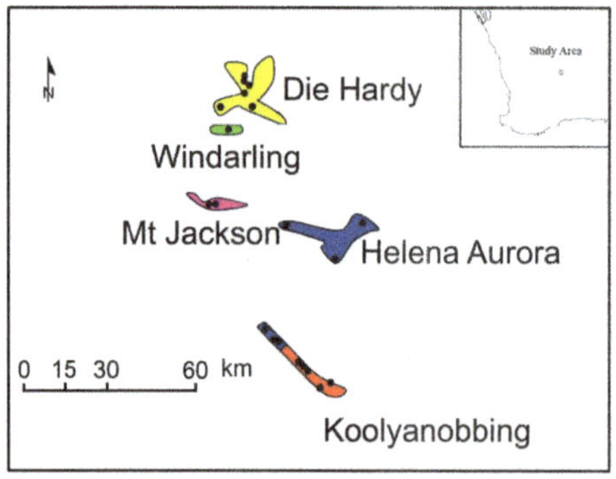

Figure 1. Schematic representation of the five *Atelomastix bamfordi* Banded Iron Formation populations, coloured according to the five genetic clades identified with mtDNA analysis. Black circles indicate sample sites. Study area shown in inset.

Pleistocene, between 0.729 Ma [0.532–0.952] and 1.078 Ma [0.792–1.433], although only the date of 1.078 for the divergence of the Mt Jackson population has high support (Fig. 2).

Isolation and migration analysis in the program IMa2 was used to determine whether the paraphyly between Koolyanobbing and Helena Aurora was due to incomplete lineage sorting (isolation) or gene flow (migration). The three models with the highest AIC support all indicated there was no historical migration between the two populations (Table S2).

Mantel tests of pairwise sequence differences against geographic distance indicated significant isolation by distance within the Koolyanobbing ($r = 0.740$ $p = 0.0001$), Mt Jackson ($r = 0.225$ $p = 0.0375$) and Helena Aurora ($r = 0.399$ $p = 0.0025$) populations.

Population genetic structure

Bayesian cluster analysis identified four genetic clusters in the microsatellite data. The first corresponded to the Die Hardy and Mt Jackson populations, and the remaining clusters to each of the three other populations (HA, WIN, KOO) (Fig. 3). Further investigation of structure in the first cluster revealed two more groupings corresponding to the separate BIF populations (DH, MTJ). There was some evidence of admixture in the Mt Jackson and Koolyanobbing populations (Fig. 3). Phylogenetic analysis of relationships between populations using Cavalli-Sforza chord distances showed distinct genetic structure amongst populations using microsatellite data, but with some clustering of Koolyanobbing and Helena Aurora (97% bootstrap support) and Die Hardy and Mt Jackson (62%) (Fig. 4). Two of the five populations, Helena

Table 2. Pairwise population nucleotide divergence based on Jukes and Cantor of five *Atelomastix bamfordi* populations using mtDNA data.

	DH	HA	MTJ	WIN	KOO
DH	0				
HA	0.016	0			
MTJ	0.017	0.022	0		
WIN	0.013	0.018	0.019	0	
KOO	0.013	0.014	0.018	0.017	0

*P value<0.005.

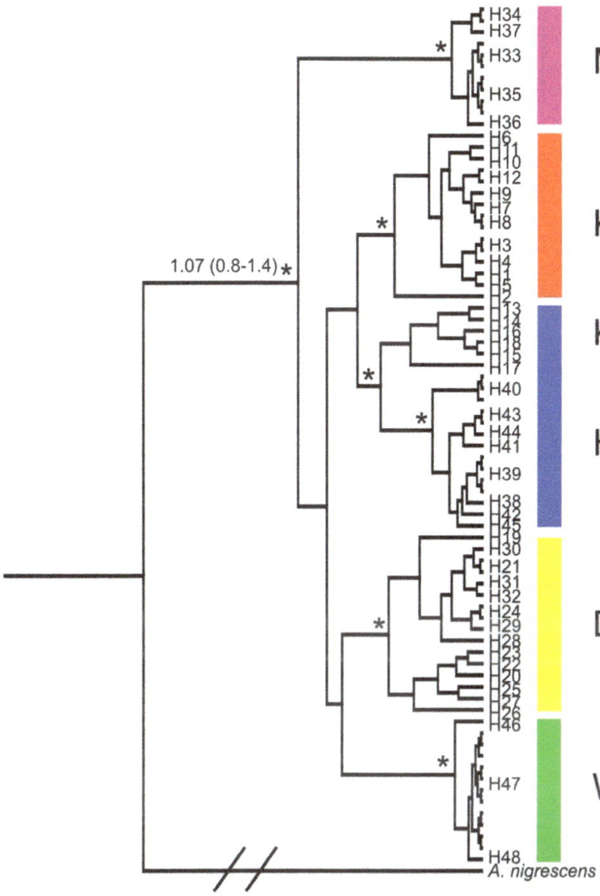

Figure 2. Bayesian cladogram of 75 *Atelomastix bamfordi* individuals based on combined CO1 and 16 S mtDNA data. Clades are designated separate colours as per Fig. 1. Numbers on terminals indicate haplotype number. The * denotes major nodes with high support (posterior probability>0.85) and the numbered node indicates divergence estimate in millions of years with confidence intervals in brackets. NB the *A. nigrescens* branch has been truncated (//) for viewing purposes.

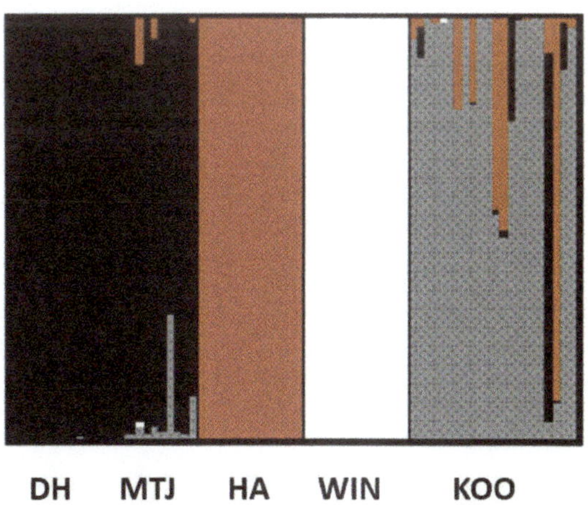

Figure 3. Bayesian population assignment of *Atelomastix bamfordi* individuals generated in TESS for the optimal grouping of K=4. Each vertical line represents an individual and the colour displays the probability of belonging to each genetic cluster.

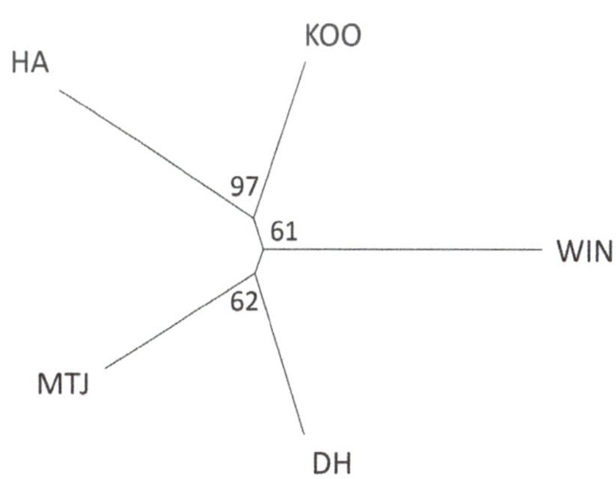

Figure 4. Unrooted neighbour-joining tree of Cavalli-Sforza chord distances between the five BIF populations of *Atelomastix bamfordi*. Distances are based on 11 microsatellite loci with bootstrap support shown at nodes.

Aurora (r = 0.30 p = 0.004) and Die Hardy (r = 0.32 p = 0.009), showed weak but significant patterns of isolation-by-distance at microsatellite loci following Mantel tests.

Discussion

Genetic differentiation of BIF populations

Spatial patterns. Phylogeographic analysis of *A. bamfordi* has revealed strong genetic differentiation between BIF populations, with mtDNA genetic lineages corresponding to each of the five BIF populations except for six individuals from Koolyanobbing that were paraphyletic with the Helena Aurora Range due to incomplete lineage sorting. Microsatellite data were broadly consistent with this result, identifying four genetic clusters, the first of which was further broken down into individual populations.

The lack of correlation between genetic divergence and geographic distance, with the two most spatially isolated populations showing the least genetic divergence, suggests there has been little, to no connectivity of BIF populations following their separation. This result was emphasized in the Bayesian analysis which showed high support for genetic clades but poor support for the relationships between clades (Fig. 2). It is also consistent with estimates of sequence neutrality (D) that indicated a lack of demographic expansion and contraction events, with the exception of Windarling, where the result can be explained by close sample proximity. This pattern of differentiation has been observed in other taxa with restricted dispersal capabilities where populations have become isolated as a response to climate change and habitat loss. Like *A. bamfordi*, these studies emphasise the role of genetic drift in driving population differentiation. For example, mitochondrial and nuclear sequence data revealed extraordinarily strong phylogeographic structure in populations of trapdoor spiders (*Moggridgea tingle*) in south-western Australia, often situated just kilometres apart. Molecular dating suggested these populations became isolated in the mesic valleys and uplands of the Porongorup and Stirling Ranges following the onset of aridity in the late Miocene [66]. Like *A. bamfordi*, reciprocally monophyletic lineages across three geographic locations highlighted the reproductive isolation of *M. tingle* populations but the pattern emphasized a more ancient division, with monophyly observed in both mtDNA and the more slowly evolving nuclear genome [66]. Studies of genetic structure in two frog species in south-western Australia, *Geocrinia lutea* and *G. rosea*, also described patterns of extremely low dispersal and long-term isolation between habitat 'islands', with divergence attributed to isolation via range changes resulting from climate change rather than geographical barriers [67]. Another striking example of genetic divergence across terrestrial islands occurred in the sky islands of south-eastern Arizona - forested mountain Ranges, separated by low-lying grassland and desert. Here, the jumping spider *Habronattus pugillis* occupies woodland habitat at altitudes above 1400 m but is absent from the lowlands. Using mtDNA, Masta [7] showed that like *A. bamfordi*, smaller populations of *H. pugillis* formed monophyletic clades whereas larger Ranges showed evidence of paraphyly. Like *A. bamfordi*, this was ascribed to incomplete lineage sorting rather than resulting from migration [7]. The isolation of populations was attributed to climate-induced contraction of woodland populations onto disjunct mountain Ranges.

Temporal patterns. Molecular dating with the arthropod clock rate has placed the separation of *A. bamfordi* clades at 0.7 to 1.1 million years ago during the Pleistocene. The timing coincides with known increases in aridity in the transitional rainfall zone (TRZ) due to climatic fluctuations associated with Milankovitch cycles during this period [68]. In particular there was a major arid shift in southern Australia corresponding to the middle Pleistocene transition (1250 ka–700 ka), a period of increased severity of glaciation, associated with greater ice volumes, lowered temperatures and increased aridity [69]. Although the TRZ remained unglaciated, this period of intense climatic instability is expected to have caused the fragmentation of species distributions due to changes in suitable habitat within this region [70]. Although the divergence estimates presented here must be treated with caution given the use of a generalised arthropod mutation rate, the errors associated with this method are lessened given we are working at shallow phylogenetic levels [71,72].

BIF population dynamics – historical and contemporary patterns

Historical patterns. The process of genetic drift acting independently on each BIF population is reflected in different patterns of genetic divergence and diversity seen in the mtDNA data. The smaller populations, Mt Jackson and Windarling, and the larger population Helena Aurora, which had less suitable millipede habitat, showed lower genetic diversity and higher levels of genetic divergence than the larger populations (KOO and DH). These three BIFs have probably harboured historically small effective population sizes, owing to the limited availability of moist sites, increasing their susceptibility to forces of genetic drift driving divergence [73]. They may also be influenced by different selective forces associated with the more arid conditions present on these BIFs. On the other hand the two large BIF populations, characterised by a greater availability of suitable habitat (KOO and DH), retained genetic diversity and had less genetic divergence, consistent with historically large population sizes [73]. The paraphyly of individuals from Koolyanobbing with those of Helena Aurora further supports the likelihood of a large historical population size at Koolyanobbing, as lineage sorting is still ongoing at this BIF. Over time this signature is likely to erode, leaving the five BIFs reciprocally monophyletic.

Contemporary patterns. Rapidly evolving microsatellite markers can provide a more contemporary view of BIF population dynamics. The information we gained from these markers was limited by a lack of Hardy-Weinberg equilibrium within our BIF populations yet we still identified patterns in genetic diversity that appear to support the impact of genetic drift on populations: lower genetic diversity in BIF's likely to have had small effective population sizes (WIN, MTJ and HA). Signals of isolation by distance (IBD) were seen in some of the BIF populations (HA, DH), but these were weak and likely to be heavily influenced by limited sample sizes. The small pairwise distances between samples at Windarling explains the lack of IBD seen in this BIF, but we were surprised to see evidence of IBD in the Helena Aurora and Mt Jackson individuals. Here, we expected limited gene flow to result in a pattern of 'islands within islands', where significant isolation has resulted in genetically divergent populations within BIFs that do not correlate with geographic distance. An increased number of samples from more locations would be required to determine whether IBD is really occurring at these spatial scales.

Conservation implications

The evolutionary significance of geographically isolated genetic lineages has long been recognised and conservation biologists have attempted to formalise this by delineating conservation units within a species such as Evolutionary Significant Units (ESUs) and Management Units (MUs) [74]. Conservation units have been defined at various levels of genetic differentiation within species and are likely to experience limited gene flow or even total reproductive isolation. For example, ESUs have been defined as a population or a group of populations that warrant separate management for conservation because of high genetic and ecological distinctiveness [75], with Mortiz [20] suggesting that ESUs might be characterised by a pattern of reciprocal monophyly amongst populations. Applying this criterion to our mtDNA data, results in the recognition of three BIF populations (DH, MTJ, WIN) as ESUs. However, Paetkau [76] and Crandall et al. [77] highlight the problems inherent in ignoring paraphyletic lineages, such as Koolyanobbing and Helena Aurora, which are often ancestral and maintain high levels of diversity and divergence within themselves. In our case, the paraphyly of Koolyanobbing with Helena Aurora is a result of a slower progression towards monophyly due to ongoing lineage sorting, probably owing to large population sizes. Given the clear evidence for restricted gene flow between these two populations, both could readily be recognised as MUs. Therefore we consider that all five BIF populations warrant designation as conservation units.

Although it is difficult to predict future climate change at such fine scales, broad-scale models for this region over the 21st century indicate increasing temperature, decreasing rainfall and a greater incidence of fire [78]. If these predictions are correct, populations are unlikely to re-connect via suitable intervening habitat in the future, further consolidating their isolation to the remaining moist pockets provided by BIF, where they will continue to be influenced by forces of genetic drift and local selection. Given their dependence on moisture, conservation efforts should focus on protecting areas of dense leaf litter, large outcrops and shaded gullies on each BIF in order to preserve the five genetic lineages identified here. To achieve this, preservation of the height and topographical complexity of BIFs is essential, in order to maintain drainage lines and water-collecting sites. Conservation of these areas is also likely to support populations of other, relict fauna and flora that are adapted to moist environments.

Conclusion

This is the first phylogeographic study conducted on a species inhabiting multiple Banded Iron Formations in this landscape and provides evidence that the disjunct populations have experienced long-term isolation following their separation during the Pleistocene. The divergence of populations during this period coincides with increases in aridity and emphasizes the role of climatic instability along with topographic features in driving differentiation and ultimately speciation in this landscape. We consider the strong phylogeographic structure and evidence of restricted gene flow between BIF populations warrant the designation of all five populations as distinct conservation units.

Supporting Information

Table S1 Specimen information, including spatial coordinates from UTM zone 50 J and haplotype number based on combined CO1 and 16 S mtDNA data. Samples from Windarling and Kooylanobbing (bold) are from specimens lodged at the Western Australian Museum, registration numbers shown (WAM_T).

Table S2 The three, best-fit models determined in IMa2 for tests of migration versus isolation between the paraphyletic populations Koolyanobbing and Helena Aurora. Results are based on mtDNA variation.

Acknowledgments

We thank M. Harvey, R. Teale, M. Rix, and K. Brennan for species information and access to WAM specimens. J. Rix and M. Beveridge for advice on DNA extraction and E. Ager for field work. Thanks to N. Harry for site access and logistical help.

Author Contributions

Conceived and designed the experiments: HN MB DC JDR. Performed the experiments: HN. Analyzed the data: HN. Contributed reagents/materials/analysis tools: HN MB DC JDR. Wrote the paper: HN.

References

1. Whittaker RJ (1998) Island biogeography: ecology, evolution and conservation. New York: Oxford University Press.
2. Raxworthy CJ, Forstner MRJ, Nussbaum RA (2002) Chameleon radiation by oceanic dispersal. Nature 415: 784–787.
3. Beheregaray LB, Gibbs JP, Havill N, Fritts TH, Powell JR, et al. (2004) Giant tortoises are not so slow: Rapid diversification and biogeographic consensus in the Galapagos. Proc Natl Acad Sci U S A 101: 6514–6519.
4. Jones AW, Kennedy RS (2008) Evolution in a tropical archipelago: comparative phylogeography of Philippine fauna and flora reveals complex patterns of colonization and diversification. Biol J Linn Soc 95: 620–639.
5. Wesener T, Raupach MJ, Sierwald P (2010) The origins of the giant pill-millipedes from Madagascar (Diplopoda: Sphaerotheriida: Arthrosphaeridae). Mol Phylogenet Evol 57: 1184–1193.
6. Robin V V, Sinha A, Ramakrishnan U (2010) Ancient geographical gaps and paleo-climate shape the phylogeography of an endemic bird in the sky islands of southern India. PLOS ONE 5: 1–13.
7. Masta SE (2000) Phylogeography of the jumping spider *Habronattus pugillis* (Araneae: Salticidae): Recent vicariance of sky island populations? Evolution 54: 1699–1711.
8. Smith C, Farrell B (2005) Phylogeography of the longhorn cactus beetle *Moneilema appressum* LeConte (Coleoptera: Cerambycidae): was the differentiation of the Madrean sky islands driven by Pleistocene climate changes? Mol Ecol 14: 3049–3065.
9. DeChaine EG, Martin AP (2005) Marked genetic divergence among sky island populations of *Sedum lanceolatum* (Crassulaceae) in the Rocky Mountains. Am J Bot 92: 477–486.
10. DeChaine EG, Martin AP (2004) Historic cycles of fragmentation and expansion in *Parnassius smintheus* (Papilionidae) inferred using mithochondrial DNA. Evolution 58: 113–127.
11. Salerno PE, Santiago RR, Senaris JC, Rojas-Runjaic FJM, Noonan BP, et al. (2012) Ancient Tepui summits harbor young rather than old lineages of endemic frogs. Evolution 66: 3000–3013.
12. Storfer A, Murphy MA, Evans JS, Goldberg CS, Robinson S, et al. (2007) Putting the "landscape" in landscape genetics. Heredity 98: 128–142.
13. Jacobi CM, Carmo FF, Vincent RC, Stehmann JR (2007) Plant communities on ironstone outcrops: a diverse and endangered Brazilian ecosystem. Biodivers Conserv 16: 2185–2200.
14. Gibson N, Yates C, Dillon R (2010) Plant communities of the ironstone ranges of south western Australia: hotspots for plant diversity and mineral deposits. Biodivers Conserv 19: 3951–3962.
15. Byrne M, Hopper SD (2008) Granite outcrops as ancient islands in old landscapes: evidence from the phylogeography and population genetics of *Eucalyptus caesia* (Myrtaceae) in Western Australia. Biol J Linn Soc 93: 177–188.
16. Baskin JM, Baskin CC (1988) Endemism in rock outcrop plant communities of unglaciated eastern United States: an evaluation of the roles of the edaphic, genetic and light factors. J Biogeogr 15: 829–840.
17. Burke A (2003) Inselbergs in a changing world-global trends. Divers Distrib 9: 375–383.
18. Dobzhansky TH (1970) Genetics and the evolutionary process. New York: Columbia University Press.
19. Avise JC, Shapira JF, Daniel SE, Aquador CF, Lansman RA (1983) Mitochondrial DNA differentiation during the speciation process in *Peromyscus*. Mol Biol Evol 1: 38–56.
20. Moritz C (1994) Defining "Evolutionary Significant Units" for conservation. Trends Ecol Evol 9: 373–375.
21. Floyd CH, Van Vuren DH, May B (2005) Marmots on Great Basin mountaintops: using genetics to test a biogeographic paradigm. Ecology 86: 2145–2153.

22. Moran GF, Hopper SD (1983) Genetic diversity and the insular population structure of the rare granite rock species, *Eucalyptus caesia* Benth. Aust J Bot 31: 161–172.
23. Gibson N, Coates DJ, Thiele KR (2007) Taxonomic research and the conservation status of flora in the Yilgarn Banded Iron Formation ranges. Nuytsia 17: 1–12.
24. Hocking RM, Langford RL, Thorne AM, Sanders AJ, Morris PA, et al. (2007) A classification system for regolith in Western Australia. West Aust Geol Surv 2007/8: 19.
25. Hopper SD (2000) Floristics of Australian granitoid inselberg vegetation. In: Porembeski S, Barthlott W, editors. Inselbergs. Berlin: Springer-Verlag. pp. 391–407.
26. McLoughlin S (2001) The breakup history of Gondwana and its impact on pre-Cenozoic floristic provincialism. Aust J Bot 49: 271–300.
27. Anand R, Paine M (2002) Regolith geology of the Yilgarn Craton, Western Australia: implications for exploration. Aust J Earth Sci 49: 3–162.
28. Martin HA (2006) Cenozoic climatic change and the development of the arid vegetation in Australia. J Arid Environ 66: 533–563.
29. Dodson JR, Macphail MK (2004) Palynological evidence for aridity events and vegetation change during the Middle Pliocene, a warm period in Southwestern Australia. Glob Planet Change 41: 285–307.
30. Hopper SD, Harvey MS, Chappill JA, Main AR, Main BY (1996) The Western Australian biota as Gondwanan heritage - a review. In: Hopper SD, Chappill JA, Harvey MS, George AS, editors. Gondwanan Heritage: Past, present and future of the Western Australian Biota. Chipping Norton, NSW: Surrey Beatty & Sons Pty limited. pp. 1–46.
31. Yates C, Gibson N, Pettit NE, Dillon R, Palmer R (2011) The ecological relationships and demography of restricted ironstone endemic plant species: implications for conservation. Aust J Bot 59: 692–700.
32. Harvey MS (2002) Short-range endemism in the Australian fauna: some examples from non-marine environments. Invertebr Syst 16: 555–570.
33. Biota (2009) Mt Jackson J1 Deposit - Summary of *Atelomastix* sampling, August 2009. Biota Environmental Sciences, Letter, 9 November 2009.
34. Bennelongia (2008) Troglofauna Survey of Mt Jackson Range, Western Australia - Report 2008/50. Bennelongia Environmental Consultants, December 2005.
35. Mesibov R (1994) Faunal breaks in Tasmania and their significance for invertbrate conservation. Mem Queensl Museum 36: 133–136.
36. Edward K, Harvey MS (2010) A review of the Australian millipede genus *Atelomastix* (Diplopoda: Spirostreptida: Iulomorphidae). Zootaxa 2371: 1–63.
37. Moir ML, Brennan KEC, Harvey MS (2009) Diversity, endemism and species turnover of millipedes within the south-western Australia global biodiversity hotspot. J Biogeogr 36: 1958–1971.
38. Kimura M, Weiss GH (1964) The stepping stone model of population structure and the decrease of genetic correlation with distance. Genetics 49: 561–576.
39. Avise JC (1989) Gene trees and organismal histories: A phylogenetic approach to population biology. Evolution 43: 1192–1208.
40. Beveridge M, Simmons LW (2004) Microsatellite loci for Dawson's burrowing bee (*Amegilla dawsoni*) and their cross-utility in other *Amegilla* species. Mol Ecol Notes 4: 379–381.
41. Folmer O, Black M, Hoeh W, Lutz R, Vrijenhoek R (1994) DNA primers for amplification of mitochondrial Cytochrome C Oxidase subunit I from diverse metazoan invertebrates. Mol Mar Biol Biotechnol 3: 294–299.
42. Simon C, Frati F, Beckenbach A, Crespi BJ, Liu H, et al. (1994) Evolution, weighting and phylogenetic utility of mitochondrial gene sequences and a compilation of conserved polymerase chain reaction primers. Ann Entomol Soc Am 87: 652–701.
43. Ji Y-J, Zhang D-X, He L-J (2003) Evolutionary conservation and versatility of a new set of primers for amplifying the ribosomal internal transcribed spacer regions in insects and other invertebrates. 3: 581–585.
44. Hillis DM, Dixon MT (1991) Ribosomal DNA: molecular evolution and phylogenetic inference. Q Rev Biol 66: 411–453.
45. Excoffier L, Lischer HEL (2010) Arlequin suite ver 3.5: A new series of programs to perform population genetics analyses under Linux and Windows. Mol Ecol Resour 10: 564–567.
46. Drummond AJ, Suchard MA, Xie D, Rambaut A (2012) Bayesian phylogenetics with BEAUti and the BEAST 1.7. Mol Biol Evol 29: 1969–1973.
47. Brower A (1994) Rapid morphological radiation and convergence among races of the butterfly *Heliconius erato* inferred from patterns of mitochondrial DNA evolution. Proc Natl Acad Sci USA 91: 6491–6495.
48. Hey J (2010) Isolation with migration models for more than two populations. Mol Biol Evol 27: 905–920.
49. Hey J, Nielsen R (2004) Multilocus methods for estimating population sizes, migration rates and divergence time, with applications to the divergence of *Drosophila psuedoobscura* and *D. persimilis*. Genetics 167: 747–760.
50. Hey J, Nielsen R (2007) Integration within the Felsenstein equation for improved Markov chain Monte Carlo methods in population genetics. Proc Natl Acad Sci 104: 2785–2790.
51. Akaike H (1974) A new look at the statistical model identification. IEEE Trans Automat Contr 19: 716–723.
52. Librado P, Rozas J (2009) DnaSP v5: A software for comprehensive analysis of DNA polymorphism data. Bioinformatics 25: 1451–1452.
53. R Core Team (2013) R: A language and environment for statistical computing. Available: http://www.r-project.org.
54. Nistelberger HM, Byrne M, Roberts JD, Coates DJ (2013) Isolation and characterisation of 11 microsatellite loci from the Western Australian Spirostreptid millipede, *Atelomastix bamfordi*. Conserv Genet Resour 5: 533–535.
55. Raymond M, Rousset F (1995) GENEPOP (version 1.2): population genetics software for exact tests and ecuminism. Heredity 86: 248–249.
56. Rice WR (1989) Analyzing tables of statistical tests. Evolution (N Y) 43: 223–225.
57. Kalinowski ST (2005) HP-RARE 1.0: a computer program for performing rarefaction on measures of allelic richness. Mol Ecol Notes 5: 187–189.
58. Chen C, Durand E, Forbes F, Francois O (2007) Bayesian clustering algorithms ascertaining spatial population structure: a new computer program and a comparison study. Mol Ecol Notes 7: 747–756.
59. Jakobsson M, Rosenberg NA (2007) CLUMPP: a cluster matching and permutation program for dealing with label switching and multimodality in analysis of population structure. Bioinformatics 23: 1801–1806.
60. Rosenberg NA (2004) Distruct: a program for the graphical display of population structure. Mol Ecol Notes 4: 137–138.
61. Cavalli-Sforza LL, Edwards AWF (1967) Phylogenetic analysis: models and estimation procedures. Am J Hum Genet 19: 233–257.
62. Langella O (2000) Populations 1.2.19: population genetic software individuals or populations distances, phylogenetic trees. Available: bioinformatics.org/~tryphon/populations/#ancre_bibliographie.
63. Goldstein DB, Pollock DD (1997) Launching microsatellites. J Hered 88: 3235–3242.
64. Page RDM (1996) TREEVIEW: an application to display phylogenetic trees on personal computers. Comput Appl Biosci 12: 357–358.
65. Jensen JL, Bohonak AJ, Kelley ST (2005) Isolation by distance, web service. BMC Genet 6. Available: http://ibdws.sdsu.edu/.
66. Cooper SJB, Harvey MS, Saint KM, Main BY (2011) Deep phylogeographic structuring of populations of the trapdoor spider *Moggridgea tingle* (Migidae) from southwestern Australia: evidence for long-term refugia within refugia. Mol Ecol 20: 3219–3236.
67. Driscoll DA (1998) Genetic structure of the frogs *Geocrinia lutea* and *Geocrinia rosea* reflects extreme population divergence and range changes, not dispersal barriers. Evolution 52: 1147–1157.
68. Bowler JM (1982) Aridity in the late Tertiary and Quaternary of Australia. In: Barker WR, Greenslade PJM, editors. Evolution of the flora and fauna of arid Australia. Frewville: Peacock Publications. pp. 35–45.
69. Clark PU, Archer D, Pollard D, Blum JD, Rial JA, et al. (2006) The middle Pleistocene transition: characteristics, mechanisms, and implications for long-term changes in atmospheric pCO2. Quat Sci Rev 25: 3150–3184.
70. Byrne M (2007) Phylogeography provides an evolutionary context for the conservation of a diverse and ancient flora. Aust J Bot 55: 316–325.
71. Buckley T, Simon C, Chambers G (2001) Exploring among-site rate variation models in maximum lielihood framework using empirical data. Effects of model assumptions on estimates of topology branch lengths and bootstrap support. Syst Biol 50: 67–86.
72. Pons J, Ribera I, Bertranpetit J, Balke M (2010) Nucleotide substitution rates for the full set of mitochondrial protein-coding genes in Coleoptera. Mol Phylogenet Evol 56: 796–807.
73. Wright S (1929) The evolution of dominance. Am Nat 63: 556–561.
74. Ryder OA (1986) Species conservation and systematics: the dilemma of subspecies. Trends Ecol Evol 1: 9–10.
75. Allendorf FW, Luikart G, Aitken SN (2013) Conservation and the Genetics of Populations. West Sussex, UK: Wiley-Blackwell.
76. Paetkau D (1999) Using genetics to identify intraspecific conservation units: a critique of current methods. Conserv Biol 13: 1507–1509.
77. Crandall KA, Bininda-Emonds ORP, Mace GM, Wayne RK (2000) Considering evolutionary processes in conservation biology. Trends Ecol Evol 15: 290–295.
78. Hughes L (2003) Climate change and Australia: Trends, projections and impacts. Austral Ecol 28: 423–443.

8

Phylogeography of *Pinus armandii* and Its Relatives: Heterogeneous Contributions of Geography and Climate Changes to the Genetic Differentiation and Diversification of Chinese White Pines

Liu Liu[1,2], Zhen-Zhen Hao[1,2], Yan-Yan Liu[1,2], Xiao-Xin Wei[1]*, Yu-Zhi Cun[3], Xiao-Quan Wang[1]

1 State Key Laboratory of Systematic and Evolutionary Botany, Institute of Botany, Chinese Academy of Sciences, Beijing, China, **2** Graduate University of the Chinese Academy of Sciences, Beijing, China, **3** Dali University, Dali, Yunnan, China

Abstract

Geographic barriers and Quaternary climate changes are two major forces driving the evolution, speciation, and genetic structuring of extant organisms. In this study, we used *Pinus armandii* and eleven other Asian white pines (subsection *Strobus*, subgenus *Pinus*) to explore the influences of geographic factors and Pleistocene climatic oscillations on species in South China, a region known to be centers of plant endemism and biodiversity hotspots. Range-wide patterns of genetic variation were investigated using chloroplast and mitochondrial DNA markers, with extensive sampling throughout the entire range of *P. armandii*. Both cpDNA and mtDNA revealed that *P. armandii* exhibits high levels of genetic diversity and significant population differentiation. Three geographically distinct subdivisions corresponding to the Qinling-Daba Mountains (QDM), Himalaya-Hengduan Mountains (HHM) and Yungui Plateau (YGP) were revealed in mainland China by cpDNA. Their break zone was located in the southeastern margin of the Qinghai-Tibetan Plateau (QTP). A series of massive mountains, induced by the QTP uplift, imposed significant geographic barriers to genetic exchange. The disjunct distribution patterns of ancestral haplotypes suggest that a large continuous population of the white pines may have existed from southwest to subtropical China. Repeated range shifts in response to the Pleistocene glaciations led to the isolation and diversification of the subtropical species. The two Taiwanese white pines share a common ancestor with the species in mainland China and obtain their chloroplasts via long-distance pollen dispersal from North Asian pines. Distinct genetic patterns were detected in populations from the Qinling-Daba Mountains, Yungui Plateau, Himalaya-Hengduan Mountains, and subtropical China, indicating significant contributions of geographic factors to the genetic differentiation in white pines. Our study depicts a clear picture of the evolutionary history of Chinese white pines and highlights the heterogeneous contributions of geography and Pleistocene climatic fluctuations to the extremely high plant species diversity and endemism in South China.

Editor: Axel Janke, BiK-F Biodiversity and Climate Research Center, Germany

Funding: This study was supported by the National Natural Science Foundation of China (Grant nos. 31270422, 30730010, 31160047). The funders had no role in study design, data collection and analysis, decision to publish, or preparation of the manuscript.

Competing Interests: The authors have declared that no competing interests exist.

* E-mail: weixx@ibcas.ac.cn

Introduction

Geographic barriers and climate change are two major forces driving the evolution, speciation and genetic structuring of extant organisms [1–2]. The Cenozoic collision between India and Asia created a highly complex montane system in southwest China, including the highest mountain range on Earth (the Himalayas) and the largest "roof of the world"-the Qinghai-Tibet Plateau (QTP). The uplift of the QTP reshaped the landscape features of central and eastern Asia and the global climate system [3–6]. In conjunction with the dramatic cooling of the global climate in the past three million years, geologic changes induced by the recent uplift of the QTP would be expected to have contributed to the development of the extremely high levels of species diversity in South China, a large region known to be centers of plant endemism and biodiversity hotspots. In this vast mountainous area, the Hengduan Mountain range, Central China, and the Nanling Mountains, which span the three-tier terrain of China from the high elevation west to the low elevation east, were recognized as plant diversity and endemism hotspots [7]. In particular, the Himalaya-Hengduan Mountains (HHM) region, which extends along the southern frontier to the southeastern margin of the QTP, encompasses 2 of the 34 biodiversity hotspots of the world, i.e., 'Mountains of Southwest China' and 'Himalaya' [8]. Although there is widespread agreement that dramatic geomorphology changes induced by the QTP uplift and/or Pleistocene glaciations profoundly affected the biota in this region, their relative influences on speciation and genetic structuring are largely unknown.

The last decade has witnessed an exponential growth in phylogeographic studies on the diverse taxa of many geographic regions in China (e.g., [9–12]). Unfortunately, most studies sampled only a single or a few closely related species, and concentrated extensively on the effects of Quaternary glaciations on genetic structure and the locations of glacial refugia. Only recently, have a few studies begun to disentangle the relative influences of the uplift of the QTP and the Pleistocene climatic fluctuations on speciation and genetic structuring in the QTP region and adjacent areas [13–16]. Moreover, there are few large-scale phylogeographic studies with sufficient coverage of distinct bioregions (for instance, warm central China, subtropical southern China, and mountainous western China) to detect the diversity of responses among closely related species to the geologic changes induced by the uplift of the QTP and Pleistocene glaciations.

The Chinese white pines, *Pinus armandii* and its relatives (*Pinus*, subgenus *Strobus*) provide an efficient system to explore these historical processes due to their extensive geographic distributions and relatively recent diversification (10~20 million years ago; Ma) [17]. Although there is no universal agreement on taxonomic relationships, most studies, including our own recent molecular phylogenies (unpublished), show a close relationship among the nine white pines distributed in the south part of mainland China and Taiwan, i.e., *P. armandii* and its variety *P. armandii* var. *mastersiana*, *P. bhutanica*, *P. dabeshanensis*, *P. fenzeliana*, *P. kwantungensis*, *P. morrisonicola*, *P. wallichiana*, and *P. wangii* [18–22]. All of these white pines are assigned to the subsection *Strobus* of Section *Quinquefoliae* [23]. Among them, *P. armandii* has the broadest geographic distribution. It is extensively and discontinuously distributed in central and southwest China, across a latitudinal range of 23°30′–36°31′N and a longitudinal range of 85°30′–113°E, with one variety, *P. armandii* var. *mastersiana*, extending to Taiwan [24]. *P. wallichiana* is a common conifer tree occurring throughout the temperate regions of the Himalayas, whereas another Himalayan pine, *P. bhutanica*, is only endemic to the eastern Himalayas. Three very closely related pines, *P. fenzeliana*, *P. kwantungensis*, and *P. wangii*, have localized and patchy distributions in the lower mountains and montane forests of southern China and possibly in adjacent northern Vietnam. *P. dabeshanensis* is a rare pine endemic to the Dabie Mountains, where it represents the easternmost distribution of soft pines in mainland China. In North China, three widespread north Asian white pines, *P. sibirica*, *P. pumila* and *P. koraiensis*, have a relatively limited distribution in the northwest and northeast (Fig. 1A).

In this study, we used a combination of phylogeography, interspecific phylogeny and molecular dating techniques with two paternally inherited chloroplast DNA fragments and three maternally inherited mitochondrial DNA markers to elucidate the effects of geography and Quaternary climatic oscillations on the genetic structure, speciation and evolution of the Chinese white pines. This study will shed further light on the mechanisms underlying plant speciation and the origin of the strikingly high plant diversity in southern China.

Materials and Methods

Ethics statement

No specific permits were required for the described locations in China because all researchers collecting the samples had introduction letters from the Institute of Botany, Chinese Academy of Sciences (IBCAS), Beijing. The samples from northern Pakistan were obtained from the international scientific research project of "Flora of Pan-Himalaya". The field studies involved collection of endangered species.

Population sampling

A total of 415 individuals from 30 populations of *P. armandii* and *P. armandii* var. *mastersiana*, representing the entire geographic range of this species, were included in our analysis. Each population had 13–23 individuals except for six from mainland China (DB, ZQ, LB, NS, SNJ, and HN) and two from Taiwan (DTS and YS), which were represented by <10 individuals. Sixty-five additional trees from the other 10 white pines distributed in China and adjacent areas (i.e., *P. wallichiana*, *P. kwangtungensis*, *P. fenzeliana*, *P. bhutanica*, *P. dabeshanensis*, *P. wangii*, *P. pumila*, *P. sibirica*, *P. morrisonicola*, and *P. koraiensis*) were also sampled for phylogeographic analysis. The detailed sampling information is shown in Table 1.

DNA extraction, PCR amplification, and sequencing

Total genomic DNA was extracted from silica gel-dried leaves using the modified CTAB method [25] or DNAsecure Plant Kit (Tiangen). We initially screened 22 primer pairs for polymorphism by using a discovery panel composed of 20 individuals that represent all major geographic subdivisions of *P. armandii*. Two chloroplast fragments, *atp*H-*atp*I and *trn*C-*trn*D, and three mitochondrial fragments, *cox*1, *nad*5 intron 1, and *nad7* intron1 were finally chosen for subsequent analysis in all sampled individuals. The five DNA segments were amplified with the primers used in previous studies [13,26–29]. The PCR mixture and amplification program followed the protocols of Wei et al. [30], except that an annealing temperature of 50–55°C was used and the elongation time was adjusted according to the different markers. The PCR products were purified with PEG8000. Both the *cox*1 and *nad*5 intron 1 strands were sequenced using the amplification primers in all samples. The *atp*H-*atp*I gene and *nad7* intron1 were sequenced by using a forward primer and *trn*C-*trn*D by using a reverse primer because all the informative polymorphisms could be covered by single primer sequencing. Sequencing was conducted with an ABI Prism 3730xl sequencer (PE Applied Biosystems). The sequences reported in this study were deposited in GenBank under accession numbers KF800948 - KF801058.

Gene genealogy

Nucleotide sequences were aligned using CLUSTAL X [31] and manually adjusted in BioEdit v7.05.3 [32]. Since organellar genomes in plants generally do not recombine and are uniparentally inherited, we treated the two cpDNA markers and the three mtDNA markers as a single locus each. A polymorphic site caused by polyG (9–12 bp) in *nad*5 was excluded due to some low quality reads. All phylogenetically informative gaps (indels) were coded as single mutation events for analyses. Unique haplotypes were determined using the program DnaSP v5 [33]. All the rare haplotypes were verified by re-extracting DNA, re-amplifying and re-sequencing both strands.

NETWORK 4.611 (http://www.fluxus-engineering.com) was used for generating haplotype networks by the median-joining network algorithm. *P. koraiensis* was used as the outgroup because of its relatively distant relationship with the southern white pines (unpublished). Substitutions and indels were considered to evolve with equal possibility.

Population genetic analyses

The spatial variance in haplotype distributions was analyzed using SAMOVA 1.0 (http://web.unife.it/progetti/genetica/Isabelle/samova.html) [34]. It implements a simulated annealing approach to define groups of populations (K) that are geographically homogenous and maximally differentiated from each other. The K values ranging from 2 to 9 were tested to search for the K that gave the highest F_{CT}.

Nuclotide diversity (π) and Watterson's parameter (θ_W) were estimated using DnaSP v5 [33]. The program PERMUTE [35] (http://www.pierroton.inra.fr/genetics/labo/software) was used

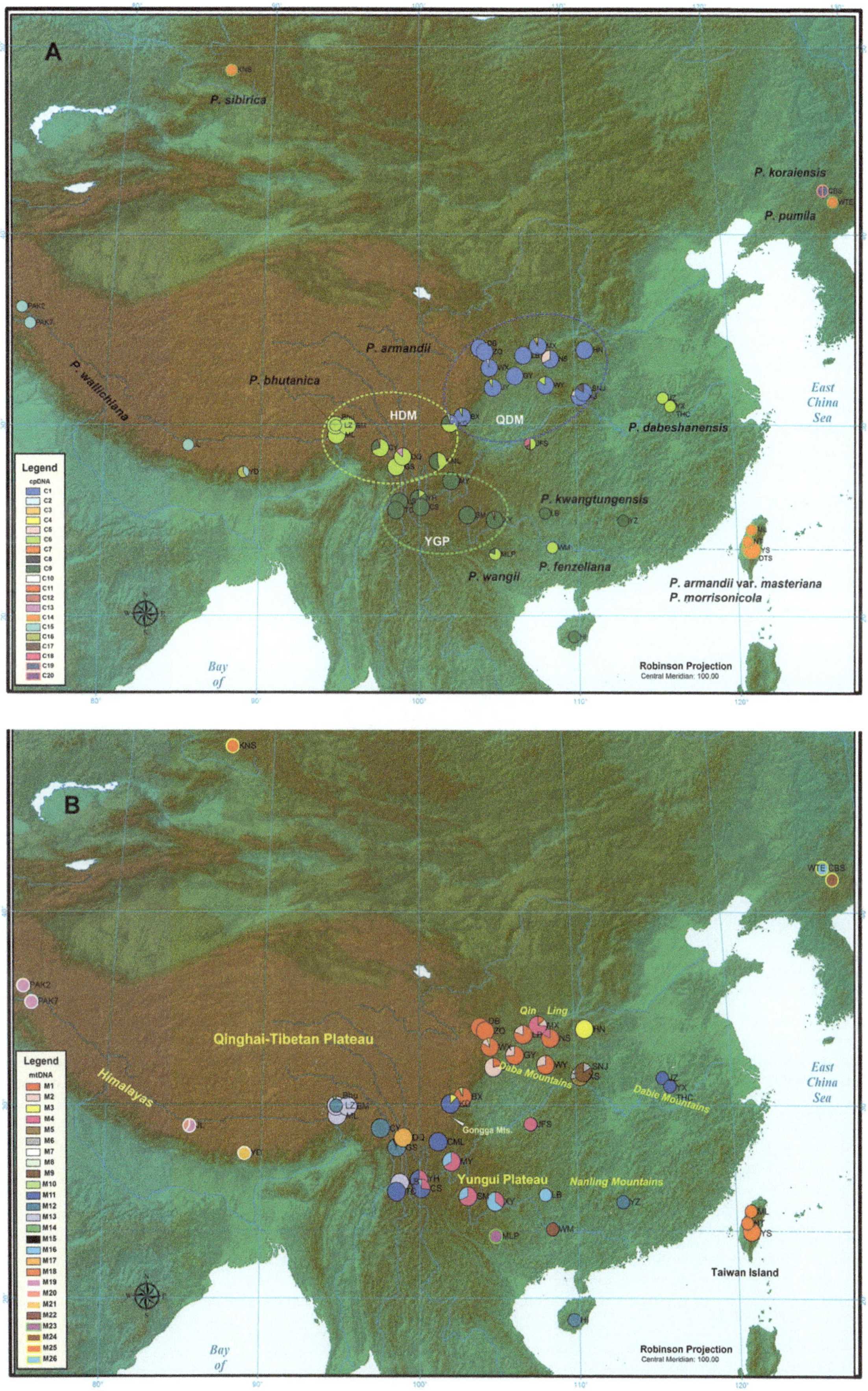
A
P. sibirica
P. koraiensis
P. pumila
P. armandii
P. bhutanica
P. wallichiana
HDM
QDM
YGP
P. dabeshanensis
East China Sea
P. kwangtungensis
P. wangii
P. fenzeliana
P. armandii var. mastersiana
P. morrisonicola
Legend
cpDNA
C1
C2
C3
C4
C5
C6
C7
C8
C9
C10
C11
C12
C13
C14
C15
C16
C17
C18
C19
C20
Bay of
Robinson Projection
Central Meridian: 100.00
B
Qinghai-Tibetan Plateau
Himalayas
Qin Ling
Daba Mountains
Dabie Mountains
Gongga Mts.
Yungui Plateau
Nanling Mountains
Taiwan Island
East China Sea
Legend
mtDNA
M1
M2
M3
M4
M5
M6
M7
M8
M9
M10
M11
M12
M13
M14
M15
M16
M17
M18
M19
M20
M21
M22
M23
M24
M25
M26
Bay of
Robinson Projection
Central Meridian: 100.00

Figure 1. Geographic distribution of chlorotypes (A) and mitotypes (B). Small circles in the pie charts indicate populations from the relatives of *Pinus armandii*. The three geographic subdivisions of *P. armandii* in mainland China are shown in A.

for calculating the average gene diversity within populations (H_S), total gene diversity (H_T), and two measures of population differentiation (G_{ST}, N_{ST}). The two estimates of population divergence were compared using a permutation test with 1000 permutations. Genetic differentiation among populations and groups (identified by SAMOVA) was estimated by the analysis of molecular variance (AMOVA) with significance tests based on 10, 000 permutations using Arlequin 3.11 [36]. Population expansion was examined using the mismatch distribution analysis (MDA). For each model, goodness-of-fit, based on the sum of squared deviations (SSD), was tested using parametric demographic expansion bootstrapping with 1000 replicates. In addition, we also calculated Tajima's D [37] and Fu's F_S [38] to infer potential expansions. All the demographic tests were performed using Arlequin 3.11. When sudden expansions were detected, the MDA-derived expansion parameter (τ) was converted to absolute expansion time (t, in number of generations) according to the equation $\tau = 2ut$ [39], where u is the mutation rate per generation for the entire analyzed sequence. We calculated u according to $u = \mu kg$, where μ is the substitution rate per nucleotide site per year, k is the average sequence length of the DNA region under study, and g is the generation time in years. We assumed that pines take 25–100 years to reach full seed production [40]. Substitution rate was obtained from the clock-calibrated BEAST (see the Results section).

Phylogenetic analyses and estimates of divergence time

The evolutionary relationships of the chlorotypes and mitotypes were reconstructed with Maximum parsimony (MP), Maximum likelihood (ML), and Bayesian inference (BI) using PAUP*4.0b10 [41], PHYML2.4.4 [42] and MrBayes 3.1 [43], respectively. All phylogenetically informative gaps (indels) were coded as single mutation events. The final alignment was 1373 bp in length. The evolutionary models of GTR+I and GTR+I+G for cpDNA and mtDNA phylogenetic analyses, respectively, were determined by the Akaike information criterion (AIC) implemented in MrModeltest v2.3 [44–45]. All characters were treated as unordered and equally weighted in the MP analysis. A heuristic search strategy with 1000 replicates of random addition of sequences, tree-bisection-reconnection (TBR) branch-swapping and MULTREES options was involved in the MP and ML analyses. To assess the relative support for monophyletic groups, bootstrap analysis was conducted with 100 replicates using the same heuristic search settings as descripted above. In the Bayesian analysis, one cold and three incrementally heated Markov chain Monte Carlo (MCMC) chains were run simultaneously for 1,000,000 generations. Trees were sampled every 100 generations with the first 2500 samples discarded as burn-in. To establish the divergence time of the Asian white pines, a subset of our initial cpDNA data and five additional GenBank sequences from *P. monophylla* (NC_011158.4) and *P. maximartinezii* (JN854184.1) in Section *Parrya*, *P. bungeana* (JN854223.1) and *P. gerardiana* (NC_011154.4) in Subsection *Gerardianae*, and a North American white pine, *P. lambertiana* (FJ899577.1), were combined and reanalyzed with the aim of obtaining a useful calibration point. Phylogenetic trees were reconstructed with MP, ML, and BI by using the methods mentioned above. In this phylogenetic analysis, the two species from Section *Parrya* were used as outgroups due to the sister relationship between section *Quinquefolia* and section *Parrya* [23].

A Bayesian approach was employed to estimate the divergence time of each major group in the chlorotype phylogeny with BEAST v1.6.1 [46], using a relaxed uncorrelated lognormal clock model and the Yule model of speciation. Due to the lack of credible fossils from *P. armandii* or closely related species, we employed the crown node age of the two monophyletic sections *Quinquefolia* and *Parrya* (subgen. *Strobus*) reported by Willyard et al. [17]. We used a normal distribution centered at 35 Ma with a standard deviation of 1.0 Ma. This gave a central 95% range of 33–37 Ma, which is within the ranges reported by the two studies [17,47] from independent fossil calibrations. The MCMC analysis was run for 30,000,000 generations. We sampled 1000 trees with the first 50% treated as burn-in. The output file was analyzed using TRACER 1.5 (http://beast.bio.ed.ac.uk/Tracer). The measures of effective sample sizes (ESS) were used to determine the Bayesian statistical significance of each parameter.

Results

Chlorotype distribution and genealogy

The lengths of the two cpDNA regions (*atp*H-*atp*I and *trn*C-*trn*D) in the white pines were found to be 674–686 bp and 893–954 bp, respectively. The combined cpDNA dataset is 1565 bp in length. Six indels and 13 substitutions were detected and resulted in 20 chlorotypes (C1–C20) among the trees surveyed. Of those, C1, C6 and C9 were the most common chlorotypes. *P. armandii* harbored 13 chlorotypes C1–C13. C14 is unique to the two Taiwanese white pines, *P. armandii* var. *masteriana* and *P. morrisonicola*, as well as the two north Asian species *P. pumila* and *P. sibirica*. C15 and C16 were fixed in *P. wallichiana*. C17 and C18 are rare chlorotypes, unique to JFS of *P. fenzeliana*. *P. koraiensis* harbored C19 and C20.

A strong geographic pattern of chlorotype variation was observed in *P. armandii*. SAMOVA showed that the value of F_{CT} reached a plateau at 0.87 for four groups (K = 4). The two populations (DTS and YS) of *P. armandii* var. *masteriana* from Taiwan were separated as a distinct group. The other three groups from mainland China also had a strong geographic basis, generally corresponding to the Qinling-Daba Mountains (DB, ZQ, WX, MX, LB, NS, XS, SNJ, HN, WY, GY, JY, and BX), Himalaya-Hengduan Mountains (KD, CML, DQ, CY, ML, LZ, BM, and GS) and Yungui Plateau (YH, LS, TC, CS, MY, SM, and XY). Hereafter, we designate these three groups as QDM, HHM, and YGP, respectively. Each group harbored a dominant chlorotype (Fig. 1A). C1 was unique to QDM. C6 and C9 dominated in HHM and YGP, respectively. In addition, C6 occurred at a low frequency in several QDM populations (BX, JY, WY, and XS) and *P. bhutanica, P. dabeshanensis, P. fenzeliana, and P. wangii*. C9 only fixed in southern populations, including the three subtropical white pines, *P. kwangtungensis*, *P. wangii*, and *P. fenzeliana*. It is notable that all the three common chlorotypes were present in the KD population with high frequency (Fig. 1A). The chlorotype network delineated three major clades of *P. armandii* populations in mainland China (Fig. 2A), which broadly correspond to the three geographic subdivisions (QDM, HHM, and YGP).

Mitotype distribution and genealogy

The sizes of the three mtDNA markers, *cox*1, *nad*5 intron 1, and *nad*7 intron 1, are 1225 bp, 1035–1042 bp and 809–1244 bp, respectively. Five individuals from populations WX, WY, and BX of *P. armandii* were excluded due to a polymorphic nucleotide site

Table 1. Geographic locations and sample sizes of the Asian white pines sampled in this study.

Species	Subdivisions[1]	Population name/location	Sample size	Latitude (°N)	Longitude (°E)
Pinus armandii	QinLing-Daba Mts.(QDM)	DB/Diebu, Gansu	3	34.01	103.92
		ZQ/Zhouqu, Gansu	4	33.82	104.27
		WX/Wenxian, Gansu	19	32.98	104.56
		MX/Meixian, Shanxi	23	34.12	107.7
		LB/Liuba, Shanxi	5	33.63	106.73
		NS/Ningshan, Shanxi	6	33.45	108.47
		XS/Xingshan, Hubei	17	31.47	110.3
		SNJ/Shennongjia, Hubei	6	31.69	110.55
		HN/Lushi, Henan	5	33.92	110.7
		WY/Wanyuan, Sichuan	13	32.07	108.12
		GY/Guangyuan, Sichuan	15	32.56	106.16
		JY/Jiangyou, Sichuan	14	31.94	104.76
		BX/Baoxing, Sichuan	15	30.41	102.76
	Himalaya-HengDuan Mts.(HHM)	KD/Kangding, Sichuan	23	30.05	101.96
		CML/Muli, Sichuan	15	28.8	101.18
		CY/Chayu, Xizang(Tibet)	23	28.78	97.53
		ML/Milin, Xizang(Tibet)	15	29.44	94.81
		LZ/Linzhi, Xizang(Tibet)	17	29.94	94.75
		BM/Bomi, Xizang(Tibet)	14	29.92	95.46
		GS/Gongshan, Yunnan	15	27.78	98.55
		DQ/Deqin,Yunnan	15	28.31	98.97
	Yungui Plateau (YGP)	YH/Yuhu, Yunnan	21	26.12	99.97
		LS/Lushui, Yunnan	15	25.96	98.75
		TC/Tengchong, Yunnan	15	25.5	98.55
		CS/Cangshan, Yunnan	15	25.66	100.12
		MY/Miyi, Sichuan	15	27.05	102.01
		SM/Songming, Yunnan	16	25.26	103.04
		XY/Xingyi, Guizhou	20	25	104.79
P. armandii var. *masteriana*	Taiwan Island	DTS/Datashan, Taiwan	9	23.53	120.82
		YS/Yushan, Taiwan	7	23.47	120.94
P. wallichiana		JL/Jilong, Xizang(Tibet)	5	28.92	85.38
		YD/Yadong,Xizang(Tibet)	5	27.5	88.97
		PAK2/Naltar Bala, Pakistan	5	36.17	74.18
		PAK7/Rama Lake, Pakistan	5	35.33	74.79
P. kwangtungensis		LB/Libo, Guizhou	3	25.35	107.95
		YZ/Yizhang, Hunan	2	24.98	112.85
		WM/Wuming Mts, Guangxi	2	23.57	108.38
P. fenzeliana		HI/Hainan Island	2	18.88	109.66
		JFS/Nanchuan, Chongqing	6	29	107.09
P. bhutanica		LZ/Linzhi, Xizang(Tibet)	5	29.94	94.75
P. dabeshanensis		YX/Yuexi, Anhui	2	31.01	116.07
		JZ/Jinzhai, Anhui	1	31.4	115.6
		THC/Taohuachong, Hubei	1	30.97	116.06
P. wangii		MLP/Molipo, Yunnan	5	23.22	104.81
P. pumila		WTE/Changbai, Jilin	5	41.73	127.86
P. sibirica		KNS/Kanasi, Xinjiang	5	48.71	87.07
P. morrisonicola		NT/Nantou, Taiwan	1	23.91	120.68
		ML/Miaoli, Taiwan	1	24.53	120.94

Table 1. Cont.

Species	Subdivisions[1]	Population name/location	Sample size	Latitude (°N)	Longitude (°E)
P. koraiensis		CBS/Changbai Mts., Jilin	4	42.33	127.27

in the *nad*5 intron 1 and 15 individuals from *P. armandii* var. *masteriana* because of failed amplification of the *nad7* intron 1. A total of 24 indels and 28 substitutions were identified for the concatenated mitochondrial DNA sequence which yielded 26 mitotypes (M1–M26). Of those, *P. armandii* harbored 17 (M1–M17), and four of them (M4, M11, M12, and M16) are shared across species. M18 was fixed in the two Taiwanese white pines (*P. armandii* var. *masteriana* and *P. morrisonicola*). M19–M26 were restricted to the remaining white pines. No more than three mitotypes were found in a single population, with the exception of XS, which harbored five (M4–M8).

QDM possessed ten mitotypes (M1–M10), and eight (M1–M2, M5–M10) were unique. Of these, M1 occurred in all populations except three in the east. M3 was only found in nine individuals from three populations (WX, HN, and KD) across a large geographic range. M4 dominated in eastern YGP, but it was also present in eastern QDM (MX, XS, NS) and in the JFS population of *P. fenzeliana*. M12 and M13 predominated in HHM. M16 occurred at high frequency in eastern YGP and shared with its adjacent population (LB) from *P. kwangtungensis*. M11 had a disjunct distribution between the southeastern margin of the QTP (*P. armandii*) and Dabie Mountains in eastern China (*P. dabeshanensis*).

Unlike the chlorotypes, the mitotype network did not show clear geographic relationships among QDM, HHM and YGP (Fig. 2B). It is compatible with the SAMOVA of mtDNA, in which no plateau value of F_{CT} was obtained. Nevertheless, two clusters linked by M3 were broadly delineated in *P. armandii*, one comprised of all but one (M10) mitotype from the QDM lineage and the other included all the mitotypes from the HHM and YGP lineages except M4, which nested in the QDM cluster. In the latter cluster, M11 (KD, CML, YH, CS, and TC) was placed in the interior position, directly linked by M3 (KD and HN), M16 (CS, MY, SM, XY, and LB), M17 (DQ), and M18 (Taiwan) (Fig. 2B). In theory, haplotypes with an interior position in a network are often older than haplotypes on the tips [48]. In addition, the close relationship of M11 with *P. koraiensis* (M26), which was used as an outgroup in the network analysis, also suggests that it is old.

Genetic diversity and population differentiation of P. armandii

Both cpDNA and mtDNA showed high levels of total genetic diversity (H_T) over populations of *P. armandii* from mainland China (0.692 for cpDNA, 0.886 for mtDNA), but the levels of average within-population diversity (H_S) were much lower (0.157 and 0.239 for cpDNA and mtDNA, respectively; Table 2). Population differentiation was remarkably high, with G_{ST} values being 0.773 and 0.730 for cpDNA and mtDNA, respectively (Table 2). The permutation tests of cpDNA showed that N_{ST} was significantly higher than G_{ST} ($P<0.01$), suggesting a strong phylogeographic structure in *P. armandii*. However, N_{ST} was not significantly greater than G_{ST} at mtDNA loci. AMOVA showed that 83.14% and 82.31% of the total cpDNA and mtDNA variation respectively resided among populations (Table 3).

At the regional level, contrasting patterns of genetic diversity and population differentiation were observed at cpDNA and mtDNA loci (Table 2, 3). For cpDNA, the level of total genetic diversity ($H_T = 0.055$–0.334) and of population differentiation ($G_{ST} = 0.048$–0.278) were relatively low in each of the three subdivisions (QDM, HHM, and YGP). Accordingly, the hierarchical analysis of molecular variance (AMOVA) showed a high level of variation among regions (85.73%) and extremely low level of variation (1.86%) among populations within regions. Taiwan populations were monomorphic. In contrast, the total genetic diversity ($H_T = 0.689$–0.833) and population differentiation ($G_{ST} = 0.557$–0.927) were very high in mtDNA. The partitioning of the mtDNA variation among regions and among populations within regions were 44.45% and 40.27%, respectively. Among the three geographic subdivisions, HHM had the highest level of genetic diversity and population differentiation at both cpDNA and mtDNA (Table 2).

Phylogenetic analyses and divergence time estimate

Both cpDNA and mtDNA phylogenies (Fig. 3A, B) indicated that the Himalayan pine, *P. wallichiana*, was distantly related to the southern pines in mainland China, i.e., *P. armandii*, *P. bhutanica*, *P. dabeshanensis*, *P. fenzeliana*, *P. kwangtungensis*, and *P. wangii*. In the cpDNA tree, three lineages corresponding to QDM, HHM, and YGP, were resolved within the clade including *P. armandii*. The two Taiwanese white pines clustered together with two North Asian pines, *P. sibirica* and *P. pumila*, and showed a sister relationship to *P. armandii*. In the mtDNA tree, however, the two white pines in Taiwan were included in the *P. armandii* clade.

Molecular dating showed that the age of the split between the Asian and North American white pines was about 20.04 Ma (95% highest posterior density (HPD) interval, 28.46 Ma – 11.09 Ma). The Asian white pines were divided into two lineages around 7.41 Ma (95% HPD interval, 12.44 Ma – 3.57 Ma). One lineage was composed of the widespread Himalayan pine *P. wallichiana*, three northern pines, *P. koraiensis*, *P. sibirica*, *P. pumila*, and two from Taiwan, *P. morrisonicola* and *P. armandii* var. *masteriana*, while the other was comprised of *P. armandii* and the other white pines in southern China. Within the *P. armandii* clade, YGP diverged at about 5.2 Ma (95% HPD interval, 9.09 Ma – 2.58 Ma). QDM and HHM were separated 4.03 Ma (95% HPD interval, 7.01 Ma – 1.84 Ma). All these estimates are shown in Fig. 4. The mean nucleotide substitution rate for the combined cpDNA region was estimated to be 0.47×10^{-9} per site per year.

Demographic analyses

Under a model of sudden population expansion ($P>0.05$), for cpDNA, each of the three regions, QDM, HHM and YGP, showed a strongly unimodal mismatch distribution and nonsignificant SSD index values (Fig. S1), indicating a historical population expansion. However, Fu's *F*s and Tajima *D* tests suggest only QDM had significantly negative Fu's F_S and Tajima D values ($F_S = -8.23581$, $P<0.001$; $D = -1.88930$, $P<0.05$) (Table S1). Therefore, we can say with confidence that QDM experienced

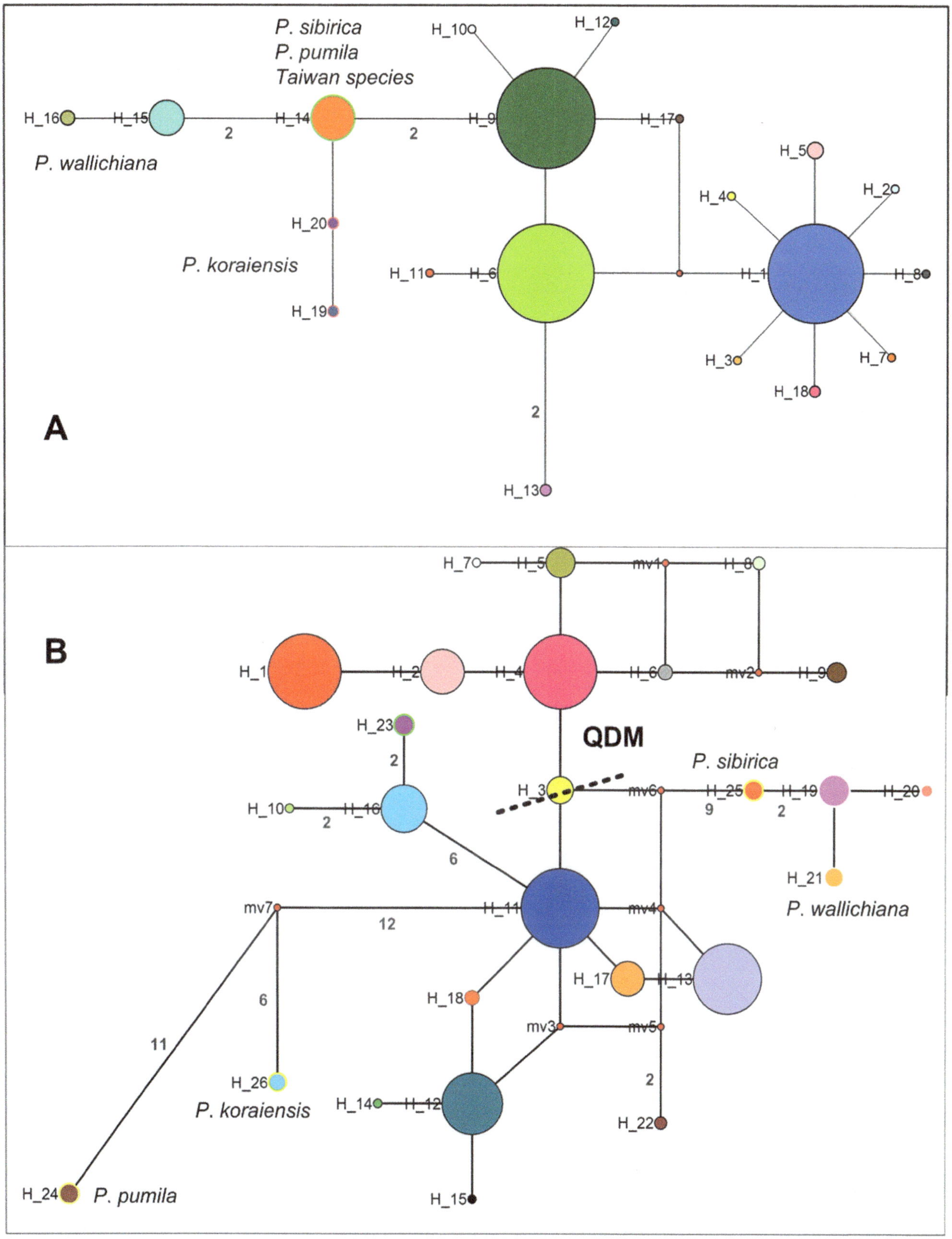

Figure 2. NETWORK-derived genealogical relationship of chlorotypes (A) and mitotypes (B). The sizes of the circles in the network are proportional to the observed frequencies of the haplotypes. The number of mutational steps larger than one is shown next to the solid line between haplotypes.

Table 2. Total genetic diversity, average genetic diversity within populations and population differentiation estimates in *Pinus armandii*.

Region	cpDNA				mtDNA			
	H_T	H_S	N_{ST}	G_{ST}	H_T	H_S	N_{ST}	G_{ST}
QDM	0.171	0.163	0.053	0.048	0.689	0.290	0.737*	0.579
HHM	0.334	0.241	0.231	0.278	0.833	0.061	0.955	0.927
YGP	0.055	0.051	0.074	0.068	0.788	0.349	0.430	0.557
Mainland China	0.692	0.157	0.860**	0.773	0.886	0.239	0.790	0.730
Entire range	0.730	0.147	0.889**	0.799	nc	nc	nc	nc

**extremely significant (Nst>Gst, P<0.01);
*significant (Nst>Gst, P<0.05); nc, not be calculated due to single individual in YS population.

expansion. The expansion of QDM was estimated at 93 ka (a generation time of 25 y applied) and 23.2 ka (a generation time of 100 y applied). For mtDNA, none of them showed strongly unimodal mismatch distribution and significantly negative values of Tajima D or Fu's F_S.

Discussion

Strong genetic differentiation of *P. armandii* and uplift of the QTP

Compared with the average genetic differentiation of conifers (cpDNA = 0.165; mtDNA = 0.764) [49] and those reported recently for other Pinaceae species (e.g., [50–55]), *P. armandii* exhibits high levels of total genetic diversity and significant population differentiation at both cpDNA ($H_T = 0.692$, $F_{ST} = 0.83$, $G_{ST} = 0.773$) and mtDNA loci ($H_T = 0.886$, $F_{ST} = 0.82$, $G_{ST} = 0.730$) (Table 2, 3). As a conifer with the capacity for long-distance pollen dispersal, the high level of the total cpDNA variation among regions (85.73%), and extremely low level among populations within regions (1.86%; Table 3), suggested that geography might play an important role in driving the genetic structure of *P. armandii* population.

The chlorotypes revealed three lineages of *P. armandii* in mainland China that geographically corresponding to the Qinling-Daba Mountains (QDM), Himalaya-Hengduan Mountains (HHM), and Yungui Plateau (YGP). Their break zone was observed along the southeastern margin of the QTP. The QDM lineage was separated from the two south lineages (HHM and YGP) at Kangding (KD) in western Sichuan, which is consistent with a previous provenance study [56]. Numerous high mountains

Table 3. Analysis of molecular variance (AMOVA) of *Pinus armandii* based on cpDNA and mtDNA haplotypes.

Source of variation	d.f.		Sum of squares		Variance components		Percentage of variation		Fixation indices	
	cpDNA	mtDNA	cpDNA	mtDNA	cpDNA	mtDNA	cpDNA	mtDNA	cpDNA	mtDNA
Mainland China										
Among populations	27	27	242.95	410.83	0.63	1.07	83.14	82.31		
Within populations	371	366	47.10	84.30	0.13	0.23	16.86	17.69	$F_{ST} = 0.83$	$F_{ST} = 0.82$
Total	398	393	290.04	495.13	0.75	1.30				
Divided into three regions (QDM, HHM, and YGP)										
Among regions	2	2	233.13	195.78	0.88	0.67	85.73	44.45	$F_{SC} = 0.13$	$F_{SC} = 0.72$
Among populations within regions	25	25	9.81	215.05	0.02	0.61	1.86	40.27	$F_{ST} = 0.88$ $F_{CT} = 0.86$	$F_{ST} = 0.85$ $F_{CT} = 0.44$
Within populations	371	366	47.10	84.30	0.13	0.23	12.42	15.28		
Total	398	393	290.04	495.13	1.02	1.51				
Divided into two regions (North and South)										
Among regions	1	1	189.55	159.46	1.01	0.82	79.38	47.60	$F_{SC} = 0.68$	$F_{SC} = 0.75$
Among populations within regions	26	26	53.39	251.37	0.14	0.68	10.69	39.10	$F_{ST} = 0.82$ $F_{CT} = 0.44$	$F_{ST} = 0.87$ $F_{CT} = 0.48$
Within populations	371	366	47.10	84.30	0.13	0.23	9.93	13.30		
Total	398	393	290.04	495.13	1.28	1.73				

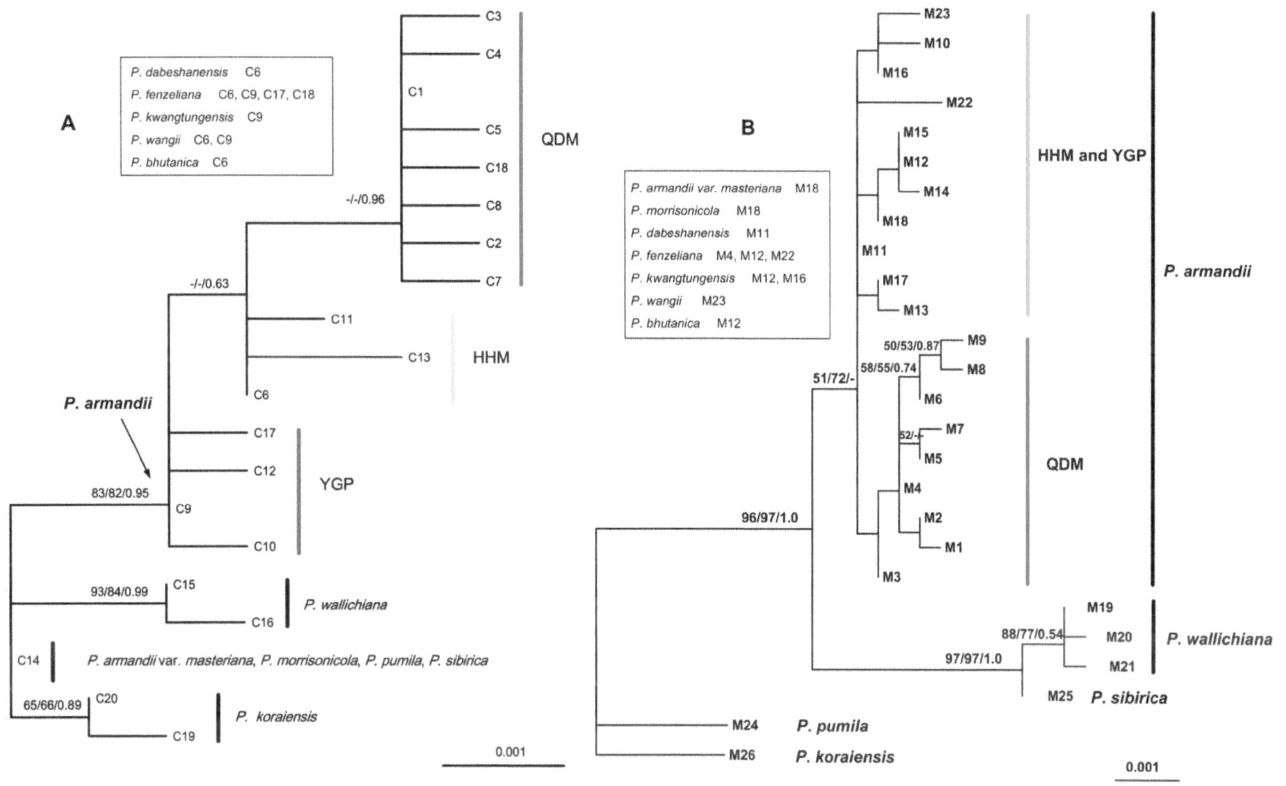

Figure 3. Phylogenetic topography obtained by analysis of chlorotypes (A) and mitotypes (B). Numbers above the branches are bootstrap values greater than 50% for ML (left) and MP (middle) analyses, and Bayesian posterior probabilities (right). A dash indicates a bootstrap value lower than 50%. The other species included in the *Pinus armandii* clade are shown in the box.

and plateaus over 6,000 m above the sea level (a.s.l.), including the highest Gongga Mountain (7,556 m a.s.l.) (Fig. 1B), are located nearby and divide the Hengduan Mountains into northern and southern floristic sub-regions [57]. The genetic isolation of the HHM and YGP populations was consistent with the geographic transition of the Hengduan Mountains and the Yungui Plateau, two large natural geographic regions characterized by a series of parallel massive mountains and deeply carved valleys running roughly north to south and mid-elevation mountains oriented from west to east, respectively. It is likely that these massive mountains at the southeastern margin of the QTP imposed significant geographic barriers to the genetic exchange among different lineages of *P. armandii*. Active mountain building, induced by the recent uplift of the southeastern margin of the QTP, might strongly affect the genetic structure of *P. armandii*.

A close link was detected between the recent rapid uplift of the QTP and the differentiation of *P. armandii*. Based on our analysis of molecular dating, *Pinus armandii* split from the Himalayan pine *P. wallichiana* around 7.41 Ma (95% HPD interval, 12.44 Ma – 3.57 Ma). The subsequent divergence of *P. armandii* was dated to early Pliocene (5.2–4.03 Ma; Fig. 4). Although we need to be cautious with these estimates due to the limited data, they are concordant with a Miocene origin (10–20 Ma) of the two subsections *Strobus* and *Cembroides* that harbor the greatest number of soft pines [58]. Several lines of evidence suggest that the rapid uplift of the eastern part of the QTP took place around 13 to 5 Ma [59–60]. Dramatic geological changes induced by the QTP uplift contributed to the extremely complicated landscape in the eastern fringe, which likely acted as a major driving force for the separation of the three lineages of *P. armandii*. Divergent selection between populations in contrasting environments would accelerate their differentiation, and eventually promote allopatric speciation.

Recent diversification of the subtropical white pines and Taiwan species

In wind-pollinated conifers with contrasting modes of inheritance, weaker population differentiation is often observed in pollen-mediated cpDNA than seed-mediated mtDNA (e.g., [30,49–50,61]). In *P. armandii*, such asymmetric differentiation between cpDNA and mtDNA was indeed observed at the regional scale. However, across its entire range, the cpDNA differentiation was surprisingly much more highly structured geographically than the mtDNA. The hierarchical AMOVA also showed that 85.73% and 44.45% of the cpDNA and mtDNA variation respectively resided among regions (Table 3), indicating more restricted pollen-mediated gene flow than seed-mediated gene flow among the three lineages. This unique pattern detected in *P. armandii* can likely be attributed to its evolutionary history and considerable gene flow via seed migration in the past.

Several lines of evidence, including the chlorotype composition of the KD population, a large number of common mitotypes and the dominant presence of the putative ancestral mitotype (M11), indicate that *P. armandii* originated in southwest China and probably in the eastern part of the QTP (Fig. 1). It is surprising that M11 and C6, which dominated in the southeastern margin of the QTP, were also found in eastern China (*P. dabeshanensis*). A similar disjunct distribution pattern was also observed in M12 and C6, between the HHM lineage of *P. armandii* and some populations of *P. fenzeliana* (HI), *P. kwantungensis* (YZ), and *P. wangii* (MLP) in southern China. It is likely that a large continuous population of

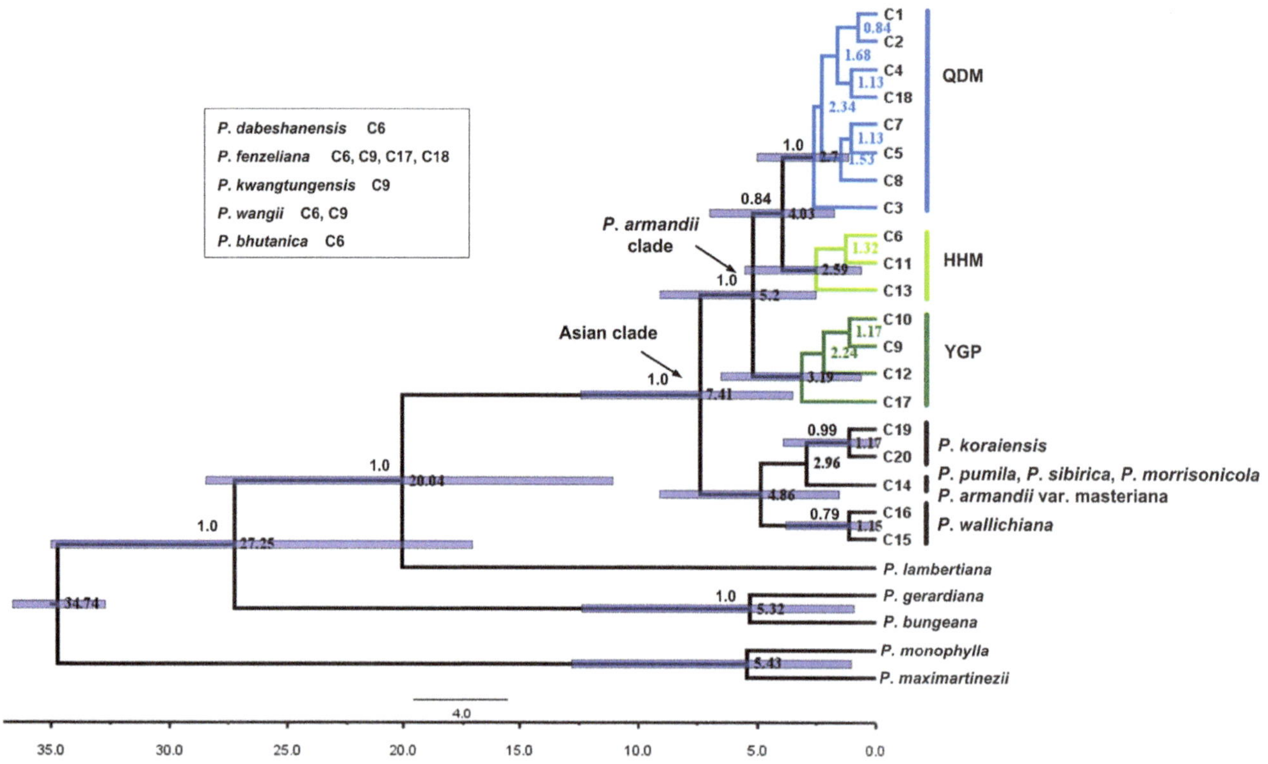

Figure 4. Phylogenetic chronogram of the chlorotypes generated from BEAST. Numbers above the branches indicate the Bayesian posterior probabilities greater than 0.5. Median ages of nodes are shown, with horizontal bars indicating the 95% highest posterior density intervals. The other species included in the *Pinus armandii* clade are shown in the box.

white pines once existed from southwest to subtropical China, but now only persists in isolated geographic aggregates. Range disjunctions for plant species are often a consequence of Quaternary climate changes [2]. A scenario could be inferred that in tracking repeated Pleistocene glaciations, these montane white pines may have migrated up and down elevation gradients. During these necessary movements into glacial refugia, they experienced severe fragmentation and isolation. In addition, population bottlenecks might have occurred in *P. dabeshanensis* and led to its extremely small population size and low genetic diversity. This scenario was supported by the phylogeographic studies of *P. kwantungensis* [54–55], as well as their current distributions and morphological similarities. These white pines are scattered in the broadleaved evergreen forest region in subtropical China, a vast region where well-known centers of plant diversity and endemism exist [7], and can also often be found on isolated mountaintops, cliffs, or slopes. Some areas, such as the Dabie Mountains, Jinfo Mountains, Hainan Island, Nanling Mountains, and eastern Yungui Plateau are often postulated as glacial refugia for warm-temperate evergreen forest species [54–55,62–64]. Long-term persistence in isolated refugia would promote their divergence and eventual speciation.

The origin of the florist affinity between Taiwan and southwestern China/Himalayas is a fascinating question regarding the biogeography of eastern Asia [30,65]. *P. armandii* var. *masteriana* and *P. morrisonicola* are endemic to Taiwan. Their primary difference is that *P. armandii* var. *masteriana* (like *P. armandii* in mainland China) has wingless seeds whereas *P. morrisonicola* has winged seeds. In addition, *P. armandii* var. *masteriana* generally occurs at higher altitudes (1800–3300 m) than *P. morrisonicola* (300–2300 m) [18]. It is interesting that the two species share a single mitotype (M18) which is closely related to M11, suggesting they evolved from a common ancestor in mainland China. The ancestor was probably in eastern China due to the presence of M11 in the Dabie Mountains and the short distance from Taiwan. The Taiwanese populations were genetically less diverse in both cpDNA and mtDNA markers, which might correspond to a rather recent colonization. Unexpectedly, the two Taiwanese white pines shared the single chlorotype C14 with two North Asian white pines, *P. pumila* and *P. sibirica*, indicating a North Asian origin of their chloroplasts. The inconsistency between the cpDNA and mtDNA markers was observed in the cytoplasmic phylogenies (Fig. 3). Introgression via long-distance pollen dispersal may be the most probable cause of the incongruence. Furthermore, a single chloroplast introgression event might occur in the two white pines followed by a loss of genetic diversity through drift effects in an island population. One may argue that the long-distance dispersal is inconsistent with the idea of limited pollen dispersal via geographic barriers among the three lineages of *P. armandii* in southwest China. However, the complicated landscapes in southwest China, particularly numerous massive mountains, likely imposed more significant barriers to pollen dispersal than the Pacific islands for wind-pollinated trees. The current distribution of *P. pumila* in Japan provides further support for this supposition. Frequent introgression and hybridization is well documented in pine species [66]. Our results suggested that interspecific introgression might also occur between the eastern YGP populations of *P. armandii* and LB population of *P. kwantungensis*, as well as between the HHM populations of *P. armandii* and *P. bhutanica*. Our study demonstrated the importance of including more closely related species in phylogeographic research and the value of a clear evolutionary background for resolving the taxonomic uncertainties in the problematic white pines.

Heterogeneous contributions of geography and climate changes to the genetic differentiation of Chinese white pines

Quaternary climate changes produced great changes in species distribution and population structure [2]. Nevertheless, the diversification patterns may be spatially varied. In the case of Chinese white pines, different genetic patterns were observed among populations from the Qinling-Daba Mountains, Himalaya-Hengduan Mountains, Yungui Plateau and southern China, which are diverse in geographic, climatologic and topographic features. The north facing Qinling Mountains oriented from east to west in central China form a natural dividing line between Northern and subtropical (Central/East/South) China and serve as a major geographic barrier to the cold, dry air coming in from the north. Thus, this region might be subjected to relatively stronger influences of Pleistocene glaciations. As expected, mismatch analysis and the indices of Fu's F^* and Tajima D suggested that only the QDM lineage of *P. armandii* has undergone a sudden population expansion which was estimated to have occurred around 23–93 ka, close to the last glacial maximum (18–24 ka BP). The large number of private haplotypes and high levels of nucleotide diversity of the SNJ and XS populations (Table S1, S2, S3) suggest that the eastern Daba Mountains likely served as a main refugium for the QDM lineage. Among the three lineages of *P. armandii*, the HHM lineage exhibits the highest level of genetic diversity and population differentiation (Table 2). This could be attributed to the extremely dissected topography of the HHM, which acted as significant barriers to gene flow.

It is noteworthy that disjunct mitotypes (e.g., M11, M12) found in the eastern margin of the QTP and the subtropical China are missing in the eastern Yungui Plateau. Instead, the eastern YGP populations (MY, SM, and XY) of *P. armandii* are dominated by M4 and M16. It is likely that the relatively less complex geomorphological configuration of the YGP not only contributed to the low genetic diversity and population differentiation in that region (Table 2), but also promoted the expansion of some putative adaptive alleles. In contrast, subtropical China is characterized by extremely rugged low mountains and relatively stable environments, which may have favored the preservation of relict lineages. Phylogeographic studies of *P. kwantungensis* have revealed many glacial refugia in this region and high levels of population differentiation at both cpDNA ($G_{st} = 0.63$ vs. 0.068 in YGP) and mtDNA loci ($G_{st} = 0.75$ vs. 0.557 in YGP) [54–55]. These findings highlighted the significance of both Quaternary climate changes and geographic factors in shaping the current genetic structures of Chinese white pines.

Supporting Information

Figure S1 Mismatch distributions for the different subdivisions of *Pinus armandii*. Capital and lower case letters indicated the distributions of chlorotypes and mitotypes, and the black and gray lines represent the expected and observed mismatch distributions, repectively. Aa: mainland China; Bb: QDM; Cc: YGP; Dd: HDM.

Table S1 Tajima's D and Fu's F_S neutrality tests and nucleotide diversity of each population and geographic subdivision of *Pinus armandii* at cpDNA and mt DNA loci.

Table S2 Chlorotype distribution in each population of *Pinus armandii* and the other white pines in this study.

Table S3 Mitotype distribution in each population of *Pinus armandii* and the other white pines in this study.

Acknowledgments

We thank two anonymous reviewers for their insightful and valuable comments and suggestions on the manuscript, and Dr. Anna W. Schoettle (USDA Forest Service, Rocky Mountain Research Station) for reading over the manuscript. We are grateful to Professor Hong Deyuan, Chen Zhiduan and Li Liangqian, Drs. Ma Xintang, Yang Fusheng, Wang Mingming, Qin Aili, Wang Qiang and Wang Hongjuan (Institute of Botany, CAS), Liu Zhanlin and Wang Xinsha (Northwest University), Kan Xianzhao, Zhang Zhongxin and Xiang Xiaoyan (Anhui Normal University) for their help in the sample collections; Miss Jin Wanqing for her assistance in the DNA sequencing.

Author Contributions

Conceived and designed the experiments: XXW. Performed the experiments: LL ZZH YYL. Analyzed the data: XXW LL ZZH YYL. Contributed reagents/materials/analysis tools: XXW YZC XQW. Wrote the paper: XXW.

References

1. Mayr E (1942) Systematics and the origin of species. New York: Columbia University Press.
2. Hewitt GM (2000) The genetic legacy of the Quaternary ice ages. Nature 405: 907–913.
3. An ZS, Kutzbach JE, Prell WL, Porter SC (2001) Evolution of Asian monsoons and phased uplift of the Himalaya-Tibetan plateau since Late Miocene times. Nature 41: 62–66.
4. Guo ZT, Ruddiman WF, Hao QZ, Wu HB, Qiao YS, et al. (2002) Onset of Asian desertification by 22 Myr ago inferred from loess deposits in China. Nature 416: 159–163.
5. Jia GD, Peng PA, Zhao QH, Jian ZM (2003) Changes in terrestrial ecosystem since 30 Ma in East Asia: stable isotope evidence from black carbon in the South China Sea. Geology 31: 1093–1096.
6. Sun XJ, Wang PX (2005) How old is the Asian monsoon system?-Palaeobotanical records from China. Palaeogeo Palaeoclim Palaeoecol 222: 181–222.
7. Ying TS (2001) Species diversity and distribution pattern of seed plants in China. Biodivers Sci 9: 393–398.
8. Mittermeier RA, Gil PR, Hoffman M, Pilgrim J, Brooks T, et al. (2005) Hotspots Revisited: Earth's Biologically Richest and Most Endangered Terrestrial Ecoregions. Cemex, Conservation International and Agrupación Sierra Madre, Monterrey, Mexico.
9. Yang SJ, Dong HL, Lei FM (2009). Phylogeography of regional fauna on the Tibetan Plateau: a review. Prog Nat Sci 19: 789–799.
10. Qiu YX, Fu CX, Comes HP (2011) Plant molecular phylogeography in China and adjacent regions: tracing the genetic imprints of Quaternary climate and environmental change in the world's most diverse temperate flora. Mol Phylogenet Evol 59: 225–244.
11. Yang FS, Qin AL, Li YF, Wang XQ (2012) Great genetic differentiation among populations of *Meconopsis integrifolia* and its implication for plant speciation in the Qinghai-Tibetan Plateau. PLoS ONE 7: e37196.
12. Qin AL, Wang MM, Cun YZ, Yang FS, Wang SS, et al. (2013) Phylogeographic evidence for a link of species divergence of Ephedra in the Qinghai-Tibetan Plateau and adjacent regions to the Miocene Asian aridification. PLoS ONE 8: e56243.
13. Cun YZ, Wang XQ (2010) Plant recolonization in the Himalaya from the southeastern Qinghai-Tibetan Plateau: Geographical isolation contributed to high population differentiation. Mol Phylogenet Evol 56: 972–982.
14. Xu TT, Abbott RJ, Milne RI, Mao KS, Du FK, et al. (2010) Phylogeography and allopatric divergence of cypress species (*Cupressus* L.) in the Qinghai-Tibetan Plateau and adjacent regions. BMC Evol Biol 10: 194
15. Wu SJ, Luo J, Li QQ, Wang YQ, Murphy RW, et al. (2013) Ecological genetics of Chinese Rhesus Macaque in response to mountain building: all things are not equal. PLoS ONE 8: e55315.

16. Yan F, Zhou WW, Zhao HT, Yuan ZY, Wang YY, et al. (2013). Geological events play a larger role than Pleistocene climatic fluctuations in driving the genetic structure of *Quasipaa boulengeri* (Anura: Dicroglossidae). Mol Ecol 22: 1120–1133.
17. Willyard A, Syring J, Gernandt DS, Liston A, Cronn R (2007) Fossil calibration of molecular divergence infers a moderate mutation rate and recent radiations for *Pinus*. Mol Biol Evol 24: 90–101.
18. Fu LG, Li N, Mill RR (1999) Flora of China 4. In: Wu CY, Raven PH, editors. Beijing & Missouri: Science Press & Missouri Botanical Garden Press.
19. Farjon A (2001) World checklist and bibliography of conifers, 2nd edn. Royal Botanic Gardens, Kew.
20. Businský R (2004) A revision of the Asian *Pinus* subsection *Strobus* (Pinaceae). Willdenowia 34: 209–257.
21. Syring J, Farrell K, Businský R, Cronn R, Liston A, et al. (2007) Widespread genealogical nonmonophyly in species of *Pinus* subgenus *Strobus*. Syst Biol 56: 163–181.
22. Tsutsui K, Suwa A, Sawada K, Kato T, Ohsawa TA, et al. (2009) Incongruence among mitochondrial, chloroplast and nuclear gene trees in *Pinus* subgenus *Strobus* (Pinaceae). J Plant Res 122: 509–521.
23. Gernandt DS, Lopez GG, Garcia SO, Liston A (2005) Phylogeny and classification of *Pinus*. Taxon 54: 29–42.
24. Wang HR, Hong JS (2004) Genetic resources, tree improvement and gene conservation of five-needle pines in East Asia. USDA Forest Service Proceedings RMRS-P 32: 73–78.
25. Rogers SO, Bendich AJ (1988) Extraction of DNA from plant tissues. Plant Mol Biol Manual A6: 1–10.
26. Ran JH, Wei XX, Wang XQ (2006) Molecular phylogeny and biogeography of *Picea* (Pinaceae): Implications for phylogeographical studies using cytoplasmic haplotypes. Mol Phylogenet Evol 41: 405–419.
27. Wang XQ, Tank DC, Sang T (2000) Phylogeny and divergence times in Pinaceae: Evidence from three genomes. Mol Biol Evol 17: 773–781.
28. Duminil J, Pemonge MH, Petit RJ (2002) A set of 35 consensus primer pairs amplifying genes and introns of plant mitochondrial DNA. Mol Ecol Notes 2: 428–430.
29. Jaramillo-Correa JP, Beaulieu J, Bousquet J (2004) Variation in mitochondrialDNA reveals multiple distant glacial refugia in black spruce (*Picea mariana*), a transcontinental North American conifer. Mol Ecol 13: 2735–2747.
30. Wei XX, Beaulieu J, Khasa DP, Vargas-Hernández J, López-Upton J, et al. (2011) Range-wide chloroplast and mitochondrial DNA imprints reveal multiple lineages and complex biogeographic history for Douglas-fir. Tree Genet Genomes 7: 1025–1040.
31. Thompson JD, Gibson TJ, Plewniak F, Jeanmougin F, Higgins DG (1997) The clustal X Windons interface: flexible strategies for multiple sequence alignment aided by quality analysis tools. Nucleic Acids Res 25: 4876–4882.
32. Hall TA (1999) BioEdit: a user-friendly biological sequence alignment editor and analysis program for Windows 95/98/NT. Nucleic Acids Symp Ser 41: 95–98.
33. Librado P, Rozas J (2009) DnaSP v.5: A software for comprehensive analysis of DNA polymorphism data. Bioinformatics 25: 1451–1452.
34. Dupanloup I, Schneider S, Excoffier L (2002) A simulated annealing approach to define the genetic structure of populations. Mol Ecol 11: 2571–581.
35. Pons O, Petit RJ (1996) Measuring and testing genetic differentiation with ordered versus unordered haplotypes. Genetics 144: 1237–1245.
36. Excoffier L, Laval G, Schneider S (2005) Arlequin version 3.0: an integrated software package for population genetics data analysis. Evol Bioinform Online 1: 47–50.
37. Tajima F (1989) Statistical method for testing the neutral mutation hypothesis by DNA polymorphism. Genetics 123: 585–595.
38. Fu YX (1997) Statistical tests of neutrality of mutations against population growth, hitchhiking and backgroud selection. Genetics 147: 915–925.
39. Rogers AR, Harpending H (1992) Population growth makes waves in the distribution of pairwise genetic differences. Mol Biol Evol 9: 552–569.
40. Ma XF, Szmidt AE, Wang XR (2006) Genetic structure and evolutionary history of a diploid hybrid pine *Pinus densata* inferred from the Nucleotide variation at seven gene loci. Mol Biol Evol 23: 807–816.
41. Swofford DL (2002) PAUP*. Phylogenetic Analysis Using Parsimony (*and other methods). Version 4. Sinauer, Sunderland, MA.
42. Guindon S, Gascuel O (2003). A simple, fast, and accurate algorithm to estimatelarge phylogenies by maximum likelihood. Syst Biol 52: 696–704.
43. Ronquist F, Huelsenbeck JP (2003) MrBayes 3: Bayesian phylogenetic inference under mixed models. Bioinformatics 19: 1572–1574.
44. Posada D, Crandall KA (1998) Modeltest: testing the model of DNA substitution. Bioinformatics 14: 917–918.
45. Nylander JAA (2004) MrModeltest v2. Program distributed by the author. Evolutionary Biology Centre, Uppsala University.
46. Drummond AJ, Rambaut A (2007) BEAST: Bayesian evolutionary analysis by sampling trees. BMC Evol Biol 7: 214.
47. Gernandt DS, Magallon S, Lopez GG, Flores OZ, Willyard A, et al. (2008) Use of simultaneous analyses to guide fossil-based calibrations of Pinaceae phylogeny. Int J Plant Sci 169: 1086–1099.
48. Posada D, Crandall KA (2001) Intraspecific phylogenetics: trees grafting into networks. Trends Ecol Evol 16: 37–45.
49. Petit RJ, Duminil J, Fineschi S, Hampe A, Salvini D, et al. (2005) Comparative organization of chloroplast, mitochondrial and nuclear diversity in plant populations. Mol Ecol 14: 689–701.
50. Jaramillo-Correa JP, Beaulieu J, Ledig FT, Bousquet J (2006) Decoupled mitochondrial and chloroplast DNA population structure reveals Holocene collapse and population isolation in a threatened Mexican-endemic conifer. Mol Ecol 15: 2787–2800.
51. Aizawa M, Yoshimaru H, Saito H, Katsuki T, Kawahara T, et al. (2007) Phylogeography of a northeast Asian spruce, *Picea jezoensis*, inferred from genetic variation observed in organelle DNA markers. Mol Ecol 16: 3393–3405.
52. Meng LH, Yang R, Abbott RJ, Miehe G, Hu TH, et al. (2007) Mitochondrial and chloroplast phylogeography of *Picea crassifolia* Kom. (Pinaceae) in the Qinghai-Tibetan Plateau and adjacent highlands. Mol Ecol 16: 4128–4137.
53. Chen KM, Abbott RJ, Milne RI, Tian XM, Liu JQ (2008) Phylogeography of *Pinus tabulaeformis* Carr. (Pinaceae), a dominant species of coniferous forest in northern China. Mol Ecol 17: 4276–4288.
54. Tian S, Luo LC, Ge S, Zhang ZY (2008) Clear genetic structure of *Pinus kwangtungensis* (Pinaceae) revealed by a plastid DNA fragment with a novel minisatellite. Ann Bot 102: 69–78.
55. Tian S, López-Pujol J, Wang HW, Ge S, Zhang ZY (2010) Molecular evidence for glacial expansion and interglacial retreat during Quaternary climatic changes in a montane temperate pine (*Pinus kwangtungensis* Chun ex. Tsiang) in southern China. Plant Syst Evol 284: 219–229.
56. Ma CG (1992) Symposium on selection of optimum seed sources for plantations of *Pinus armandii* Franch. Beijing: Beijing Agricultural University press.
57. Zhang DC, Boufford DE, Ree RH, Sun H (2009) The 29°N latitudinal line: an important division in the Hengduan Moutains, a biodiversity hotspot in southwest China. Nordic J Bot 27: 405–412.
58. Willyard A, Syring J, Gernandt D, Liston A, Cronn R. (2006) Molecular evolutionary rates indicate a recent and rapid diversification for modern pine lineages. Mol Biol Evol 23:1–12.
59. Harris N (2006) The elevation history of the Tibetan Plateau and its implications for the Asian monsoon. Palaeogeogr Palaeoclimatol Palaeoecol 241: 4–15
60. Clark MA, House MA, Royden LH, Whipple KX, Burchfiel BC, et al. (2005) Late Cenozoic uplift of southeastern Tibet. Geology 33: 525–528.
61. Wang B, Mao JF, Gao J, Zhao W, Wang XR (2011) Colonization of the Tibetan Plateau by the homoploid hybrid pine *Pinus densata*. Mol Ecol 20: 3796–3811.
62. Wang HW, Ge S (2006) Phylogeography of the endangered *Cathaya argyrophylla* (Pinaceae) inferred from sequence variation of mitochondrial and nuclear DNA. Mol Ecol 15: 4109–4122.
63. Gao LM, Moeller M, Zhang XM, Hollingsworth ML, Liu J, et al. (2007) High variation and strong phylogeographic pattern among cpDNA haplotypes in *Taxus wallichiana* (Taxaceae) in China and North Vietnam. Mol Ecol 16: 4684–4698.
64. Wang J, Gao P, Kang M, Lowe A J, Huang H W (2009) Refugia within refugia: the case study of a canopy tree (*Eurycorymbus cavaleriei*) in subtropical China. J Biogeogr 36: 2156–2164.
65. Chen ZD, Ying JS, Lu AM (2012) Disjunct distribution of seed plants between Southwestern China and Taiwan Island of China. Chinese Bull Bot 47: 551–570.
66. Critchdield WB (1984) Impact of the Pleistocene on the genetic structure of North American conifers. In: Lanner RM, editor. Proceedings of the 8th North America forest biology workshop. Logan, Utah. pp. 70–118.

9

Pasture Characteristics in Three Different Ecotypes at Khovd Aimag, Western Mongolia

Jutta Beher*

ARC Centre of Excellence for Environmental Decisions, University of Queensland, Brisbane, Queensland, Australia

Abstract

The transition of nomadic pastoralism to more sessile forms of rangeland utilization and increased stocking rates can result in the degradation of pasture. After political changes in the 1990s in Mongolia, population growth and missing alternative livelihoods intensified the grazing pressure on pastures, and further decreased the condition of the fragile arid ecosystems. To learn more about the productivity and quality of pasture land in Khovd Aimag in the western region of Mongolia, standing biomass was measured in the alpine region, mountain steppe and semi-desert. Plant samples were analyzed for nitrogen and fiber contents by wet chemistry and Near Infrared Spectroscopy (NIRS). Results show clear differences in distribution of biomass with reduced biomass in the vicinity of temporary settlements. From July to early September plant nitrogen contents decreased in the alpine region, remained unchanged in the mountain steppe and increased in the semi-desert. Nitrogen concentrations were elevated in vegetation close to temporary settlements. For fiber contents (ADF) no clear patterns were found. Neither biomass/m^2 nor vegetation cover were appropriate indicators for food quality.

Editor: Bazartseren Boldgiv, National University of Mongolia, Mongolia

Funding: DAAD funded student travel stipend (https://www.daad.de/deutschland/stipendium/en/); VW foundation funded field work expenses within the project Changing Pastoral Ecosystems in Western Mongolia (Khovd Aimag): A Landscape-Ecological Approach for Assessing Grazing Land Degradation and Grazing Capacity (http://www.volkswagenstiftung.de/de/foerderung/beendet/partnerschaftsprogramm/bewilligungen-2002.html). The funders had no role in study design, data collection and analysis, decision to publish, or preparation of the manuscript.

Competing Interests: The author has declared that no competing interests exist.

* Email: j.beher@uq.edu.au

Introduction

Almost half of the Mongolian human population depends on livestock in a direct or indirect way. About 75% of the vast country is used as pasture for goats, sheep, cattle, horses and camels [1]. For hundreds of years pastoralists pursued a semi-nomadic lifestyle, moving livestock and households among seasonal pastures about four times per year to meet the nutritional demands of their animals, and to conserve pasture resources [2–4]. Climate on the Mongolian Plateau has been getting warmer and drier since the early 1960s [5,6]. The number of droughts and winter snowstorms has increased in Mongolia over the last 50 years, particularly in the last decade [7]. Due to lack of effective resource institutions, conflicts among herders over grassland use have increased since the early 1990s [3] and degradation of many areas has become an issue due to socio-economic changes and increased stocking rates [2,6]. At the same time, distance and frequency of nomadic movement have declined [8]. The transition from ancient tradition to centralized organization and back to individually organized rangeland organization, for example in parts of Russia and China, has led to severe degradation of pasture ecosystems in the recent past. The same is now probable in rural Mongolia, and coherent management plans still do not exist. The growth of the human population of Mongolia requires intensified productivity in the agricultural sector, which increases the pressure on pastures [4,9–11]. Still, relatively little is known about most Mongolian steppe ecosystems and their dynamics. Opinions on correct classification of underlying general dynamics of Mongolian rangelands differ immensely [12–15]. For sustainable livestock management, data of seasonal and site specific pasture conditions and response to climate and grazing are needed [12]. The suitability of an area depends not only on food biomass but also on food quality. Most ongoing research focuses on vegetation change and plant species but little is known about the nutritional quality of the pasture systems in time and space.

Apart from quantitative differences in food availability, the chemical composition of plants changes during the growing season. Besides the secondary plant chemicals (most important in herbs and woody plants), food quality for livestock is characterized by nitrogen and fiber concentrations with the ratio of nitrogen-to-fiber used as a measure of food quality for ruminants [16]. In general, concentrations of nitrogen decrease while fibers increase as the plants grow older [17–22].

In order to come to a better understanding of the spatial and temporal pattern of food for the livestock of nomadic herdsmen in Mongolia, I analyzed mixed vegetation samples from pastures in three different ecotypes. I addressed the following questions: (1) What are the quantitative and qualitative changes of vegetation during growing season? (2) How does the quantity and quality of the vegetation differ between temporary inhabited valleys and areas further away from these settlements?

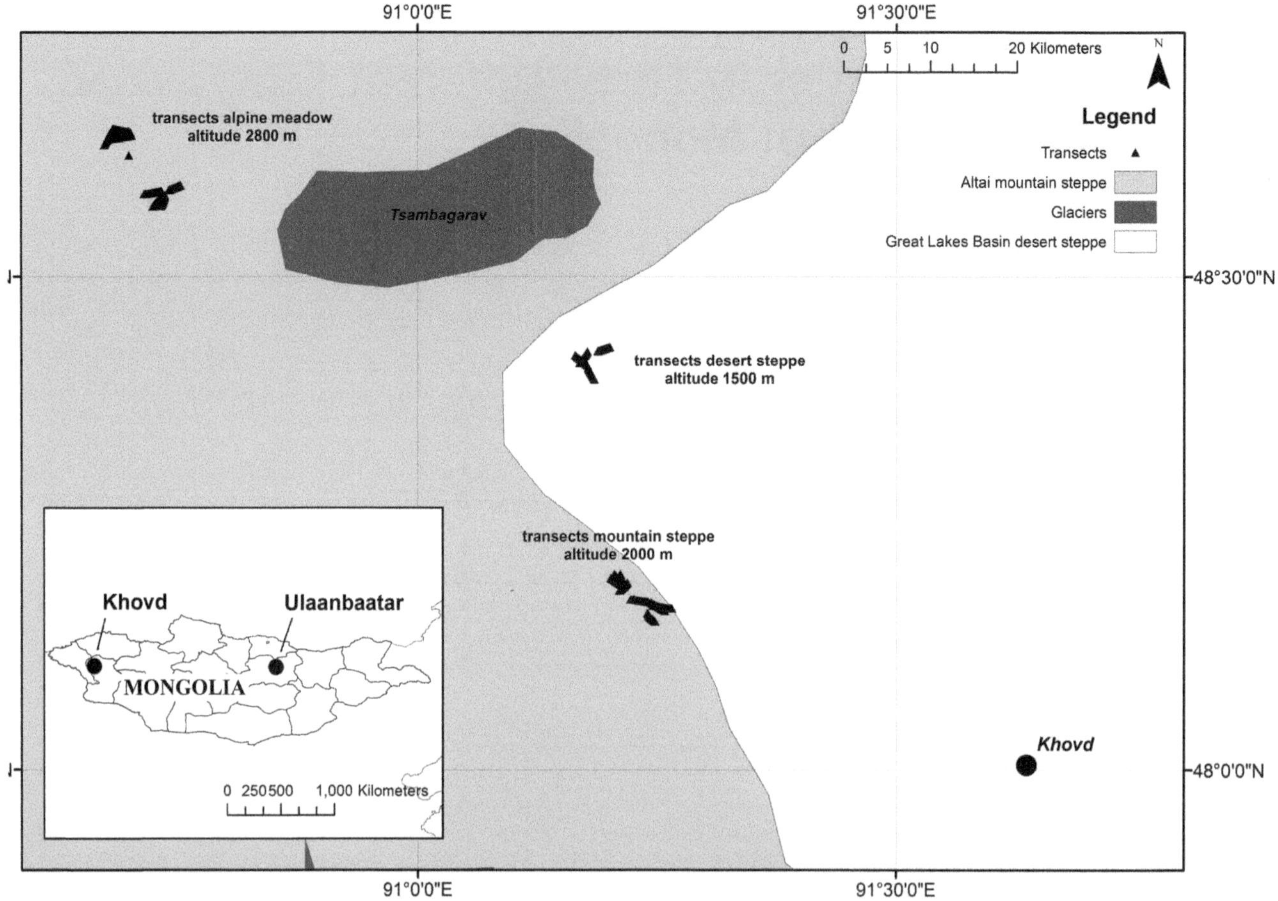

Figure 1. Study area in the northwest of Khovd City. Study sites at desert steppe in the area of the Great Lakes Basin, alpine region close to mount Tsambagarav and mountain steppe.

Materials and Methods

Study area

The study was carried out in the north-west of Khovd city in Khovd Aimag in Western Mongolia (48°01′N, 91°34′E) (Figure 1).

The area extends across three different ecotypes from the semi-deserts of the Great Lake Basin to mountain steppe and alpine regions in the Altai Mountains [23,24]. These ecotypes are three major ecological zones in Mongolia and their close vicinity at the Khovd region simplifies comparative studies. The distance between the studied areas was about 30–80 kilometers (Figure 1).

Annual rainfall in Mongolia ranges from 50 mm/year in the southern regions and 500 mm/year in the northern mountains. Up to 96% of the rains fall in summer. Data from the weather station at Khovd (1405 m above sea level) show a clear peak between May and September (Figure 2).

The time of increased precipitation represents the general growth period of vegetation, however, the time span of climate conditions that allow growth of plants is much shorter in higher regions, with a later start and an earlier ending than in lower regions.

Extreme temperatures, drought and short transitions between seasons result in limited soil moisture. Therefore, soil formation occurs only in thin layers [25], and strong winds can denude the sparsely covered soils. The alpine region and upper mountain steppe show kastanozems, semi-desert regions mostly burozems. The alpine region above 2700 m altitude is dominated by sedgegrass and kobresia meadows, mountain steppe (1800–2500 m) and semi-deserts (1200–1700 m) by scant graminoid communities dominated by *Stipa* sp. and often assorted with dwarf bushes like *Caragana* sp. The whole area is used for pasture by nomad families and their animals who settle for short spans of time, averaging several weeks, at different riverbanks. Gers are set up on their arrival and deconstructed again shortly before the next migration. This way, temporary villages of 20–50 gers are built, demolished and rebuilt over time and place. Rivers are used as a watersource for human beings and animals. The use of wells was not observed in the studied area. Most people own the five common livestock species of yaks, camels, horses, sheep and goats. The former three are low in numbers and graze on their own, while sheep and goats dominate in numbers and are kept in mixed herds which are driven by a herdsman with daily path-lengths of about 10 km. Every family in the study area owns at least sheep and goats in a mixed herd of a minimum of 100 animals. Most families follow their own schedule coming back to the same places for many years. Nomads in the study area used the alpine region and mountain steppe for pasture from July 15 to August 20, and semi-desert from August 13 until September 2 [26].

Sampling design

Biomass and nutrient dynamics were studied in relation to: (1) seasonality, (2) ecotype and (3) distance to a river in valleys inhabited temporarily during the summer months.

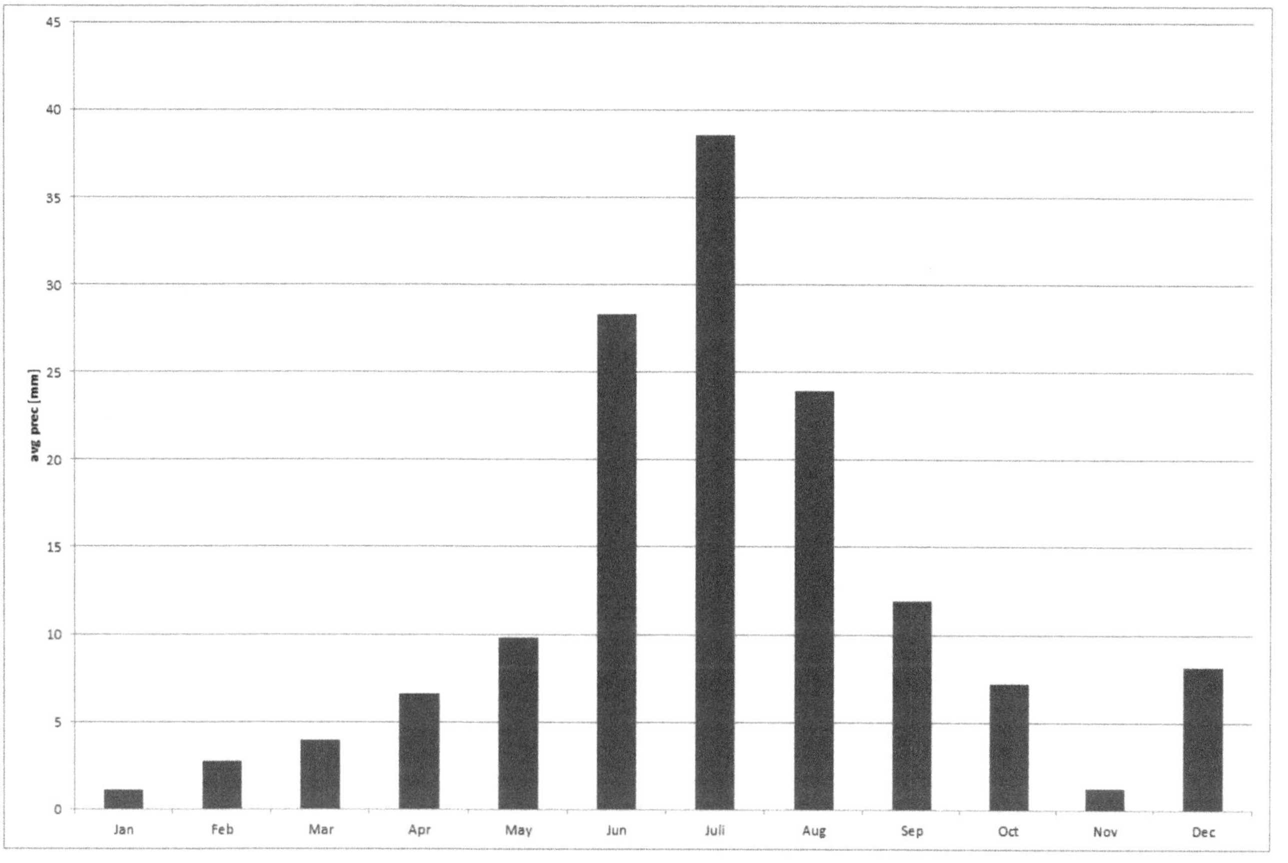

Figure 2. Mean amount of rainfall at Khovd Region from 1983 to 2004 [47].

Seasonality is defined as time span of active plant physiology and was assessed by sampling in July, August and September, which are the main months of the growth period. Ecotype is defined as semi-desert (sampling site at 1700 m, 48°25′N, 91°10′E), mountain steppe (sampling site at 2000 m, 48°11′N, 91°12′E and 48°9′N, 91°15′E) and alpine region (sampling site at 2700 m, 48°38′N, 90°40′E and 48°34′N, 90°43′E). Distance to a river implies two sets of plots within an ecotype differing in distance to the next temporarily inhabited riverbank. Ger numbers at the study sites were about 50 at alpine region and mountain steppe and 25 at semi-desert.

Distance to the river with the associated settlements was assumed to indicate utilization intensity by livestock. This was verified by tracking of herds with GPS collars: A range of 2 km on both sides of the rivers seemed to be used more frequently by livestock than areas outside this zone [26]. Distance classes are referred to as "distant" and "close". Close plots have less than 2 km distance to the riverbank, distant ones are more remote but not more than 5 km away.

Arrangement and distance between plots should create a representative sample for the studied pasture. To avoid geomorphological or topographical effects [15,27], large homogenous areas were chosen that had no difference in slope or aspect and were typical examples of the ecotype [25]. The areas directly adjacent to a river were classified as a different ecotype and were not sampled. In the mountain steppe and the alpine region one transect of 1.5 km length was established in a random direction at the two distance classes each month. Starting from a random point

Table 1. Sampling design: 20 plots of 1 m^2 were sampled per month and distance class in the alpine region and the mountain steppe; and 20 plots were sampled per month in the semi-desert.

	Alpine region	Mountain steppe	Semi-desert
distant	July: 1 transect with 20 plots	July: 1 transect with 20 plots	July: 1 transect with 20 plots
	Aug: 1 transect with 20 plots	Aug: 1 transect with 20 plots	Aug: 1 transect with 20 plots
	Sept: 1 transect with 20 plots	Sept: 1 transect with 20 plots	
close	July: 1 transect with 20 plots	July: 1 transect with 20 plots	
	Aug: 1 transect with 20 plots	Aug: 1 transect with 20 plots	
	Sept: 1 transect with 20 plots	Sept: 1 transect with 20 plots	Sept: 1 transect with 20 plots

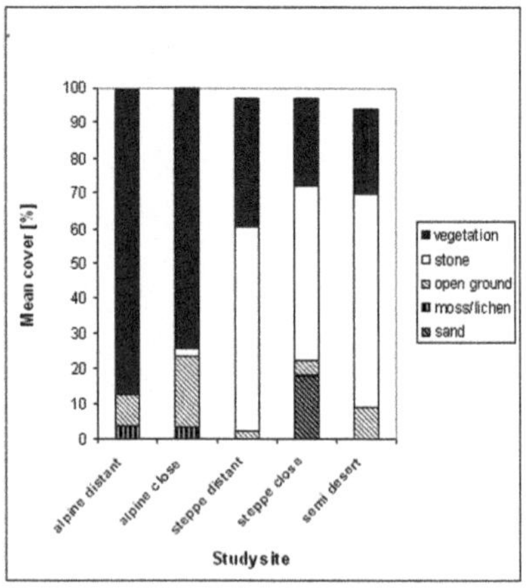

Figure 3. Mean cover of vegetation and particle size of the ground at study sites of alpine region, mountain steppe and semi-desert.

and using a GPS device for bearing, every 75 m a 1 m^2 plot was sampled, resulting in 20 different plots along each transect (Table 1) Exposure and slope where consistent for all three transects within each ecotype. For each 1 m^2 plot the vegetation composition was described on the species level following Braun-Blanquet [28]. Vegetation descriptions included all herbaceous plants, graminoids, and shrubs to fully describe the ecotype and its plant communities (Table S1 "Species List"). Shrub species were excluded in the biochemical analysis to avoid bias, because the probability that they would frequently occur in the plots along transects was low. In the rare occasion that a caragana bush occurred inside of a plot, the plot was established 1 m behind the bush to avoid effects of shading or root competition. The percentage cover of moss/lichen, stone, sand, bare soil and vegetation were estimated. Stone, sand, and bare soil were classified according to particle size: stone>2 mm, sand 0.1–2 mm and bare soil<0.1 mm (Figure 3).

At the semi-desert only one transect was established per month. Due to the late arrival of nomad families it was unknown where they would install their camps at the riverbank. Therefore, classification of close and distant sites was possible only afterwards and data for the semi-desert is incomplete. This data is included in correlations but not in the multivariate analysis.

Plant sampling for biomass determination and chemical analyses

Plant material was collected every third/fourth week from the beginning of July through to early September. All plant material was cut at 1 cm height above ground in each of the twenty 1 m^2 plots per transect. The field work was conducted in collaboration with the University of Khovd and no additional specific permission was required for any sampling at the study sites. The accessed land was neither privately owned nor protected and no endangered or protected species were involved in the sampling. All locations were subject to temporary livestock grazing. In total 300 plots were sampled (Table 1).

Biomass was defined as above-ground Dry Organic Matter (DOM [g/m^2]) measured after 24 h of drying at 60°C in an oven. Food quality was defined as the ratio of Acid Detergent Fiber and nitrogen content (N/ADF).

Of the 300 samples of 1 m^2 plots 291 plant samples were analyzed for total nitrogen content using Kjeldahl digestion with

Table 2. Effects of the ecotype, seasonality and distance to the river on the biomass, nitrogen, ADF and food quality in the alpine meadow and the mountain steppe according to ANOVA.

Factor	df	Biomass		Nitrogen		ADF		Quality	
		F	p	F	p	F	p	F	p
Ecotype	1	0.1622	0.686	**239.8722**	<0.0001	0.0796	0.778	**127.1525**	<0.0001
Seasonality	2	1.9426	0.146	**38.6886**	<0.0001	**14.8750**	<0.0001	**37.2654**	<0.0001
Distance	1	**196.7121**	<0.0001	**52.1571**	<0.0001	0.0027	0.959	**27.1747**	<0.0001
Ecotype * Seasonality	2	**6.6454**	0.002	**40.6727**	<0.0001	**3.0549**	0.049	**15.6000**	<0.0001
Ecotype * Distance	1	0.5521	0.459	1.5799	0.210	0.1462	0.702	1.2844	0.25834
Seasonality * Distance	2	**5.7054**	0.004	2.1194	0.123	**3.5532**	0.030	**3.3430**	0.037
Ecotype * Seasonality * Distance	2	**12.3303**	<0.0001	**14.6348**	<0.0001	**4.4605**	0.013	**13.6231**	<0.0001

N = 229. A * indicates interaction terms.

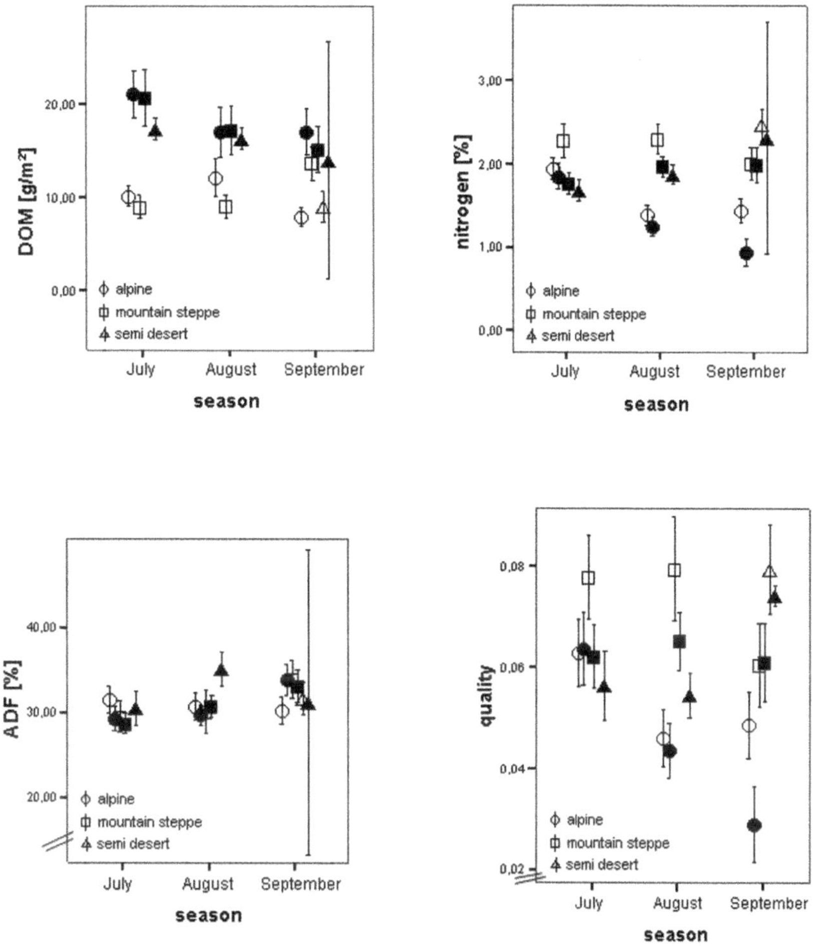

Figure 4. DOM, nitrogen, ADF and quality (N/ADF) of vegetation of 1 m² plots in relation to ecotype, season and distance class of study sites. Error bars indicate the 95% confidence interval of the mean. Empty symbols = sites close to rivers, filled symbols = sites away from rivers. Data for the semi-desert were not assigned to close or distant sites.

sulphuric acid at 400°C and Near-Infra-Red-Spectroscopy (NIRS) [29–31]. In addition, fiber contents were measured as acid detergent fiber (ADF) extraction according to Van Soest et al. [18].

Relationships between nitrogen, fiber, quality and biomass using ecotype, season and distance class as independent variables were analyzed with ANOVA. Data were log-transformed prior to analysis to meet ANOVA assumptions. Relationships between biomass and chemical characteristics were checked with simple correlations. Statistics are based on the values of the single plots.

Results

Ecotype

The number of plant species is highest in the mountain steppe and lowest in the semi-desert (Table S1 "Species List"). Biomass does not differ between ecotypes although vegetation cover is highly different across ecotypes. Alpine meadows show a very high cover in comparison to the sparse grass communities of semi-desert and mountain steppe. Altitudinal differences were primarily found for nitrogen and quality, with lower values in alpine region (Table 2, Figure 4).

Seasonality effects on biomass, nitrogen, ADF and food quality are different for the two ecotypes. Effects of distance class are the same for the alpine region and mountain steppe (check interaction terms in Table 2).

Seasonality

Seasonality shows significant effects on nitrogen, ADF and food quality (Table 2). Effects depend on spatial scale (Table 3). Effects on biomass are more pronounced for close than for distant sites. Nitrogen concentration and quality decrease from July to September in the alpine region but show less differences in mountain steppe. For ADF, distant sites seem more affected by seasonal changes than close sites: significant changes are more pronounced for both distant sites than for the close site in the mountain steppe. At the close site in the alpine region, ADF does not change significantly.

Distance from a river

Biomass, nitrogen and quality differ between distant and close sites in both alpine region and mountain steppe (Table 3, Figure 4).

Biomass is higher at distant sites while N and quality is higher in the vicinity to temporary human settlements. Interaction of distance class and seasonality is significant for biomass, ADF and quality in the alpine region and the mountain steppe.

Table 3. Single factor analyses for the main factors ecotype, seasonality and distance class in the alpine meadow (alpine) and mountain steppe (steppe).

Factor	Spatial scale	N	df	Biomass		Nitrogen		ADF		Quality	
				F	p	F	p	F	p	F	p
Season.	alpine dis.	58	2	3.0859	0.053	**47.15**	<0.0001	**11.776**	<0.0001	**40.202**	<0.0001
	alpine cl.	59		**9.4752**	<0.001	**22.226**	<0.0001	0.6739	0.514	**9.2926**	<0.001
	steppe dis.	57		3.3982	0.041	2.8519	0.07	**10.607**	<0.001	0.6073	0.549
	steppe cl.	55		**12.665**	<0.0001	**3.603**	<0.05	**5.31**	<0.01	**6.2688**	<0.01
Distance	alpine	117	1	**96.774**	<0.0001	**14.012**	<0.001	0.0952	0.758	**9.4725**	<0.01
	steppe	112		**68.904**	<0.0001	**20.016**	<0.0001	0.0362	0.8495	**8.7536**	<0.01

At close sites in the alpine meadow and mountain steppe as well as in all plots in the semi-desert there is a significant negative correlation of DOM and quality. At distant sites of the alpine meadow and mountain steppe there is no such correlation. For the Pearson correlations between biomass and quality, data of the semi-desert were included (Table 4).

Discussion

In the Western Mongolian pasture systems food quantity and quality change in time and space. At all three ecotypes seasonal changes were found for nitrogen and food quality, while the distance to a temporary settlement affected both quantity and quality. Different ecotypes in Khovd Aimag provide fodder of different quality at the same time of year. The natural temporal dynamics of main pasture plants is well known, e.g. Tserendash and Erdenebaatar [32] describe an early summer flowering with maximal nitrogen contents of species of *Koeleria*, *Poa* and *Agropyron* at steppe ecosystems. At the same time of the year species of *Carex* in meadow ecosystems show late flowering. Plants serving as fodder during the growing season can show compensatory growth after grazing producing new material rich in nitrogen and poor in fiber [33]. As a consequence the quality of the pasture is improved. All study sites were used as pasture to some extent during the summer. Thus compensatory growth as suggested in literature [34–36] could be a possible reaction of plants to grazing. A study from a meadow steppe in Inner Mongolia within a similar plant community shows a clear increase in total nitrogen contents under increasing grazing intensity for e.g. *Koeleria cristata*, *Artemisia frigida* and species of *Stipa*, *Thalictrum*, and *Pulsatilla* [37]. Nitrogen contents in grasses and sedges were generally lower than of forbs, and leguminous plants showed highest nitrogen contents in the same study.

Negative correlations between biomass and the chemical quality of the vegetation as shown in Table 4 have been found in other studies on grazing and its impact on the vegetation [38]. The negative correlation was found only for close study sites but not for the distant sites. This could indicate more active plant production and new growth after grazing in these areas. In all ecotypes nitrogen contents were higher close to the river than at more distant sites.

However, this picture of compensatory growth should not be generalized. Whether plants can compensate or not and how they compensate can depend on the type of damage, abiotic conditions such as rainfall or nutrient availability, and the effect of nutrient influx caused by the feces of livestock itself [33,38,39]. In addition, plants seem to have the option of differential compensation within the same individual, resulting in plant material of diverging quality within the same plant [31].

At the studied sites nitrogen contents are significantly higher for soils in the alpine region than for soils of mountain steppe or semi-desert (analyzed soil parameter from data of [40]). In the alpine region the availability of nitrogen in the soil could facilitate compensatory growth, or allow different plant reactions than soil conditions in the mountain steppe or semi-desert.

The decreasing quality in alpine vegetation as a result of reduced nitrogen concentrations is likely due to the seasonal changes in weather. Temperature dropped and any precipitation resulted in snowfall. Many plants started to wither in late August, which follows the described decrease in nutritive value until October in the literature [32]. Compensatory growth under these conditions was unlikely.

Changing patterns in nitrogen contents during the season were very different in the three ecotypes. A clear decrease was found in

Table 4. Pearson's correlation analysis between food quality (measured as the ratio of nitrogen/ADF) and DOM for pooled data, single pooled ecotypes and each distance class per ecotype.

Data Set	N	r coefficient	p
alpine region, mountain steppe and semi-desert	286	−0.3115764	>0.0001***
alpine region	115	−0.2951415	>0.01**
mountain steppe	112	−0.3664732	>0.0001***
Semi-desert	59	−0.6854251	>0.0001***
alpine region, distant	59	−0.06731023	0.622
alpine region, close	56	−0.3052649	0.019*
mountain steppe, distant	57	−0.1258348	0.355
mountain steppe, close	55	−0.3886314	0.003**

***$p \leq 0.0001$,
**$p \leq 0.01$,
*$p \leq 0.05$.

alpine meadow, mostly at the distant site. In contrast, nitrogen concentrations increase in the semi-desert between July and September. One reason for the different dynamics is surely the different climatic conditions and the shift in season. Winter was starting in the alpine regions while summer had reached its peak in the semi-desert. While the vegetation in the alpine region had already dried up, the vegetation of the semi-desert had just started growing after heavy rainfalls in August. The vegetation of the mountain steppe showed no clear signs of change, neither in nitrogen contents nor in outer appearance. Thus, seasonality in the three ecotypes cannot be defined through the Julian calendar but should be defined phenologically or climatically. Given the successive changes over the growing season found in this and many other studies, the decline of nitrogen content in semi-desert and mountain steppe is to be expected later during the year, with the change of climate during autumn and winter [17,21,32,41]. Cattle and sheep are likely to avoid food with nitrogen content below 1.12%; dairy cows prefer food with nitrogen contents above 1.9% [42]. Nitrogen contents in the alpine region are about 1.2% in August and September. Thus, leaving this region in August is advantageous from a biochemical point of view.

The distance to temporary human settlements seems to be an important factor for the nitrogen and even more for DOM in the alpine region and the mountain steppe. Although vegetation cover is lower in semi-desert and mountain steppe than in alpine region, differences in DOM are much clearer between distance classes than between ecotypes. This might be due to the taller habit of grass communities that compensate sparse cover as far as total biomass is concerned.

Reduced biomass as a consequence of grazing is often found, but intermediate grazing can increase plant production [42–44]. Thus, the differences in biomass found between the two distance classes could be due to reduced growth at close sites or increased plant production at distant sites or even both. The study in Inner Mongolia [37] reveals consistent decrease of plant biomass with increasing grazing intensity for species of *Stipa*, *Bupleurum* and *Pulsatilla* and increasing biomass for *Artemisia frigida*. This can only be revealed by exclosure experiments or longterm data.

The present study did not indicate negative effects of grazing such as a shift in species composition, decrease in fodder quality due to damage or a shift to unpalatable species or intensive erosion. Only very few species showed a significantly different coverage in close and distant plots: In the alpine meadow *Thalictrum alpinum* and *Oxytropis altaica* showed higher coverage in distant plots while *Androsace fedtschenkoi*, *Oxytopis oligantha*, *Artemisia argyrophylla*, *Stellaria pulvinata and Festuca ovina* showed significant higher cover in close plots [40]. In the steppe, *Koeleria cristata* and *Poa attenuata* showed higher coverage in distant plots whereas *Artemisa rutifolia* and *Convolvulus ammanni* showed higher coverage in close plots. Still, the significances were low and the community structure seemed not to be influenced heavily by the current pasture usage. In contrast to that, Lkhagva et al found up to 50% of the biomass consisting of unpalatable species in boreal steppe ecosystems under high grazing pressure and grazing related increase of grazing tolerant species was observed in steppe and riparian zones under high grazing pressure [27]. However, conclusions based on short-term studies have to be interpreted with care as pasture characteristics in dry ecosystems can be highly variable between years. Here, differences of more than 275% in biomass production are not uncommon between years, especially for arid and highly seasonal regions where the importance of abiotic factors can override the effect of grazing [45] and plant productivity is much more variable than in ecosystems with higher precipitation [46].

Supporting Information

Table S1 Species List: List of species found in more than 5% of all plots in an ecotype.

Code S1 R code for statistical analysis.

Data S1 Data table used for ANOVA in R (semi-desert excluded).

Data S2 Data table used for Pearson's correlation analysis (semi-desert included).

Acknowledgments

The project has been carried out as part of the collaboration between the universities of Hamburg, Germany and Khovd, Mongolia. The author would like to thank Jörg Ganzhorn and Udo Schickhoff for their help and inspiration, the project "Changing Pastoral Ecosystems in Western Mongolia (Khovd Aimag): A Landscape-Ecological Approach for Assessing Grazing Land Degradation and Grazing Capacity" for the possibility to conduct this study, Dr. Oyunchimeg for her help in the field, and the

Mongolian and Kazakh families for their outstanding hospitality. I would also like to thank the Center of Excellence of Environmental Decision for the support.

Author Contributions

Conceived and designed the experiments: JB. Performed the experiments: JB. Analyzed the data: JB. Contributed reagents/materials/analysis tools: JB. Contributed to the writing of the manuscript: JB.

References

1. NSO (National Statistical Office of Mongolia) (2007) Mongolian statistical yearbook 2007. National Statistical Office of Mongolia NSO, editor National Statistical Office of Mongolia, Ulaanbaatar.
2. Janzen J, Bazargur D (2003) Wandel und Kontinuität in der mobilen Tierhaltung der Mongolei. Petermanns Geogr Mitt 147: 50–57.
3. Upton C (2009) Custom and Contestation: Land Reform in Post-Socialist Mongolia. World Dev 37: 1400–1410. Available: http://dx.doi.org/10.1016/j.worlddev.2008.08.014.
4. World Resource Institution (2001) Grassland Ecosystems: Sustaining the Steppe: The Future of Mongolias Grasslands. Available: http://pdf.wri.org/wr2000_grasslands_mongolia.pdf. Accessed 30 April 2014.
5. Batima P (2006) Climate Change Vulnerability and Adaptation in the Livestock Sector of Mongolia. Washington D.C.
6. Wang J, Brown DG, Chen J (2013) Drivers of the dynamics in net primary productivity across ecological zones on the Mongolian Plateau. Landsc Ecol 28: 725–739. doi:10.1007/s10980-013-9865-1.
7. UNDP (2011) National Climate Risk Management Strategy and Action Plan of Mongolia 2009, revised draft v5 August 2011.
8. Fernandez-Gimenez M (1999) Sustaining the Steppe: A Geographical History Pastoral Land Use in Mongolia. Geogr Rev 3: 315–342.
9. NSO (National Statistical Office of Mongolia) (2011) Mongolian Statistical Yearbook 2011. NSO (National Statistical Office of Mongolia), editor NSO (National Statistical Office of Mongolia), Ulaanbaatar.
10. UNDP (2011) From vulnerability to sustainability: Environment and human development. Mongolia Human Development Report. Ulaanbaatar.
11. UN (United Nations); DPADM (Division for Public Administration and Development Management; DESA (Department of Economic and Social Affairs (2004) Mongolia - Public Administration Country Profile. UN; DPADM; DESA.
12. Sternberg T (2012) Piospheres and Pastoralists: Vegetation and Degradation in Steppe Grasslands. Hum Ecol 40: 811–820. doi:10.1007/s10745-012-9539-7.
13. Fernandez-Gimenez M, Allen-Diaz B (1999) Testing a non-equilibrium model of rangeland vegetation dynamics in Mongolia. J Appl Ecol 36: 871–885.
14. Zemmrich A, Manthey M, Zerbe S, Oyunchimeg D (2010) Driving environmental factors and the role of grazing in grassland communities: A comparative study along an altitudinal gradient in Western Mongolia. J Arid Environ 74: 1271–1280. Available: http://dx.doi.org/10.1016/j.jaridenv.2010.05.014.
15. Sasaki T, Okayasu T, Shirato Y, Jamsran U, Okubo S, et al. (2007) Can edaphic factors demonstrate landscape-scale differences in vegetation responses to grazing? Plant Ecol 194: 51–66. Available: http://link.springer.com/10.1007/s11258-007-9274-0. Accessed 1 May 2014.
16. Robbins C (1983) Wildlife Feeding and Nutrition. New York: Academic Press.
17. Savage D (1947) Nutritional Qualities of Range Forage Plants in Relation to Grazing With Beef Cattle on the Southern Plains Experimental Range. Technical Bulletin. Washington D.C.: United States Department of Agriculture, Vol. 943.
18. Van Soest P, Robertson J, Lewis B (1997) Methods for Dietary fiber, neutral detergent fiber, and Nonstarch Polysaccharides in Relation to Animal Nutrition. J Dairy Sci 74: 3583–3597.
19. Vallentine J (2001) Grazing Management. 2. ed. San Diego: Academic Press.
20. Kilcher M (1981) Plant Development, Stage of Maturity and Nutrient Composition. J range Manag 34: 363–364.
21. Tuku A (1991) Evaluation of dry matter yield, chemical composition and in vitro digestibility of different forage crops as influenced by cutting age, site (altitude) and year. Berlin: Technical University of Berlin.
22. Iason S, Van Wieren G (1999) Digestive and ingestive adaptations of mammalian herbivores to low-quality forage. In: Olff H, Brown V, Drent R, editors. Herbivores, between Predators and Plants. Oxford: Blackwell Science. 337–363.
23. Hilbig W (1995) The vegetation of Mongolia. Amsterdam: SPB Academic Publishing.
24. Zemmrich A (2005) Die Steppengliederung der Mongolei aus Sicht der russischen und mongolischen Geobotanik. Arch für Naturschutz und Landschaftsforsch: 17–35.
25. Gunin P, Vostovoka E, Dorofeyuk N, Tarasov P, Black C (1999) Vegetation Dynamics of Mongolia. Dordrecht.: Kluwer Academic Publishers.
26. Beher J (2008) Qualität und Quantität westmongolischer Weidegebiete in Abhängigkeit von Zeit, Raum und Nutzungsintensität Hamburg University.
27. Lkhagva A, Boldgiv B, Goulden CE, Yadamsuren O, Lauenroth WK (2013) Effects of grazing on plant community structure and aboveground net primary production of semiarid boreal steppe of northern Mongolia. Grassl Sci 59: 135–145. Available: http://doi.wiley.com/10.1111/grs.12022. Accessed 1 May 2014.
28. Traxler A (1997) Handbuch des Vegetationsökologischen Monitorings - Methoden, Praxis, angewandte Projekte, Teil A: Methoden. Bundesministerium für Umwelt Jugend und Famlie, editor Östereich.
29. Horowitz W (1970) Official Methods of Analysis of the Association of Official Analytical Chemists. 11th ed. Washington DC: AOAC [Association of Official Analytical Chemists].
30. Bruker Analytik GmbH (1998) Benutzerhandbuch Vector 22/N.
31. Stolter C, Julkunen-Tiitto R, Ganzhorn J (2006) Application of Near Infrared Reflectance Spectroscopy (NIRS) to assess some properties of a sub-arctic ecosystem. Basic Appl Ecol 7: 167–187.
32. Tserendash S, Erdenebaatar B (1993) Performance and management of natural pasture in Mongolia. Nomad People 33: 9–15.
33. Karban R, Baldwin I (1997) Induced Responses to Herbivory. Chicago: The University of Chicago Press.
34. Van der Graaf AJ, Stahl J, Bakker JP (2005) Compensatory growth of Festuca rubra after grazing: can migratory herbivores increase their own harvest during staging? Funct Ecol 19: 961–969. Available: http://doi.wiley.com/10.1111/j.1365-2435.2005.01056.x Accessed 15 May 2014.
35. McIntire EJB, Hik DS (2002) Grazing history versus current grazing: leaf demography and compensatory growth of three alpine plants in response to a native herbivore (Ochotona collaris). J Ecol 90: 348–359. Available: http://doi.wiley.com/10.1046/j.1365-2745.2001.00672.x.
36. McNaughton S (1984) Grazing lawns: Animals in herds, plant form and coevolution. Am Nat 124: 863–886.
37. Han G, Hao X, Zhao M, Wang M, Ellert BH, et al. (2008) Effect of grazing intensity on carbon and nitrogen in soil and vegetation in a meadow steppe in Inner Mongolia. Agric Ecosyst Environ 125: 21–32. Available: http://linkinghub.elsevier.com/retrieve/pii/S0167880907002629. Accessed 28 May 2014.
38. Drent R, van der Wal R (1999) Cyclic grazing in vertebrates and the manipulation of the food resource. In: Olff H, Brown V, Drent R, editors. Herbivores, between plants and predators. Oxford: Blackwell Science. 271–295.
39. Jeffries R (1999) Herbivores, nutrients and trophic cascades in terrestrial environments. In: Olff H, Brown V, Drent R, editors. Herbivores, between plants and predators. Oxford: Blackwell Science. 301–321.
40. Goepfert L (2006) Vegetationsökologische Untersuchungen zum Einfluss von Beweidung in drei unterschiedlichen Steppenökosystemen der Westmongolei. Univ Hamburg, Inst für Geogr.
41. Hendricks H, Novellie P, Bond W, Midgley J (2002) Diet selection of goats in the communally grazed Richtersveld National Park. African J Range Forage Sci 19: 1–11.
42. FAO (1991) Guidelines: land evaluation for extensive grazing. FAO Soils Bull 58.
43. Rhodes B, Sharrow S (1990) Effect of grazing by sheep on the quantitiy and quality of forage available to big game in Oregons Coast Range. J Range Manag 43: 235–237.
44. Bakker E, Olff H, Boekhoff M, Gleichmann J, Berendse F (2004) Impact of herbivores on nitrogen cycling: contrasting effects of small and large species. Oecologia 138: 91–101.
45. Fernandez-Gimenez M, Allen-Diaz B (2001) Vegetation change along gradients from water sources in three grazed Mongolian ecosystems. Plant Ecol 157: 101–118.
46. Wehrden H Von, Wesche K (2007) Relationships between climate, productivity and vegetation in southern Mongolian drylands. Basic Appl Dryland Res 1: 100–120.
47. Institute for Hydrology and Metereology (2004) Climate Change Studies in Mongolia. Ulaanbaatar.

10

Recent Changes in Glacial Area and Volume on Tuanjiefeng Peak Region of Qilian Mountains, China

Junli Xu[1], Shiyin Liu[1], Shiqiang Zhang[1,2]*, Wanqin Guo[1], Jian Wang[1]

1 State Key Laboratory of Cryosphere Science, Cold and Arid Regions Environmental and Engineering Research Institute, Chinese Academy of Sciences, Lanzhou, PR China, **2** The Land in the Earth System, Max Planck Institute for Meteorology, Hamburg, Germany

Abstract

Glaciers' runoff in the Qilian Mountains serves as a critical water resource in the northern sections of the Gansu province, the northeastern sections of the Qinghai province, and the northeastern fringe of the Tibetan Plateau. Changes in the glacial area and volume around the highest peak of the Qilian Mountains, i.e., Tuanjiefeng Peak, were estimated using multi-temporal remote-sensing images and digital elevation models, and all possible sources of uncertainty were considered in detail. The total glacier area decreased by 16.1 ± 6.34 km^2 ($9.9\pm3.9\%$) during 1966 to 2010. The average annual glacier shrinkage was -0.15% a^{-1} from 1966 to 1995, -0.61% a^{-1} from 1995 to 2000, -0.20% a^{-1} from 2000 to 2006, and -0.45% a^{-1} from 2006 to 2010. A comparison of glacier surface elevations using digital elevation models derived from topographic maps in 1966 and from the Shuttle Radar Topography Mission in 1999 suggests that 65% of the grid cells has decreased, thereby indicating that the glacier thickness has declined. The average change in glacier thickness was -7.3 ± 1.5 m (-0.21 ± 0.04 m·a^{-1}) from 1966 to 1999. Glaciers with northeastern aspects thinned by 8.3 ± 1.4 m from 1966 to 1999, i.e., almost twice as much as those with southwestern aspects (4.3 ± 1.3 m). The ice volume decreased by $11.72\pm2.38\times10^8$ m^3 from 1966 to 1999, which was about 17.4% more than the value calculated from the statistical relationship between glacier area and volume. The relationship between glacier area change and elevation zone indicates that glacier change is not only dominated by climate change but also affected by glacier dynamics, which are related to local topography. The varied response of a single glacier to climate change indicates that the glacier area change scheme used in some models must be improved.

Editor: Qinghua Sun, The Ohio State University, United States of America

Funding: This work was supported by the CAS Strategic Priority Research Program-Climate Change: Carbon Budget and Relevant Issues (Grant No. XDA05090302, http://www.cas.cn/), and the Innovation Project of the Chinese Academy of Sciences (Grant No. KZCX2-YW-GJ04, http://www.cas.cn/). The China National Natural Science Foundation (Grants No. 41030638, 41271090, and 41121001, http://www.nsfc.gov.cn/Portal0/default152.htm).The funders had no role in study design, data collection and analysis, decision to publish, or preparation of the manuscript.

Competing Interests: The authors have declared that no competing interests exist.

* E-mail: zhangsq@lzb.ac.cn

Introduction

Glaciers can serve as a sensitive barometer of climate change and play an important role in the climate system by growing or shrinking in response to changes in temperature and snowfall [1]. Glaciers are also important fresh water resources in many arid and semi-arid regions, like western China [2]. In recent decades, changes in the lengths and areas of glaciers have been documented through geodetic surveying, geographic information systems, and remote-sensing data. Most research has indicated that glaciers are rapidly retreating in terms of length and shrinking in terms of area in most glaciated areas of western China [3–6] and other regions of the world [7–9]. Consequently, glacier runoff is expected to have changed as well [10–12].

Changes in ice thickness and volume are important and direct evidence of glacial changes in studies on basin water resources. The traditional methods of measuring changes in ice thickness is to repeatedly measure glacier thickness using ground penetrating radar or repeatedly measure glacier mass balance via snow pits. The impact of regional and global glacier changes has been estimated based on the extrapolation of measurements of mass balance for limited benchmarks glaciers in a particular region [13,14]; however, the number of benchmark glaciers in china is very limited. About 34 glaciers in China have been measured for several months, i.e., up to one or two years. Only the Meikuang Glacier, the Dongkemadi Glacier and Glacier No. 1 at the headwaters of the Urümqi River have been measured consecutively for more than 10 years [15].

Changes in ice thickness also can be derived from difference observed in glacier surface elevation at different time periods in the field survey. This has been accomplished through repeated measuring by real-time kinematic global positioning system surveys in western China [6,16,17], but these surveys are difficult to carry out for a large number of glaciers due to the high costs of the surveys and the hostile environment on the glacier. An alternative method of measuring the changes in glacier surface elevation involves the use of multiple temporal digital elevation models (DEMs). Indeed, DEMs generated from topographical maps and/or aerial photographs have often been used as reference data [18–20]. Recently, the Shuttle Radar Topography Mission (SRTM), and the Geoscience Laser Altimeter System which carried onboard the Ice, Cloud, and Land Elevation Satellite (ICESat), were used in studies on glaciers changes in Antarctica, Greenland, Alaska Svalbard, the Alps, and the central Tien Shan

Mountains [20–25]. Comparisons between DEMs suggest that glacier surface elevation changes varied greatly among the different regions, and more data and results are needed in order to understand glacier response to global warming.

Glaciers runoff in the Qilian Mountains is a critical water resource for northern sections of the Gansu province and northeastern sections of the Qinghai province. Glaciers around Tuanjiefeng (TJF, also known as Kangze'gyai), i.e., the highest peak in the Qilian Mountains, are the main water resource for the rivers in the Qilian mountains. The glaciers in the area of the western Qilian Mountains, including TJF, have receded by about 16.9% from the end of Little Ice Age (LIA, from 1622 to 1740) to 1990 [26]. The glacier area on the Su-lo Mountains, part of western Qilian Mountains, including TJF, reported to decrease from 429.1 km^2 in 1966 to 458.2 km^2 in 1999, and the loss in the ice volume were estimated at 1.4 km^3 from 1966 to 2000 via a comparison of the glacier surface elevation between SRTM and DEM in 1966 [27]. However, this study did not account for the error associated with the shift between DEMs and the different resolution [28,29].

On the other hand, recent research has suggested that glacier shrinkage began accelerating in 2000 in the central Qilian Mountains [17,30]. Thus, this study was designed to investigate the glacier length and area changes in TJF during 1966 to 2010 via topographic maps from 1966 and Landsat Multispectral Scanner (MSS)/Thematic Mapper (TM)/Enhanced Thematic Mapper Plus (ETM+) imagery from 1973, 1995, 1999, 2006, and 2010; moreover, the ice surface elevation and volume changes in the glaciers around TJF from 1966 to 1999 were investigated by comparing SRTM DEM and the DEM from 1966, and the accuracy of the results were considered in detail.

Study Area

The TJF region (36.5°–39°N, 93.5°–103°E) is located on the northeastern fringe of the Tibetan Plateau (Figure 1). The peak of the TJF region is 5826 m above Sea level (ASL), and the average firn line is at about 4862 m ASL during late summer [31]. The average annual precipitation and temperature at 4000 m ASL are about 300 mm and −6.3°C, respectively [31]. According to the China Glacier Inventory [32], 98 glaciers with a total area of 162.7 km^2 can be found in the TJF region. Most of the glaciers in this area (i.e., 90.8%) are smaller than 5 km^2, and the combined area of these smaller glaciers account for 54.2% of the total glacier area. There are only nine glaciers larger than 5 km^2. Glaciers around TJF are separated by a main ridge line from northwest to southeast, thus, all the glaciers have southwestern or northeastern aspects. Note that glaciers with a northeasterly aspect cover an area of 105.57 km^2 (i.e., 64.9% of the total area for all glaciers in the region). All glaciers in the TJF region are free of debris.

Datasets

SRTM

SRTM imagery was used to divide glacier basins into individual glaciers and to study ice surface elevation changes. The SRTM has a specially modified radar system that was used onboard the space shuttle *Endeavour* during an 11-day mission in February of 2000. SRTM elevation data are available between the latitudes of 60°N and 57°S. The geodetic reference for SRTM data is the World Geodetic System 1984 (WGS84), defined as the 1996 Earth Gravitational Model (EGM96) geoid.

The horizontal resolution of the SRTM DEM is 90 m, and the absolute vertical accuracy is better than 9 m [33]. A previous study [34] found that the average elevation difference between the SRTM DEM and the topographic map was about 6 m in the Tien Shan Mountains.

The penetrating depth of the C band radar on snow is highly dependent on the wetness of the snow. Wet snow will show no significant penetration. It increases from 0 to 4 m for Alaskan glaciers and up to 10 m on the dry, cold firn of the Greenland ice sheet [35].

The climate for an extensive region of Asia involves highly concentrated summer precipitation climate [36]. In fact, the total winter precipitation (i.e., for December of 1999 and January and February of 2000) was only 4.5 mm, which accounted for 1.2% of the annual precipitation in 2000 in the TJF region. Therefore, the measurable accumulated snow in February of 2000 was very small. On the other hand, the average air temperature during the Winter of 2000 was about −15.9°C in the TJF region, and the average air temperature over the glacier was lower than −15.9°C due to a lapse rate of about 0.6°C/100 m, which indicates that the small amount of accumulated snow nearly had no chance to melt in the Winter of 2000 in the TJF region. Therefore, the snow should be relatively dry, thereby allowing the C band radar to penetrate into dry snow and perhaps even reach 10 meters of depth. Thus, the elevation of the SRTM data referred to the 1999 glacier surface at the end of the melt season [28,37].

SRTM DEM version filled finished_B was downloaded from the Global Land Cover Facility (http://glcf.umd.edu/data/srtm/). This DEM data was adjusted for voids of high quality and to address minimal gaps in the data. The quality of the SRTM DEM in the TJF region was also checked via visual examination.

We tried to find other stereo image pairs in the TJF region to construct the DEM ourselves or other DEM products after 1999, but no qualified satellite images or products available due to snow cover and cloudy cover.

DEM 1966

Four 1:50,000 topographic maps, which were compiled from 1:40,000 aerial stereo pairs taken in 1966 by the State Bureau of Surveying and Mapping of China (SBSMC), were collected and merged into one DEM called DEM 1966. The geographic projection of the topographic maps was based on the Beijing Geodetic Coordinate System 1954 (BJ54) geoid and the Yellow Sea 1956 datum (i.e., the mean sea level of the Yellow Sea at Qingdao Tidal Observatory in 1956).

The contour lines of digital elevation was obtained from SBSMC and were digitized from topographic maps and re-projected as the Xian Geodetic Coordinate System 1980 (GDZ80) geoid using the Yellow Sea 1985 datum. The errors in the 1966 contours were reported as 3–5 m over flat and hilly areas and 8–14 m over mountains [38]. Note that the digital elevation of the contours on the snow probably contains larger errors than digital elevations of areas on an ice-free mountain due to the lack of visible contrast on the surface of the snow. The data were mainly collected in the late summer or early autumn, i.e., when most of the snow has melted. On the other hand, a large scale field survey was conducted in which part of the survey took place on glacier areas with snow covered surfaces, and some experts provided visual interpretation for the snow covered areas. Because we did not have enough original field survey data in 1966 around the TJF, this error is difficult to measure and was ignored in this study.

The contour lines of DEM 1966 were re-projected to UTM zone 47N using a seven parameter transformation method that will be described subsequently and interpolated to 30 m resolution by employing the Thiessen polygon method. More than 12 triangulation points around the TJF, the GDZ80 geoid and

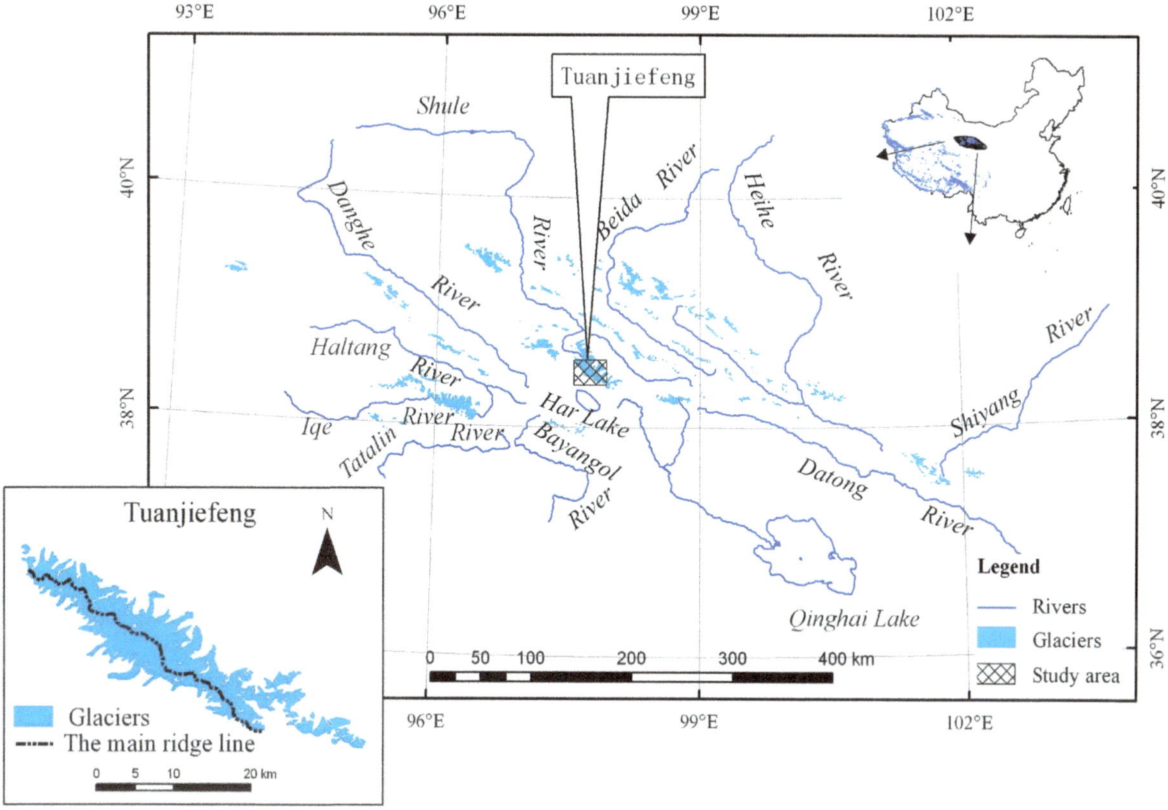

Figure 1. Location of the Tuanjiefeng region in the Qilian Mountains on the northeastern Tibet Plateau, China. The lower left map is the distribution of glaciers in the region.

WGS84 ellipsoid elevations obtained from the National Geomatics Center of China were used for seven parameter transformation.

Landsat satellite imagery

All Landsat images, i.e., the MSS, TM, and ETM+ images, were used in this study (Table 1). Landsat MSS images have a spatial resolution of 57 m and five spectral bands (i.e., two visible bands, two near-infrared bands, and one thermal band). The TM images have a spatial resolution of 30 m, and include seven spectral bands (i.e., blue, green, and red visible bands, one near-infrared band, two short-infrared bands, and one thermal band). The ETM+ images have a panchromatic band with a spatial resolution of 15 m plus the TM bands. Because of the low resolution and snow cover, Landsat MSS data were not used for g the whole area; instead, only the data for four typical glaciers were used. The orthorectified Landsat MSS 1973 and TM 1995 (L1G) were collected from the Global Land Cover Facility (GLCF: http://www.landcover.org). The other orthorectified images (L1T) were obtained from the United States Geological Survey (USGS: http://landsat.usgs.gov). For all of the Landsat images, UTM zone 47N projection was used.

Table 1. Metadata of Landsat satellite images around the Tuanjiefeng region, Qilian Mountain, China.

Sensor	Date (mm-dd-yyyy)	Path/row	Resolution/m
MSS	10-28-1973	145/33	57
TM	8-19-1995	135/33	30
ETM+	16-8-1999	135/33	15
TM	8-1-2006	135/33	30
TM	9-13-2010	135/33	30

Methods

Glacier boundary extraction

Four topographic maps of the TJF scanned in 600 DPI resolution were geo-registered, and the Beijing 1954 GK zone 17N projection was used. The root mean square error (RMSE) of the geo-registration was 18.6 m. Glacier outlines from 1966 were digitized on these maps using the ArcGIS 9.3 software and were transformed to UTM zone 47N projection based on a seven parameter transformation model.

Several glacier boundaries for 1973 were digitized using the Landsat MSS false-color composite 4, 3, 2 (i.e., R, G, B, respectively) imagery. For 1995, 1999, 2006, and 2010, the glacier outlines were obtained via the automatic delineation of a Landsat TM/ETM+ band ratio image of TM3/TM5 with a threshold of 2.1. This approach has been successfully applied in other studies to extract the outlines of many glaciated areas [39]. We use the method [7] to divide glacier boundaries into individual glaciers. All of the boundaries of each glacier were visually checked after extraction.

Uncertainty in extracting glacier boundaries

The best way to assess the accuracy of glacier boundary extraction via Landsat TM/ETM+ images is to compare the extracted boundaries with a higher resolution satellite image [40]. However, obtaining high resolution images during similar time periods as the Landsat TM/ETM+ images was difficult. We attempted to assess the uncertainty associated with the glacier boundaries for each period by looking at the influencing factors (e.g., image resolution), errors in co-registration, and scene quality (e.g., clouds, seasonal snow, and shadow).

Because the accurate detection glacier changes that are smaller than the resolution of the measurement is impossible, we assumed that all of the outermost pixels over the glacier in question were uncertain. Therefore, the number of uncertain grids was assumed to be the number of grids passing through the glacier boundary perimeter. In general, three kinds of glacier boundaries are extracted from raster to vector, i.e., the stair-shape, the straight line, and the tooth-shape. However, most boundaries are stair-shaped, and only a few boundaries are like the latter two. The tooth-shape is usually formed in some sharp terminals of glaciers, and the influence of these sharp terminals on the number of pixels can be omitted. The straight line is found in a few glaciers. Therefore, the number of uncertain grids can be estimated by the following equation:

$$Number = (1/2 * Perimeter_1 + Perimeter_2)/resolution \quad (1)$$

$$\sigma_r = (1/2 * Perimeter_1 + Perimeter_2) * resolution \quad (2)$$

where $Perimeter_1$ is the total length of the stair-shapes boundaries, $Perimeter_2$ is the total length of the straight line boundaries, and the resolution is the image resolution. Thus, the length uncertainty is the one pixel of images, and the area uncertainty σ_r is the total area of the "number" glaciers.

To estimate errors in the co-registration of topographic maps as compared to that of the Landsat images, 20 independent tie points were selected. The co-registration accuracy of between topographic map and MSS 1973, TM 1995, ETM+ 1999, TM 2006 and TM 2010 was 30.6 m, 15.4 m, 11.8 m, 16.9 m, and 14.7 m, respectively. In all cases, the error was less than the resolution of the image (i.e., one pixel), and the error was assosiated with the uncertainty of the outermost glacier pixels.

The influence of cloud and shadow was considered negligible. No seasonal snow was involved except with the Landsat MSS images. The MSS generated a 1% difference in snow cover as compared with the other TM images. This error was added to the overall error.

The total error is calculated using the flowing equation

$$\sigma = \sqrt{\sigma_1^2 + \sigma_2^2}, \quad (3)$$

where σ is the total error, σ_1 is the error from the first image and σ_2 is the error from the second image.

Seven parameter transformation

Shifts and deformations occur in three-dimensional directions between the coordinates of the DEM 1966 and the SRTM DEM due to the different geoid and datum used in each of them. A seven parameter transformation (i.e., so named for the three translations, one scaling operation, and three rotations) called the Helmert transformation was used to transform the Xian 80 data into the WGS84 data in this study. The datum transformation model is written as:

$$\begin{bmatrix} X_{84} \\ Y_{84} \\ Z_{84} \end{bmatrix} = M * \begin{bmatrix} 1 - \omega_Z + \omega_Y \\ \omega_Z + 1 - \omega_X \\ -\omega_Y + \omega_X + 1 \end{bmatrix} * \begin{bmatrix} X_{80} \\ Y_{80} \\ Z_{80} \end{bmatrix} + \begin{bmatrix} X_0 \\ Y_0 \\ Z_0 \end{bmatrix}, \quad (4)$$

where X_{84}, Y_{84} and Z_{84} are the WGS84 datum coordinates, X_{80}, Y_{80} and Z_{80} are the Xian 80 datum coordinates, X_0, Y_0, and Z_0 are the translation parameters, ω_X, ω_Y, and ω_Z are the rotation parameters; and M is the scale factor. These parameters for each map were estimated from three known national trigonometric points. The mean error is 0.5 m in the vertical direction when validated by 8 kinematic Global Positioning System points. The accuracy of this seven parameter transformation has also proved to be less than 0.5 m by other studies [4,41].

DEMs accuracy and change correction

The surface elevation accuracy of SRTM DEM and DEM 1966 for the TJF was tested by comparing the over barren areas with the GLA14 product of the National Aeronautics and Space Administration's (NASA) Ice, Cloud, and Land Elevation Satellite (ICESat) Geoscience Laser Altimeter System (GLAS) (http://nsidc.org/data/icesat/). Four hundred and thirty GLAS footprints over barren areas in 2007 were found in the TJF region. The surface elevation difference over barren areas between the GLAS and DEM 1966 was 2.8 m±9.6 m (mean value ± standard deviation), and −0.67±9.3 m between the GLAS and SRTM. This is less than that found in Tien Shan Mountains [34]. The available number of ICEsat GLAS data limited the spatial correction of the DEM 1966 and SRTM. Thus, the corrections were made on the difference between the DEMs rather than the DEM itself.

Systemic biases can be found among different DEMs [29,42–44]. These biases are a result of the combined effect of a co-registration shift [29], different resolution [45], and the C-band radar penetration of the SRTM signals [46]. A coarse resolution will underestimate elevation in a glacier area [45] (Figure 2a), and the co-registration shift leads to a clear sinusoidal relationship between the elevation difference and the aspect [29,44].

Studies have shown that a sinusoidal relationship between the elevation difference and the aspect occurs across the entire aspect range [29,44]. Thus, biases related to the aspect were corrected for the elevation difference between the SRTM and the DEM 1966. The co-registration shift between the SRTM DEM and the DEM 1966 varies from 10 to 60 m. This was assessed in detail by comparing 19 tie points extracted from the SRTM DEM and DEM 1966. This uneven error distribution probably attributed to the uneven distribution of the triangulation points. The transform points that are closer to the triangulation points will be more accurately transformed [47].

However, an ideal the sinusoidal relationship between the elevation difference and aspect is not ideal for the whole aspect range in the TJF region (Figure 2c). A clear sinusoidal relationship was found between 0 and 124 degrees, and two linear relationships identified between 125 and 181 degrees and between 182 and 360 degrees (Figure 2c). Therefore, three different models were applied to correct aspect bias.

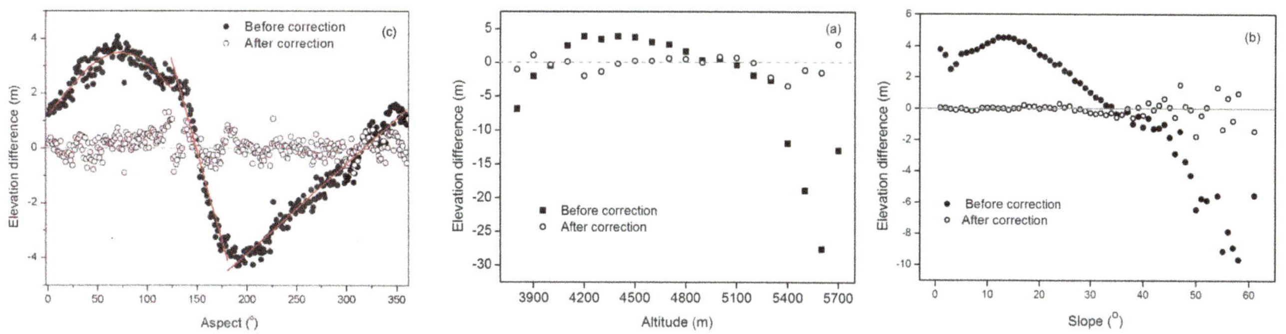

Figure 2. The before and after correction average surface elevation differences between the SRTM and the DEM 1966 over the barren terrain of the Tuanjiefeng region. (a) elevation zone, (b) slope, and (c) aspect.

$$y=\begin{cases} 3.23462\times\sin(\frac{2\pi\phi}{360}-0.15)+0.0126, 0\leq\phi\leq124 \\ -0.1315\times\phi+19.646, 124<\phi\leq181 \\ 0.0332\times\phi-10.518, 181<\phi\leq360 \end{cases} \quad (5)$$

where φ is the aspect in degree and y is the elevation difference.

After the aspect-related bias correction, the bias over the barren areas in the TJF region was -0.67 ± 12.63 m. A nearly linear relationship was still observed between the elevation difference and slope after the aspect-related bias correction. This bias is probably due to different resolutions. It was corrected via the method [45], which used the relationship between the elevation differences and the maximum curvature of each pixel taken from the 30 resolution DEM with a 5×5 kernel size established on ice-free terrain. After the aspect-related and slope-related bias correction, the bias over the barren areas in the TJF region decreased to 0.06 ± 8.73 m.

The relationship between the elevation difference between the SRTM and the DEM 1966 and the altitude (Figure 2a), slope (Figure 2b), and aspect (Figure 2c) over barren areas before and after the aspect-related and slope-related bias correction were compared. The relationship between the surface elevation difference and the altitude zones (Figure 2a) suggested that after the correction elevation difference was not altitude dependent. The relationship between the surface elevation and slope (Figure 2b) and the aspect (Figure 2c) before and after correction suggested that the correction significantly improved the accuracy of elevation difference calculation. The average surface elevation difference in glacier area is 8.34 m and 7.27 m before and after the correction, which means that a 12.8% error was corrected. Therefore, the ice surface elevation difference and ice volume change reported [27] were probably overestimated.

Elevation change uncertainty

In general, the best way to estimate the absolute errors in elevation data is to use ground control points (GCPs) from a more accurate elevation data set [48,49]. Due to the fact that no higher resolution elevation datasets were available, we used relative error estimated from residuals of the difference between the DEM 1966 and the SRTM DEM over stable barren terrain. Standard principles of error propagation were used for estimating the errors between the two DEMs [50]. Following the standard principles of error propagation, the mean standard error of the linear model is calculated using the following equation:

$$\bar{\sigma}=\sqrt{\sum_{i=1}^{n}k_i^2\sigma_i^2} \quad (6)$$

where $\bar{\sigma}$ is the mean standard error, k_i is the linear model slope, and σ_i is the standard error of the regression model. Here, the parameter k_i for 3 linear models is -0.1315, 0.0332, and 3.6036, and σ_i for the three linear models is 0.47, 0.32, and 0.11. Then:

$$\sigma=\sqrt{\bar{\sigma}^2+3.23462^2\times\cos(\varphi)^2\times\sigma_S^2}, \quad (7)$$

where $\bar{\sigma}$ is the value obtained from Equation 6, 3.23462 is the regression parameter mentioned in Equation 6, and σ_S is the standard error of the sinusoidal model (i.e., 0.62). The minimum error was 0.59 m, and the maximum error was 2.09 m. When analyzing and quantifying glacier elevation changes, the spread of elevation changes should be considered, and the mean of the elevation changes over a particular area (e.g., a glacier or glacial zone) should also be considered. The standard error equation for the mean is calculated using the following equation:

$$\varepsilon=\sqrt{\sum_{i=1}^{n}\sigma_i^2\Big/n}, \quad (8)$$

where ε is the mean error of the measurement area, n is the total number of measurements, and σ_i is standard error of the ith pixel obtained from Equation 8.

Ice volume change and uncertainty

Ice volume change was calculated based on the glacier surface elevation change using the following three methods: 1) The arithmetic mean of the glacier elevation differences multiplied by the total glacier area, 2) An area-weighted average per 100 m of elevation zone, and 3) An area-weighted average for each glacier.

The change in ice volume was determined by changes in the glacier area and the ice surface elevation change as expressed in equation 8. The error can be estimated using Equation 9 according to the standard principles of error propagation. More specifically, the change in ice volume is expressed as follows:

$$\Delta V=S\times\Delta Z, \quad (9)$$

where S is the total glacier area obtained from measurements, and

ΔZ is the elevation change. The error is estimated using the following equation:

$$\sigma_v = \sqrt{(S \times \sigma_z)^2 + (\Delta Z \times \sigma_s)^2}, \quad (10)$$

where σ_v is the error for the ice volume change, σ_s is the error for area, and σ_z is the error for the elevation change.

Results

Changes in glacier area in the TJF region

The total glacier area in the TJF region was 162.70±3.07 km^2 in 1966; interestingly, it decreased to 146.60±5.55 km^2 by 2010. The overall decrease in the glacier area was 16.1±6.34 km^2 (9.9±3.9%) from 1966 to 2010, accounting for 0.37 km^2·a^{-1} (Table 2). The average annual glacier shrinkage was 0.15% a^{-1} from 1966 to 1995, 0.61% a^{-1} from 1995 to 2000, 0.20% a^{-1} from 2000 to 2006, and 0.45% a^{-1} from 2006 to 2010.

The relationship between glaciers' annual area change and elevation zone (Figure 3a) suggested that glacier's area change is much larger when the glacier is located below 4800 m ASL than above this line, and glacier's annual area change is the most pronounced around 4500 m ASL. In fact, the area of all glaciers above 4500 m ASL elevation shrunk during all study periods.

Also note that the annual area change of glacier is typically not consistent when comparing the periods 1995–1999, 1999–2006, and 2006–2010. For example, the annual area change of glaciers is about 11% a^{-1} in 4500 m elevation zone from 1999 to 2006, but it is about 4.8% a^{-1} during the period from 1995 to 1999 at the 4400 m elevation zone; this suggests that the glacier area even expanded during some periods. These results indicate that the relationship between the expansion and contraction of the glacier area and the elevation zone is complex.

The relationship between the glacier area and the elevation zone (Figure 3b) suggested that the maximum area change occurred around 5000 m ASL during 1966 to 2010, and the area of the glaciers located above 5500 m ASL are stable. The maximum absolute glacier area change occurred at 4800 m ASL from 1966 to 1995, and shifted to 5100 m ASL from 2006 to 2010. Note that the glacier area between 4400 m ASL and 4600 m ASL, i.e., near the glacier terminus, has been nearly a constant (i.e., above zero) since 1995.

Changes in glacier area and length of larger glaciers

Four larger glaciers (i.e., greater than 5 km^2), including 5Y445G0020, 5Y445G0026, 5Y445G0028, and 5Y445G0031, were selected for analysis of the changes in glacier area (Table 3) and length (Table 4) during the periods from 1966 to 1973, 1973 to 1995, 1995 to 1999, 1999 to 2006, and 2006 to 2010.

Table 2. Changes in the glacier areas from 1966 to 2010 in the Tuanjiefeng region, Qilian Mountain, China.

Year	Glacier area (km^2)	Change in percentage from 1966 (%)	
		Total	Annual
2010	146.60±5.55	−9.9±3.9	−0.45
2006	149.31±5.51	−8.2±3.9	−0.20
2000	151.11±2.83	−7.1±2.6	−0.61
1995	155.84±5.68	−4.2±4.0	−0.15
1966	162.70±3.07	-	-

The glacier area of 5Y445G0028 receded 1.36±0.30 km^2 (i.e., −16.43±5.31%) from 1966 to 2010, which was greater than the average of the entire TJF region (i.e., −9.9%), and the recession rates have been greater than 0.40% a^{-1} from 1973 to 1995 and 1995 to 2000, (Table 3). The largest recession rate was 1.83% a^{-1} for 5Y445G0028 from 2006 to 2010. The largest glacier area expansion speed is 0.67% a^{-1} for 5Y445G0020 from 2000 to 2006.

Great differences in the glacier area change were observed during the same time periods. For example, the glacier area changes varied from −0.64% a^{-1} for 5Y445G0026 to 0.67% a^{-1} for 5Y445G0026 during 2000 to 2006 (Table 3), and the average for the whole TJF region was −0.20% a^{-1} (Table 2). This indicates that the change in glacier area is not only affected by climate change but also other factors, such as glacier size, slope, aspect, etc. It also indicates that great uncertainty probably exists when presenting some "benchmark" glacier area changes as representative for regional glacier area changes.

Glacier length fluctuations are not always in accordance with that of glacier area. For 5Y445G0020, the glacier area contracted or expanded at the rate of −0.09, −0.23, −0.46, 0.67, and −0.58% a^{-1} while the glacier length retreated or advanced at the rate of −0.79, −0.27,−0.33, 1.88, and 0% a^{-1} during the periods of 1966 to 1973, 1973 to 1995, 1995 to 2000, 2000 to 2006, and 2006 to 2010, respectively. However, for 5Y445G0026, the glacier area contracted by 0.35% a^{-1} during 1995 to 2000, but the glacier length advanced by 0.16% a^{-1} during the same time period (Table 4).

Great variations in glacier length changes were also observed within the same period. For example, the glacier length change was −0.69% a^{-1} for 5Y445G0028, and 1.88% a^{-1} for 5Y445G0020 during the period from 2000 to 2006 (Table 4). This evidence also indicates that glacier length change is not only affected by climate change but also other factors.

Changes in glacier elevation from 1966 to 1999

The mean glacier surface elevation change for the entire TJF region was −7.3±1.5 m from 1966 to 1999 (Figure 4), thereby indicating an annual change of 0.21±0.04 m·a^{-1}. The surface elevation for about 65% of grid area decreased, i.e., mostly in the glacier ablation zone and in areas of accumulation on mountain crests. We also observed extreme thinning (i.e., locally exceeding 90 m) at the low-lying terminus of glacier 5Y445G0028, i.e., the lowest point in the TJF site, which is believed to be to the result of melting of ice. Extreme thinning was also observed on the steep ridges of glacier 5Y571E0007 (Figure 4), which was probably caused by an avalanche. Glacier thickening occurred mostly in firn areas.

Note that some measurable mass displacements were observed down the glacier, possibly as a result of sliding, avalanches, or other forces, all of which can change the ice thickness distribution but not the ice volume loss or gain. For example, the terminus of glacier 5Y445G0026 had advanced 14.02% (Table 4), while the glacier area of glacier 5Y445G0026 contracted by 1.04% during the period from 1966 to 2000 (Table 3), and the average ice thickness change −7.9±1.5 m from 1966 to 1999.

Glaciers in the study area are separated into northeasterly and southwesterly aspects by the main ridgeline (Figure 1), and we found a regional difference in the glacier surface elevation changes. Glaciers with a northeastern aspect thinned by 8.3±1.4 m, i.e., almost twice as much as those with a southwestern aspect, which thinned by 4.3±1.3 m. Interestingly, 76% of the

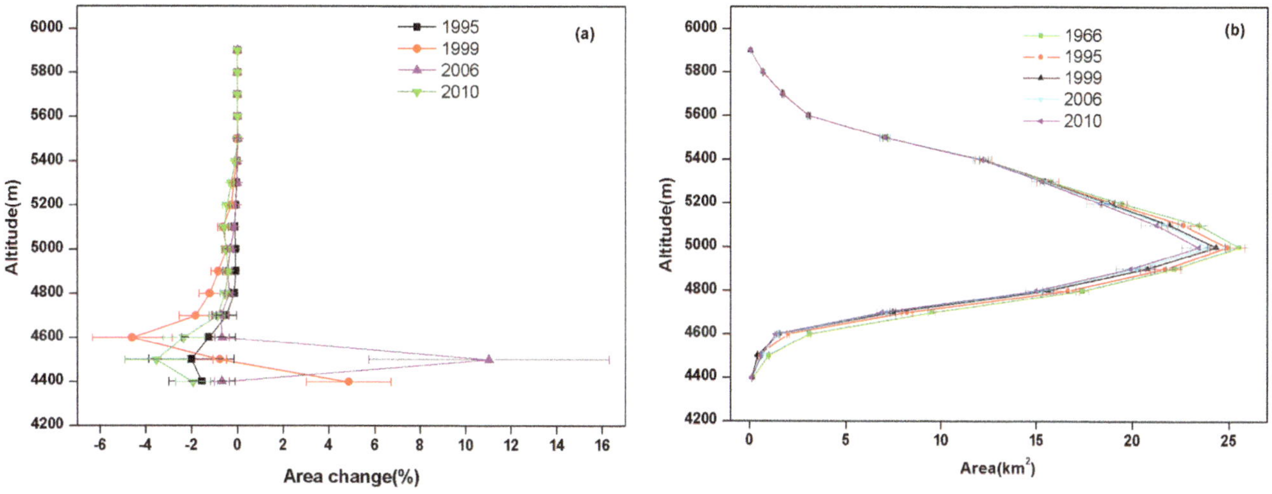

Figure 3. The average annual changes in glacier area relative to the area measured in 1966 along the altitude zone over the time periods of 1966–1995, 1995–1999, 1999–2006, and 2006–2010 in the Tuanjiefeng region. (a) is relative change(%), and (b) is absolute change (km^2).

Table 3. Changes in areas of four typical glaciers (km^2) around Tuanjiefeng region in 1966, 1973, 1995, 2000, 2006, and 2010, respectively.

Glacier ID	Year	Area(km^2)	Change (%) relative to 1966	
			Total	Annual
5Y445G0020	2010	15.26±0.58	−6.27±4.04	−0.58
	2006	15.62±0.58	−4.05±4.04	0.67
	2000	15.02±0.28	−7.74±2.57	−0.46
	1995	15.37±0.56	−5.59±3.93	−0.23
	1973	16.18±1.31	−0.61±8.27	−0.09
	1966	16.28±0.31		
5Y445G0026	2010	5.07±0.19	−11.83±3.82	−1.83
	2006	5.47±0.20	−4.87±3.97	−0.64
	2000	5.69±0.11	−1.04±2.7	−0.35
	1995	5.79±0.21	0.7±4.12	−0.05
	1973	5.86±0.47	1.91±8.39	0.27
	1966	5.75±0.11		
5Y445G0028	2010	6.92±0.26	−16.43±5.31	−0.21
	2006	6.98±0.26	−15.7±5.31	−0.28
	2000	7.1±0.13	−14.25±3.59	−0.49
	1995	7.28±0.27	−12.08±5.49	−0.46
	1973	8.10±0.65	−2.17±11.64	−0.31
	1966	8.28±0.16		
5Y445G0031	2010	6.41±0.24	−5.48±4.75	0.08
	2006	6.39±0.12	−5.75±3.08	−0.33
	2000	6.52±0.24	−3.83±4.75	−0.36
	1995	6.64±0.24	−2.06±4.75	−0.09
	1973	6.78±0.55	0	0
	1966	6.78±0.13		

Table 4. Changes in lengths of four typical glaciers (m) around Tuanjiefeng region in 1966, 1973, 1995, 2000, 2006, and 2010, respectively.

Glacier ID	Year	Length (km)	Change (%) relative to 1966	
			Total	Annual
5Y445G0020	2010	6724.3±30	−2.82±0.48	0
	2006	6724.3±30	−2.82±0.48	1.88
	2000	6042.1±15	−12.68±0.31	−0.33
	1995	6143.6±30	−11.21±0.48	−0.27
	1973	6538.8±57	−5.5±0.85	−0.79
	1966	6919.2±15		
5Y445G0026	2010	5678.9±30	13.54±0.48	0.003
	2006	5678.1±30	13.52±0.48	−0.07
	2000	5703.1±15	14.02±0.31	0.16
	1995	5657.0±30	13.1±0.48	0.42
	1973	5177.9±57	3.52±0.85	0.5
	1966	5001.7±15		
5Y445G0028	2010	5011.5±30	−32.51±0.48	−0.97
	2006	5214.6±30	−29.78±0.48	−0.69
	2000	5441.0±15	−26.73±0.31	−0.96
	1995	5716.2±30	−23.02±0.48	−0.92
	1973	7176.1±57	−3.36±0.85	−0.48
	1966	7425.6±15		
5Y445G0031	2010	3858.1±30	−16.92±0.48	0.48
	2006	3785±30	−18.49±0.48	0
	2000	3785±15	−18.49±0.31	−0.78
	1995	3939.7±30	−15.16±0.85	−0.69
	1973	4643.6±57	0	0
	1966	4643.6±15		

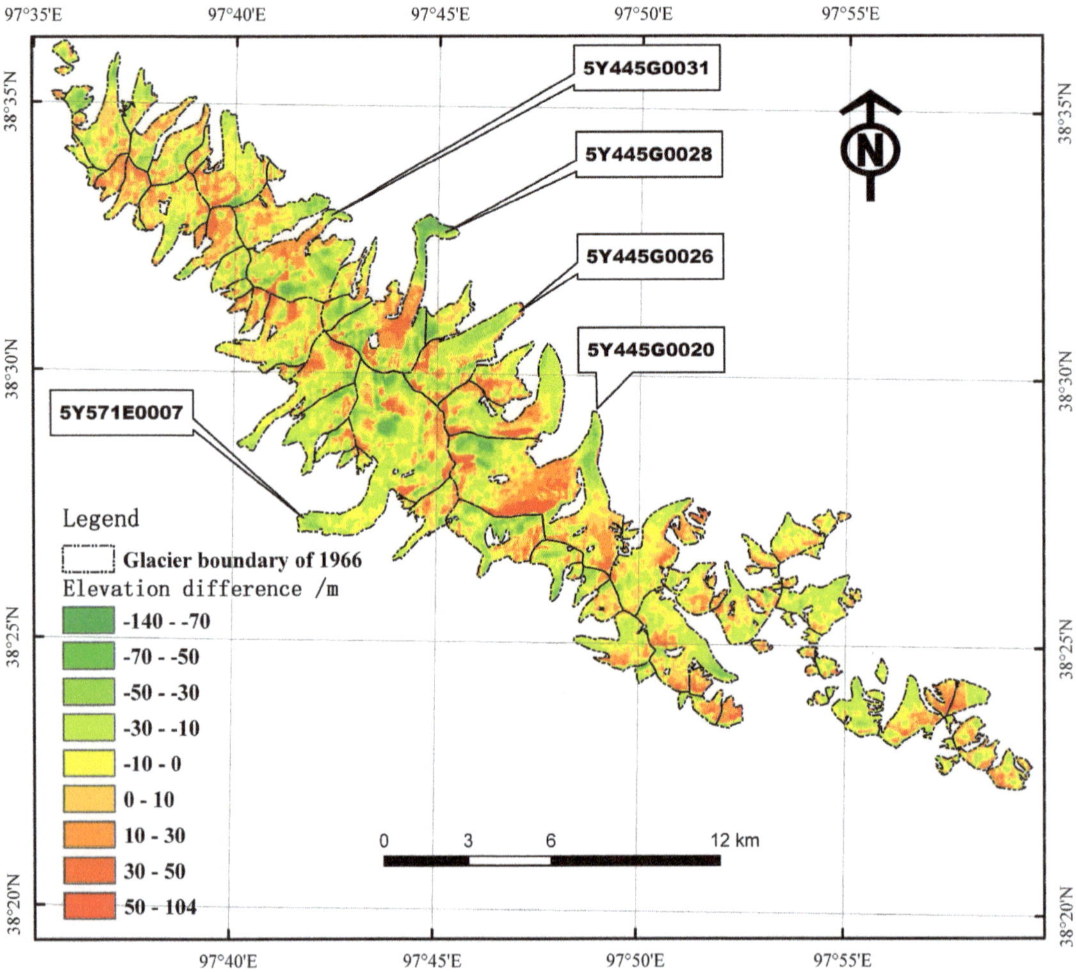

Figure 4. Spatial distribution of the difference in glacier surface elevation between the corrected SRTM data and the DEM 1966 of the Tuanjiefeng region.

glaciers with ice thickening have a northeasterly aspect. However, the area change for glaciers with a northeasterly aspects is only a little larger than that of the glaciers with the southwesterly aspects (i.e., −7.5% and −6.4%, respectively). This agrees with results from previous research [51], which reflect the climatic conditions in the Qilian Mountains where the annual precipitation and temperature decreases from northeast towards the southwest.

Ice volume changes from 1966 to 1999

As determined by method 1 for the measurement of ice volume changes as described in the methods section, the change in ice volume for the entire study area was about $-11.88 \pm 2.45 \times 10^8$ m^3. For method 2 and method 3, the calculated change was about $-11.71 \pm 2.37 \times 10^8$ and $-11.56 \pm 2.32 \times 10^8$ m^3, respectively. The difference between the results of the three methods was less than 1.5%. A previous study [23] found the difference between an area-weighted average per 100 m of elevation zone and the arithmetic mean of ice thickness was only 1.7%. The results from method 2 showed that the maximum change occurred at 5200 to 5300 m ASL, which include the mass migration area shown in Figure 4.

Mass loss still occurs above 5500 m ASL as a result of glacier surface elevation changes without area change. This change is represents real change and is easily underestimated by SRTM DEM due to the low resolution with these images. Similar findings were described by previous study [52], and other explanations, including heterogeneous radar penetration and instrumental biases, have been suggested [42].

The relationship between ice volumes changes for each glacier and the glacier area (Figure 5a) suggests that the decrease in ice volume is nearly linear with glacier area, and the negative correlation is significant (i.e.,$\alpha = 0.01$). Therefore, large glaciers experienced the greatest mass loss with no dependence on geographic location assuming that they are located in the same climate zone. The same phenomenon has been also reported by Paul and Haeberli [19] for glaciers in Swiss Alps.

A Previous study [26] found a strong statistical relationship between the area and volume of a single glacier in western China, which is expressed as follows:

$$V = 0.04 \times S^{1.35}, \tag{11}$$

where S is area (km^2) and V is volume (km^3). We used glacier area in 1966 and 1999 to calculate ice volume change during the period from 1966 to 1999. The relationship between ice volume change and glacier area (Figure 5b) suggests that the decrease in ice volume change is nearly linear with glacier area, and the relationship is significant (i.e.,$\alpha = 0.01$). The slope in Figure 5b is

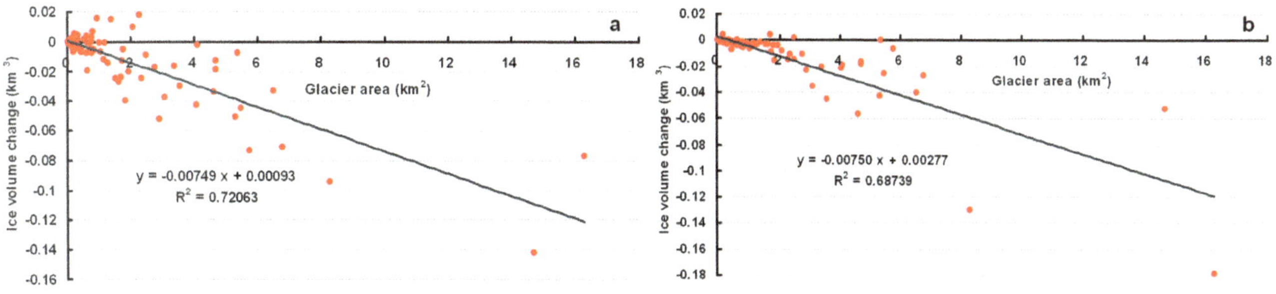

Figure 5. Relationship between changes in ice volume for each glacier and glacier size. (a) ice volume calculated from the glacier surface elevation change between the corrected SRTM and DEM 1966, and (b) ice volume derived from the statistical relationship between the glacier area and ice volume in the Tuanjiefeng region.

nearly the same with that in Figure 5a (i.e., −0.00750 and −0.00749, respectively), thereby indicating that the ice volume change difference between estimated by the two methods is not strongly correlated with glacier area. Thus, the error for the total ice volume as estimated by the glacier area-volume relationship can be approximated as the differences between the two interceptions (0.00184 km^3). The total ice volume change was $-9.53\pm2.54\times10^8$ m^3 in the TJF region during the period from 1966 to 1999, which is 17.4±4.6% less than the ice volume change estimated by the glacier surface elevation change. This indicates that the ice volume probably underestimated when using Equation 11.

Discussion

Glacier advances or surges

Glacier surge are short-lived events where a glacier can advance substantially, moving at velocities up to 100 times faster than normal. In some glaciers, surges can occur in fairly regular cycles and may include 15 to 100 or more surge events per year. In other glaciers, surging is unpredictable [53]. A previous study found that mass displacement to down-glacier is an important signal that occurs before a glacier surge [54]. Such mass displacements were found in three glaciers, i.e., 5Y445G0020, 5Y445G0028, and 5Y445G0031 during the period from 1966 to 1999. 5Y445G0020 advanced from 1999 to 2006, and 5Y445G0031 and 5Y445G0020 advanced from 2006 to 2010, thereby indicating that the prediction of the actual surge time is still difficult to even mass displacement downward was found.

This kind of mass displacement was also found in other Asian glaciers, and these observations were more frequently noted after 1999 than before [28,37]. They were all assumed to potentially surging glaciers. Surging glaciers also were also found in the Tibet plateau based on glacier surface velocities measurement [55]. Sudden glacier surges are difficult to predict via direct monitor glacier area/length changes.

Most surge glaciers, including our study glaciers, retreated due to the climate warming during the period from 1980 to 2000 before they actually surged or advanced [6]. A previous study [28] suggested that surging glaciers in Karakoram showed slight mass gain. Another previous study [37] assumed surging glaciers correspond with mass gain, but this cannot be confirmed due to a lack of upper part elevation change data. Furthermore, we must consider whether the climate warming from 1980 to 2000 accelerated glacier advanced. Among the four typical glaciers in the TJF region, one glacier advanced from 1999 to 2006, whereas two glaciers advanced from 2006 to 2010. More glaciers seemingly exhibited advances after the increased climate warming than before. However, the number of studied glaciers is rather small, thereby preventing any type of generalizability of the results. However, this phenomenon was also found in Tien Shan [37], but the mechanism for this phenomenon is still unclear.

Glaciers response to climate change

Climate change, especially temperature change, is the fundamental factor underlying changes observed in glacier [56,57]. Thus, meteorological data obtained from the Tuole weather station (38.80°N, 98.42°E; 3367 m ASL), which is located about 6.7 km northeast of the TJF, were used to analyze the driver of glacier change in the TJF region. The Tuole station reported a clear increasing trend by 0.023°C a^{-1} in the average summer temperature from 1957 to 2003, whereas the annual precipitation varied significantly and showed a slightly increasing trend of 0.87 mm·a^{-1} during the same period [27]. The increase in the summer average temperature will lead to an increase in the annual positive degree-day. The increase in the annual positive degree-day when the summer air temperature will result in more glacier ablation and runoff and greater glacial mass balance sensitivity to temperature [58].

A previous study suggested that the climate of northwest China had shifted from warm and dry to warm and wet since 1987 [59]. Detection of transition points in annual air temperature at the Tuole station via the Mann-Kendall method indicated that the tipping point occurred in 1986 [60]. Thus, the time series of annual precipitation, average air temperature, positive cumulative temperature, and positive days during which daily air temperature was equal to or more than 0°C during 1957 to 2011 were divided into two periods (i.e., before and after 1987) for analysis (Figure 6).

A slight increase in precipitation of 2.61 mm·a^{-1} (i.e., $p<0.05$) from 1957 to 1986 was observed, while the annual average temperature was showed no significant trend (Figure 6a); moreover, the positive cumulative temperature and positive days also showed no significant trend (Figure 6b). These conditions would have contributed to higher accumulation and the corresponding positive glacier mass balance, which may perhaps lead to some glacial advance. For example, the annual glacier area was only −0.15% a^{-1} during the period from 1966 to 1995 (Figure 6c), but the glacier 5Y445G0026 advanced during this period (Figure 6d).

Precipitation was nearly stable, and a rapid significant (i.e., $p<0.001$) increase in annual average temperature by 0.048°C a^{-1} (Figure 6a) was observed during 1987 to 2011; moreover, a significant increase in positive days by 0.59 day a^{-1} and a significant (i.e. $p<0.001$) increase in positive cumulative temperature by 11.17°C a^{-1} (Figure 6b) was observed during the same period. These conditions would contribute to intense ablation and corresponding negative glacier mass balance, which would lead to

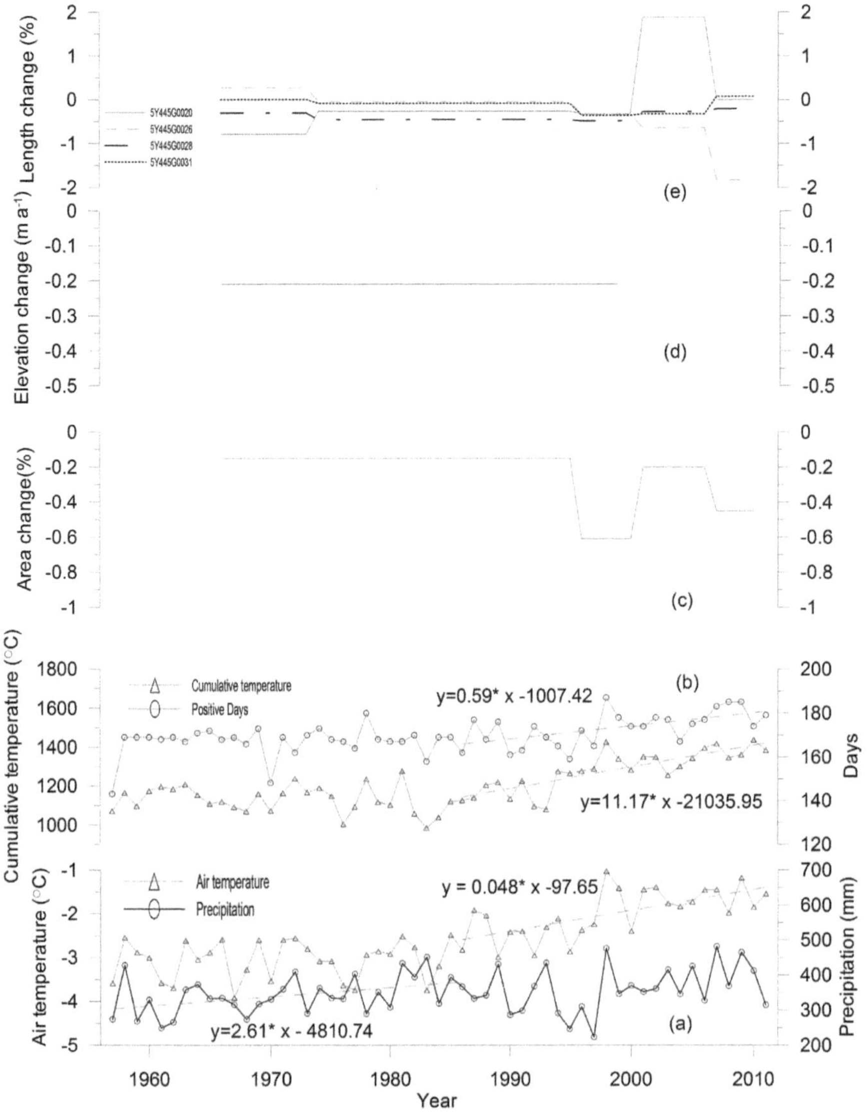

Figure 6. Comparison of changes in precipitation and temperature and glacier surface elevation and length. (a) annual precipitation (mm) and annual temperature (°C), (b) positive cumulative temperature (°C) and positive days at the Tuole meteorological station, (c) average annual glacier area change (% a^{-1}), (d) average annual glacier surface elevation (m a^{-1}), and (e) average annual glacier length of typical larger glaciers (i.e., 5Y44G0020, 5Y44G0026, 5Y44G0028, and 5Y44G0031) from 1957 to 2011 in the Tuanjiefeng region. All the linear fits are at significant (i.e., $p<0.05$).

accelerated glacier area shrinkage and glacier thinning. The annual glacier area change reached −0.61% a^{-1} during 1995 to 2000 (Figure 6c).

On the other hand, the responses by individual glaciers to climate change varied greatly, which is probably due to different response times. For example, the area of glacier 5Y445G0020 expanded by 0.67% a^{-1} from 2000 to 2006, while the area of glaciers 5Y445G0026, 5Y445G0028, and 5Y445G0031 contracted by 0.64, 0.28, and 0.33% a^{-1}, respectively (Table 3). The length of glacier 5Y445G0020 advanced by 1.88% a^{-1} from 2000 to 2006, while the length of glaciers 5Y445G0026, 5Y445G0028, and 5Y445G0031 retreated by 0.07, 0.69, and 0.00% a^{-1}, respectively (Figure 6e). This indicates that great uncertainties are likely still present when using the observed mass balance of a typical glacier to represent regional glacial changes.

Glacier response time and implications to glacier area change simulation

A previous study [61] suggested that glacier surface elevation change can depend not only on current but also multi-decadal average accumulation rate. Another study [58] implied that the area change of a glacier probably has a time lag (i.e., about 10 years) when compared to the mass balance. The response time depends on the glacier size and ice thickness and the slope of the glacier surface [62]. The varied glacier response time leads to more uncertainty when modeling the glacier area changes.

Recently, some models [12,58,63,64] have developed schemes for simulating glacial area changes using annual mass balance; but they considered all single glaciers in one grid box or sub-basin as one single glacier, and they assumed that the expansion or recession of a glacier occurred at the lowest glacier elevation zone. However, this is not always the case when we consider all glaciers as one glacier due to glacier dynamics. For example, in our study, the most pronounced glacier area contraction occurred around 5000 m ASL from 1966 to 2010 (Figure 3b), while the glacier terminus elevation ranged from 4400 to 4600 m ASL in the TJF region. Intense ice thinning was also observed around 5000 m ASL. These results indicate that the glacier area change and ice thickness change was not only caused by climate change, but also glacier dynamics, which are affected by local factors (e.g., slope, aspects, and glacier size). Glacier area change schemes should consider different responses to climate change by various individual glaciers in each grid box or sub-basin, especially when the resolution of the model is relative large or includes many individual glaciers.

Conclusions

Topographic maps, multi-temporal remote-sensing images, and SRTM DEM were used to monitor changes in glacier area and ice surface elevation/volume from 1966 to 2010 in the TJF region. The following conclusions can be reached:

(1) The overall change in the glacier area in the TJF region is -16.1 ± 6.34 km^2 ($-9.9\pm3.9\%$) from 1966 to 2010, which translates to an average annual change of -0.37 km^2·a^{-1}. The average annual glacier change was -0.15% a^{-1} from 1966 to 1995, -0.61% a^{-1} from 1995 to 2000, -0.20% a^{-1} from 2000 to 2006, and -0.45% a^{-1} from 2006 to 2010. The annual area change for the studied glaciers was the most pronounced around 4500 m ASL. Glacier length and glacier area change varied greatly among different individual glaciers.

(2) The mean glacier surface elevation change for the entire TJF region was -7.3 ± 1.5 m from 1966 to 1999, which suggests an annual change of -0.21 ± 0.04 m·a^{-1}. The surface elevation for about 65% of the grid area decreased, mostly in the glacier ablation zone and in areas of accumulation on mountain crests.

(3) Glaciers with a northeastern aspect thinned almost twice as much as those with a southwestern aspect. However, the area change for glaciers with northeasterly aspects was only a little larger than that of glaciers with southwesterly aspects.

(4) The change in ice volume as estimated from ice surface elevation changes for the TJF region was about $-11.88\pm2.45\times10^8$ m^3, which is about 17.4% larger than that estimated by ice volume-area relationship.

Acknowledgments

We thank our reviewers for their valuable comments and suggestions, which significantly improved this manuscript. Our appreciation also is extended to NASA for the provision of the GLA14 and SRTM data.

Author Contributions

Conceived and designed the experiments: JLX SYL SQZ. Performed the experiments: JLX. Analyzed the data: JLX SQZ. Contributed reagents/materials/analysis tools: WQG JW. Wrote the paper: JLX SQZ.

References

1. Benn ID, Evans JAD (1998) Glaciers and glaciation. London: Hodder Arnold Publication. 4–93 p.
2. Yao T, Wang Y, Liu S, Pu J, Shen Y, et al. (2004) Recent glacial retreat in High Asia in China and its impact on water resource in Northwest China. Science in China Series D: Earth Sciences 47: 1065–1075.
3. Du W, Qing X, Liu Y, Wang X (2008) Variation of the Laohugou Glacier No. 12 in the Qilian Mountains. Journal of Glaciology and Geocryology 30: 373–379.
4. Liu GK, Zhao YQ, Wang ZL, Du MW (2006) The conversion method and accuracy analysis between Xi'an 1980 and WGS-84 coordinate system. Geomatics & spatial information technology 29: 40–41.
5. Liu Q, Liu SY, Zhang Y, Wang X, Zhang YS, et al. (2010) Recent shrinkage and hydrological response of Hailuogou glacier, a monsoon temperate glacier on the east slope of Mount Gongga, China. Journal of Glaciology 56: 215–224.
6. Shangguan D, Liu SY, Ding YJ, Ding LF, Xu JL, et al. (2009) Glacier changes during the last forty years in the Tarim Interior River basin, northwest China. Progress in Natural Science 19: 727–732.
7. Bolch T, Menounos B, Wheate R (2010) Landsat-based inventory of glaciers in western Canada, 1985–2005. Remote Sensing of Environment 114: 127–137.
8. Narama C, Kaab A, Duishonakunov M, Abdrakhmatov K (2010) Spatial variability of recent glacier area changes in the Tien Shan Mountains, Central Asia, using Corona (similar to 1970), Landsat (similar to 2000), and ALOS (similar to 2007) satellite data. Global and Planetary Change 71: 42–54.
9. Silverio W, Jaquet JM (2005) Glacial cover mapping (1987–1996) of the Cordillera Blanca (Peru) using satellite imagery. Remote Sensing of Environment 95: 342–350.
10. Casassa G, López P, Pouyaud B, Escobar F (2009) Detection of changes in glacial run-off in alpine basins: examples from North America, the Alps, central Asia and the Andes. Hydrological Processes 23: 31–41.
11. Liu S, Zhang Y, Zhang Y, Ding Y (2009) Estimation of glacier runoff and future trends in the Yangtze River source region, China. Journal of Glaciology 55: 353–361.
12. Zhang S, Ye B, Liu S, Zhang X, Hagemann S (2012) A modified monthly degree-day model for evaluating glacier runoff changes in China. Part I: model development. Hydrological Processes 26: 1686–1696.
13. Arendt A, Echelmeyer K, Harrison W, Lingle C, Zirnheld S, et al. (2006) Updated estimates of glacier volume changes in the western Chugach Mountains, Alaska, and a comparison of regional extrapolation methods. Journal of Geophysical Research-Earth Surface 111.
14. Haeberli W, Hoelzle M, Paul F, Zemp M (2007) Integrated monitoring of mountain glaciers as key indicators of global climate change: the European Alps. Annals of Glaciology, Vol 46, 2007 46: 150–160.
15. Liu SY, Xie ZC, Liu CH (2000) Mass balance and fluctuations of glaciers. In: Shi YF, editor. Glaciers and their environments in China—the present, past and future. Beijing: Science Press. pp. 101–131.
16. Ma LL, Tian LD, Pu JC, Wang PL (2010) Recent area and ice volume change of Kangwure Glacier in the middle of Himalayas. Chinese Science Bulletin 55: 2088–2096.
17. Shangguan DH, Liu SY, Ding YJ, Zhang YS, Li XY, et al. (2010) Changes in the elevation and extent of two glaciers along the Yanglonghe river, Qilian Shan, China. Journal of Glaciology 56: 309–317.
18. Miller PE, Kunz M, Mills JP, King MA, Murray T, et al. (2009) Assessment of Glacier Volume Change Using ASTER-Based Surface Matching of Historical Photography. Ieee Transactions on Geoscience and Remote Sensing 47: 1971–1979.
19. Paul F, Haeberli W (2008) Spatial variability of glacier elevation changes in the Swiss Alps obtained from two digital elevation models. Geophysical Research Letters 35: L21502.
20. Schenk T, Csatho B, van der Veen CJ, Brecher H, Ahn Y, et al. (2005) Registering imagery to ICESat data for measuring elevation changes on Byrd Glacier, Antarctica. Geophysical Research Letters 32: L23S05.
21. Farinotti D, Huss M, Bauder A, Funk M (2009) An estimate of the glacier ice volume in the Swiss Alps. Global and Planetary Change 68: 225–231.
22. Howat IM, Smith BE, Joughin I, Scambos TA (2008) Rates of southeast Greenland ice volume loss from combined ICESat and ASTER observations. Geophysical Research Letters 35: L17505.
23. Kääb A (2008) Glacier Volume Changes Using ASTER Satellite Stereo and ICESat GLAS Laser Altimetry. A Test Study on Edgeoya, Eastern Svalbard. Ieee Transactions on Geoscience and Remote Sensing 46: 2823–2830.
24. Sauber J, Molnia B, Carabajal C, Luthcke S, Muskett R (2005) Ice elevations and surface change on the Malaspina Glacier, Alaska. Geophysical Research Letters 32: L23S01.
25. Surazakov AB, Aizen VB (2006) Estimating volume change of mountain glaciers using SRTM and map-based topographic data. Ieee Transactions on Geoscience and Remote Sensing 44: 2991–2995.
26. Liu SY, Sun WX, Shen YP, Li G (2003) Glacier changes since the Little Ice Age maximum in the western Qilian Shan, northwest China, and consequences of glacier runoff for water supply. Journal of Glaciology 49: 117–124.
27. Wang Y, Hou S, Hong S, Hur S, Liu Y (2008) Glacier extent and volume change (1966~2000) on the Su-lo Mountain in northeastern Tibetan Plateau, China. Journal of Mountain Science 5: 299–309.
28. Gardelle J, Berthier E, Arnaud Y (2012) Slight mass gain of Karakoram glaciers in the early twenty-first century. Nature Geoscience 5: 322–325.
29. Nuth C, Kaab A (2011) Co-registration and bias corrections of satellite elevation data sets for quantifying glacier thickness change. Cryosphere 5: 271–290.

30. Tian HZ, Yang TB, Liu QP (2012) Relation between climate change and Dunde glacier retreat studied by using remote sending data, 1970–2010. Journal of Glaciology and Geocryology 34: 277–283.
31. Wang ZT, Liu CH, You GX, Pu JC, Yang HA, et al. (1981) Glacier inventory of China I Qilian mountains. Lanzhou: Lanzhou institute of glaciaology and cryopedology, Academia Sicica. 1–35 p.
32. Shi Y (2008) Concise Glacier Inventory of China. Shanghai, China: Shanghai Popular Science Press.
33. Farr TG, Rosen PA, Caro E, Crippen R, Duren R, et al. (2007) The shuttle radar topography mission. Reviews of Geophysics 45: RG2004.
34. Bolch T, Kamp U, Olsenholler J (2005) Using ASTER and SRTM DEMs for studying geomorphology and glaciation in high mountain areas. New Strategies for European Remote Sensing: 119–127.
35. Rignot E, Echelmeyer K, Krabill W (2001) Penetration depth of interferometric synthetic‐aperture radar signals in snow and ice. Geophys Res Lett 28: 3501–3504.
36. Fujita K (2008) Effect of precipitation seasonality on climatic sensitivity of glacier mass balance. Earth and Planetary Science Letters 276: 14–19.
37. Pieczonka T, Bolch T, Wei J, Liu S (2012) Heterogeneous mass loss of glaciers in the Aksu-Tarim Catchment (Central Tien Shan) revealed by 1976 KH-9 Hexagon and 2009 SPOT-5 stereo imagery. Remote Sensing of Environment 130: 233–244.
38. State Bureau of Surveying and Mapping (2007) Technical rules for producing digital products of 1: 10000 1:50000 fundamental geographic information, CH/T 1015.1-2007. Beijing: State Bureau of Surveying and Mapping.
39. Rott H (1993) Thematic Studies in Alpine Areas by Means of Polarimetric Sar and Optical Imagery. Remote Sensing of Earths Surface and Atmosphere 14: 217–226.
40. Paul F, Huggel C, Kääb A, Kellenberger T (2003) Comparison of TM-derived glacier areas with higher resolution data sets. EARSeL eProceedings 2 (Observing our cryosphere from space); pp. 15–21.
41. Gao YJ, Zang DF (2009) The conversion method and accuracy analysis between WGS-84 coordinate system and Xi'an 80 coordinate system. Engineering of surveying and mapping 18: 55–57.
42. Berthier E, Arnaud Y, Vincent C, Remy F (2006) Biases of SRTM in high-mountain areas: Implications for the monitoring of glacier volume changes. Geophysical Research Letters 33: L08502.
43. Schiefer E, Menounos B, Wheate R (2007) Recent volume loss of British Columbian glaciers, Canada. Geophysical Research Letters 34: L16503.
44. Van Niel TG, McVicar TR, Li LT, Gallant JC, Yang QK (2008) The impact of misregistration on SRTM and DEM image differences. Remote Sensing of Environment 112: 2430–2442.
45. Gardelle J, Berthier E, Arnaud Y (2012) Impact of resolution and radar penetration on glacier elevation changes computed from DEM differencing. Journal of Glaciology 58: 419–422.
46. Rignot E, Echelmeyer K, Krabill W (2001) Penetration depth of interferometric synthetic-aperture radar signals in snow and ice. Geophysical Research Letters 28: 3501–3504.
47. Wang YC, Hu WS (2008) Influence Caused by Public Point Selection on Accuracy of Coordinate Conversion. Modern Surveying and Mapping 31: 13–15.
48. Fujita K, Suzuki R, Nuimura T, Sakai A (2008) Performance of ASTER and SRTM DEMs, and their potential for assessing glacial lakes in the Lunana region, Bhutan Himalaya. Journal of Glaciology 54: 220–228
49. Zhang Y, Fujita K, Liu SY, Liu QA, Wang X (2010) Multi-decadal ice-velocity and elevation changes of a monsoonal maritime glacier: Hailuogou glacier, China. Journal of Glaciology 56: 65–74.
50. Burrough P, McDonnell R (1998) Principles of geographical information systems. Oxford: Oxford University Press. 8 p.
51. Wang Z, Liu C (1983) Development conditions distribution features and regional division of present glaciers in Qilian shan. Acta Geographica Sinica 38: 141–153.
52. Paul F (2008) Calculation of glacier elevation changes with SRTM: is there an elevation-dependent bias? Journal of Glaciology 54: 945–946.
53. Summerfield MA (1991) Global Geomorphology, an introduction to the study of landforms. Harlow, England: Pearson, Prentice Hall.
54. Sund M, Eiken T, Hagen JO, Kaab A (2009) Svalbard surge dynamics derived from geometric changes. Annals of Glaciology 50: 50–60.
55. Guo W, Liu S, Xu J, Wei J (2012) Monitoring recent surging of the Yulinchuan glacier on north slope of Muztag range by remote sensing. Journal of Glaciology and Geocryology 34: 765–774.
56. Hoelzle M, Haeberli W, Dischl M, Peschke W (2003) Secular glacier mass balances derived from cumulative glacier length changes. Global and Planetary Change 36: 295–306.
57. Oerlemans J (2005) Extracting a climate signal from 169 glacier records. Science 308: 675–677.
58. Zhang S, Gao X, Ye B, Zhang X, Hagemann S (2012) A modified monthly degree-day model for evaluating glacier runoff changes in China. Part II: application. Hydrological Processes 26: 1697–1706.
59. Shi Y, Shen Y, Kang E, Li D, Ding Y, et al. (2007) Recent and Future Climate Change in Northwest China. Climatic Change 80: 379–393.
60. Li Z, Xu Z (2011) Detection of Change Points in Temperature and Precipitation Time Series in the Heihe River Basin over the Past 50 Years (in Chinese with English abstract). Resources Science 33: 1877–1882.
61. Helsen MM, van den Broeke MR, van de Wal RSW, van de Berg WJ, van Meijgaard E, et al. (2008) Elevation changes in Antarctica mainly determined by accumulation variability. Science 320: 1626–1629.
62. Cuffey KM, Paterson WSB (2010) The physics of glaciers. Pergamon: Pergamon press. 453–510 p.
63. Luo Y, Arnold J, Liu S, Wang X, Chen X (2013) Inclusion of glacier processes for distributed hydrological modeling at basin scale with application to a watershed in Tianshan Mountains, northwest China. Journal of Hydrology 477: 72–85.
64. Möller M, Finkelnburg R, Braun M, Hock R, Jonsell U, et al. (2010) Climatic mass balance of the ice cap Vestfonna, Svalbard: A spatially distributed assessment using ERA-Interim and MODIS data. J Geophys Res 116: F03009.

11

Vigorous Dynamics Underlie a Stable Population of the Endangered Snow Leopard *Panthera uncia* in Tost Mountains, South Gobi, Mongolia

Koustubh Sharma[1,2]*, Rana Bayrakcismith[3], Lkhagvasumberel Tumursukh[1,4], Orjan Johansson[1,3,5], Purevsuren Sevger[1,4], Tom McCarthy[3], Charudutt Mishra[1,2]

1 Snow Leopard Trust, Seattle, Washington, United States of America, **2** Nature Conservation Foundation, Mysore, Karnataka, India, **3** Panthera, New York, New York, United States of America, **4** Snow Leopard Conservation Foundation, Ulaan Baatar, Mongolia, **5** Grimso Wildlife Research Station, Swedish University of Agricultural Sciences, Uppsala, Sweden

Abstract

Population monitoring programmes and estimation of vital rates are key to understanding the mechanisms of population growth, decline or stability, and are important for effective conservation action. We report, for the first time, the population trends and vital rates of the endangered snow leopard based on camera trapping over four years in the Tost Mountains, South Gobi, Mongolia. We used robust design multi-season mark-recapture analysis to estimate the trends in abundance, sex ratio, survival probability and the probability of temporary emigration and immigration for adult and young snow leopards. The snow leopard population remained constant over most of the study period, with no apparent growth ($\lambda = 1.08+-0.25$). Comparison of model results with the "known population" of radio-collared snow leopards suggested high accuracy in our estimates. Although seemingly stable, vigorous underlying dynamics were evident in this population, with the adult sex ratio shifting from being male-biased to female-biased (1.67 to 0.38 males per female) during the study. Adult survival probability was 0.82 (SE+−0.08) and that of young was 0.83 (SE+−0.15) and 0.77 (SE +−0.2) respectively, before and after the age of 2 years. Young snow leopards showed a high probability of temporary emigration and immigration (0.6, SE +−0.19 and 0.68, SE +−0.32 before and after the age of 2 years) though not the adults (0.02 SE+−0.07). While the current female-bias in the population and the number of cubs born each year seemingly render the study population safe, the vigorous dynamics suggests that the situation can change quickly. The reduction in the proportion of male snow leopards may be indicative of continuing anthropogenic pressures. Our work reiterates the importance of monitoring both the abundance and population dynamics of species for effective conservation.

Editor: Roscoe Stanyon, University of Florence, Italy

Funding: The authors are grateful to the Whitley Fund For Nature, BBC Wildlife Fund, Kolmarden Zoo, AZA Conservation Endowment Fund, Disney, Worldwide Conservation Fund, David Shepherd Wildlife Foundation, Woodland Park Zoo, Columbus Zoo and Aquarium, Nysether Family Foundation and Cheyenne Mountain zoos for financial support. The funders had no role in study design, data collection and analysis, decision to publish, or preparation of the manuscript.

Competing Interests: The authors have declared that no competing interests exist.

* Email: koustubhsharma@gmail.com

Introduction

Population monitoring programs form an important component of conservation efforts. They provide information on the status of wildlife populations and can help evaluate the effectiveness of conservation actions, thereby allowing for adaptive management. In addition to population trends, population monitoring programs can help better understand the biology of the species, especially the fundamental processes of survival, reproduction and temporary emigration and immigration, along with their associated vital rates [1]. Empirical estimates of various vital rates such as survival, recruitment, and migration can provide insights into population performance, connectivity, and vulnerability to external threats [2,3].

Large carnivores play an important functional role in ecosystems, are often the focus of conservation programs, and yet, monitoring their population dynamics has remained challenging. Their large home ranges tend to make them useful umbrella species, whose conservation can help protect several other species and large habitats [4]. Additionally, the charismatic image of large carnivores often incites strong emotions, making them suitable symbols for management of large ecosystems. With a growing human population and its ecological footprint, anthropogenic activities come into direct conflict with the ecological needs of large wide-ranging carnivores [5]. Low densities [6], large home ranges [7], and elusive behaviour make it difficult to monitor large carnivore populations in their natural environment. Non-invasive methods, such as camera trapping [8] and fecal genetics [9] can provide statistically robust population estimates of large carnivores. For felid species with individually distinct fur patterns, such as tigers *Panthera tigris* [8], jaguars *Panthera onca* [10–12], snow leopards *Panthera uncia* [13], leopards *Panthera pardus* [14], cheetahs *Acinonyx jubatus* [15] and ocelots *Leopardus pardalis* [16,17], data from camera-trapping can be analysed using capture-recapture models to estimate abundances and population dynamics [2].

The snow leopard is a globally threatened large carnivore distributed over an area of c. 2 million km^2 in the mountain

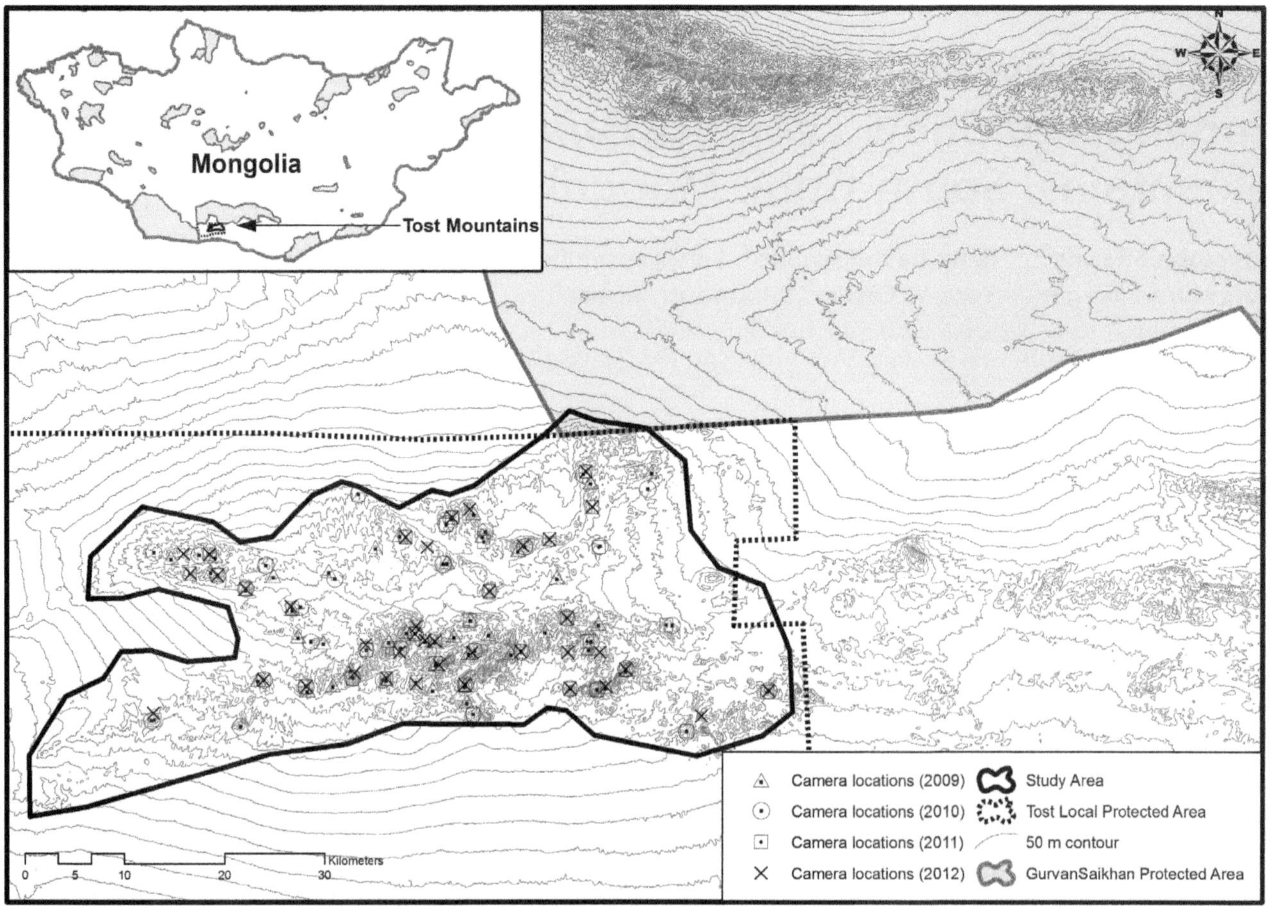

Figure 1. Location of camera traps from 2009 to 2012 in the Tost Mountains in South Gobi, Mongolia, where the population monitoring study of snow leopards was conducted. The camera locations were adjusted every year within the study area after conducting fresh surveys to ensure best camera placements for the year.

ecosystems of 12 countries in South and Central Asia. It is believed that their global population may be between 4,000 and 6,500 in the wild, though there is no robust basis for these estimates [18]. The species is classified as endangered by the IUCN. Retaliatory persecution by pastoralists, depletion of wild prey, habitat degradation, poorly planned developmental activities, extractive industries such as mining, and poaching pose the main threats to the species' survival. The snow leopard enjoys the highest legal protection in the 12 range countries where it is found and considerable conservation efforts are being made towards improving their status.

A few studies have been conducted on snow leopards to estimate their population using mark-recapture framework on camera trapping data (e.g. [13,19,20]), but these have been single-season efforts, and robust design population analyses based on time-series data are not available for any snow leopard population (but see [13]). Based on four years of camera trapping in the Tost Mountains of South Gobi, Mongolia, we describe, for the first time, the abundance and dynamics of a wild snow leopard population over multiple years. We examine the trends in population abundance and sex ratio, and estimate the detectability, survival probability and probabilities of temporary emigration and immigration for adult and young snow leopards. Our results bring to light vigorous underlying dynamics in an otherwise 'stable' population of an endangered large carnivore.

Study Area

Tost mountain range is situated in the Tost *Bag* (smallest administrative unit in Mongolia) of the South Gobi province in Southern Mongolia (43.2°N, 100.5°E) and comprises of Tost and Tosonbumba Mountains. Temperatures here range between −35°C in the winter and +35°C in the summer. Water is a limiting resource during summer as most waterholes dry up leaving only a few perennial springs to sustain wildlife. Some man-made wells support livestock and people, though even they often rely on springs especially through summer. The altitude ranges between 1600 m and 2400 m above mean sea level and although not very high, the mountain slopes are characterized by steep cliffs and crevices. The study area is surrounded by steppe on three sides with a few isolated rugged hillocks potentially creating narrow corridors to habitats in the north and west. The Nemegt Mountains, part of the Gurvan Saikhan National Park, are separated by 60 km of steppe north of the study area. Towards the west, the study area is connected to the Great Gobi 'A' Strictly Protected Area through a broken network of small hillocks in the Trans Altai Gobi desert, with the largest gaps in between the hillocks being c. 25 km. The nearest mountain range in the south is separated by the international border with China, delineated by a trench and an approximately 400 km wide steppe of the Alashan Gobi desert. Towards the east, there is prominent human presence in the form of roads, mining establishments, and the Gurvantes township, followed by the Noyon Mountains.

Legally, the entire study area now lies within the 7,000 km^2 Tost Local Protected Area that was declared in 2011 (Fig. 1). The human population of Tost Bag was 935, according to a census in 2012. Goats and sheep are the most abundant livestock with 31,400 heads, followed by camel (1,100 heads), horse (120 heads), and cattle (47 heads). The Snow Leopard Trust and Snow Leopard Conservation Foundation have been implementing community-based conservation programs in the study area for several years. Ibex *Captra sibirica* is the primary prey of snow leopards in this region, followed by domestic goat *Capra hircus*, argali *Ovis ammon*, domestic sheep *Ovis aries* and other species including Tolai hare *Lepus tolai*, Chukar partridge *Alectoris chukar*, and several species of rodents ([21], Johansson, unpublished data). Other predators include lynx *Lynx lynx*, wolf *Canis lupus*, red fox *Vulpes vulpes* and at least two species of mustelids.

Methods

Field

We defined our study area as the entire mountainous region of Tost that was separated from other mountains by at least 20 km of steppe (Fig. 1). This comprised a surface area of 1,684 km^2. We sought all necessary permits from the Mongolian Ministry of Environment and Green Development to conduct research in the Tost Mountains. The methods for the study were approved by the Mongolian Institute of Biology and the work detailed here did not involve any direct sampling methods or the collection of any specimens.

Data from the on-going study in Tost involving 25,297 locations from 19 radio-collared snow leopards indicated that snow leopards did not use the steppe, except for occasional forays or to migrate to neighbouring mountain ranges. We overlaid a grid of 5×5 km on the mountainous part of the study area to aid placement of one camera trap station in each grid cell. This size was deemed appropriate allowing about 2 camera stations per home range [22] since the mean minimum summer and annual ranges of snow leopards in Tost are estimated to be c. 55 km^2 and 70 km^2, respectively for 9 adult breeding females using fixed Kernel estimators on 12,498 locations over a period of five years (Johansson et. al. *unpublished data*). In each grid cell we searched along saddles on the ridgelines, valley or cliff bases and near overhanging boulders for fresh signs of snow leopard presence such as urine spray markings and scrapes. Within each cell, the site with most signs of recent activity was chosen and a camera trap was installed. Camera placements were rarefied if certain grid cells did not have ample suitable habitat or signs of snow leopard presence. The resultant density varied between 2 and 4 camera trapping stations/100 km^2. Overall, 40–42 camera trapping stations were operated each year with one camera trap (Reconyx RM45, RC40 or HC500) placed at each (Fig. 2). Camera traps were left in the field for at least 62 days each summer, except in 2009 when due to logistic constraints, cameras were operational for 32 days. Camera traps were checked once every two weeks to ensure battery and memory card adequacy. We used at least 2 Gigabyte cards, allowing each camera to store more than 10,000 pictures. In total we invested a sampling effort of 8,490 camera-trap-nights over 4 years.

We installed camera traps in sites where snow leopards were expected to spend more time (for marking or investigating) than just passing through. We programmed the camera traps to take 5 pictures on each trigger with an interval of about 0.5 seconds between each picture. The time period between two triggers was kept as zero. This meant that if the activity that triggered the first set of pictures continued after the first five pictures were taken, then another set of five pictures was taken immediately. This allowed us to get multiple pictures (range 1 to 115) of individual snow leopards for most captures (99%) and on many occasions we were able to photograph combinations of face, tail, foreleg, hind leg, and/or flank during the same encounter (Fig. 3). The trigger sensor on Reconyx cameras are calibrated to take pictures using a combination of heat and motion sensors that prevent unnecessary triggers due to heated rocks or movements of grass blades in front of the camera's field of view.

Analysis

Snow leopard identification. Images were downloaded at the end of each sampling period. Markings on the face, rump, tail and flanks were used to identify individuals using pictures obtained during camera trapping.

Snow leopards have thick fur that tends to get ruffled up, which can lead to misidentification. Misidentification of an individual as two different individuals can result in an underestimation of detection probability and increase the minimum population count (number of individuals captured). Together, this can lead to an exponential overestimation of the population. To prevent this, we did not consider any two snow leopard images as belonging to different individuals unless at least three marking patterns were confirmed to be different. Conversely, images with three or more similarities were considered to belong to the same individual. In cases where we failed to find at least three markings that were either similar or different, the images and the particular encounter were discarded from the analyses. Images from different years were also compared with each other to identify individuals. Many images that remained unidentified in the initial years could be identified by the end of the fourth year.

Abundance estimation. Single season mark-recapture analysis allows for sampling designs where a study area is sampled in parts and analysed as a whole (e.g. [23,24]). Although we sampled the entire area during the same period, this flexibility allowed us to use the entire data from a camera trap right from the day of installation, irrespective of whether or not the other camera traps had become operational. We used single season models to estimate the adult population each year using a 32 days' dataset in 2009 and up to 62 days' datasets for the remaining years using software CAPTURE [25]. We tested for possible effects of the shorter sampling duration in 2009 by randomly truncating the datasets from the subsequent years to 32 days. We found that the shorter sampling duration increased the variance of the estimates but did not influence the means, and therefore, used the longer data-sets starting 2010.

We created capture histories in the standard X-matrix format [25] by pooling the one-day sampling occasion capture histories into two-day sampling occasions capture histories. Population closure was tested using program CAPTURE [25]. The snow leopard population size was estimated independently for each year. To assess the accuracy of our population estimates, we separately analysed the photographic capture data of radio-collared snow leopards in the same study site, and compared our mark-recapture estimates for radio-collared animals each year with the known number of individuals wearing collars during the respective sampling period.

We tested whether the population had remained constant over the four primary sampling periods using robust design as well as Pradel's models [26]. The robust design model enabled us to test the hypothesis that the population remained constant, while the

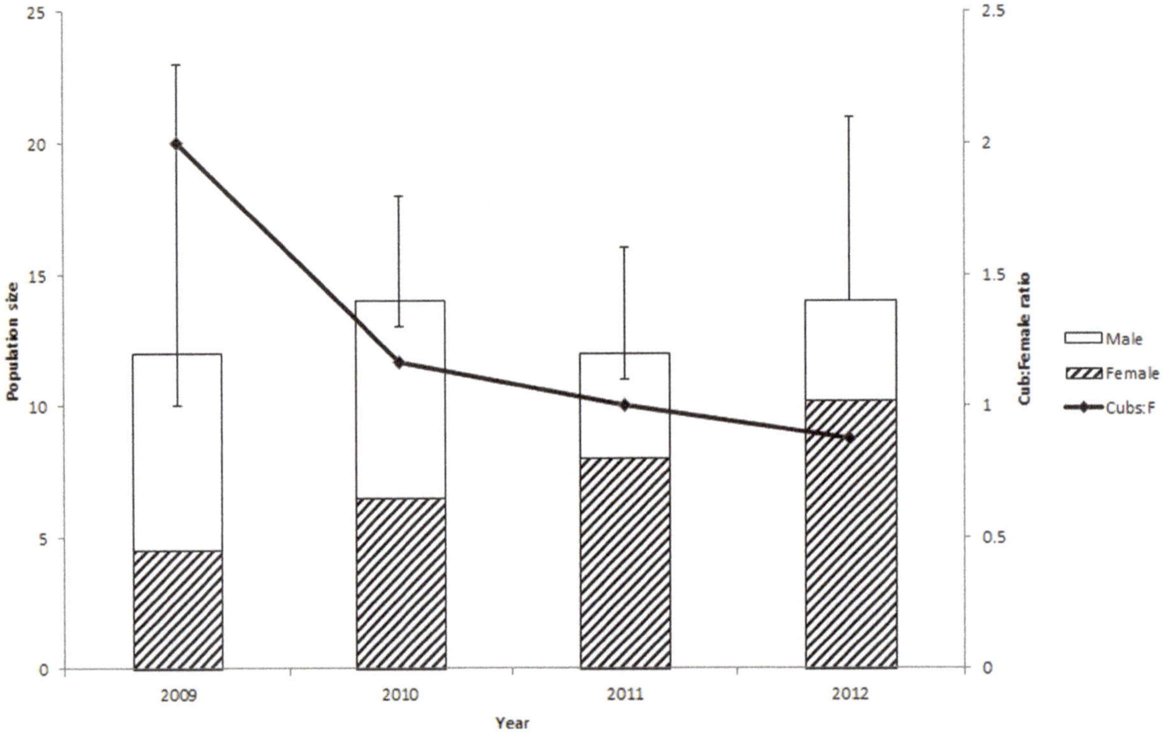

Figure 2. Trends in sex ratio and size of the adult population of snow leopards in the Tost Mountains, South Gobi, Mongolia, between 2009 and 2012. Population size was estimated independently over 4 years using the Jackknife Heterogeneity Model (Mh). Also shown on the secondary axis is the ratio of cubs per adult female. It is worth noting that the number of cubs recorded remained constant even though the ratio declined.

Pradel's model allowed us to assess the rate of change in abundance (λ) over the four annual intervals (2009–2012).

Estimating life history parameters. Robust design multi-season mark-recapture modelling [27] was used to estimate life history parameters such as survival, temporary emigration and immigration, and changes in abundance. Since camera setup in the field took up to 15 days, and about a week for retrieval, we truncated the data to the days when all cameras were operational for the multi-season analysis. This resulted in a secondary sampling period of 16 days for 2009, and 32 days each for 2010, 2011 and 2012. For each primary sampling period (year) we thus obtained 8, 16, 16 and 16 secondary sampling periods, respectively, by pooling the one-day sampling occasion capture histories into two-day sampling occasions capture histories. The snow leopard populations were expected to be open to changes between primary sampling periods (i.e. years) while we assumed

Figure 3. Multiple images with more than one body part of the snow leopard could be photographed using a single camera trap capable of taking multiple images rapidly (<1 second intervals). Here we show 3 out of the 30 different photos of a male snow leopard taken on the same encounter during the 2 minutes that he lingered in front of the camera. These images were used to create references for identifying 8 body parts, including face, tail, and both flanks, hindlegs and forelegs of this particular snow leopard.

Table 1. Abundance of snow leopards estimated using the Jackknife estimator on camera trapping data in Tost Mountains, South Gobi, Mongolia, between 2009 and 2012.

Year			Adult			Adult+Young		
				95% Confidence Interval			95% Confidence Interval	
	z	p_z	N	LCL	UCL	N	LCL	UCL
2009	−0.38	0.35	12	10	21	20	17	35
2010	−1.41	0.13	14	13	17	20	19	26
2011	−0.42	0.34	12	11	15	19	17	23
2012	0.36	0.64	14	14	21	21	21	31

Abundance of Adult+Young snow leopards was estimated by multiplying the estimates of Adult abundance by mean group size of each identified snow leopard. z = Population closure test statistic, p_z = closure test significance, N = Abundance, LCL = lower confidence limit, and UCL = upper confidence limit.

the population to be closed to changes between secondary sampling periods.

Survival probability was estimated as the probability that an individual survived between two primary sampling periods. Modelling of capture-recapture data from multiple seasons using robust design allows estimation of probability of temporary emigration and immigration [27,28]. This addresses the issue of individuals that survived and were unavailable for camera trapping during the subsequent primary sampling period(s) (considered temporarily emigrated from the study area), but were available again at a later primary sampling period (considered immigrated back into the study area). We tested the probabilities of temporary emigration and immigration to be random as well as Markovian. Markovian models expect that the probability of an individual being in a particular state (within the population or as a migrant) is dependent on its previous state, whereas random models expect that the probabilities of temporary emigration and immigration are independent of the previous state [28]. Detection probability was estimated for individuals present in an area but not encountered (captured on cameras) during the secondary sampling periods. Unlike the proportion of known individuals that were assumed to have temporarily emigrated and unavailable for capture, this addressed the proportion of individuals that were encountered (or not encountered), given presence.

We grouped encounter histories into two categories for multi-season analysis of survival and temporary emigration and immigration - individuals first time marked as young (cubs accompanying mother) and first time marked as adult. For individuals marked as young, the survival and emigration-immigration probabilities were modelled to change after two years, by which time they were expected to have grown into independent adults (or sub-adults). Although it is relatively difficult to sex snow leopards from their pictures given their thick fur, over the years of sampling, we were able to correctly sex all adult snow leopards based on visible testicles, presence of young cubs, and through information gathered during snow leopard captures for our telemetry study. Model based inferences from a separate analysis using adult snow leopards' data, grouped as male and female on program Mark indicated that the detection probability did not vary between sexes. We therefore used this information for estimating annual adult sex ratios, but did not use it to estimate additional population parameters due to further fragmentation of the small sample size.

Using robust design analysis of multi-season data, we developed several models to understand the effects of different variables on the modelled parameters that included survival probability, temporary emigration and immigration, and capture and recapture probabilities. Models were developed using various combinations of constraints including time, behaviour after first capture, and age. For age models, we considered that young snow leopards turned into adults after the age of 2 years. We tested if the parameters for these young snow leopards remained the same after 2 years, became similar to other known adults, or were entirely different from young and adult snow leopards. Models were compared using the corrected Akaike Information Criteria (AICc) and the final estimates were obtained using model averaging following recommendations by [29] on model selection. Variances of the model-averaged results were obtained using methods prescribed by [30].

Results

Of 381 snow leopard encounters over the four primary sampling periods, we could identify the individuals in 361 (95%).

Table 2. Top 10 models (cumulative AIC weight, CW = 0.98) in the robust design analysis of multi-season camera trapping of snow leopards for variables affecting survival (S), abundance (N), temporary immigration (G1), temporary emigration (G2), capture probability (p), and recapture probability (c) for snow leopards in the Tost Mountains, South Gobi, Mongolia, between 2009 and 2012.

Model	AICc	Δ AICc	W	Model Likelihood	K	Model Deviance	CW
{S(.) G1 = G2 (AgeGrp) p = c(AgeGrp) N(.)}	659.4	0	0.237	1	6	638.289	0.24
{S(.) G1 = G2 (Yng-Ad_2yr&diff) p = c(AgeGrp) N(.)}	659.44	0.0239	0.234	0.9881	7	636.084	0.47
{S(.) G1 = G2 (AgeGrp) p = c(AgeGrp) N(Age)}	660.12	0.6998	0.167	0.7048	7	636.760	0.64
{S(Yng-Ad_2yr&diff) G1 = G2 (Yng-Ad_2yr&diff) p = c(AgeGrp) N(.)}	661.50	2.0813	0.084	0.3532	8	635.876	0.72
{S(Yng→Ad_2yr) G1 = G2 (AgeGrp) p = c(AgeGrp) N(.)}	661.60	2.1781	0.080	0.3365	7	638.238	0.80
{S(Yng→Ad_2yr) G1 = G2 (AgeGrp) p = c(AgeGrp) N(Age)}	662.33	2.9139	0.055	0.233	8	636.709	0.86
{S(.) G1 = G2 (Yng→Ad_2yr) p = c(AgeGrp) N(.)}	662.63	3.2157	0.047	0.2003	6	641.505	0.90
{S(.) G1 = G2 (Yng→Ad_2yr) p = c(AgeGrp) N(Age)}	663.04	3.6172	0.039	0.1639	7	639.677	0.94
{S(.) G1 = G2 (AgeGrp) p = c(AgeGrp) N(Session)}	663.57	4.1541	0.030	0.1253	9	635.647	0.97
{S(Yng→Ad_2yr) G1 = G2 (AgeGrp) p = c(AgeGrp) N(Session)}	665.86	6.4427	0.009	0.0399	10	635.596	0.98

Included in the table are: AICc = corrected Akaike Information Criteria, ΔAICc = differences in AICc values between each model and the best fitting model, *W* = model weight, Model Likelihood, K = number of model parameters, and Model Deviance. Constraints tested included Yng→Ad_2yr = snow leopards marked as young turn into adults after 2 years with parameters similar to the latter, Yng-Ad_2yr&diff = snow leopards marked as young turn into sub-adults after 2 years with parameters different from young as well as adults; AgeGrp = snow leopards marked as young have parameters different than for those marked as adult, Session or Period = conditions when parameters are considered to be different for each year.

Table 3. A comparison of the known 'population' of radio-collared snow leopards and their population estimated using mark-recapture analysis in the Tost Mountains, South Gobi, Mongolia, between 2009 and 2012.

Year	Known population of collared individuals	Camera trapped collared individuals	Mark-Recapture estimates of collared population	Std. Err.	95% LCL	95% UCL
2009	7	5	7.33	2.32	5.45	16.85
2010	10	9	9.83	1.43	9.08	17.3
2011	8	7	7.52	1.24	7.04	14.75
2012	10	10	10	0.29	10	10

In total we used data from 13 individuals marked as adults and 21 marked as young in our analysis across four years. The adult population was estimated separately for each year using photographic data of 8, 11, 13 and 13 adult snow leopards identified respectively for each year.

For the multi-season models, heterogeneity in detection of individuals did not rank high among the most parsimonious models. However the models indicated that detection probability for secondary sampling occasions of individuals marked as adults (0.16+−SE 0.02) was greater than that of young individuals (0.11+−SE 0.02). AIC model weights indicated little difference between capture and recapture probabilities, suggesting no behavioural effect of camera trapping.

Populations closure test for all the four years were consistent with the assumption that the snow leopard population was closed during the primary sampling periods (Table 1). Jackknife estimation of abundance using the heterogeneity models [25] showed that the estimated mean adult population ranged from 12 (95% CI = 11–19) to 14 (95% CI = 14–21) during the four years (Table 1). AIC weights from the robust design analysis models (Table 2) indicated there was greater than 95% probability that the adult population remained constant over four years. Pradel's population growth rate model also suggested that the population had remained stable across the four years ($\lambda = 1.08+-0.25$). Our mark-recapture estimate for the population subset comprising of radio-collared individuals was identical to the actual 'known population' in each of the four years (7, 10, 8 and 10 individuals, respectively), indicating that our overall population estimates may have been accurate (Table 3).

Estimation of male and female ratios in the four primary sampling periods showed a significant change in adult sex ratio ($\chi^2 = 5.8$; $p = 0.02$) from 1.67 males per female in 2009 to 0.38 males per female in 2012 (Fig. 2). The proportion of females in the adult population increased from 0.38 (estimated 4 individuals) to 0.67 (10), whereas that of males declined from 0.63 (6) to 0.33 (4). The total number of cubs recorded on camera traps each year remained between 8 and 9, while the number of cubs per female declined from 2 in 2009 to 0.88 in 2012 as the female population increased (Fig. 2).

We found no evidence in support of models assuming variation in survival probability, indicating that snow leopard survival probability did not vary over the years (Table 2). This enabled us to use the reduced parameters approach and estimate survival probability of snow leopards for the entire study period. The model averaged estimate of survival probability of adults was 0.83 (SE+−0.08), and that of the individuals marked when young were 0.83 (SE+−0.15) and 0.77 (SE +−0.2) respectively, before and after the age of 2 years (Table 4). This meant that approximately 17% of the adult and young and 23% of the sub-adult snow leopards either died or permanently emigrated out of the study area annually. While we could estimate temporary emigration and immigration, our study design did not allow us to differentiate between permanent migration and mortality. However, assuming individuals that migrated permanently were lost from the study population (died), we estimated the life expectancy of adult snow leopards to be 5 years (95%CI: 2.05–13.78 years).

For the entire dataset, the probability of temporary emigration and immigration appeared to be random rather than Markovian, indicating that it was not dependent on the previous state of the individuals. For individuals marked as young, the probability of random temporary emigration and immigration was estimated to be 0.6 (SE +−0.19) and 0.68 (SE +−0.32) respectively, before and after the age of 2 years, suggesting that an estimated 60–68% of young individuals emigrated from the study area, potentially returning subsequently (Table 4). However, there was little evidence of temporary emigration and immigration in the adult population (0.01 SE+−0.07).

Discussion

In this first ever multi-year monitoring of snow leopards, we found the population in Tost Mountains of South Gobi to have remained almost constant, and the estimated mean adult population remained between 12 and 14, and the total population (including young) between 19 and 21. That our overall abundance estimates were reliable was supported by the accurate estimation of the known population of radio-collared snow leopards during the study period. The population also appeared fecund as we recorded confirmed births of at least 21 cubs in the study area over the 4 years, and estimated a total of 32 cubs after incorporating detection probability. Our estimate of annual survival of adult snow leopards (0.82±0.08) was comparable to annual adult survival probabilities reported for tigers in South India (0.77, [2]) and leopards in Kruger National Park, South Africa (0.82, [31]), both characterizing stable populations. Other studies with higher level of human induced mortality have reported relatively lower survival probabilities, such as in the case of male Amur tigers in southeast Russia (0.72, [32]), male cougars in the Pacific north-west America (0.55, [33]), and male leopards in South Africa (0.45, [34]), and predicted negative population growths. From these perspectives, the snow leopard population in Tost thus appeared to be stable and not in decline.

Yet, while the abundance estimates indicate a stable snow leopard population in Tost, a closer examination reveals vigorous underlying dynamics. For instance, the significant change in adult sex ratio of snow leopards over the sampling period we recorded was unexpected. We started with a male-biased population, and over the 4 years, it became female-biased with the proportion of males becoming almost half and that of the females in the population almost doubling. Indeed, in our radio-collaring efforts

Table 4. Probabilities of annual survival and temporary emigration and immigration of adult and young snow leopards in the Tost Mountains, South Gobi, Mongolia, between 2009 and 2012.

	Survival (S)	SE	LCL	UCL	Emigration/Immigration (γ)	SE	LCL	UCL
Adult	**0.83**	0.075	0.614	0.926	**0.01**	0.07	0.00	0.15
Young (<2 yrs)	**0.83**	0.154	0.328	0.960	**0.60**	0.19	0.24	0.87
Young (>2 yrs)	**0.77**	0.201	0.264	0.968	**0.68**	0.32	0.11	0.97

SE = Standard Error, LCL = 95% Lower Confidence Limit, UCL = 95% Upper Confidence Limit.

on the same population during the same study period, we physically captured only 1 adult female and 8 male snow leopards during the first two years, whereas in the subsequent years, 8 females and 6 males (including some that had been captured earlier) were captured [35], which may support our finding of the change in sex ratio. Disproportionate mortality of males, presumably an important cause of the change in sex ratio of the snow leopard population in Tost, is reported in several carnivores as resulting from human-induced factors such as poaching and retaliatory killing, including Amur tigers in southeast Russia [32], leopards in South Africa [34], and cougars in the Pacific Northwest [33], although there are exceptions (e.g. tiger population in Panna, Central India, that had turned male-biased prior to extinction due to poaching; [36,37]). Males of large carnivores are known for their tendency to disperse further and range over much larger areas as compared to females who may remain in close vicinity to their mother's range after dispersal e.g. see [38,39], exposing the males to greater anthropogenic risks. However in a fragmented habitat where dispersal distance could be limited by habitat restrictions hindering connections with a neighbouring population, the difference might not be so striking. During our 4 year study period, we recorded the mortality of 4 adult snow leopards, all of which were males. We could ascertain the cause of mortality only in one case, where the individual had been shot by a herder. This may be indicative of anthropogenic pressure on the Tost population.

The change in sex ratio of snow leopards in Tost was accompanied by a potential decline in the number of cubs per female over the study period as the female population increased. It is likely that young females dispersing around their natal ranges may have resulted in the proportion of females in the study area to increase while males suffered disproportionate mortality or dispersed to areas beyond the study area. The younger females perhaps do not reproduce in the first year or so after dispersal, possibly explaining the reduced number of cubs per female. With more data in the future on age at first reproduction, sexual differences in dispersal strategies, and connectivity between neighbouring habitats, we hope to be able to better understand the mechanisms behind these population dynamics in Tost.

The much lower random probability of temporary emigration and immigration we recorded for adult snow leopards points to their greater site fidelity and perhaps territoriality. Expectedly, as compared to the adults, we recorded a relatively higher rate of temporary emigration and immigration by young snow leopards, presumably due to dispersal attempts and efforts to establish their territories outside their natal areas as observed in other cat species [40,41]. Although the Tost Mountains appear relatively isolated, it appears that young individuals use the surrounding landscape for dispersal, despite the absence of contiguous, large mountain chains. Two young collared snow leopards illustrated this by dispersing to the nearby Nemegt Mountains after crossing about 80 km of steppe. The movements between landscapes through steppe underscore the need for paying attention to the intervening steppe habitats of Central Asia while planning for snow leopard conservation.

From a methodological perspective, we achieved high (95%) success in identification of individual snow leopards from photographs, even though we used only one camera per trapping station. As long as trapping stations are set up in sites where snow leopards are likely to linger (e.g. marking locations), thereby allowing multiple images and both flanks to be photographed, one camera trap appears adequate, especially when conducting long-term monitoring and using cameras that are capable of taking multiple images rapidly (<1 second intervals). This enables twice the area to be sampled with a given number of cameras. Indeed many of our snow leopard encounters that remained unidentified during the initial years could be identified in due course over multi-year sampling. For studies of a shorter duration, or one-time sampling of the population, one camera might not always be adequate. Although all snow leopards were identified as unique individuals based on camera trap pictures, the sexing of adults would perhaps not have been possible for many individuals (except females with young), had we not been capturing them for collaring.

Our work, apart from providing the first estimates of vital rates and other population characteristics of the endangered snow leopard, reiterates the value of long-term monitoring of both abundance and population dynamics for conservation planning and action.

Acknowledgments

We would like to thank the Mongolian Ministry of Environment and Green Development for permitting us to conduct research in the Tost Mountains. We are grateful to Dr. Gustaf Samelius for his comments on the manuscript and to Bayarjargal A. for managing the snow leopard research and conservation programs in Mongolia. We thank Purevjav Lkhagvajav, Tserennadmid Mijiddorj, Dorj, Kim Murray, Louis McNutt, Ashley Spearing, Mijid Ayush, Oyungerel Baatar, and the student interns Munkhoo (B. Bolormunkh), Nasaa (O. Nasantogtokh), Buma (N. Bumantuya), Khishigee (N.Batkhishig), Ulzii (Ts. Enkh-Ulzii), Pugi (O. Purevnyam), Ganaa (J. Ganchimeg) from the National University of Mongolia who participated in camera setup and retrieval. We are grateful to the people living in Tost Bag for supporting our work and ensuring the safety of the cameras. The editor and two anonymous reviewers are thanked for their constructive comments that helped us improve the manuscript.

Author Contributions

Conceived and designed the experiments: KS CM. Performed the experiments: KS LT OJ PS. Analyzed the data: KS RB. Contributed reagents/materials/analysis tools: CM TM. Wrote the paper: KS CM TM RB PS LT OJ. Project managers, helped coordinate field work and analysis: TM CM.

References

1. Williams BK, Nichols JD, Conroy MJ (2002) Analysis and management of animal populations. 1st ed. San Diego, California: Academic Press.
2. Karanth KU, Nichols JD, Kumar NS, Hines JE (2006) Assessing tiger population dynamics using photographic capture-recapture sampling. Ecology 87: 2925–2937. Available: http://www.ncbi.nlm.nih.gov/pubmed/17168036.
3. Kerley LL, Goodrich JM, Miquelle DG, Smirnov EN, Quigley HB, et al. (2003) Reproductive parameters of wild female Amur (Siberian) tigers (Panthera tigris altaica). J Mammal 84: 288–298. Available: http://www.asmjournals.org/doi/abs/10.1644/1545-1542(2003)084<0288:RPOWFA>2.0.CO;2. Accessed 8 November 2013.
4. Caro TM (2003) Umbrella species: critique and lessons from East Africa. Anim Conserv 6: 171–181. Available: http://doi.wiley.com/10.1017/S1367943003003214. Accessed 30 July 2012.
5. Treves A, Karanth KU (2003) Human-Carnivore Conflict and Perspectives on Carnivore Management Worldwide. Conserv Biol 17: 1491–1499. Available: http://doi.wiley.com/10.1111/j.1523-1739.2003.00059.x.
6. Karanth KU, Chellam R (2009) Carnivore conservation at the crossroads. Oryx 43: 1. Available: http://www.journals.cambridge.org/abstract_S003060530843106X. Accessed 8 November 2013.
7. Gittleman JL, Harvey PH (1982) Carnivore home-range size, metabolic needs and ecology. Behav Ecol Sociobiol 10: 57–63. doi:10.1007/BF00296396.
8. Karanth K (1995) Estimating tiger Panthera tigris populations from camera-trap data using capture—recapture models. Biol Conserv 71: 333–338. Available: http://www.sciencedirect.com/science/article/pii/000632079400057W. Accessed 30 January 2014.
9. Mills LS, Citta JJ, Lair KP, Schwartz MK, Tallmon DA (2000) Estimating animal abundance using noninvasive DNA sampling: promise and pitfalls. Ecol Appl 10: 283–294. Available: http://www.esajournals.org/doi/abs/10.1890/1051-0761(2000)010[0283:EAAUND]2.0.CO;2. Accessed 8 November 2013.
10. Wallace RB, Gomez H, Ayala G, Espinoza F (2003) Camera trapping for jaguar (Panthera onca) in the Tuichi Valley, Bolivia. J Neotrop Mammal 10: 133–139. Available: http://www.sarem.org.ar/wp-content/uploads/2012/11/SAREM_MastNeotrop_10-1_10_Wallace.pdf. Accessed 12 January 2014.
11. Silver S, Ph D (2004) Assessing jaguar abundance using remotely triggered cameras. Wildl Conserv.
12. Maffei L, Cuéllar E, Noss A (2004) One thousand jaguars (Panthera onca) in Bolivia's Chaco? Camera trapping in the Kaa-Iya National Park. J Zool 262: 295–304. Available: http://onlinelibrary.wiley.com/doi/10.1017/S0952836903004655/abstract. Accessed 8 November 2013.
13. Jackson RM, Roe JO, Wangchuk R, Hunter DO (2006) Estimating Snow Leopard Population Abundance Using Photography and Capture – Recapture Techniques. Wildl Soc Bull 34: 772–781.
14. Athreya V, Odden M, Linnell JDC, Krishnaswamy J, Karanth U (2013) Big Cats in Our Backyards: Persistence of Large Carnivores in a Human Dominated Landscape in India. PLoS One 8: e57872. Available: http://dx.plos.org/10.1371/journal.pone.0057872. Accessed 7 March 2013.
15. Marnewick K, Funston PJ, Karanth KU (2008) Evaluating camera trapping as a method for estimating cheetah abundance in ranching areas. South African J Wildl Res 38: 59–65. Available: http://www.bioone.org/doi/abs/10.3957/0379-4369-38.1.59.
16. Trolle M, Kéry M (2003) Estimation of ocelot density in the Pantanal using capture-recapture analysis of camera-trapping data. J Mammal 84: 607–614. Available: http://www.asmjournals.org/doi/abs/10.1644/1545-1542(2003)084<0607:EOODIT>2.0.CO;2. Accessed 8 November 2013.
17. Maffei L, Noss AJ, Ceullar E, Rumiz DI (2005) Ocelot (Felis pardalis) population densities, activity, and ranging behaviour in the dry forests of eastern Bolivia: data from camera trapping. J Trop Ecol 21: 1–6.
18. Mccarthy TM, Chapron G (2003) Snow Leopard Survival Strategy. Seattle, USA: ISLT and SLN.
19. Mccarthy KP, Fuller TK, Ming M, M MT, Waits L, et al. (2008) Assessing Estimators of Snow Leopard Abundance. J Wildl Manage 72: 1826–1833. Available: http://onlinelibrary.wiley.com/doi/10.2193/2008-040/abstract. Accessed 8 November 2013.
20. Janečka JE, Munkhtsog B, Jackson RM, Naranbaatar G, Mallon DP, et al. (2011) Comparison of noninvasive genetic and camera-trapping techniques for surveying snow leopards. J Mammal 92: 771–783. Available: http://www.bioone.org/doi/abs/10.1644/10-MAMM-A-036.1. Accessed 2 March 2013.
21. Shehzad W, Mccarthy TM, Pompanon F, Purevjav L, Coissac E, et al. (2012) Prey preference of snow leopard (Panthera uncia) in South Gobi, Mongolia. PLoS One 7: 1–8. Available: http://dx.plos.org/10.1371/journal.pone.0032104. Accessed 12 January 2014.
22. Karanth K, Nichols J (1998) Estimation of tiger densities in India using photographic captures and recaptures. Ecology 79: 2852–2862. Available: http://www.esajournals.org/doi/abs/10.1890/0012-9658(1998)079 [2852: EOTDII]2.0.CO;2. Accessed 12 January 2014.
23. Nichols JD, Karanth KU (2002) Statistical concepts: estimating absolute densities of tigers using capture–recapture sampling. In: Nichols JD, Karanth KU, editors. Monitoring tigers and their prey: amanual for researchers,managers and conservationists in tropical Asia. Bangalore: Centre for Wildlife Studies. pp. 121–137.
24. Karanth KU, Chundawat RS, Nichols JD, Kumar NS (2004) Estimation of tiger densities in the tropical dry forests of Panna, Central India, using photographic capture–recapture sampling. Anim Conserv 7: 285–290. Available: http://onlinelibrary.wiley.com/doi/10.1017/S1367943004001477/abstract. Accessed 8 November 2013.
25. Otis DL, Burnham KP, White GC, Anderson DR (1978) Statistical inference from capture data on closed animal populations. Wildl Monogr 62: 3–135.
26. Pradel R (1996) Utilization of capture-mark-recapture for the study of recruitment and population growth rates. Biometrics 52: 703–709.
27. Kendall WL, Pollock KH, Brownie C (1995) A Likelihood-Based Approach to Capture-Recapture Estimation of Demographic Parameters under the Robust Design A Likelihood-Based Approach to Capture-Recapture Estimation of Demographic Parameters Under the Robust Design. Biometrics 51: 293–308.
28. Kendall WL, Bjorkland R (2001) Using Open Robust Design Models to Estimate Temporary Emigration from Capture-Recapture Data Using Open Robust Design Models to Estimate Temporary Emigration from Capture-Recapture Data. Biometrics 57: 1113–1122.
29. Burnham KP, Anderson DR, Huyvaert KP (2010) AIC model selection and multimodel inference in behavioral ecology: some background, observations, and comparisons. Behav Ecol Sociobiol 65: 23–35. Available: http://link.springer.com/10.1007/s00265-010-1029-6. Accessed 22 May 2013.
30. Cooch E, White G (2008) Program Mark: a gentle introduction. 7th ed. Cooch E, White G, editors.
31. Bailey TN (1993) The African leopard: Ecology and Behaviour of a solitary felid. New York: Columbia University Press.
32. Goodrich JM, Kerley LL, Smirnov EN, Miquelle DG, McDonald L, et al. (2008) Survival rates and causes of mortality of Amur tigers on and near the Sikhote-Alin Biosphere Zapovednik. J Zool 276: 323–329. Available: http://doi.wiley.com/10.1111/j.1469-7998.2008.00458.x. Accessed 13 August 2013.

33. Lambert CMS, Wielgus RB, Robinson HS, Katnik DD, Cruickshank HS, et al. (2006) Cougar population dynamics and viability in the Pacific Northwest. J Wildl Manage 70: 246–254. Available: http://www.bioone.org/doi/abs/10.2193/0022-541X(2006)70[246:CPDAVI]2.0.CO;2. Accessed 8 November 2013.
34. Balme G, Hunter L (2004) Mortality in a protected leopard population, Phinda Private Game Reserve, South Africa: A population decline. Ecol J 6: 1–6.
35. Johansson Ö, Malmsten J, Mishra C, Lkhagvajav P, McCarthy T (2013) Reversible immobilization of free-ranging snow leopards (panthera uncia) with a combination of medetomidine and tiletamine-zolazepam. J Wildl Dis 49: 338–346. Available: http://www.ncbi.nlm.nih.gov/pubmed/23568909. Accessed 7 November 2013.
36. Chundawat RS, Gruisen J (2009) Panna's last tiger. New Delhi: BAAVAN-Baagh Aap Aur Van.
37. Gopal R, Qureshi Q, Bhardwaj M, Jagadish Singh RK, Jhala Y V (2010) Evaluating the status of the endangered tiger Panthera tigris and its prey in Panna Tiger Reserve, Madhya Pradesh, India. Oryx 44: 383–389.
38. Smith JLD (1993) The Role of Dispersal in Structuring the Chitwan Tiger Population. Behaviour 124: 165–195. doi:10.1163/156853993X00560.
39. Gour DS, Bhagavatula J, Bhavanishankar M, Reddy PA, Gupta J a, et al. (2013) Philopatry and dispersal patterns in tiger (Panthera tigris). PLoS One 8: e66956. Available: http://www.pubmedcentral.nih.gov/articlerender.fcgi?artid = 3699573 &tool = pmcentrez&rendertype = abstract.
40. Samelius G, Andren H, Liberg O, Linnel JDC, Odden J, et al. (2012) Spatial and temporal variation in natal dispersal by Eurasian lynx in Scandinavia. J Zool 286: 120–130.
41. Vangen KM, Persson J, Landa A, Anderson R, Segerstrom P (2001) Characteristics of dispersal in wolverines. Can J Zool 79: 1641–1649.

12

Molecular Phylogeography and Population Genetic Structure of *O. longilobus* and *O. taihangensis* (*Opisthopappus*) on the Taihang Mountains

Yiling Wang, Guiqin Yan*

College of Life Sciences, Shanxi Normal University, Linfen, China

Abstract

Historic events such as the uplift of mountains and climatic oscillations in the Quaternary periods greatly affected the evolution and modern distribution of the flora. We sequenced the *trn*L–*trn*F, *ndh*J-*trn*L and ITS from populations throughout the known distributions of *O. longilobus* and *O. taihangensis* to understand the evolutionary history and the divergence related to the past shifts of habitats in the Taihang Mountains regions. The results showed high genetic diversity and pronounced genetic differentiation among the populations of the two species with a significant phylogeographical pattern ($N_{ST}>G_{ST}$, $P<0.05$), which imply restricted gene flow among the populations and significant geographical or environmental isolation. Ten chloroplast DNA (cpDNA) and eighteen nucleus ribosome DNA (nrDNA) haplotypes were identified and clustered into two lineages. Two corresponding refuge areas were revealed across the entire distribution ranges of *O. longilobus* and at least three refuge areas for *O. taihangensis*. *O. longilobus* underwent an evolutionary historical process of long-distance dispersal and colonization, whereas *O. taihangensis* underwent a population expansion before the main uplift of Taihang Mountains. The differentiation time between *O. longilobus* and *O. taihangensis* is estimated to have occurred at the early Pleistocene. Physiographic complexity and paleovegetation transition of Taihang Mountains mainly shaped the specific formation and effected the present distribution of these two species. The results therefore support the inference that Quaternary refugial isolation promoted allopatric speciation in Taihang Mountains. This may help to explain the existence of high diversity and endemism of plant species in central/northern China.

Editor: Igor Mokrousov, St. Petersburg Pasteur Institute, Russian Federation

Funding: The study was financially supported by Natural Science Foundation of Shanxi Normal University (ZR1106), Natural Science Foundation of Shanxi Province, China (2011011031-2) and High-tech Industrialization project of Shanxi Province, China (20110014). The funders had no role in study design, data collection and analysis, decision to publish, or preparation of the manuscript.

Competing Interests: The authors have declared that no competing interests exist.

* Email: guiqiny@dns.sxnu.edu.cn

Introduction

During the last three million years, repeated glacial and interglacial cycles have greatly affected the landform and fauna of the total earth [1–3]. Historical ecological and biogeographic factors are often considered to have played significant roles in shaping global biodiversity by influencing regional differences in speciation, extinction, and migration [4–6]. The genetic structures of the current species record the simultaneous consequences of two fundamental processes: population dynamics in response to past geological or climatic changes, and lineage sorting within a species under natural selection. Thus, knowledge of the evolutionary history of many plant species is central to the identification of divergence and speciation processes [3,7–8].

Phylogeographical analyses can provide an understanding of how paleo-environmental changes in landscape and climate have influenced species distributions and population demography [9]. Such analyses play an important role in understanding the evolutionary history of species with changes [3]. Over the last decade, molecular research on the evolutionary history of plant species, both in China and throughout East Asia, has been conducted with reference to past climatic oscillations [10–18]. One of the results of such research is the proposal by some authors that glacial refugia were maintained in both the northern and the southern regions, or at different spatial-temporal scales in China, during these glacial periods. These refugia are thus suggested to have acted as sites for subsequent range expansion during the interglacial (or postglacial) periods [12–14,19–20]. Some plant species have extended their distribution from southwestern China to central/northern China [21]. Reduction of genetic diversity is expected under a scenario of rapid postglacial expansion, as has been found in northern Europe and America [2]. Because central/ northern China is believed not to have been glaciated during the Quaternary [2,22–23], it is an open question whether species experienced past northern expansion and southern retreat during the Quaternary climatic oscillations.

Taihang Mountains locates on 35°19′–40°51′N and 113°10′–115°48′E. As a natural boundary mountain of the east, southeast of Shanxi province and Hebei and Henan provinces, it is the important mountain range and geographical boundary for eastern China. The northern part of Taihang Mountains is higher than its southern part. Most areas of Taihang Mountains are higher than 1200 meters. The uplift of Taihang Mountains, followed by the formation of high mountains and deep valleys within the plateau [24–25], was one of the most important geological events. Several lines of evidence suggest that the rapid uplift of Taihang

Mountains took place after the late Pleistocene [26–27]. Rich sources of species, most of which are endemic, are found in this region [24]. The high species richness in Taihang Mountains has led to the hypothesis that this region is a distribution and diversity center for many plant genera [24].

Molecular techniques have provided powerful tools for studying the phylogeography or migratory footprints of species [7–9,28–29]. Maternally inherited chloroplast DNA lineages in natural populations often display geographical structure [9], which is useful for deciphering the evolutionary history of species. The cpDNA markers are thought to be the most appropriate candidates because of their slow evolution and lack of recombination [30]. Nevertheless, the joint use of molecular markers derived from different genomes provides a more complete description of population structure and insights into population history and dynamics, particularly for comparisons of maternally inherited organelle and bi-parentally inherited nuclear markers [31]. Several studies about inter-specific and intra-specific divergence from China, using different molecular markers, have played an important role in discovering phylogeographical patterns in East Asian flora. This is particularly true for the mechanisms of plant speciation, along with the production of the high plant biodiversity and endemism found in East Asian flora [12–13,15,20,32]. However, most of the studied species from China have been trees [13,33–34]. Herbaceous vegetation has received much less attention [16]. Therefore, it would be of great interest to study the phylogeography of an herbaceous species to understand the evolution and modern distribution of the vegetation, especially in Taihang Mountains of central/northern China.

Opisthopappus longilobus Shih and *Opisthopappus taihangensis* (Ling) Shih belong to *Opisthopappus* (Asteraceae). The genus *Opisthopappus* is endemic to China, and its wild distribution is mainly restricted to Taihang Mountains across the provinces of Shanxi, Hebei, and Henan, occurring on the slopes at an elevation of about 1000 m or in the cracks of the steep cliffs [35]. *O. longilobus* is mainly distributed in the province of Hebei, whereas *O. taihangensis* is found in the provinces of Shanxi and Henan [35–36]. Both species are diploid (2n = 18) [25]. A strict morphological differentiation occurred between *O. longilobus* and *O. taihangensis*. The leaf blade of *O. longilobus* is smooth and subcylindrical, and most stem and leaves are pinnatifid, with one pair of bracteal leaf. *O. taihangensis*, apprised pubescent on both surfaces of leaf blade, stem, and leaves, is bipinnatifid, with no bracteal leaf [35–36]. *O. longilobus* and *O. taihangensis* have been listed among the Class II State-Protected Endangered Plant Species [36] due to the narrow distribution range, unique ecological environment, and serious artificial plucking.

O. longilobus and *O. taihangensis* are an ideal candidate for investigating the influences of Quaternary climate change in Taihang Mountains for the following reasons: (i) Geographically, *O. longilobus* and *O. taihangensis* are distributed only in Taihang Mountains, including Shanxi, Hebei, and Henan provinces. This geographic range may be advantageous for enabling investigation into whether the glacial refuges were maintained in Taihang Mountains during the last glacial maximum, or earlier cold periods. (ii) *O. longilobus* and *O. taihangensis* have intra-specific and endemic taxa in China that display visible phenotypic differences, which are important in exploring the incipient speciation caused by isolation provided by a glacial refuge. Finally, (iii) *O. longilobus* and *O. taihangensis* are an herbaceous plant and sensitive to environmental changes. In research on the effects of climate oscillations on plant evolution, it can represent the phylogeography of an herb, and can serve as a model to reveal the historical and evolutionary processes of such plants since the period of Quaternary climate change.

Thus, our objectives included the following: (1) explore the phylogeographical structure and the evolutionary history of *O. longilobus* and *O. taihangensis*; (2) elucidate the effects of the uplift of Taihang Mountains and the Quaternary climatic oscillations on the genetic structure, divergence and distribution of *O. longilobus* and *O. taihangensis*; (3) infer the possible refugia of *O. longilobus* and *O. taihangensis* across their whole distribution range.

Materials and Methods

Ethics statement

This study was conducted in accordance with all People's Republic of China laws. No specific permits were required for the described field studies because all researchers collecting the samples had introduction letters from the College of Life Science, Shanxi Normal University, Linfen.

Plant sampling

Thirteen populations of *Opisthopappus* were investigated in this study, covering the entire distribution area on Taihang Mountains. The samples comprised five populations of *O. longilobus* and eight populations of *O. taihangensis* (Table 1, Figure 1). Each population included 10 to 25 individuals that were collected at least 5 m apart. Healthy leaves were collected and dried in silica gel until total DNA was extracted. In each of the 13 populations, the ecological factors of each location were determined and recorded (Table 1).

DNA extraction, amplification and sequencing

Total DNA was extracted from the silica gel–dried leaves using the modified 2×CTAB procedure [37]. DNA quality was checked by electrophoresis on 0.8% agarose gels, and DNA concentration was determined using an Eppendorf biophotometer protein nucleotide analyzer (Eppendorf China Ltd., Beijing, Germany). The DNA samples were diluted to 10 ng·μL^{-1} and stored at 20°C for subsequent use.

The *trn*L-*trn*F sequences were amplified using *trn*L and *trn*F [38].The primers *ndh*J and *trn*L [39] were used to amplify the *ndh*J-*trn*L sequences. ITS sequences were then amplified using the primers ITS4 and ITS5 [40]. A polymerase chain reaction (PCR) was then performed in a 25 μL volume, with 50 ng plant DNA, 2×MasterMix (0.2 mM dNTPs, 3 mM $MgCl_2$, 1×PCR buffer, and 0.1 unit Taq DNA polymerase), and 0.6 mM of both forward and reverse primers. The PCR parameters for all amplification programs of ITS and cpDNA were as follows: 4 min of pre-denaturation at 94°C, followed by 34 cycles of 30 s of denaturation, 40 s of annealing (49.2°C for ITS or 58.8°C for *trn*L-*trn*F or 59.8°C *ndh*J–*trn*L), 1 min 20 s of elongation at 72°C, and a final elongation step of 7 min at 72°C. PCR products were purified using the Wizard PCR Preps DNA purification system (Promega, Madison, WI, USA) following the manufacturer's instructions. Cycle sequencing reactions were conducted using the purified PCR product, AmpliTaq DNA polymerase, and fluorescent BigDye terminators. The sequencing products were analyzed using an ABI Prism 310 DNA sequencer (Applied Biosystems Inc., Foster City, CA, USA).

Data analysis

The sequences were edited manually based on the chromatograms and aligned by CLUSTAL X [41] and then adjusted manually. Inserts and indels within all cpDNA and nrDNA sequences were firstly treated as a single character resulting from

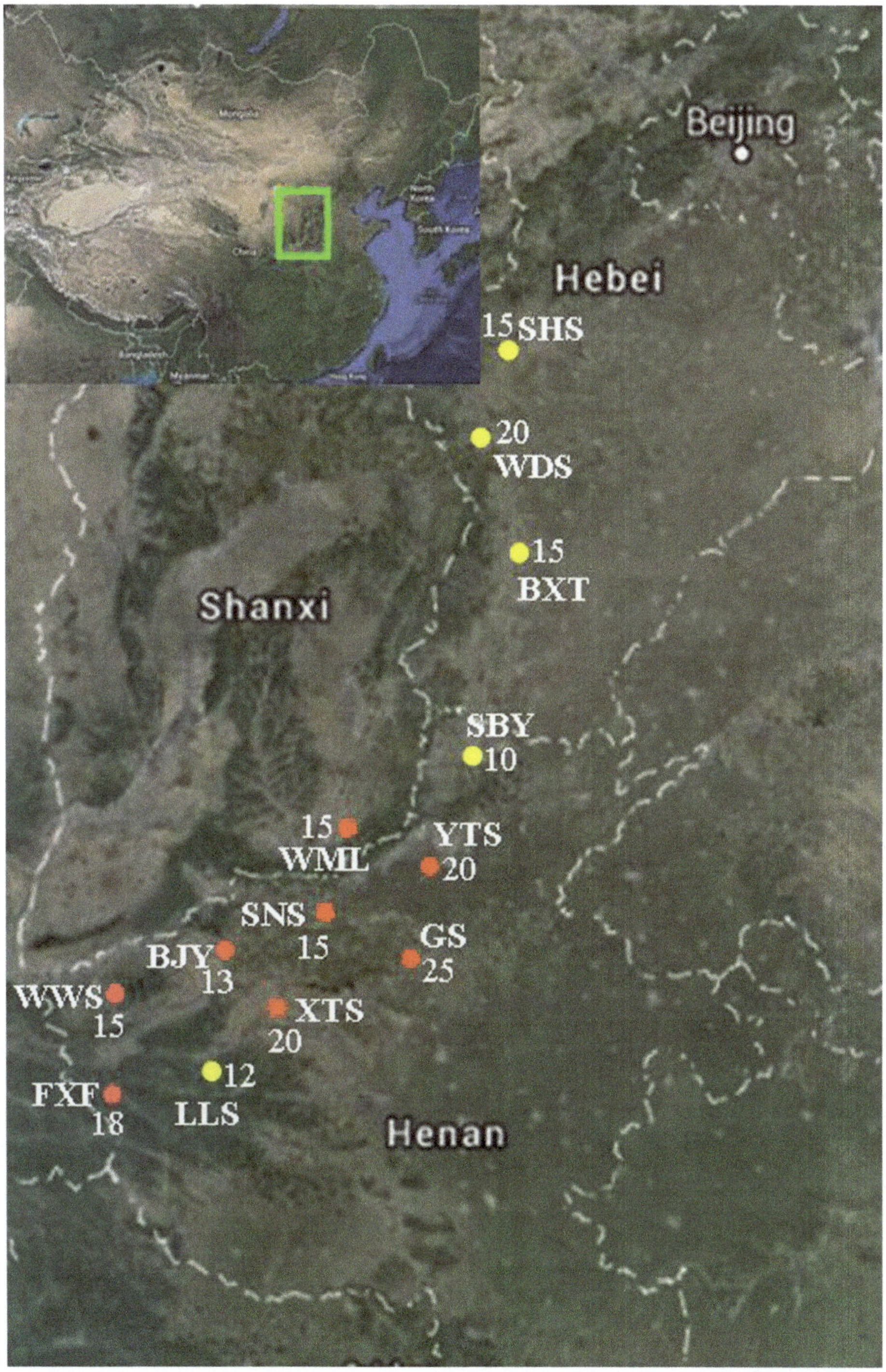

Figure 1. Map of sample sites for *Opisthopappus* (Asteraceae) on the Taihang Mountains. Location details are given in Table 1, and locality numbers correspond to those in Table 1. Yellow circle dots represent the populations of *O. longilobus*; red circle dots represent the populations of *O. taihangensis*.

Table 1. Location of the sampled *Opisthopappus* populations and the estimated diversity indexes.

Population		Geographic origin	Sample size	Latitude	Longitude	Number hapelotypes cpDNA (ITS)	Haplotype diversity cpDNA (ITS)	Nucleotide diversity cpDNA (ITS)
O. longilobus	WDS	Wudangshan, Hebei	20	113°47′	36°57′	6(7)	0.854(0.819)	0.00194(0.00240)
	BXT	Beixiangtang, Hebei	15	114°09′	36°02′	1(3)	0.000(0.556)	0.00000(0.00130)
	SHS	Shuanghuangshan, Hebei	15	113°57′	36°54′	2(3)	0.500(0.833)	0.00119(0.00391)
	LLS	Linlvshan, Henan	12	113°02′	34°43′	1(5)	0.000(0.857)	0.00000(0.00196)
	SBY	Shibanya, Henan	10	113°32′	36°03′	1(1)	0.000(0.000)	0.00000(0.00000)
Species level						8(15)	0.838(0.908)	0.00175(0.00537)
O.taihangensis	GS	Guanshan, Henan	25	113°34′	34°45′	1(3)	0.000(0.714)	0.00000(0.00251)
	YTS	Yuntaishan, Henan	20	113°25′	35°25′	2(2)	0.500(0.400)	0.00089(0.00059)
	XTS	Xiantaishan, Henan	20	113°49′	34°11′	2(1)	0.600(0.000)	0.00107(0.00000)
	SNS	Shennongshan, Henan	15	112°44′	35°16′	1(3)	0.000(0.618)	0.00000(0.00155)
	FXF	Foxifeng, Henan	18	112°62′	35°11′	1(3)	0.000(0.607)	0.00000(0.00162)
	WWS	Wangwushan, Henan	15	112°26′	35°38′	1(3)	0.000(0.700)	0.00000(0.00176)
	BJY	Baijiayan, Henan	13	112°92′	35°46′	2(2)	0.667(0.667)	0.00040(0.00293)
	WML	Wangmangling, Shanxi	15	113°35′	35°41′	2(2)	0.533(0.667)	0.00063(0.00098)
Species level						5(6)	0.562(0.558)	0.00073(0.00158)

one mutation. Haplotype diversity (h) and nucleotide diversity (π) were calculated for each population (h, π) and at the species level (h_d, π_d) using DNAsp [42]. The program PERMUT [44] was used to calculate the within-population diversity (h_S), total diversity (h_T), geographical total haplotype diversity (V_T), geographical average haplotype diversity (V_S), level of population differentiation at the species level (G_{ST}), and an estimate of population subdivisions for phylogenetically ordered alleles (N_{ST}). We further tested the phylogeographical structure at the species range between (N_{ST}) and (G_{ST}) by using the U-statistic. An N_{ST} value higher than the G_{ST} value indicates that closely related haplotypes are observed more often in a given geographical area than would be expected by chance [43].

To quantify the genetic differentiation partitioned among different groups and total genetic variance, analyses of molecular variance (AMOVA) were carried out using ARLEQUIN [44], with 1000 random permutations to test for significance of partitions. The spatial genetic structure of haplotypes was analyzed by spatial analysis of molecular variance using SAMOVA [45]. This program uses a simulated annealing approach to define groups of populations (K) that are geographically homogenous and maximally differentiated from each other. In this analysis of haplotypes, K varied from 2 to 13, with each simulation starting from 100 random initial conditions. An F_{CT} index of genetic differentiation among initial K groups was computed, followed by an iterative simulated annealing process to obtain the optimal configuration of groups and final F_{CT} values. The simulated annealing process was repeated 1000 times. The configuration with the largest F_{CT} value among the 100 tested was retained as the best grouping of populations.

Genealogical relationships between haplotypes were inferred from a maximum parsimony median-joining network calculated in NETWORK 4.5.0.2 [46]. To complement the results of NETWORK, TCS 1.21 [47] was also used to construct haplotype relationships. Phylogenetic analyses were conducted for the nrDNA and cpDNA sequence data using maximum likelihood (ML) and Bayesian inference (BI), respectively.

The time of the most recent common ancestor (TMRCA) of all haplotypes was estimated via a Bayesian approach implemented in *BEAST (Star BEAST) [48–49] using a GTR+I substitution model. The best substitution model was determined according to the Akaike Information Criterion (AIC) in jModeltest [50]. These models were applied in ML, BI, and subsequent *BEAST analyses. Bayesian inference was performed using MrBayes 3.1.2 [51]. Two independent runs of Metropolis-coupled Markov chain (MCMC) analysis were executed, each including one cold chain and three incrementally heated chains that started randomly in the parameter space. Two independent runs of 10^8 generations were carried out, with sampling at every 1000 generations. The first 25% of sampled trees were discarded as burn-in, and the remaining trees were used to construct a Bayesian consensus tree. The convergence of chains was checked using Tracer 1.5 [52]. The remaining trees were pooled to estimate the posterior probabilities (PPs).

No fossil records or biogeographic events isolating distinct populations are available to calibrate a cpDNA-IGS substitution rate for *O. longilobus* and *O. taihangensis*. Therefore, for our combined chloroplast non-coding regions, we assumed minimum and maximum values of a range of average mutation rates reported for synonymous sites of plant chloroplast genes [i.e., 1.2 and 1.7×10^{-9} substitutions per site per year (s/s/y)] [53]. These rates were then used for estimating TMCRA in *BEAST under a relaxed molecular clock assumption.

To detect whether population groups experienced recent population expansion and satisfied the assumption of neutrality, a mismatch distribution analysis was performed using the DNAsp program. Tajima's D and Fu and Li's D* were calculated for the entire genus and groups of populations. The statistical significance of D and D* was estimated with coalescent simulations as implemented in this program [54–55]. To further infer demographic processes, the null hypotheses of a spatial expansion and of a pure demographic expansion were tested in ARLEQUIN by comparing observed and expected distributions of pairwise sequence differences (mismatch distributions). For each expansion model, goodness-of-fit was tested using the sum of squared differences (SSD) and Harpending's [56] raggedness index (HRag).

Finally, isolation by distance (IBD) for both nrDNA and cpDNA data was tested based on pairwise geographical and genetic distances (F_{ST}) with the Isolation by Distance Web Service [57] (http://ibdws.sdsu.edu/~ibdws/), running 10,000 Mantel permutations.

Results

Patterns of variability in cpDNA and ITS sequences

The *trn*L-*trn*F and *ndh*J-*trn*L intergenic spacers varied from 800 to 866 bp and from 790 to 835 bp, respectively. The *trn*L-*trn*F intergenic spacer, which exhibits considerable nucleotide polymorphism, is characterized by 10 substitutions and a 4 bp insertions and deletions (Table S1). The 4 bp insertions and deletions region was mainly observed in *O. longilobus* populations. The *ndh*J-*trn*L spacer was characterized by one indel, three substitutions, and a 10 bp insertions and deletions (Table S1). The total cpDNA combined matrix comprised 1701 sites, of which 13 positions were variable and 18 harbored gaps. Ten haplotypes (Figure 2A) were observed by combining cpDNA data: eight haplotypes were within *O. longilobus* populations and five were within *O. taihangensis* populations. The haplotype diversity (h) ranged from 0 to 0.854, and the nucleotide diversity (π) from 0 to 0.00194 in the 5 populations of *O. longilobus* (Table 1). At species level, $h_d = 0.838$ and $\pi_d = 0.00175$ for *O. longilobus*. Within 8 *O. taihangensis* populations, h ranged from 0 to 0.667, π from 0 to 0.00107, h_d was 0.562 and π_d was 0.00073. The highest haplotype diversity was observed in population WDS from *O. longilobus* and BJY from *O. taihangensis* (Table 1).

The full ITS sequences, including ITS1+5.8S+ITS2, were 682 bp in length, comprising of 254 bp and 224 bp for ITS1 and ITS2, respectively. Eighteen haplotypes (Figure 2B) and 17 polymorphic sites, the latter consisting of 15 parsimony informative sites and 2 singleton variable sites, were detected at ITS sequences. Two populations, SBY and XTS, were monomorphic, whereas the remaining populations were polymorphic (Table 1). The haplotype diversity (h) ranged from 0 to 0.857 and the nucleotide diversity (π) from 0 to 0.00391 in the 5 populations *O. longilobus*. Within 8 *O. taihangensis* populations, the haplotype diversity h ranged from 0 to 0.714 and the nucleotide diversity π from 0 to 0.00293. For *O. longilobus*, $h_d = 0.908$, $\pi_d = 0.00537$; for *O. taihangensis* populations, $h_d = 0.558$, $\pi_d = 0.00158$ (Table 1).

Genetic diversity and population structure

A genetic diversity analysis revealed high genetic diversity harbored in *O. longilobus* and *O. taihangensis* populations. The parameters h_T, V_T, h_S, and V_S were 0.908, 0.916, 0.669, and 0.581, respectively, for *O. longilobus* based on cpDNA sequences. By ITS sequences, $h_T = 0.944$, $V_T = 0.937$, $h_S = 0.824$, and $V_S = 0.889$ within *O. longilobus* populations. For *O. taihangensis*

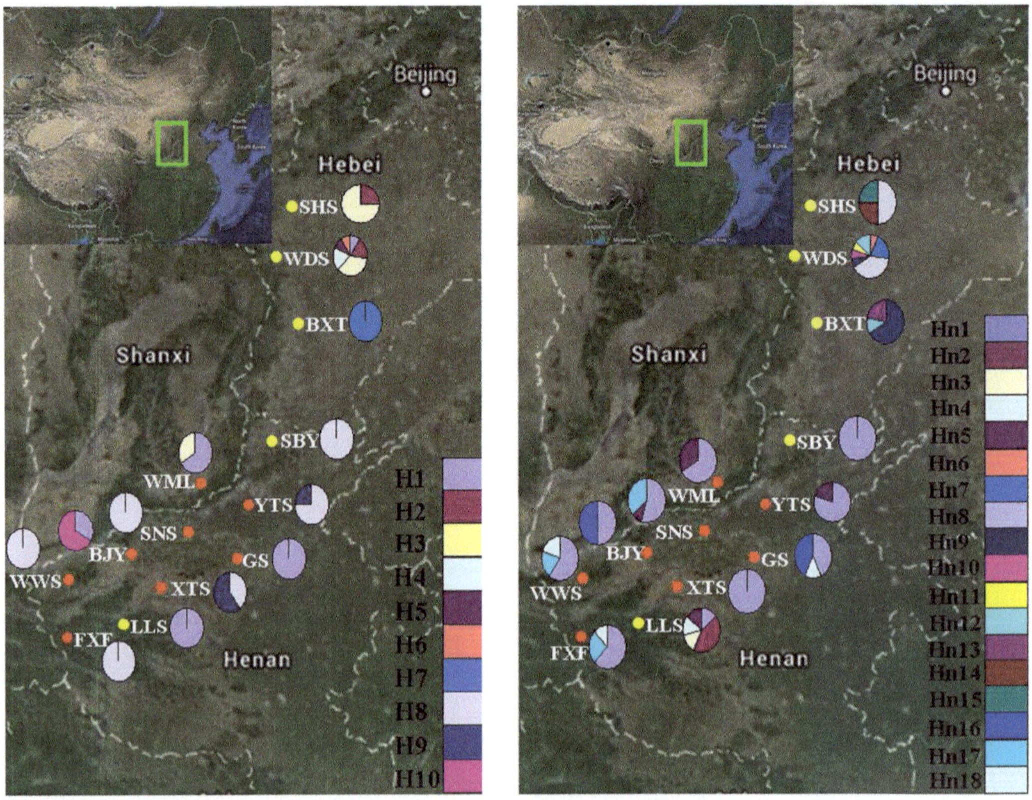

Figure 2. Geographical distribution of 10 cpDNA (A) and 18 (B) nrDNA haplotypes. The pie charts reflect the frequency of haplotype occurrence in each population. Haplotype colours correspond to those shown in panel.

populations, $h_T = 0.828$, $V_T = 0.833$, $h_S = 0.531$, and $V_S = 0.455$ by cpDNA sequences; $h_T = 0.806$, $V_T = 0.810$, $h_S = 0.704$, and $V_S = 0.592$ with nrDNA sequences.

Based on ITS sequences, phylogenetic analyses were carried (Figure S1). All studied populations made up a single monophyletic lineage, indicating that all of these divergent populations were derived within *Opisthopappus* itself, rather than the result of introgression by other species. Spatial genetic analyses of cpDNA and nrDNA haplotypes in all 13 populations using SAMOVA indicated that F_{CT} increased to a maximal value of 0.921 when $K = 2$ (K, the number of groups). Thus, the division of all the 13 sampled populations approximately into two groups is appropriate.

AMOVA analysis revealed that 60.58% and 67.70% ($P<0.01$) of the total variation was due to differences among the cpDNA and nrDNA populations for all studied populations, respectively (Table 2). When the populations were grouped by taxonomic variety, the cpDNA showed 27.28% variation among populations within species ($P<0.01$; Table 2). Similarly, for the nrDNA, AMOVA revealed that 26.68% of the variation was attributed to differences among populations within the species ($P<0.01$; Table 2). There was a significant ($P<0.01$) variation between species, suggesting that the two separately distributed groups have a genetic differentiation. In addition, 60.71% of the total cpDNA variation and 68.24% of the total nrDNA variation existed among *O. longilobus* populations (Table 2). The genetic structure of *O. taihangensis* populations showed a similar trend with that of *O. longilobus* populations, with most of the genetic differentiation occurring among populations (Table 2).

Results of the Mantel test showed a significant correlation between nrDNA data differentiation among populations and the natural logarithm of the geographical distances throughout the sampled range of *O. longilobus* ($r = 0.442$; $P = 0.001$) and *O. taihangensis* ($r = 0.385$; $P = 0.001$). Likewise, a significant correlation was detected among populations from cpDNA data of *O. longilobus* ($r = 0.401$; $P = 0.001$) and *O. taihangensis* ($r = 0.399$; $P = 0.001$). This indicates a significant correlation between genetic differentiation and geographical distance, implying strong IBD.

Haplotype and phylogeographical structure

In this study, similar topologies were obtained from the two different network approaches used to infer the relationships among the cpDNA and nrDNA haplotypes. Only the median-joining network is shown. The haplotype H8 was the most frequent haplotype in cpDNA data, occurring in 6 of the 13 populations. Three haplotypes, H1, H3, and H8, were shared by *O. longilobus* and *O. taihangensis* populations (Table 1 and Figure 3). In nrDNA data, Hn1 was the most frequent haplotype, occurring in 10 of the 13 populations (Figure 3 and Figure 4). The haplotypes Hn1, Hn4, and Hn5 were shared by *O. longilobus* and *O. taihangensis* populations (Table 1 and Figure 4). The distributions of haplotypes that were restricted to a single population were extremely skewed. Four cpDNA and nine nrDNA haplotypes were found in *O. longilobus* single populations. Only one cpDNA haplotype was identified for *O. taihangensis* populations.

Based on cpDNA data, we estimated a G_{ST} value (0.649, $P<0.05$), which is significantly smaller than the N_{ST} value (0.761). When the nrDNA was examined, the G_{ST} value (0.253) also

Table 2. Analyses of Molecular Variance (AMOVA) based on the nrDNA (ITS) and cpDNA sequences.

Source of variation	d.f	Sum of squares	Variance components	Variation percentage	Fixation indices
ITS(all population)					
Among populations	19	105.101	1.218	67.70	$F_{ST}=0.677$ ($P<0.01$)
Within populations	178	44.179	0.581	32.30	
ITS (species)					
Among species	1	54.511	1.113	48.16	$F_{CT}=0.482$($P<0.01$)
Among populations within species	18	50.590	0.617	26.68	$F_{ST}=0.748$($P<0.01$)
Within populations	178	44.179	0.581	25.16	$F_{SC}=0.515$($P<0.01$)
ITS(*O. longilobus*)					
Among populations	12	44.767	1.499	68.24	$F_{ST}=0.682$ ($P<0.01$)
Within populations	76	23.022	0.698	31.76	
ITS(*O. taihangensis*)					
Among populations	14	21.157	0.492	90.04	$F_{ST}=0.900$ ($P<0.01$)
Within populations	95	5.823	0.054	9.96	
ITS (SMOVA species)					
Among species	7	57.023	2.456	50.21	$F_{CT}=0.502$($P<0.01$)
Among populations within species	12	55.147	2.078	28.34	$F_{ST}=0.786$($P<0.01$)
Within populations	178	40.563	1.789	21.45	$F_{SC}=0.497$($P<0.01$)
cpDNA(all population)					
Among populations	19	78.452	2.187	60.58	$F_{ST}=0.606$ ($P<0.01$)
Within populations	178	50.234	1.023	39.42	
cpDNA (species)					
Among species	1	60.231	1.586	46.59	$F_{CT}=0.466$($P<0.01$)
Among populations within species	18	58.427	0.789	27.28	$F_{ST}=0.738$($P<0.01$)
Within populations	178	45.126	0.567	26.13	$F_{SC}=0.533$($P<0.01$)
cpDNA (*O. longilobus*)					
Among populations	12	29.313	1.032	60.71	$F_{ST}=0.607$ ($P<0.01$)
Within populations	76	19.364	0.668	39.29	
cpDNA (*O. taihangensis*)					
Among populations	14	21.601	0.467	68.64	$F_{ST}=0.686$ ($P<0.01$)
Within populations	95	9.183	0.214	31.36	
cpDNA (SMOVA species)					
Among species	7	56.273	1.756	52.34	$F_{CT}=0.523$($P<0.01$)
Among populations within species	12	55.012	1.111	27.55	$F_{ST}=0.799$($P<0.01$)
Within populations	178	41.603	0.654	22.36	$F_{SC}=0.498$($P<0.01$)

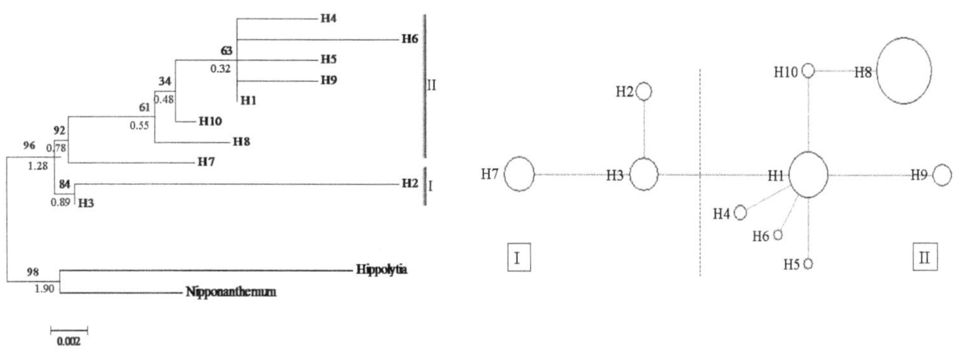

Figure 3. The evolutionary relationships among cpDNA haplotypes. (A) NJ phylogenetic tree for the 10 cpDNA haplotypes. Numbers above branches are support values from bootstrap resampling/Bayesian inference. (B) Median-joining network. Sizes of the circles are proportional to the overall frequency of the haplotypes in the entire sample of all species.

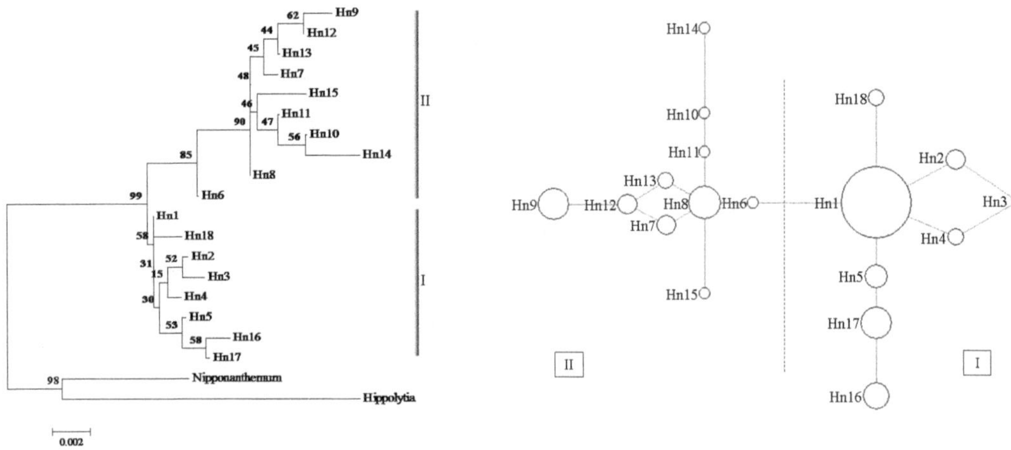

Figure 4. The evolutionary relationships among nrDNA haplotypes. (A) NJ phylogenetic tree for the 18 nrDNA haplotypes. Numbers above branches are support values from bootstrap resampling. (B) Median-joining network. Sizes of the circles are proportional to the overall frequency of the haplotypes in the entire sample of all species.

differed significantly from the N_{ST} value (0.641). A significant phylogeographic structure was indicated by both cpDNA and nrDNA data ($N_{ST}>G_{ST}$, $P<0.05$).

The phylogeographical relationships of the 10 cpDNA haplotypes and 18 nrDNA haplotypes were assessed under Neighbor-joining analyses (NJ) and Bayesian inferences drawn using *Hippolytia* and *Nipponanthemum* as outgroups (Figure 3 and Figure 4). Similar topology to phylogeographical relationships indicated a single phylogenetic split. All haplotypes were clustered into two lineages: I and II (Figure 3 and Figure 4). Lineage I contained some populations of *O. longilobus* with a high bootstrap support. The remaining haplotypes formed lineage II, containing some *O. longilobus* populations and all *O. taihangensis* populations, also with a high bootstrap support. This unrooted network of cpDNA haplotypes is consistent with the strict consensus tree produced by NJ and Bayesian inference and furthermore displays the relationship of interior (ancestral) and tip (derived) haplotypes (Figure 3). For lineage I (Figure 3), the haplotype H3 located in the center and fixed in three populations, being an ancestral haplotype. In clade II, H1 was shared by five populations. It appears to be an ancestral haplotype distributed in the center of clade II. The haplotype H8 was located in the tip of the network despite though contained a higher frequency of haplotypes. In the nrDNA haplotype network (Figure 4), Hn8 was ancestral for the lineage I; Hn1 was the predominant and widespread one in the clade II.

To test whether the current expansion had occurred in different populations, we calculated the frequency distribution of pairwise nucleotide differences among *O. longilobus* populations and *O. taihangensis* populations respectively. The resultant mismatch distribution consisted of multiple multimodal curves (Figure S2). Neutrality tests revealed nonsignificant positive values in considering all populations (Table 3). The observed mismatch distributions of cpDNA haplotypes were multimodal for all groups except for *O. taihangensis*. The mismatch distributions based on nrDNA haplotypes were present a similar pattern with cpDNA data (Table 3). Thus, a recent population expansion for *O. longilobus* is unlikely. *O. taihangensis* perfectly fits the expected expansion distribution (Figure 3).

Divergence times were estimated using the program *BEAST. The *ESS* values were from 150 to 240 for all the nodes discussed as follows. All sampled haplotypes of *Opisthopappus* coalesced at about 1.90 Ma (95% highest posterior density, HPD) (0.50–1.72), which indicated the origin of *Opisthopappus* during the late Tertiary periods. The divergence times, estimated at approximately 1.28 Ma (95% HPD) (0.70–2.34) between *O. longilobus* and *O. taihangensis*, are shown in the haplotype phylogenic tree (Figure 3A). *O. longilobus* and *O. taihangensis* thus appear to have diverged from each other during the early Pleistocene.

Discussion

Taxonomic implications

At the morphological character levels, the evidences that illustrated *O. taihangensis* as a separate species from *O. longilobus* were only leaf form and involucres. So, these two species were once regarded as a species. In this study, the analyses of the combined cpDNA and nrDNA matrix support the monophyly of two clades (Figure 3A, Figure 4A, and Figure S1). The phylogenetic relationships among all populations show an overall congruence with the result of the haplotype network (Figure 3B and Figure 4B). All haplotypes are split into two major groups according to cpDNA and nrDNA data. The populations from two distinct clades approximately exhibited specific geographical distributions in Taihang Mountains. SAMOVA analysis identified two approximately defined groups corresponding to *O. longilobus* and *O. taihangensis* populations at every level of divergence. Within DNA sequences of cpDNA and ITS, the total of thirty polymorphic sites harbored in populations. Unique haplotypes found in *O. longilobus* and *O. taihangensis* populations respectively. Moreover, a significant variation based on AMOVA occurred between *O. longilobus* and *O. taihangensis* ($F_{ST cpDNA}=0.482$ and $F_{ST nrDNA}=0.466$, Table 2). By used SRAP and cpSSR markers [58–59], AMOVA indicated that a significant genetic differentiation between *O. longilobus* and *O. taihangensis*. Accordingly, our molecular results support that *O. longilobus* and *O. taihangensis* were regarded as two species rather than a species. That result is consistent with the point of Hu [25]. The divergence times between *O. longilobus* and *O. taihangensis* (Figure 3A) indicate that genetic divergence between two species occurred in early Pleistocene. This split was related to the ecological and geographical habitat changes resulting from climate oscillations during the Quaternary glacial periods and complicated topography with the uplift of Taihang Mountains.

Table 3. Parameters of mismatch distribution analysis.

	Tajima's D	P	Fu's Fs	P	Fu and Li's D*	P	Fu and Li's F*	P	SSD	P	H_{RAG}	P
ITS data												
O. longilobus	0.085	>0.10	−3.988	0.012	0.283	>0.10	0.257	>0.10	0.111	<0.05	0.461	<0.05
O. taihangensis	0.483	>0.10	−0.787	0.157	1.002	>0.10	0.984	>0.10	0.106	<0.05	0.350	<0.05
cpDNA data												
O. longilobus	0.017	>0.10	0.271	0.182	0.012	>0.10	0.015	>0.10	0.168	<0.05	0.424	<0.05
O. taihangensis	0.261	>0.10	0.495	0.217	1.095	>0.10	0.977	>0.10	0.254	<0.05	0.806	<0.05

Geographic patterns of genetic diversity and structure

The high levels of genetic diversity were observed in *O. longilobus* and *O. taihangensis* populations (Table 1). These results are similar to previous research on herbaceous plants in East Asia and significantly exceed the average value of 0.67 for the 170 plant species documented [8,12,19,31,60]. In *O. longilobus*, the WDS population obtained the most haplotypes and highest genetic diversity, followed by the SHS population. This suggests the presence of two genetic diversity centers for *O. longilobus*, i.e., the regions where WDS and SHS populations are located. For *O. taihangensis*, the BJY and WML populations, which obtained the high genetic diversity, were genetic diversity centers for *O. taihangensis*. Significant genetic divergence and a highly structure were illustrated in all studied populations (Figure 3, Figure 4, and Table 2). For *O. longilobus* ($F_{STcpDNA} = 0.607$, $F_{STnrDNA} = 0.682$) and *O. taihangensis* ($F_{STcpDNA} = 0.686$, $F_{STnrDNA} = 0.805$), the results of our analysis suggested that significant variation was detected from among populations rather than from variation within groups (Table 2). Moreover, very high genetic differentiation in *O. longilobus* and *O. taihangensis* greatly exceeding the mean value ($G_{ST} = 0.22$) reported by Nybom [61].

Species genetic diversity and structure can be affected by life history traits (e.g., life cycle, breeding system, pollination mechanism) and environmental effects (e.g., geographical range, climate, topography) [61–62]. *O. longilobus* and *O. taihangensis* possess a sexual reproductive mode and a relatively long life cycle as perennial herbs, all of which are traits associated with low total genetic diversity [63–64]. However, high genetic diversity was observed in our study, and this diversity appears to result from high among-population genetic variation. It has been widely documented that the dispersal of pollen and seeds in plant species is strongly linked to the development of a population genetic structure in bi-parentally inherited (nuclear-encoded) and maternally inherited (cytoplasmic) genetic markers [12,31,61,62,64]. Generally, outcrossing taxa with high seed dispersal capacity retain the majority of both types of genetic marker variation within populations, whereas selfing (and/or asexual) taxa with restricted seed dispersal allocate the majority of such variation among populations. *O. longilobus* and *O. taihangensis* are self-compatible plant species [25]. These two species, therefore, might be expected to have a heterogeneous distribution of both cpDNA and nrDNA variation among populations. Taken at face value, the above results may suggest that restricted gene flow via pollen and seed has resulted in significant population genetic differentiation in *O. longilobus* ($N_{mcpDNA} = 0.12$; $N_{mnrDNA} = 0.10$) and *O. taihangensis* ($N_{mcpDNA} = 0.19$; $N_{mnrDNA} = 0.14$). The physiographic complexity of Taihang Mountains and the deeply carved valleys and ravines between the inhabit areas probably would imposes significant barriers to gene flow among populations of *O. longilobus* and *O. taihangensis*. The genetic divergence between the two lineages occurred at 1.28 Ma, corresponding to the early Pleistocene. This divergence falls within the glaciations during the Taku Glaciation (1.05–1.20 Ma). The glaciers and/or extremely low temperature in the mountains might also have created barriers to gene flow between geographically isolated populations, which therefore promoted the divergence. For both marker systems, we generally observed a good fit of these populations to an IBD model. When coupled with high levels of population subdivision, such conditions strongly suggest that geographical isolation had a larger historical role in extant population structure compared with limited pollen and seed flow alone. The habitats of *O. longilobus* and *O. taihangensis* are either rocky or massifs, and these species grow discontinuously in different habitats along the altitudinal gradient from 700 to 1500 m, in which the variable micro-surrounding, complex topography, and great altitudinal variability of the region might promote the high degree of genetic population differentiation [65–66]. Limited migration coupled with a heterogeneous environment could promote local adaptation and fixation of different alleles in *O. longilobus* and *O. taihangensis* populations.

Phylogeographical structure and inference of demographic history

Both *O. longilobus* and *O. taihangensis* populations have a significant phylogeographic structure. The current distribution of haplotypes could be a result of past climatic changes related to the advance/retreat of glaciations [11]. During the Quaternary periods, climatic oscillations resulted in repeated drastic environmental changes, which further caused massive range shifts of most plants and animals, leading to accumulated genetic differences and particular phylogeographic patterns [2–3].

Our analyses suggested that *Opisthopappus* originated during the Pliocene. Before the Pleistocene, the current range of Taihang Mountains would have been covered by grassland with roughly similar geographic conditions [67]. As herbs, *O. longilobus* and *O. taihangensis* can disperse following grassland migration. On this basis, we suspect that *O. longilobus* and *O. taihangensis* were once distributed throughout Taihang Mountains and its adjacent region. With uplift of Taihang Mountains [27] and climate fluctuation of Quaternary glacial periods, *O. longilobus* and *O. taihangensis* were fragmented and forced into the refuge areas. The flora would migrate and expand during interglacial and glacial periods. This migration and expansion resulted in ecological displacement, which might have resulted in populations situated at different spatial-temporal scales. Different habitats might enhance the isolation of plant populations and prevent gene flow, which in turn could lead to incipient allopatric speciation or further allopatric speciation [12,18,19,68,69]. During the Quaternary periods, the paleovegetation in Taihang Mountains repeatedly appear the replacement of grasslands and forests cycles [67]. Those multiple replacements likely fragmented and isolated the habitat of herbs. Therefore, the refuge areas for *O. longilobus* and *O. taihangensis* would have remained separated and fragmented at different spatial-temporal scales for a long time. Moreover, dramatic geological changes induced by the Taihang Mountains uplift contributed to the extremely complicated landscape, which likely acted as a major driving force for the separation of the *O. longilobus* and *O. taihangensis*. Divergent selection between populations in contrasting environments, longer temporal and spatial separation and fragmentation would accelerate their differentiation, and eventually promote allopatric speciation.

Following this divergence, *O. longilobus* and *O. taihangensis* underwent strikingly different histories. For *O. longilobus*, long-distance dispersal and colonization may be a major historical process. *O. longilobus* was divided into different lineages around 0.89 Ma, which fell within the interglacial stage between the Taku Glaciation and Lushan Glaciation. The paleovegetation grasslands of Taihang Mountains [67] might be regarded as a corridor connecting the currently isolated populations. Populations of *O. longilobus* in Hebei province may have extended across this corridor and reached Henan province. Haplotypes H1 and Hn1 located in populations LLS and WDS or LLS and SBY imply this long dispersal (Table 2). The 10 bp insertions of *ndh*J-*trn*L occurring in population LLS also supported this viewpoint (Table S1). Subsequently, with the advent of inter-glaciations or post-glaciations, the climate changed, which led to the paleovegetation of Taihang Mountains experiencing a transition from grasslands to forests. The corridor for *O. longilobus* migration gradually disappeared, which would have resulted in isolation of the population [11,12,68]. Additionally, we did not detect any population expansion due to the clear multimodel mismatch distribution (Figure S2). The complex geological conditions of Taihang Mountains would limit *O. longilobus* population expansion on a macro scale and thus restricting *O. longilobus* to sustain in situ during the Quaternary periods, perhaps by moving upwards and downwards in their mountain ranges. For *O. taihangensis*, a pattern of population expansion was observed, because the observed mismatch distribution for cpDNA and nrDNA haplotypes was unimodal (a pattern consistent with expansion) (Figure S2). This expansion was dated to be 0.78 Ma, which corresponding the interglacial stage between the Taku Glaciation and Lushan Glaciation. The above results suggest an expansion consistent with a range expansion and re-colonization before the mainly uplift of Taihang Mountains. With the uplift of Taihang Mountains, large geographical distances and significant topographical barriers could restrict gene flow and limit any large-scale population expansion [12,70]. The cpDNA sequences characteristic of the *trn*L-*trn*F and *ndh*J-*trn*L intergenic spacers further confirmed this hypothesis (Table S1).

Crandall and Templeton [71] suggest that ancestral haplotypes are located in a central position within the haplotype phylogeographical network and that potential refugia are usually characterized by widely dispersed ancestral haplotypes and other tip and unique haplotypes. Geographic areas displaying increased levels of genetic diversity are thus good candidates in the search for past refuges [8,34]. These regions are characterized by relatively stable ecological conditions during environmental fluctuations, which foster the accumulation of genetic diversity [72]. Most *O. longilobus* individuals or populations gathered in lineage I (Figure 3 and Figure 4). Haplotypes H3 and Hn8 are located in the center of the haplotype network. And Haplotype H3 is detected in three populations (WDS, SHS and WML), Hn8 in two populations (WDS and SHS) (Figure 2). WDS and SHS are located in the southern part of Hebei province and have higher haplotype diversity (Table 1). Thus, we suggest that the WDS and SHS populations' regions were two major potential refugia for *O. taihangensis*. These regions may therefore have been characterized by a relatively mild climate, potentially containing microclimatic environments able to impart relative stability to a range of habitats. Lineage II consisted of all populations of *O. taihangensis* and some individuals of *O. longilobus*. Haplotypes H1 and Hn1, which are located in the center of the haplotype network, were the most dominant haplotypes found in five and ten populations, respectively. Moreover, some populations in these haplotypes (H1 and Hn1) had higher haplotype diversity (Table 1, e.g., BJY, WML, and YTS). Thus, we suggest that there were multiple small refuges located in Shanxi and Henan provinces during the last glacial maximum or earlier cold periods for *O. taihangensis*. The above populations of *O. taihangensis*, distributed in southern Taihang Mountains, would have been provided with a complex and stable environment. Thus, the potential refugia area may have been located in those regions.

However, our comprehensive sampling showed that assessing the species-level relationships in our study is complicated by the fact that not all cpDNA and nrDNA haplotypes are species-specific. Incongruence between the species boundaries and the genealogy of their cpDNA and nrDNA sequences is illustrated by the sharing of haplotypes (e.g., H1 and Hn8) between the two species. On the other hand, several haplotypes were found within a single species (Table 1 and Figure 2). Haplotype distribution does not strictly follow species circumscription. Moreover, an association of haplotypes with geographically circumscribed regions rather than with taxonomic boundaries is a phenomenon observed in *O. longilobus* and *O. taihangensis*. The sharing of haplotypes and lack of reciprocal monophyly might be explained by the persistence of ancestral polymorphisms during speciation events and/or exchange of genes by inter-specific hybridization (Table S1 and Figure S1). These two species have similar habitats and the same period of flowering [25], and possible visitation by the same pollinators in the absence of physiological or genetic reproductive barriers might have enabled hybridization events. Nonetheless, further work on chromosomes and the sequencing of additional genomic regions are needed to make more in-depth conclusions about the hybrid status of these taxa.

Conclusions

In summary, *O. longilobus* and *O. taihangensis* have a high level of genetic diversity. The following phylogeographical structure and historical scenario of *O. longilobus* and *O. taihangensis* can also be drawn. The vicariance following the uplift of Taihang Mountains and a transition of paleovegetation of Taihang Mountains mainly shaped the present distribution of these two species. The above findings imply that multiple small refugia could have been maintained in the Taihang Mountains regions, which allowed *O. longilobus* and *O. taihangensis* to persist in situ and maintain sizable haplotype and nucleotide diversity at the lineage-wide scale during the glaciations. The current distribution pattern of *O. longilobus* and *O. taihangensis* refuges is similar to that of other East Asian plants (e.g., *Cathaya argyrophylla*, *Picea crassifolia*, *Sinopodophyllum hexandrum*, *Lagochilus Bunge* ex Bentham) [8,20,73], with multiple refuges sustained across their distribution ranges. Furthermore, complex topography rendered these refuge areas both isolated and fragmented in different geographic units. Geographical and ecological isolation likely restricted seed and pollen migration ability. The isolation and fragmentation further promoted the intra-specific split. These results support the conclusion that the high plant diversity and endemism found in central/northern China and throughout eastern Asia has mainly resulted from allopatric speciation due to the complex topography of Mountains and allopatric fragmentation during the late Tertiary and Quaternary periods.

Author Contributions

Conceived and designed the experiments: GQY YLW. Performed the experiments: YLW. Analyzed the data: YLW. Contributed reagents/materials/analysis tools: YLW. Wrote the paper: YLW. Revised the manuscript: GQY.

References

1. Avise JC, Walker D, Johns GC (1998) Speciation durations and Pleistocene effects on vertebrate phylogeography. Proceedings of the Royal Society of London Series B Biological Sciences 265: 1707–1712.
2. Hewitt GM (2000) The genetic legacy of the Quaternary ice ages. Nature 405: 907–913.
3. Hewitt GM (2004) Genetic consequences of climatic oscillations in the Quaternary. Philosophical Transactions of the Royal Society of London Series B Biological Sciences 359: 183–195.
4. Willis KJ, Whittaker RJ (2002) Species diversity-scale matters. Science 1245–1248.
5. Turchetto-Zolet AC, Cruz F, Vendramin GG, Simon MF, Salgueiro F, et al. (2012) Large-scale phylogeography of the disjunct Neotropical tree species *Schizolobium parahyba* (Fabaceae-Caesalpinioideae). Molecular Phylogenetics and Evolution 65: 174–182.
6. Xu JW, Chu KH (2012) Genome scan of the mitten crab *Eriocheir sensu stricto* in East Asia: Population differentiation, hybridization and adaptive speciation. Molecular Phylogenetics and Evolution 64: 118–129.
7. Cao MM, Jin YT, Liu NF, Ji WH (2012) Effects of the Qinghai–Tibetan Plateau uplift and environmental changes on phylogeographic structure of the *Daurian Partridge* (Perdix dauuricae) in China. Molecular Phylogenetics and Evolution 65: 823–830.
8. Meng HH, Zhang ML (2013) Diversification of plant species in arid Northwest China: Species-level phylogeographical history of *Lagochilus Bunge* ex Bentham (Lamiaceae). Molecular Phylogenetics and Evolution 68: 398–409.
9. Avise JC (2000) Phylogeography: The History and Formation of Species. Harvard University Press, Cambridge.
10. Yu G, Chen X, Ni J, Cheddadi R, Guiot J, et al. (2000) Palaeovegetation of China: a pollen databased synthesis for the mid-Holocene and last glacial maximum. Journal of Biogeography 27: 635–664.
11. Zhang YH, Volis S, Sun H (2010) Chloroplast phylogeny and phylogeography of *Stellera chamaejasme* on the Qinghai-Tibet Plateau and in adjacent regions. Molecular Phylogenetics and Evolution 57: 1162–1172;
12. Zhao YP, Qi ZC, Ma WW, Dai QY, Li P, et al. (2013) Comparative phylogeography of the *Smilax hispida group* (Smilacaceae) in eastern Asia and North America – Implications for allopatric speciation, causes of diversity disparity, and origins of temperate elements in Mexico. Molecular Phylogenetics and Evolution 68: 300–311.
13. Xie KQ, Zhang ML (2013) The effect of Quaternary climatic oscillations on *Ribes meyeri* (Saxifragaceae) in northwestern China. Biochemical Systematics and Ecology 50:39–47.
14. Tian B, Liu RR, Wang LY, Qiu Q, Chen KM, et al. (2009) Phylogeographic analyses suggest that a deciduous species (Ostryopsis davidiana Decne., Betulaceae) survived in northern China during the Last Glacial Maximum. Journal of Biogeography 36: 2148–2155.
15. Jia DR, Liu TL, Wang LY, Zhou DW, Liu JQ (2011) Evolutionary history of an alpine shrub *Hippophaë tibetana* (Elaeagnaceae): allopatric divergence and regional expansion. Biological Journal of Linnean Society 102: 37–50.
16. Chen KM, Abbott RJ, Milne RI, Tian XM, Liu JQ (2008) Phylogeography of *Pinus tabulaeformis* Carr. (Pinaceae), a dominant species of coniferous forest in northern China. Molecular Ecology 17: 4276–4288.
17. Bai WN, Liao WJ, Zhang DY (2010) Nuclear and chloroplast DNA phylogeography reveal two refuge areas with asymmetrical gene flow in a temperate walnut tree from East Asia. New Phytologist 188: 892–901.
18. Harrison SP, Yu G, Takahara H, Prentice IC (2001) Palaeovegetation: diversity of temperate plants in East Asia. Nature 413: 129–130.
19. Qiu YX, Guan BC, Fu CX, Comes HP (2009) Did glacials and/or interglacials promote allopatric incipient speciation in East Asian temperate plants? Phylogeographic and coalescent analyses on refugial isolation and divergence in *Dysosma versipellis*. Molecular Phylogenetics and Evolution 51: 281–293.
20. Li Y, Stocks M, Hemmilä S, Källman T, Zhu HT, et al. (2010) Demographic histories of four spruce (Picea) species of the Qinghai-Tibetan Plateau and neighboring areas inferred from multiple nuclear loci. Molecular Biology and Evolution 27: 1001–1014.
21. Wu CY (1995) Vegetation of China. Science Press, Beijing.
22. Zhao SL, Li GG (1990) Desertization on the shelves adjacent China in the Later Pleistocene. Oceanologia et Limnologia Sinica 8: 289–298.
23. Liu KB (1988) Quaternary history of the temperate forests of China. Quaternary Sciences Reviews 7: 1–20.
24. Zhu LM (2008) Spider community structure in fragmented habitats of Taihang Mountain area, China. Thesis for Master' Degree, Hebei Univiersity.
25. Hu X (2008) Preliminary studies on inter-generic hybridization within *Chrysanthemum* in broad sense (III). Thesis for Master' Degree, Beijing Forestry University.
26. Gong MQ (2010) Uplifting process of southern Taihang Mountain in Cenozoic. Chinese Academy of Geological Science Thesis for Doctor' Degree.
27. Wu C, Zhang XQ, Ma YH (1999) The Taihang and Yan Mountains rose mainly in Quarteranary. Norht China Earthquake Sciences 17(3): 1–7.
28. Gao LM, Möller M, Zhang XM, Hollingsworth ML, Liu J, et al. (2007) High variation and strong phylogeographic pattern among cpDNA haplotypes in *Taxus wallichiana* (Taxaceae) in China and North Vietnam. Molecular Ecology 16: 4684–4698.
29. Zhou TH, Li S, Qian ZQ, Su HL, Huang ZH, et al. (2010) Strong phylogeographic pattern of cpDNA variation reveals multiple glacial refugia

for *Saruma henryi* Oliv. (Aristolochiaceae), an endangered herb endemic to China. Molecular Phylogenetics and Evolution 57: 176–188.
30. Wolfe KH, Li WH, Sharp PM (1987) Rates of nucleotide substitution vary greatly among plant mitochondria, chloroplast, and nuclear DNAs. Proceeding of the National Academy of Sciences of the United States of America 84: 9054–9058.
31. Petit RJ, Duminil J, Fineschi S, Salvini D, Vendramin GG (2005) Comparative organization of chloroplast, mitochondrial and nuclear diversity in plant populations. Molecular Ecology 14: 689–701.
32. Zhang RY, Song G, Qu YH, Alström P, Ramos R, et al. (2012) Comparative phylogeography of two widespread magpies: Importance of habitat preference and breeding behavior on genetic structure in China. Molecular Phylogenetics and Evolution 65: 562–572.
33. Zhang Q, Chiang TY, George M, Liu JQ, Abbott RJ (2005) Phylogeography of the Qinghai-Tibetan Plateau endemic *Juniperus przewalskii* (Cupressaceae) inferred from chloroplast DNA sequence variation. Molecular Ecology 14: 3513–3524.
34. Zhao C, Wang CB, Ma XG, Liang QL, He XJ (2013) Phylogeographic analysis of a temperate-deciduous forest restricted plant (*Bupleurum longiradiatum* Turcz.) reveals two refuge areas in China with subsequent refugial isolation promoting speciation. Molecular Phylogenetics and Evolution 68: 628–643.
35. Wang FZ, Tang J, Chen XQ, Liang SJ, Dai LK, et al. (1978) Liliaceae. In: Editorial Board of the Flora of China of the China Science Academy. Flora of China 73–74 p.
36. Ding BZ, Wang SY (1998) Flora of Henan. Henan Science & Technology Press, Zhengzhou.
37. Doyle JJ, Doyle JL (1987) A rapid DNA isolation procedure for small quantities of fresh leaf material. Phytochemistry Bulletin 19: 11–15.
38. Taberlet PT, Gielly L, Patou G, Bouvet J (1991) Universal primers for amplification of three noncoding regions of chloroplast DNA. Plant Molecular Biology 17: 1105–1109.
39. Vijverberg K, Bachmann K (1999) Molecular evolution of a tandemly repeated trnF (GAA) gene in the chloroplast genomes of *Microseris* (Asteraceae) and the use of structural mutations in phylogenetic analyses. Molecular Biology and Evolution 16(10): 1329–1340.
40. White TJ, Bruns T, Lee S, Taylor J (1990) Amplification and direct sequencing of fungal ribosomal RNA genes for phylogenetics. In: Innis MA, Gelfand DH, Shinsky JJ, White TJ (eds), PCR Protocols: A Guide to Methods and Applications. Academic Press, San Diego 315–322 p.
41. Thompson JD, Gibson TJ, Plewniak F, Jeanmougin F, Higgins DG (1997) The ClustalX windows interface: flexible strategies for multiple sequence alignment aided by quality analysis tools. Nucleic Acids Research 25: 4876–4882.
42. Rozas J, Sanchez-DelBarrio JC, Messeguer X, Rozas R (2003) DnaSP, DNA polymorphism analyses by the coalescent and other methods. Bioinformatics 19: 2496–2497.
43. Pons O, Petit RJ (1996) Measuring and testing genetic differentiation with ordered versus unordered alleles. Genetics 144: 1237–1245.
44. Excoffier L, Laval G, Schneider S (2005) Arlequin ver. 3.0: an integrated software package for population genetics data analysis. Evolutionary Bioinformatics Online 1: 47–50.
45. Dupanloup I, Schneider S, Excoffier L (2002) A simulated annealing approach to define the genetic structure of populations. Molecular Ecology 11: 2571–2581.
46. Polzin T, Daneshmand SV (2003) On Steiner trees and minimum spanning trees in hypergraphs. Operations Research Letters 31: 12–20.
47. Clement M, Posada D, Crandall KM (2000) TCS: a computer program to estimate gene genealogies. Molecular Ecology 9: 1657–1660.
48. Drummond AJ, Rambaut A (2007) BEAST: Bayesian evolutionary analysis by sampling trees. BMC Evolutionary Biology 7: 214.
49. Heled J, Drummond AJ (2010) Bayesian inference of species trees from multilocus data. Molecular Biology and Evolution, 27, 570–580.
50. Posada D (2008) JModelTest: phylogenetic model averaging. Molecular Biology and Evolution 25: 1253–1256.
51. Ronquist F, Huelsenbeck JP (2003) MrBayes 3: Bayesian phylogenetic inference under mixed models. Bioinformatics 19: 1572–1574.
52. Rambaut A, Drummond AJ (2009) Tracer v1.5. <http://tree.bio.ed.ac.uk/software/tracer/>.
53. Graur D, Li WH (200) Fundamentals of Molecular Evolution, second ed., Sinauer Associates Inc., Sunderland, Massachusetts.
54. Tajima F (1989) Statistical method for testing the neutral mutation hypothesis by DNA polymorphism. Genetics 123: 585–595.
55. Fu YX (1997) Statistical tests of neutrality of mutations against population growth, hitchhiking and background selection. Genetics 147: 915–925.
56. Harpending HC (1994) Signature of ancient population growth in a low-resolution mitochondrial DNA mismatch distribution. Human Biology 66: 591–600.
57. Jensen JL, Bohonak AJ, Kelley ST (2005) Isolation by Distance, Web Service. BMC Genet. 6, 13 (Version 3.16, <http://ibdws.sdsu.edu/>).
58. Wang YL (2013) Chloroplast microsatellite diversity of *Opisthopappus* Shih. Plant Systematic and Evolution 299: 1849–1858.
59. Wang YL, Yan GQ (2013) Genetic diversity and population structure of *Opisthopappus longilobus* and *Opisthopappus taihangensis* (Asteraceae) in China determined using sequence related amplified polymorphism markers. Biochemical Systematics and Ecology 49: 115–124.
60. Wang YL, Li X, Guo J, Guo ZG, Li SF, et al. (2010) Chloroplast DNA phylogeography of *Clintonia udensis* Trautv. & Mey. (Liliaceae) in East Asia. Molecular Phylogenetics and Evolution 55: 721–732.
61. Nybom H (2004) Comparison of different nuclear DNA markers for estimating intraspecific genetic diversity in plants. Molecular Ecology 13: 1143–1155.
62. Loveless MD, Hamrick JL (1984) Ecological determinants of genetic structure in plant populations. Annual Review of Ecology and Systematics 15: 65–95.
63. Hamrick JL, Godt MJW (1989) Allozyme diversity in plant species. In: Brown AHD, Clegg MT, Kahler AL, Weir BS (eds.), Plant Population Genetics. Breeding and Genetic Resources. Sinauer, Sunderland MA 43–63 p.
64. Hamrick JL, Godt MJW (1996) Effects of life history traits on genetic diversity in plant species. Philosophical Transactions of the Royal Society of London Series B Biological Sciences 351: 1291–1298.
65. Till-Bottraund I, Gaudeul M (2002) Intraspecific genetic diversity in alpine plants. In: Körner C, Spehn EM (eds.), Mountain Biodiversity: A Global Assessment. Parthenon Publishing, New York 23–34 p.
66. Liu L, Hao ZZ, Liu YY, Wei XX, Cun YZ, et al. (2014) Phylogeography of *Pinus armandii* and its relatives: heterogeneous contributions of geography and climate changes to the genetic differentiation and diversification of Chinese white pines. PLoS one 9(1): e85920.
67. Yang XL, Xu QH, Zhao HP (1999) The vegetation succession of Taihang Mountains during the late glaciations. Geography and Territorial Research 15(1): 81–88.
68. Qiu YX, Fu CX, Comes HP (2011) Plant molecular phylogeography in China and adjacent regions: tracing the genetic imprints of Quaternary climate and environmental change in the world's most diverse temperate flora. Molecular Phylogenetics and Evolution 59: 225–244.
69. Qian H, Ricklefs RE (2001) Diversity of temperate plants in East Asia–reply. Nature 413: 130.
70. Chou YW, Thomas PI, Ge XJ, LePage BA, Wang CN (2011) Refugia phylogeography of Taiwaniain East Asia. Journal of Biogeography 38: 1992–2005.
71. Crandall KA, Templeton AR (1993) Empirical tests of some predictions from coalescent theory with applications to intraspecific phylogeny reconstruction. Genetics 134: 959–969.
72. Gong W, Chen C, Dobeš C, Fu CX, Koch MA (2008) Phylogeography of a living fossil: Pleistocene glaciations forced *Ginkgo biloba* L. (Ginkgoaceae) into two refuge areas in China with limited subsequent postglacial expansion. Molecular Phylogenetics and Evolution 48:1094–1105.
73. Wang HW, Ge S (2006) Phylogeography of the endangered *Cathaya argyrophylla* (Pinaceae) inferred from sequence variation of mitochondrial and nuclear DNA. Molecular Ecology 15: 4109–4122.

13

Flying between Sky Islands: The Effect of Naturally Fragmented Habitat on Butterfly Population Structure

Sandhya Sekar*, Praveen Karanth

Centre for Ecological Sciences, Indian Institute of Science, Bangalore, India

Abstract

High elevation montane areas are called "sky islands" when they occur as a series of high mountains separated by lowland valleys. Different climatic conditions at high elevations makes sky islands a specialized type of habitat, rendering them naturally fragmented compared to more continuous habitat at lower elevations. Species in sky islands face unsuitable climate in the intervening valleys when moving from one montane area to another. The high elevation shola-grassland mosaic in the Western Ghats of southern India form one such sky island complex. The fragmented patches make this area ideal to study the effect of the spatial orientation of suitable habitat patches on population genetic structure of species found in these areas. Past studies have suggested that sky islands tend to have genetically structured populations, possibly due to reduced gene flow between montane areas. To test this hypothesis, we adopted the comparative approach. Using Amplified Fragment Length Polymorphisms, we compared population genetic structures of two closely related, similar sized butterfly species: *Heteropsis oculus*, a high elevation shola-grassland specialist restricted to the southern Western Ghats, and *Mycalesis patnia*, found more continuously distributed in lower elevations. In all analyses, as per expectation the sky island specialist *H. oculus* exhibited a greater degree of population genetic structure than *M. patnia*, implying a difference in geneflow. This difference in geneflow in turn appears to be due to the natural fragmentation of the sky island complexes. Detailed analysis of a subset of *H. oculus* samples from one sky island complex (the Anamalais) showed a surprising genetic break. A possible reason for this break could be unsuitable conditions of higher temperature and lower rainfall in the intervening valley region. Thus, sky island species are not only restricted by lack of habitat continuity between montane areas, but also by the nature of the intervening habitat.

Editor: Daniele Canestrelli, Tuscia University, Italy

Funding: This work used funding from the Department of Biotechnology (DBT), Government of India. Website: http://dbtindia.nic.in/index.asp. Grant number: BT/24/NE/TBP/2010. The funders had no role in study design, data collection and analysis, decision to publish, or preparation of the manuscript.

Competing Interests: The authors have declared that no competing interests exist.

* E-mail: sandysek@gmail.com

Introduction

"Sky islands" are a sequence of high elevation mountain areas separated by lowland valleys [1]. The high elevation areas are characterized by different climate, and hence different vegetation. Sky islands can be regarded as continental equivalents of oceanic islands, with mountain tops acting as cradles of evolution, and the intervening valleys with very different climate, akin to a sea of alien vegetation [1]. Species that inhabit the high elevation montane areas have been shown to be physiologically incapable of dispersing through low lying valleys [2].

There are many examples of restricted gene flow between sky islands from the well studied Madrean archipelago of sky islands. One of the most extreme examples is that of genetic diversification over a small spatial scale in the jumping spider *Habronattus pugillis* [3]. Reduced gene flow has also been reported in other taxa, like the longhorn cactus beetle [4], the New Mexico ridge-nosed rattlesnake, *Crotalus willardi obscurus* [5] and the black bear *Ursus americanus* [6]. Similar results have been obtained in other sky island systems like the Rocky mountains (for instance, reduced gene flow in the alpine plant *Sedum lanceolatum* [7] and the montane grasshopper *Melanoplus oregonensis* [8]).

The high elevation areas of the Western Ghats mountain range in southern India form one such sky island group. The Western Ghats (WG) is a 1600 km long mountain chain, which extends from north to south along the west coast of peninsular India. The mountain chain is broken by only one large gap: the 40 km long Palghat Gap. The southern Western Ghats are on an average much higher than the northern Western Ghats, with Anaimudi in south Kerala being the highest peak, with an elevation of 2295 m. These high elevation montane regions have an altitude of over 1900 m and are characterized by a special type of vegetation - the shola-grassland complex. The sholas are stunted forests with an upper story of trees, a twisted second story and a dense shrub layer of saplings; they occupy ridges between mountaintops. The shola forests are surrounded by undulating grasslands, consisting of fire and frost-resistant species of grasses [9].

The sky islands of the Western Ghats exhibit a high degree of endemism, with fifty endemic plant genera, half of which are monotypic, and with almost 50% of India's 206 amphibian species endemic to this region [10]. This makes it more crucial to study the gene flow patterns among populations of species isolated in such a naturally fragmented system. The phylogeographic structure of the white bellied shortwing (*Brachypteryx major*), a bird endemic to the shola forests of the Western Ghats, showed maximal genetic differentiation across the widest valley in the Western Ghats - the 40 km wide Palghat Gap [11]. Some montane plant species have also been shown to exhibit genetic differentiation among sky island populations [12,13].

These studies suggest that species confined to sky islands tend to have structured populations, ostensibly due to reduced gene flow across the sky islands. However, most of these studies have looked at a single sky island species, and more importantly, have not compared it with non-sky island species. Comparing species which inhabit the naturally disjunct sky island habitat, with similar species that inhabit the more continuously distributed low and mid elevation areas, can offer some insights into whether sky island species are capable of maintaining connectivity between different populations.

Butterflies in the Western Ghats make an interesting system to address such questions, because the composition of the butterfly community changes drastically between the mid-elevation and montane areas. Genetic structure as a result of habitat specialization has been shown in butterflies earlier, especially in the case of montane species, like in the high altitude meadow butterflies *Parnassius smintheus* [14], and *Erebia melampus* [15]. However, these studies have not compared sky island species with continuously distributed species from lower elevations.

In this study, we looked at the population genetic structure of two closely related butterfly species from peninsular India, both from the subtribe Mycalesina (Family Nymphalidae, subfamily Satyrinae). *Heteropsis oculus* is an endemic species found only to the south of the Palghat Gap in the Western Ghats, and restricted to montane areas above 1200 m. The closely related *Mycalesis patnia* is a generalist woodland species that occupies lower and mid elevation areas. The two species are similar with respect to all traits that affect dispersal in butterflies, except for the kind of habitats they occupy [16]: both species are of approximately the same size, their larvae are generic grass feeders, they have multiple generations throughout the year, and are on the wing for most of the year. Both species are weak fliers, and do not undertaking noticeable long distance flights [17]. The relatedness and high similarity between the two species makes this a robust comparison; thus a difference in genetic structure between them can be attributed to the difference in the spatial orientation of their habitat with greater confidence.

In this study, we predict that the sky island specialist *Heteropsis oculus* will have more geographically structured populations, because of the spatial orientation of montane forests (disjunct habitat) along the Western Ghats. *Mycalesis patnia*, being more continuously distributed in the lower elevations, will have a more homogenous population structure. Such a result will provide greater support to the hypothesis that sky island species tend to have structured populations due to reduced gene flow across suitable habitat patches. We sampled both species along the Western Ghats, and also sampled *Heteropsis oculus* intensively across one mountain range, the Anamalais, to enumerate gene flow between sky island populations separated by short distances.

Materials and Methods

Study Species

Both *M. patnia* and *H. oculus* belong to subtribe Mycalesina, which is part of subfamily Satyrinae and Family Nymphalidae. The subfamily Satyrinae in peninsular India has about 20 species in the low and mid elevation areas, which are almost completely replaced by just two species in the higher elevation areas [17]. Even if other species do occur, their densities are very low. The sub tribe Mycalesina has about six species in low and mid elevation areas, which are almost completely replaced above 1200 m by *Heteropsis oculus* (henceforth HO) to the south of the Palghat Gap, and *Heteropsis adolphei* to the north of the Palghat Gap. Both *Heteropsis* species are endemic to the sky islands to the north and south of the Palghat Gap respectively [18]. *Mycalesis patnia* (henceforth MP) is a generalist woodland species endemic to the Western Ghats and Sri Lanka that occupies evergreen, deciduous and scrub forests in the lower and mid elevation areas. HO belonged to the genus *Mycalesis* until recently, when a molecular phylogenetic study placed it in *Heteropsis*, another genus in the same subtribe [19].

Sampling

Both study species were collected from across their range in the Western Ghats: MP from northern Karnataka to south Kerala, and HO from the south of the Palghat Gap in Kerala and Tamil Nadu. We conducted active searches and also set up traps using fermented bait made from a mixture of over-ripe bananas, sugarcane juice and rum. For intensive sampling of HO, three hill complexes that are high enough for shola forests south of the Palghat Gap were demarcated, using an elevation map of the Western Ghats. These included the Anamalai-Palni complex (ANA), the Meghamalai-High Wavies complex (MEGH) and the Agastyamalai complex (AGA) (Figure 1 panel C). The ANA complex was sampled most intensively for HO, resulting in 58 individuals over a distance of about 60 km. The full HO dataset had 99 individuals sampled over a straight-line distance of about 200 km.

MP was sampled from the low elevation areas corresponding to the above shola complexes (see Figure 1 panel B). A total of 57 samples, including 45 from the south of the Palghat Gap, and 12 from the north, were collected. In all areas, we took care to sample from many different patches to capture maximal genetic diversity within the area. All captured individuals were killed immediately and stored in paper envelopes as voucher specimens. We would like to thank the forest departments of Kerala, Karnataka and Tamil Nadu for collection permits. For a full list of collected specimens, and the GPS locations of where they were collected, refer to Table S1 in the Supporting Information.

Lab Techniques

We first tried out sequencing of mitochondrial DNA markers (cytochrome oxidase b – cyt b, cytochrome *c* oxidase subunit I - COI, and control region). All these markers did not show any intraspecific variation within the specimens we had from the Western Ghats. So, we decided to use the Amplified Fragment Length Polymorphisms (AFLPs) technique instead. The AFLP protocol is known to be sensitive to contamination, especially from microbial DNA, which is abundant in the digestive system of the insect [20]. Hence, care was taken to use only two legs for the extractions. The extraction protocol was adapted from Reineke *et al* [21]. All extracts were quantified using a spectrophotometer NanoDrop ND-1000 (NanoDrop Technologies, Wilmington, DE, USA), and then diluted to 20 ng/µl to ensure uniform amounts of DNA in the starting material.

The restriction and ligation reactions were accomplished in a single step. The total DNA was digested with 1 unit of MseI (New England Biolabs) and 5 units of EcoRI (New England Biolabs), along with the adaptors. The MseI adaptors were 5′-GAC GAT GAG TCC TGA G-3′ (forward) and 5′-TAC TCA GGA CTC AT-3′ (reverse). The EcoRI adaptors were 5′-CTC GTA GAC TGC GTA CC-3′ (forward) and 5′-AAT TGG TAC GCA GTC TAC-3′ (reverse). The restriction-ligation reaction was carried out at 37 degrees for 4 hours. The products were diluted with 90 µl Tris-EDTA before the next step. The pre-selective PCR used the EcoRI primer E/A (5′-GACTGCGTACCAA TTCA-3′) and the

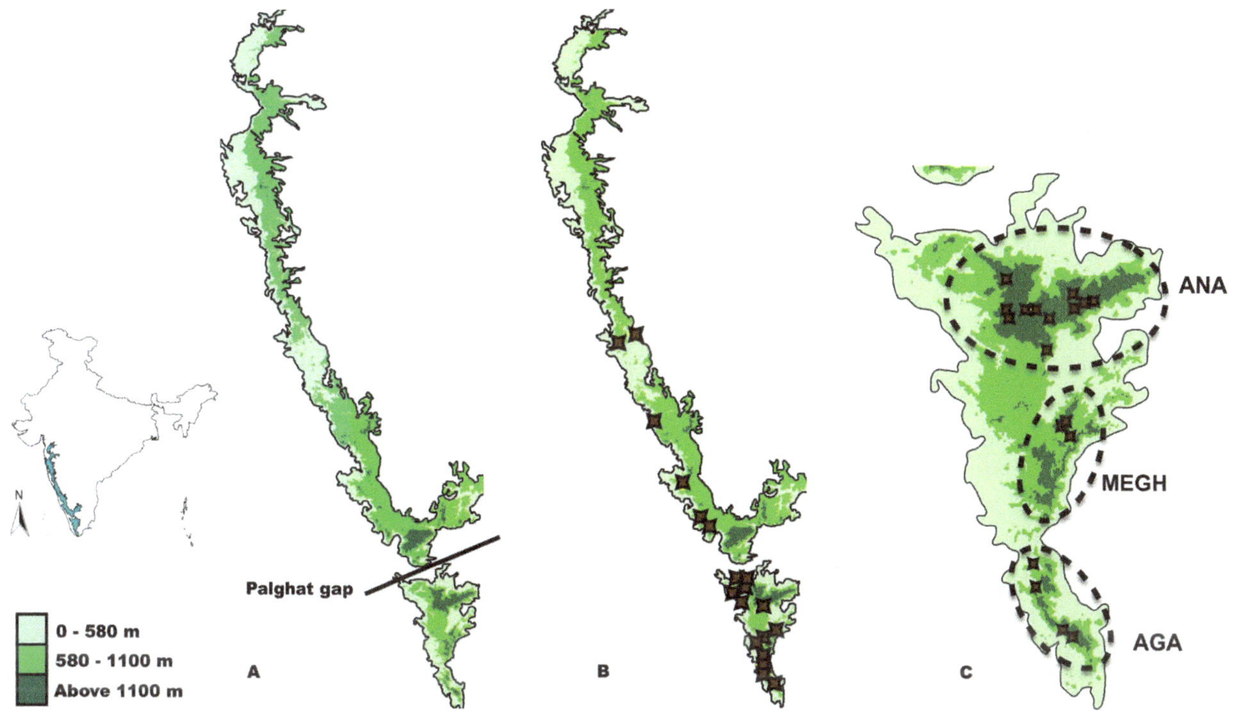

Figure 1. Maps showing sampling locations. Inset: A map of India showing the position of the Western Ghats. Panel A is an outline of the Western Ghats, showing the altitudinal gradient. A scale shows the altitudes represented by the different shades of green. The largest break in the mountain chain, the 40 km long Palghat Gap, is also shown. Panel B shows sampling locations for *Mycalesis patnia*. Panel C is the outline of the Western Ghats south of the Palghat Gap, showing sampling locations for *Heteropsis oculus*. The three shola complexes sampled in this study are indicated with dashed ellipses. (ANA – Anamalai complex; MEGH – Meghamalai complex; and AGA – Agastyamalai complex).

MseI primer M/C (5′GATGAGTCCTGAGTAAC-3′), with the PCR amplification parameters as follows: 72°C for 2 min, 30 cycles of denaturing at 94°C for 30 sec, annealing at 56°C for 30 sec, extension at 72°C for 2 min, and final elongation at 60°C for 10 min.

The product of the pre-selective PCR was run on a 2% agarose gel to check for amplification, indicated by a smear in the 100 to 1000 bp range. Successful samples were diluted with 100 µl using Tris-EDTA, and used in the selective amplification. AmpliTaq Gold (Applied Biosystems) was used during the selective PCR to minimize amplification error. The PCR parameters were: 94°C for 2 min, and 36 cycles of denaturing at 94°C for 30 sec, annealing for 30 sec and extension at 72°C for 2 min. Annealing was initiated at 65°C and then reduced by 0.7°C per cycle for the next 12 cycles, and maintained at 56°C for the next 23 cycles.

Four selective primer pairs were chosen, after testing over 20 combinations of primer pairs. Selective primers with both 2 and 3 extra nucleotides than the preselective primers (+2 and +3 primers [22]) were tested. The nucleotides to be added were chosen based on the relative numbers of A and C [23]. The EcoRI primers were labeled with fluorescent dyes. The final four primer pairs for each species are given in Table 1. All oligonucleotides were procured from Eurofins Genomics India Private Limited. All PCR reactions were carried out in a BioRad thermocycler.

After the selective PCR, the samples were diluted by 1:15, and run on the 16 capillary Applied Biosystems 3130xl sequencer at the National Centre for Biological Sciences, Bangalore. 0.6 µl of the sample was mixed with 0.5 µl of GeneScan-500 ROX size standard (Applied Biosystems) and 12 µl of Formamide. The plate was centrifuged to mix the contents, and then the DNA denatured by heating at 95°C for 3 min, followed by a quick chill on ice. The Applied Biosystems Standard Dye Set D was used to detect peaks. The electropherograms were visualized, and edited where necessary, using Peak Scanner (Applied Biosystems). The sizing table, containing information on the intensity and size of the peaks, was obtained from Peak Scanner, and imported into RawGeno [24] for binning and scoring.

In order to minimize error, the same person (SS) did all the lab work, from extraction to loading the samples into the sequencer. All the samples could not be worked on at the same time, because the field work was done in batches. To ensure that between run variability did not affect results, positive and negative controls were run and error rates checked at the start of each batch of runs. When needed, all the electropherograms from one primer pair were normalized before analyses using the R package AFLPScore [25]. About 10% of the total number of samples (both species combined) were used as blind replicates to estimate error rates (using AFLPScore), which came to about 4%. The estimated error rate of 4% is considered acceptable for publication [26].

Analyses

During the analysis, we compared two AFLP datasets: the MP dataset, with all the individuals of *M.patnia* ($n = 57$), and the HO dataset, with all the individuals of *H.oculus* ($n = 99$). Additionally, since we sampled the Anamalais region intensively for HO, we performed all the analyses for a reduced dataset of HO, which included only individuals collected from the Anamalais (the HO-ANA dataset, $n = 58$; see Figure 2 Panel A for a map showing sampling locations). For all datasets, we carried out 1) individual based analyses to obtain overall patterns in the dataset - genetic diversity, spatial autocorrelation, isolation by distance and clustering; and 2) population based analyses after assigning individuals into populations - genetic differentiation and assignment tests.

Overall Patterns

a) Spatial autocorrelation. The spatial autocorrelation statistic *r*, which is closely related to Moran's *I*, was estimated using the methodology from [25], implemented in GenAlex [26]. Like Moran's *I*, *r* also assumes values ranging from −1 to +1. Eight uniform distance classes of 100 km each was used for the MP dataset, five distance classes of 50 km each for the HO dataset, and six distance classes of 10 km each for the HO-ANA dataset. The significance of *r* in each distance class was established using 1000 permutations [27]. The result of the permutation tests was analyzed using both one-tailed and two-tailed tests. The differences between the two datasets were estimated using three statistical tests, following the methodology described in Banks (2012) [27].

b) Isolation by distance. Associations between geographic and genetic distance in both species were tested using a Mantel's test, implemented in the package ade4 version 1.4 [28] in the statistical language R v2.14 [29]. Jaccard's distance measure was used to compute a genetic distance matrix, which is the best at dealing with homoplasy in binary data [30]. A simple Euclidean distance matrix was calculated for geographic distance. Both distance matrices were calculated using the R package "proxy" v 0.6 [31]. The Mantel test was run for 1000 permutations to test for significance.

c) Clustering analyses. Since Bayesian clustering algorithms like STRUCTURE make a lot of assumptions (Hardy-Weinberg equilibrium, linkage equilibrium and regular sampling) while performing the clustering of genotypes [32], multivariate clustering approaches were adopted. A recently developed protocol, the spatial principal component analyses (sPCA) technique, was used [33]. sPCA uses a synthetic variable derived from variation in allele frequencies and spatial information. Spatial information is usually provided in the form of a connection network, chosen based on the sampling protocol. Since sampling was clumped for MP and HO datasets, the inverse distances connection network was used; for HO-ANA, since the sampling was more evenly placed, a Gabriel graph connection network was used [33].

Table 1. List of selective primers used in the AFLP analysis of MP and HO.

Mycalesis patnia			Heteropsis oculus		
Labeled	Unlabeled	Number of loci	Labeled	Unlabeled	Number of loci
E4	M1	126	E4	M1	136
E4	M7	152	E4	M3	113
E7	M1	121	E5	M7	140
E7	M7	139	E7	M7	127
Total loci: 538			Total loci: 516		

E4 (FAM): GACTGCGTACCAATTC-CAA; E5 (HEX): GACTGCGTACCAATTC-CAT; E7 (FAM): GACTGCGTACCAATTC-CT. Here, fluorescent dye used for labeling is mentioned in brackets. FAM: 6-FAM Fluorescein (blue dye) and HEX: hexachlorofluorescein (green dye). M1: GATGAGTCCTGAGTAAACAT; M3: GATGAGTCCTGAGTAAA CTA; M7: GATGAGTCCTGAGTAAA-AT.

Two permutation tests were used to test for positive and negative spatial autocorrelation: the Gtest (test for global structures, positive spatial autocorrelation), and Ltest (test for local structures, negative spatial autocorrelation). The PCA was performed in the package ade4 version 1.4 [28], and sPCA in the package adegenet version 1.3 [34], both implemented in the statistical language R v2.14 [29].

Population Based Analyses

In order to carry out the population-based analyses, we grouped samples from each dataset into populations by putting individuals from the same area together. For deciding the cutoff for placing individuals to a group, we used the first quartile from a distribution of geographic distances between sampling coordinates. This gave rise to three groups in the HO dataset (see panel A of Figure 3): Anamalai group (ANA, $n = 58$), Agastyamalai (AGA, $n = 21$) and Meghamalai (MEGH, $n = 20$). The HO-ANA dataset clustered into seven groups (see panel A of Figure 2): Pambadam shola (PAM, $n = 9$), Mannavan shola (MANN, $n = 7$), Akkamalai (AKK, $n = 4$), Rajmala (RAJ, $n = 6$), Vaguvaraiyar shola (VAG, $n = 10$), Mathikettan shola (MATH, $n = 11$) and Kodaikanal (KODI, $n = 11$). The MP dataset had five groups (see panel A of Figure 3): samples from Anamalais hill complex and the surrounding plains (ANA, $n = 15$), samples from Meghamalais and Periyar (MEGH, $n = 15$), samples from Agastyamalais (AGA, $n = 15$), samples from northern Karnataka (NKAR, $n = 7$) and samples from Wayanad and Coorg (WAC, $n = 5$).

a) Genetic diversity. Genetic diversity parameters were estimated using the program AFLPSurv [35]. Allele frequencies were estimated using a Bayesian method with non-uniform prior distribution of allele frequencies [36], under the assumption of Hardy-Weinberg genotypic proportions. This method has been demonstrated to be the best available for dominant markers as it has the least bias [37]. Heterozygosities were further calculated using a method specifically for dominant markers [38].

b) Genetic differentiation. The genetic differentiation between the groups in each dataset was estimated using both band-based and allele-frequency based F_{ST} [37,39]. AFLPSurv was used to estimate allele frequencies, as described earlier, and the allele frequency values were used in the calculation of F_{ST}. Pairwise genetic distances (Nei's D, after [38]) were also estimated. The significance of the F_{ST} values was evaluated using a permutation test, with 1000 permutations. The program Arlequin [40] was used to perform the Analysis of Molecular Variance (AMOVA), in order to calculate the molecular variance found among groups, within groups, within populations and within individuals.

c) Assignment tests. Assignment tests were carried out in AFLPOP [41]. A relatively high minimum log-likelihood difference (MLD) of 1 was used, to reduce chances of misassignment [42]. The value for ε was chosen as 0.001, as per the default settings. When individuals were allocated to a population other than the one where they were sampled from, the AFLPOP simulation option was used to assess the probability of incorrect assignment, using 1000 random specimens. The simulation gives a *P* value of allocation to the second population; when this *P* value is low (<0.01), the chance of correct allocation to the second population is high. In some cases, the software cannot place an individual into any of the populations with high confidence; in such cases, the individual is considered an immigrant from outside the sampled metapopulations.

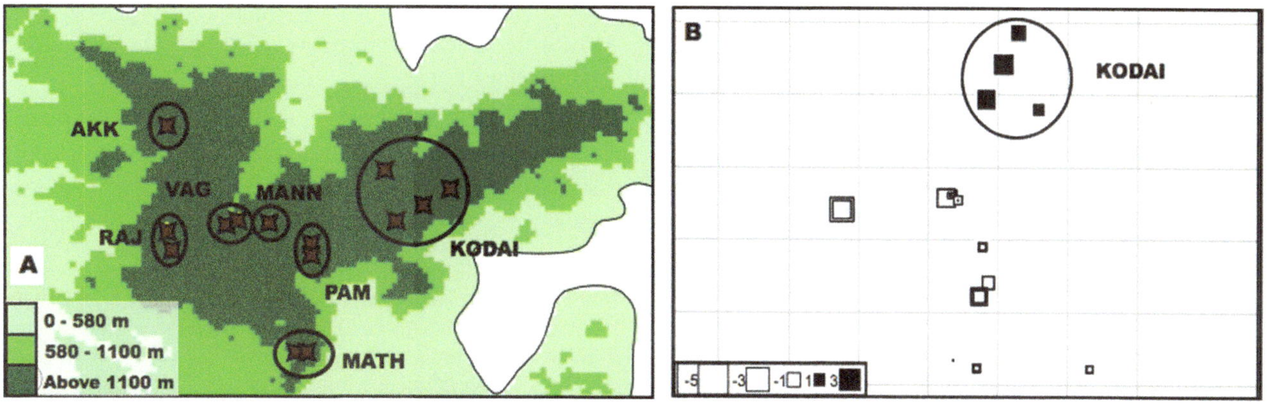

Figure 2. sPCA plots for the Anamalai population of *Heteropsis oculus.* Panel A shows the populations created for the population based analyses. Panel B is a representation of the first axis sPCA scores. The graph denotes the scores using squares; colour of the squares denotes positive (black) or negative (white) spatial autocorrelation, and the size of the squares denotes the magnitude of genetic variance. The squares are overlaid on a connection graph that represents the sampling design. If there was genetic differentiation between populations, squares of similar colour will clump together in one population. The individuals from Kodaikanal (denoted with a circle, named KODAI) are well differentiated from the other individuals in the Anamalais.

Results

Overall Patterns

a) Spatial autocorrelation. The MP dataset did not show significant spatial autocorrelation in the first distance class (MP first distance class, 0–100 km, $r = 0.004$, $P =$ NS; HO first distance class, 0–50 km, $r = 0.003$, $P = 0.003$) while the HO dataset did, though of very low magnitude. The third distance class (100–150 km) was significant in the HO dataset, but the r was low (0.008). However, the spatial autocorrelation correlograms of MP and HO did not differ significantly (r MP = 0.004, 95% bootstrap CI = −0.005 to 0.006, r HO = 0.003, 95% bootstrap CI = −0.002 to 0.003; T2 (first distance class) = 0.253, $P =$ NS; Omega = 11.08, $P =$ NS). In the HO-ANA dataset, the first four distance classes do not show significant spatial autocorrelation. However, the fifth and sixth distance classes (above 40 km) showed significant negative

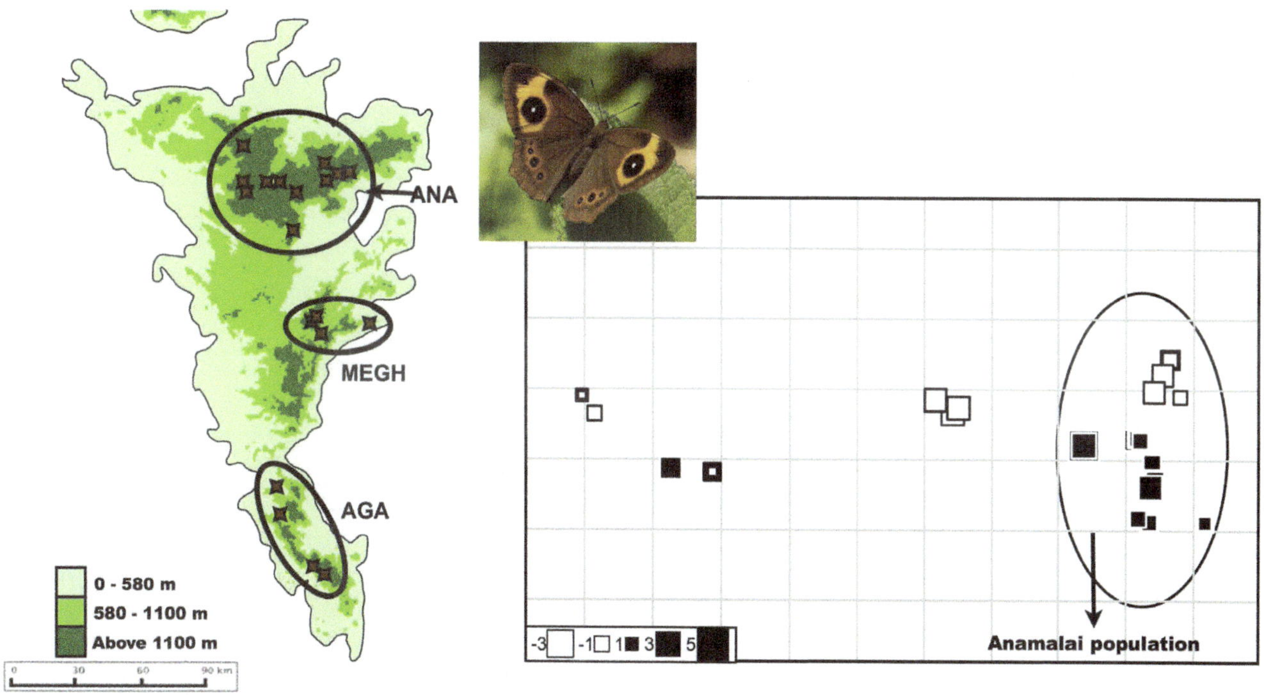

Figure 3. sPCA plots for *Heteropsis oculus.* Panel A shows the populations created for the population based analyses. Panel B is a representation of the first axis sPCA scores. The graph denotes the scores using squares; colour of the squares denotes positive (black) or negative (white) spatial autocorrelation, and the size of the squares denotes the magnitude of genetic variance. The squares are overlaid on a connection graph that represents the sampling design. If there was genetic differentiation between populations, squares of similar colour will clump together in one population. There is some clumping in the Anamalai population (shown in a circle), which is further explored in the next figure. Inset: photograph of *H. oculus.* Credit: Balakrishnan Valappil.

spatial autocorrelation. The complete results of the spatial autocorrelation analysis are given in Table S2.

b) Isolation by distance. Both the MP and HO datasets exhibited significant IBD; however, the correlation value for HO was more than twice that of MP (MP Mantel $r = 0.1168$, $P = 0.014$; HO Mantel $r = 0.272$, $P = 0.0001$). The graphs for the Mantel test are presented in Figure S1. IBD is significant in HO-ANA (Mantel $r = 0.218$, $P = 0.0005$), with an elevated r value at a Euclidean distance of 0.3 – 0.4.

c) Clustering analysis. Based on an examination of the screeplot of sPCA values of the MP dataset (see Figure S2, part A), only the first sPCA eigenvalue (variance = 4.302, Moran's $I = 0.655$) was retained. There were no significant clusters retrieved in this analysis (see Figure 4), and both global and local tests were not significant (Gtest max(t) = 0.034, P = NS; Ltest max(t) = 0.044, P = NS).

In the case of the HO dataset, only the first sPCA eigenvalue was retained (variance = 3.211, Moran's $I = 0.868$) (see Figure S2, part B for the screeplot). The Gtest was significant while the Ltest was not (Gtest max(t) = 0.0205, $P = 0.001$; Ltest max(t) = 0.017, $P = 0.189$). However, the significant global structure did not translate into any geographic structuring (see Figure 3). Close examination of the Anamalais group (encircled in Figure 3), shows the presence of some pattern, which was further explored in the HO-ANA dataset.

For HO-ANA, the first sPCA eigenvalue (variance = 3.158, Moran's $I = 0.887$) was retained, since it was far higher than all the others (see Figure S2, part C for the screeplot). Global structures were significant, but not local structures (Gtest max(t) = 0.034, $P = 0.029$; Ltest max(t) = 0.030, $P = 0.498$). In Figure 2, the samples represented by black squares inside the circle (which encompasses sampling locations from the Kodaikanal subpopulation) are different from the other samples, which are all lighter colored; this indicates a progressive genetic differentiation between Munnar and Kodaikanal.

Population Based Analyses

a) Genetic diversity. The genetic diversity parameters for all the datasets are summarized in Table 2. The percentage polymorphic loci (PLP) was similar across all the populations of the MP dataset, ranging from 47.8% in WAC to 57.9% in NKAR. The expected heterozygosity (H_e) was similarly uniform, ranging from 0.15 to 0.17; the total heterozygosity was 0.162. In HO, the PLP was about 42% in the ANA and MEGH populations, but as high as 51.6% in the AGA population. The total heterozygosity was 0.115, with H_e ranging from 0.105 to 0.12. There was higher variation in genetic diversity among sub populations in the HO-ANA dataset: the PLP varied from 38.5% in AKK, to about 71% in MATH. The total heterozygosity was 0.167, with H_e varying from 0.14 to 0.19.

b) Genetic differentiation. Genetic differentiation among populations in MP dataset was not significant in the allele-based method (see Table 3a) and only marginally significant in the band-based method (see Table S3 a). Nei's genetic distance was less than 0.001 among all populations (see Table 3a). The HO-dataset exhibited low, but significant differentiation among its three constituent populations, with a low overall allele-based F_{ST} of 0.0217, P<0.0001 (see Table 3b) and an overall band-based F_{ST} of 0.043, P<0.0001 (see Table S3 b). Nei's D values were also very low, around 0.002 (see Table 3b)).

In HO-ANA, the overall F_{ST} value was 0.002, at P<0.0001; see Table 3c for more details. The F_{ST} between KODI and the other sub populations is the major contributor to the overall F_{ST} value. F_{ST} is 0.034 between KODI and MATH, and KODI and VAG,

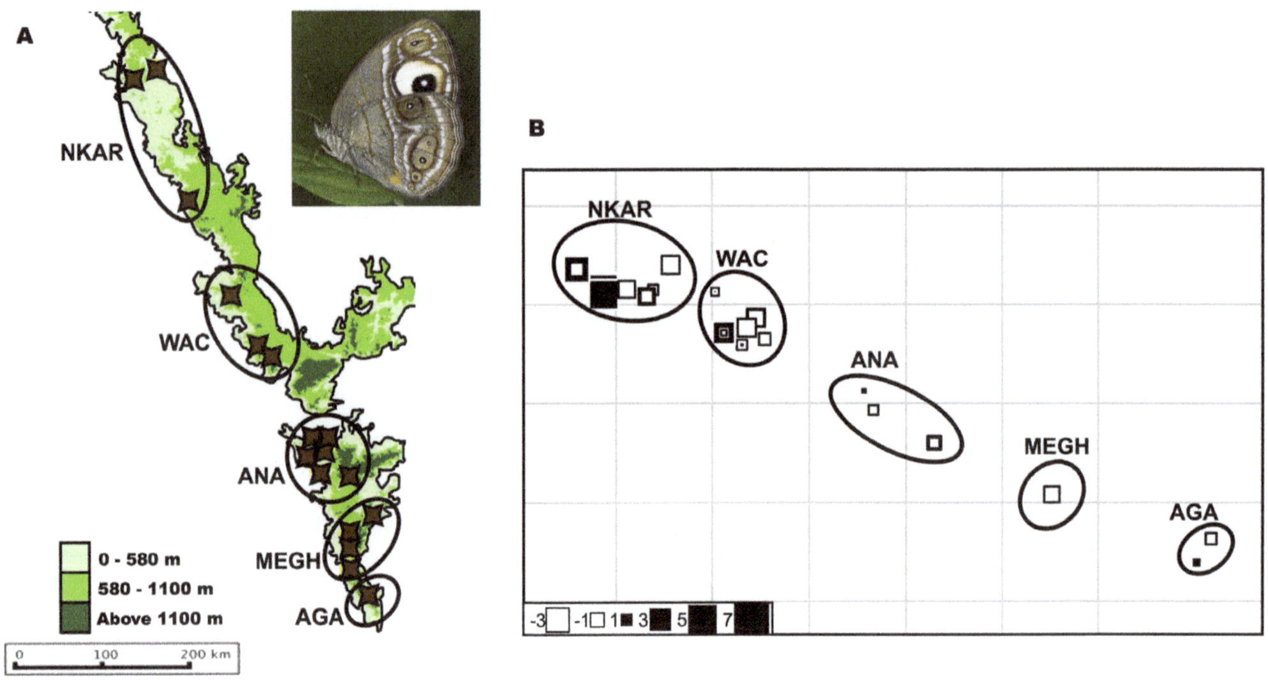

Figure 4. sPCA plots for *Mycalesis patnia*. Panel A shows the populations created for the population based analyses. Panel B is a representation of the first axis sPCA scores. The graph denotes the scores using squares; colour of the squares denotes positive (black) or negative (white) spatial autocorrelation, and the size of the squares denotes the magnitude of genetic variance. The squares are overlaid on a connection graph that represents the sampling design. If there was genetic differentiation between populations, squares of similar colour will clump together in one population. This is clearly not the case for *M. patnia*. Inset: photograph of *M. patnia*. Credit: Krushnamegh Kunte.

Table 2. Genetic diversity in the MP, HO and HO-ANA datasets.

Dataset	Population	PLP	H_e (± SD)
MP	AGA	52.2	0.167 (0.005)
	ANA	49.6	0.154 (0.005)
	MEGH	56.6	0.170 (0.005)
	NKAR	57.9	0.166 (0.005)
	WAC	47.8	0.155 (0.005)
HO	AGA	51.6	0.112 (0.004)
	ANA	41.8	0.121 (0.004)
	MEGH	42	0.105 (0.004)
HO-ANA	AKK	38.5	0.157 (0.007)
	KODI	65.6	0.147 (0.006)
	MANN	54.9	0.156 (0.006)
	MATH	70.9	0.194 (0.006)
	PAM	66	0.177 (0.006)
	RAJ	46.7	0.158 (0.007)
	VAG	66	0.179 (0.006)

Here, PLP is the percentage of polymorphic loci, and H_e is the expected heterozygosity (± standard deviation). Population names are abbreviated as follows: AGA – Agastyamalais, ANA – Anamalais, MEGH – Meghamalais, NKAR – North Karnataka, WAC – Wayanad/Coorg complex, AKK – Akkamalai, KODI – Kodaikanal, MANN – Mannavan shola, MATH –Mathikettan shola, PAM – Pambadam shola, RAJ – Rajmala, VAG – Vaguvaraiyar.

while the F_{ST} among other sub-populations is 0. Similar patterns can be seen with Nei's D (Table 3c)) and band-based F_{ST} (see Table S3 c). The only other pair of sub-populations showing differentiation are MATH and MANN, though not as high as the values between KODI and other sub-populations.

c) Assignment tests. In the MP dataset, 15 out of the 57 individuals could not be assigned to any population, and 15 were assigned to the population from which they were sampled. Simulations were performed for the 27 individuals that were not assigned to the populations from which they were sampled. 22 out of these 27 individuals were assigned with high probability to the population they were sampled from, making the total number of individuals correctly assigned, 37 (65%). The remaining five individuals were assigned with high probability to populations they were not sampled from: two from ANA were assigned to MEGH, two from WAC to MEGH and one from MEGH to NKAR.

Of the 99 individuals in the HO dataset, 16 could not be assigned to any one population with confidence. 70 samples were assigned to the population from which they were sampled. However, simulations showed that the remaining 13 individuals had the highest probability of being assigned to the population they were sampled from.

In the HO-ANA subset, 9 out of the 58 individuals could not be assigned to one population with confidence, and only 15 of the remaining 49 individuals were correctly assigned: 9 to KODI, and 6 to MATH. But, simulations showed that all the individuals except one were assigned with high probability to the population they were sampled from. Only one individual from AKK was assigned to MATH with high confidence.

Discussion

In this study, we compare the population genetic structure of two butterfly species that occupy habitats with different levels of continuity. Among these species, HO is a sky-island specialist distributed in naturally disjunct habitat, and MP is distributed in relatively more continuous lowland forests. The two species are similar in all other traits that affect dispersal ability; thus a difference in genetic structure can be attributed to a difference in habitat continuity. We first established that the overall genetic diversity in both species was not significantly different. The two AFLP datasets can thus be compared meaningfully. Importantly, the H_e values from this study are comparable to the values obtained in other studies that have used AFLP [43–45].

The Effect of Spatial Orientation of Suitable Habitat Patches

Spatial autocorrelation was not observed in MP, while HO exhibits significant spatial autocorrelation, even though the correlation value is very low. The clustering analysis sPCA did not retrieve clusters in either MP or HO. However, the sPCA analysis of HO showed a positive global structure (significant G test), indicating significant positive spatial autocorrelation, as opposed to the lack of global structures in MP. Genetic differentiation was not significant in MP, and significant, though low, between the three populations of HO. The same pattern was observed in the assignment tests, with the algorithm unable to confidently assign individuals to the correct populations in MP, because of a homogenous genetic structure. There were, however, a high percentage of correct assignments in HO, showing more restricted dispersal. Taken together, these results indicate that HO, the sky island species, showed a higher population genetic structure than MP, the mid and low elevation species. A summary of the analyses is given in Table 4.

The difference between the genetic structure of MP and HO can be attributed to the spatial orientation of the habitats they occupy, because the two species are similar in all other respects. HO, the sky island specialist inhabiting a naturally disjunct habitat exhibits more geographic structuring than MP, which inhabits the more continuously distributed low and mid elevation forests in the Western Ghats. Thus our results support the hypothesis that sky islands tend to have structured populations due to reduced gene flow across the sky islands.

Similar results of reduced gene flow among sky islands have been observed in the montane butterflies, *Erebia melampus* [15] and *Erebia Euryale* [46] in Europe. A case in point is *Parnassius smintheus* from the Canadian rockies [14], where the patches of suitable high altitude meadows have been pushed further apart by habitat fragmentation, resulting in a drop in gene flow. The landscape level changes in the Western Ghats need to be investigated further, but further reduction in connectivity between sky islands is a likely scenario.

As mentioned above, MP seems to posses a uniform genetic structure across the Western Ghats. More intense sampling may show different patterns; but, MP has been sampled across most of its range in the Western Ghats, and the genetic diversity values are comparable to global values. This indicates that sampling has not been a problem. Such a lack of genetic structure has been observed in other butterfly species, especially among Satyrines: examples are *Chazara briseis* [47] and *Melanargia galathea* [48].

Also, some butterfly species can have an innate ability to disperse over long distances, depending on the spatial distribution of resources necessary for their survival: sites where males and females can find each other and mate, for oviposition where larval

Table 3. Genetic differentiation matrices among populations in a) MP b) HO and c) HO-ANA datasets.

a) MP dataset

Populations	**AGA**	**ANA**	**MEGH**	**NKAR**	**WAYCOO**
AGA	–	0.0016	0.0011	0.0019	0.0008
ANA	0.0085	–	0	0	0
MEGH	0.0056	0	–	0	0
NKAR	0.0096	0	0	–	0
WAYCOO	0.0043	0	0	0	–
Overall $F_{ST}=0.0015$, $P=0.064$					

b) HO dataset

Populations	**AGA**	**ANA**	**MEGH**
AGA	–	0.0024	0.002
ANA	0.0183	–	0.004
MEGH	0.0162	0.0304	–
Overall $F_{ST}=0.0217$, $P<0.0001$			

c) HO-ANA dataset

Populations	**AKK**	**KODI**	**MANN**	**MATH**	**PAM**	**RAJ**	**VAG**
AKK	–	0.0044	0	0	0	0	0
KODI	0.0239	–	0.0018	0.0071	0.0023	0.0045	0.0067
MANN	0	0.0098	–	0	0	0	0
MATH	0	0.0349	0.0074	–	0	0	0
PAM	0	0.0123	0	0	–	0	0
RAJ	0	0.0246	0	0	0	–	0
VAG	0	0.0341	0	0	0	0	–
Overall $F_{ST}=0.002$, $P<0.0001$							

Nei's D genetic distance is presented in the upper triangle, and allele-based F_{ST} values in the lower triangle.

food plants are proximately distributed, and where adults can find food and survive to be able to perform both these activities [49]. This has been demonstrated in *Parnassius apollo* [50]. Some species can travel through patches of unfavorable habitat even if they prefer a particular habitat type. A study on butterflies of the subfamily Ithomiinae (Family Nymphalidae) in the cloud forest fragments in the Columbian Andes showed that butterflies moved through the coffee plantations around forests, flying faster through sunny rather than shady patches, but still being able to move through unsuitable habitats [51].

Genetic Differentiation among Sky Islands in HO

Within HO, an interesting pattern was revealed. In the HO-ANA dataset, KODI samples formed a distinct cluster in the PCA plot when the first and third axes were used. Also, KODI samples formed a distinct cluster in the sPCA analyses, indicating differentiation between KODI and all other sub-populations

Table 4. Comparison of population genetic analyses performed for MP and HO datasets.

Analysis	**Support for hypothesis**
Spatial autocorrelation	Not significant in MP, significant in HO
IBD	Mantel's r 2.3 times greater in HO
sPCA	Gtest not significant in MP, significant in HO
Band-based F_{ST}	Not significant in MP, significant in HO
Allele-based F_{ST}	Not significant in MP, significant in HO
AMOVA	Variation between populations explained two times more genetic variation in HO
	2.4 times higher percentage of correct assignment in HO

(which are all situated in Munnar, see Figure 2). The uniqueness of the Kodaikanal population within the Anamalai subset of HO is also evident by the other analysis parameters. Spatial autocorrelation and isolation by distance within the Anamalais samples are evident at about 40 km, the distance of KODI from other populations. Genetic differentiation (both F_{ST} measures and Nei's D) in the HO-ANA dataset is low, but the overall values are contributed by the differentiation between KODI and the other sub-populations. Even in assignment tests, only one individual was assigned to a population from which it was not sampled.

The presence of such genetic structuring over a short distance of about 40–50 km is further evidence of the reduced connectivity between sky islands, which is limited in comparison to connectivity between lower elevations. In the white-bellied shortwing, a bird species also found in the sky islands of the Western Ghats, a study of song structure [52] demonstrated that birdsong was beginning to diverge between Kodaikanal and the nearby Grass Hills. The samples of *H.oculus* from Akkamalai (which is adjacent to Grass Hills) are closely related to the individuals from Munnar, and well differentiated from the Kodaikanal samples. It is possible that similar forces are at play here. Anthopogenic habitat fragmentation is probably causing songs to diverge because populations are getting isolated (though with no corresponding genetic differentiation yet) in the white-bellied shortwing, and genetic differentiation in *H.oculus*.

The reasons for a break in gene flow between Kodaikanal and Munnar can only be speculated upon. The following information was obtained from the high resolution vegetation map [53] and from the bioclimatic maps [54] of the Western Ghats, both developed by the French Institute. The altitude of the areas sampled from Munnar (average 1730 m) and Kodaikanal (average 1845 m) are similar, though Kodaikanal is slightly higher. Temperatures are thus similar, with the mean temperature in the coldest month being less than 13°C in both places. The geology is also similar, with the bedrock composed of metamorphic rocks from the Precambrian, with gneiss and granite; soils are "ferralitic (laterites) to fersialitic (red soils), with a massive development of kaolinite as a product of rock weathering where the annual soil water balance is consistently positive (i.e. above 1,200 mm rainfall)" [53]. The vegetation is also similar, with patches of montane shola-grasslands, interspersed with plantations of tea, eucalyptus, wattle and pine.

The main difference is the amount and seasonality of rainfall in both places, which manifests as a strong west-east gradient in the number of dry months, from Munnar to Kodaikanal. Munnar experiences the Allepey-Mangalore rainfall regime, characterized by heavy monsoon showers in June, July and August, with 2000 to 5000 mm rainfall annually, (mean of 3594 mm), spread over 144 rainy days, and an average dry season length of 3 months per year. As one moves east to Kodaikanal, there is a succession of different rainfall regimes: 1) a strip of the Shimoga regime (1200 to 1500 mm annual rainfall, with 3–6 dry months, less than 100 rainy days and temperature of coldest month between 13.5°C and 16°C), and 2) the Madurai-Pollachi regime (mean annual rainfall 875 mm, with 4–8 dry months, about 66 rainy days and temperature of coldest month between 16°C and 23°C), until Kodaikanal, which is a special case of the Madurai-Pollachi regime due to increased rainfall over a few days (mean annual rainfall 1571 mm, with 0–4 dry months and about 92 rainy days, and temperature again about 13.5°C). This probably means that HO, being adapted to the high rainfall and low temperature in Kodaikanal, is unable to disperse to Munnar due to the intervening dry area.

This brings to light an important fact: apart from the spatial orientation of suitable habitat patches, the nature of the intervening habitat is equally important in determining whether dispersal (and thus gene flow) can occur in a particular species. Replicating studies of this kind with other taxa across the Munnar – Kodaikanal gradient, and in other areas of the Western Ghats that show such a strong gradient, like the Nilgiris, will help further our understanding of habitat connectivity between sky islands. This assumes further importance given the high degree of endemism in the sky islands of the Western Ghats, especially in the case of plants. While there may be a difference between species based on species biology, comparative studies such as these lay the ground for making generalizations regarding the ability of a species to maintain genetically viable populations.

There is now information regarding two species that inhabit the sky islands of the Western Ghats: an endemic bird, and an endemic butterfly. Both studies show that gene flow is restricted at these high elevations. The high elevation areas of the Western Ghats have been systematically converted into plantations from British times: tea plantations in Munnar started in the 1870s, and the replacement of grasslands with timber trees (pine, wattle and eucalyptus) in Kodaikanal started in the early 1900s. The montane shola forests and grasslands have been swamped with invasive exotic species that have changed the composition of native vegetation [55,56]. Human occupation at high densities (about 261 people per square km) has resulted in extensive modification of both the montane landscape, and the intervening 'valley' areas. There is a strong need to conserve unmodified areas in the Western Ghats as a whole, and the unique shola grassland mosaic in particular; for this endeavor, establishing the connectedness between sky islands across taxa is important.

Acknowledgments

The authors would like to thank Krushnamegh Kunte, Ullasa Kodandaramaiah, S.P. Vijayakumar and Kartik Shanker for discussions on the manuscript. A big thanks to Jahnavi Joshi for helping with the maps. We would also like to thank the forest departments of Kerala, Karnataka and Tamil Nadu for collection permits. Voucher specimens have been deposited at the Centre for Ecological Sciences, Bangalore, and at the Kerala Forest Research Institute at Peechi, Kerala. Also, the people who provided logistical support during field work: drivers (Sekar and Kumar), and numerous field assistants at each collection site.

Author Contributions

Conceived and designed the experiments: SS PK. Performed the experiments: SS. Analyzed the data: SS PK. Contributed reagents/materials/analysis tools: SS PK. Wrote the paper: SS PK. Collected specimens: SS.

References

1. Warshall P (1994) The Madrean Sky Island Archipelago: a planetary overview. In: DeBano FL, Ffolliott PF, Ortegae-Rubio A, Gottfried RH, Hamre GJ, editors. Biodiversity and the Management of the Madrean Archipelago: The Sky Islands of Southwestern United States and Northwestern Mexico. Fort Collins, Colorado: United States Department of Agriculture. 6–18.
2. Mayr E, Diamond JM (1976) Birds on islands in the sky: Origin of the montane avifauna of Northern Melanesia. Proceedings of the National Academy of Sciences 73: 1765–1769.
3. Maddison W, McMahon M (2000) Divergence and reticulation among montane populations of a jumping spider (*Habronattus pugillis* Griswold). Systematic biology 49: 400–421. doi:10.1080/10635159950127312.
4. Smith CI, Farrell BD (2005) Phylogeography of the longhorn cactus beetle *Moneilema appressum* LeConte (Coleoptera: Cerambycidae): was the differentiation of the Madrean sky islands driven by Pleistocene climate changes? Molecular Ecology 14: 3049–3065. doi:10.1111/j.1365-294X.2005.02647.x.
5. Holycross A, Douglas M (2007) Geographic isolation, genetic divergence, and ecological non-exchangeability define ESUs in a threatened sky-island rattlesnake. Biological Conservation 134: 142–154. doi:10.1016/j.biocon.2006.07.020.
6. Varas-Nelson AC (2010) Conservation Genetics of Black Bears in Arizona and Northern Mexico. The University of Arizona.
7. DeChaine EG, Martin AP (2005) Marked genetic divergence among sky island populations of *Sedum lanceolatum* (Crassulaceae) in the Rocky Mountains. American Journal of Botany 92: 477–486. doi:10.3732/ajb.92.3.477.
8. Knowles LL (2001) Did the Pleistocene glaciations promote divergence? Tests of explicit refugial models in montane grasshopprers. Molecular Ecology 10: 691–701. doi:10.1046/j.1365-294x.2001.01206.x.
9. Wikramanayake E, Dinerstein E, Loucks C, Olson D, Morrison J, et al. (2001) Terrestrial ecoregions of the Indo-Pacific: a conservation assessment. Washington, D.C.: Island Press.
10. Bawa KS, Das A, Krishnaswamy J, Karanth U, Sambakumar N, et al. (2007) Ecosystem profile: Western Ghats & Sri Lanka biodiversity hotspot - Western Ghats region.
11. Robin V, Sinha A, Ramakrishnan U (2010) Ancient geographical gaps and paleo-climate shape the phylogeography of an endemic bird in the sky islands of southern India. PLoS ONE 5: e13321.
12. Deshpande AU, Apte GS, Bahulikar RA, Lagu MD, Kulkarni BG, et al. (2001) Genetic diversity across natural populations of three montane plant species from the Western Ghats, India revealed by intersimple sequence repeats. Molecular Ecology 10: 2397–2408.
13. Bahulikar R, Lagu M, Kulkarni B, Pandit S, Suresh H, et al. (2004) Genetic diversity among spatially isolated populations of *Eurya nitida* Korth. (Theaceae) based on inter-simple sequence repeats. Current Science 86: 824–832.
14. Keyghobadi N, Roland J, Strobeck C (2005) Genetic differentiation and gene flow among populations of the alpine butterfly, *Parnassius smintheus*, vary with landscape connectivity. Molecular Ecology 14: 1897–1909.
15. Haubrich KAS, Schmitt T (2007) Cryptic differentiation in alpine-endemic, high-altitude butterflies reveals down-slope glacial refugia. Molecular Ecology 16: 3643–3658.
16. Sekar S (2012) A meta-analysis of the traits affecting dispersal ability in butterflies: can wingspan be used as a proxy? The Journal of Animal Ecology 81: 174–184. doi:10.1111/j.1365-2656.2011.01909.x.
17. Wynter-Blyth MA (1957) Butterflies of the Indian Region. Mumbai: Bombay Natural History Society.
18. Kehimkar I (2008) The Book of Indian Butterflies. Mumbai, India: Bombay Natural History Society and Oxford University Press. 1 pp.
19. Kodandaramaiah U, Lees DC, Müller CJ, Torres E, Karanth KP, et al. (2010) Phylogenetics and biogeography of a spectacular Old World radiation of butterflies: the subtribe Mycalesina (Lepidoptera: Nymphalidae: Satyrini). BMC Evolutionary Biology 10: 172. doi:10.1186/1471-2148-10-172.
20. Conord C, Lempérière G, Taberlet P, Despres L (2006) Genetic structure of the forest pest *Hylobius abietis*on conifer plantations at different spatial scales in Europe. Heredity 97: 46–55. doi:10.1038/sj.hdy.6800837.
21. Reineke A, Karlovsky P, Zebitz CPW (1998) Preparation and purification of DNA from insects for AFLP analysis. Insect Molecular Biology 7: 95–99.
22. Meudt HM, Clarke AC (2007) Almost Forgotten or Latest Practice? AFLP applications, analyses and advances. Trends in Plant Science 12: 106–117.
23. Althoff DM, Gitzendanner MA, Segraves KA (2007) The utility of Amplified Fragment Length Polymorphisms in phylogenetics: a comparison of homology within and between genomes. Systematic biology 56: 477–484. doi:10.1080/10635150701427077.
24. Arrigo N, Tuszynski J, Ehrich D, Gerdes T, Alvarez N (2009) Evaluating the impact of scoring parameters on the structure of intra-specific genetic variation using RawGeno, an R package for automating AFLP scoring. BMC Bioinformatics 10: 33.
25. Whitlock R, Hipperson H, Mannarelli M, Butlin RK, Burke T (2008) An objective, rapid and reproducible method for scoring AFLP peak-height data that minimizes genotyping error. Molecular Ecology Resources 8: 725–735.
26. Bonin A, Bellemain E, Eidesen PB, Pompanon F, Brochmann C, et al. (2004) How to track and assess genotyping errors in population genetics studies. Molecular Ecology 13: 3261–3273.
27. Banks SC, Peakall R (2012) Genetic spatial autocorrelation can readily detect sex-biased dispersal. Molecular Ecology 21: 2092–2105. doi:10.1111/j.1365-294X.2012.05485.x.
28. Dray S, Dufour A-B (2007) The ade4 package: implementing the duality diagram for ecologists. Journal of statistical software 22: 1–20.
29. R Development Core Team (2011). R: A language and environment for statistical computing. R Foundation for Statistical Computing, Vienna, Austria. ISBN 3-900051-07-0, URL http://www.R-project.org/. Accessed 2013 July 7.
30. Kosman E, Leonard KJ (2005) Similarity coefficients for molecular markers in studies of genetic relationships between individuals for haploid, diploid, and polyploid species. Molecular Ecology 14: 415–424.
31. Meyer D, Buchta C (2010) "proxy: Distance and Similarity Measures". R package version 0.4-6. http://CRAN.R-project.org/package = proxy. Accessed 2013 July 7.
32. Frantz AC, Cellina S, Krier A, Schley L, Burke T (2009) Using spatial Bayesian methods to determine the genetic structure of a continuously distributed population: clusters or isolation by distance? Journal of Applied Ecology 46: 493–505.
33. Jombart T, Pontier D, Dufour A-B (2009) Genetic markers in the playground of multivariate analysis. Heredity 102: 330–341.
34. Jombart T (2008) adegenet: a R package for the multivariate analysis of genetic markers. Bioinformatics 24: 1403–1405. doi:10.1093/bioinformatics/btn129.
35. Vekemans X, Beauwens T, Lemaire M, Roldan-Ruiz I (2002) Data from amplified fragment length polymorphism (AFLP) markers show indication of size homoplasy and of a relationship between degree of homoplasy and fragment size. Molecular Ecology 11: 139–151.
36. Zhivotovsky LA (1999) Estimating population structure in diploids with multilocus dominant DNA markers. Molecular Ecology 8: 907–913.
37. Bonin A, Ehrich D, Manel S (2007) Statistical analysis of amplified fragment length polymorphism data: a toolbox for molecular ecologists and evolutionists. Molecular Ecology 16: 3737–3758.
38. Lynch M, Milligan BG (1994) Analysis of population genetic structure with RAPD markers. Molecular Ecology 3: 91–99. doi:10.1111/j.1365-294X.1994.tb00109.x.
39. Irwin DE, Irwin JH, Smith TB (2011) Genetic variation and seasonal migratory connectivity in Wilson's warblers (*Wilsonia pusilla*): species-level differences in nuclear DNA between western and eastern populations. Molecular Ecology 20: 3102–3115. doi:10.1111/j.1365-294X.2011.05159.x.
40. Excoffier L, Lischer H (2010) Arlequin suite ver 3.5: a new series of programs to perform population genetics analyses under Linux and Windows. Molecular Ecology Resources 10: 564–567. doi:10.1111/j.1755-0998.2010.02847.x.
41. Duchesne P, Bernatchez L (2002) aflpop: a computer program for simulated and real population allocation, based on AFLP data. Molecular Ecology Notes 2: 380–383.
42. He T, Krauss SL, Lamont BB, Miller BP, Enright NJ (2004) Long-distance seed dispersal in a metapopulation of *Banksiahookeriana* inferred from a population allocation analysis of amplified fragment length polymorphism data. Molecular Ecology 13: 1099–1109. doi:10.1111/j.1365-294X.2004.02120.x.
43. Takami Y, Koshio C, Ishii M, Fujii H, Hidaka T, et al. (2004) Genetic diversity and structure of urban populations of *Pieris* butterflies assessed using amplified fragment length polymorphism. Molecular Ecology 13: 245–258.
44. Timm AE, Geertsema H, Warnich L (2006) Gene flow among *Cydia pomonella* (Lepidoptera: Tortricidae) geographic and host populations in South Africa. Journal of Economic Entomology 99: 341–348.
45. Groot AT, Classen A, Inglis O, Blanco CA, López J, et al. (2011) Genetic differentiation across North America in the generalist moth *Heliothis virescens*and the specialist *H. subflexa*. Molecular Ecology 20: 2676–2692. doi:10.1111/j.1365-294X.2011.05129.x.
46. Vila M, Marí Mena N, Guerrero A, Schmitt T (2011) Some butterflies do not care much about topography: a single genetic lineage of *Erebia euryale* (Nymphalidae) along the northern Iberian mountains. Journal of Zoological Systematics and Evolutionary Research 49: 119–132.
47. Johannesen J, Schwing U, Seufert W, Seitz A, Veith M (1997) Analysis of gene flow and habitat patch network for *Chazara briseis* (Lepidoptera: Satyridae) in an agricultural landscape. Biochemical Systematics and Ecology 25: 419–427.
48. Vandewoestijne S, Martin T, Liegeois S, Baguette M (2004) Dispersal, landscape occupancy and population structure in the butterfly *Melanargia galathea*. Basic and Applied Ecology 5: 581–591.
49. Wiklund C (1977) Oviposition, feeding and spatial separation of breeding and foraging habitats in a population of *Leptidea sinapis* (Lepidoptera). Oikos 28: 56–68.
50. Brommer JE, Fred MS (1999) Movement of the Apollo butterfly *Parnassius apollo* related to host plant and nectar plant patches. Ecological Entomology 24: 125–131.

51. Muriel SB, Kattan GH (2009) Effects of patch size and type of coffee matrix on Ithomiine butterfly diversity and dispersal in cloud-forest fragments. Conservation Biology 23: 948–956.
52. Robin V, Katti M, Purushotham C, Sancheti A, Sinha A (2011) Singing in the sky: song variation in an endemic bird on the sky islands of southern India. Animal Behaviour.
53. Data Paper – High Resolution Vegetation Cover Data for the Southern Western Ghats of India. (IFP_ECODATA_VEGETATION) (2010). Available: http://hal.archives-ouvertes.fr/hal-00481614.Accessed 11 July 2012.
54. Pascal J-P (1982) Explanatory notes on the bioclimate maps of the Western Ghats.
55. Muniappan R, Viraktamath CA (1993) Invasive alien weeds in the Western Ghats. Current Science 64: 555–558.
56. Reddy CS (2008) Catalogue of invasive alien flora of India. Life Science Journal 5: 84–89.

Examining the Extinction of the Barbary Lion and Its Implications for Felid Conservation

Simon A. Black[1]*, Amina Fellous[2], Nobuyuki Yamaguchi[3], David L. Roberts[1]

1 Durrell Institute of Conservation and Ecology, School of Anthropology and Conservation, University of Kent, Canterbury, Kent, United Kingdom, 2 Agence Nationale pour la Conservation de la Nature, Algiers, Algeria, 3 Department of Biological and Environmental Sciences, University of Qatar, Doha, Qatar

Abstract

Estimations of species extinction dates are rarely definitive, yet declarations of extinction or extirpation are important as they define when conservation efforts may cease. Erroneous declarations of extinctions not only destabilize conservation efforts but also corrode local community support. Mismatches in perceptions by the scientific and local communities risk undermining sensitive, but important partnerships. We examine observations relating to the decline and extinction of Barbary lions in North Africa. Whilst the extinction predates the era of the scientific conservation movement, the decline is relatively well documented in historical records. Recently unearthed accounts suggest Barbary lions survived later than previously assumed. We use probabilistic methods to estimate a more recent extinction date for the subspecies. The evidence presented for a much later persistence of lions in North Africa, including generations when sightings were nil, suggests caution when considering felid populations as extinct in the wild. The case raises the possibility that captive animals descended from the Moroccan royal collection are closer contemporaries to wild Barbary lions. Furthermore, our results highlight the vulnerability of very small lion populations and the significance of continued conservation of remnant lion populations in Central and West Africa.

Editor: Alfred L. Roca, University of Illinois at Urbana-Champaign, United States of America

Funding: The authors have no support or funding to report.

Competing Interests: Co-author David L Roberts is a PLOS ONE Editorial Board member. This does not alter the authors' adherence to all the PLOS ONE policies on sharing data and materials.

* E-mail: S.Black@kent.ac.uk

Introduction

Several statements of extinction relating to felids have been proven unreliable, such as the recent rediscovery of the Barbary leopard [1] and the late persistence of the Caspian tiger, recognized as extinct since the early 1970s, yet later found present in Turkey where a local trade in hunted skins persisted into the 1980s suggesting survival into at least the early 1990s [2]. In both cases local people's observations remained unknown to science and therefore had no impact on conservation policy.

Wild populations of lion (*Panthera leo*), like other large mammalian carnivores, are suffering severe decline and in Africa have contracted sharply over the past 50 years [3], [4]. Outside East Africa, lion populations are fragmented, with remnants in Central and Western Africa threatened with extinction [5], [6]. The present-day status in sub-Saharan Africa mirrors the situation north of the Sahara a century ago where the once extensive lion distribution from North Africa to India was reduced to fragmented populations by the 20th century [7].

The North African 'Barbary lion' or 'Atlas lion' occupied the Maghreb, the region isolated from the rest of non-arid Africa by the Sahara [8]. Until the 18th century, Barbary lions ranged from the Atlas Mountains to the Mediterranean [7]. Extensive persecution in the 19th century reduced populations to remnants in Morocco in the west, and Algeria and Tunisia further east, all of which were extirpated during the 20th century [9], [10].

The Barbary lion (*P. l. leo*) is considered distinct from the six commonly defined sub-species in the rest of Africa [11], owing to its geographic separation, morphology and unique montane habitat with cold winters. Although the phylogenetic status of lion populations remains unclear [12], [13], recent morphological and genetic studies consistently differentiate northern lions (India to North Africa) from sub-Saharan African lions [9], [14], [15].

The IUCN's assessments of species status, threats and extinctions have been established over the past four decades [4], which post-dates the extinction of the Barbary lion, so Red List declarations on the sub-species are based on reviews of literature [11]. The IUCN Wild Cats Status Survey and Action Plan recognized that certain captive lions may be descendants of the Barbary lion, designating the subspecies as 'extinct in the wild' [11]. Currently all extant African populations are grouped together as a single taxon, *Panthera leo*, with the species listed as 'regionally extinct' across its former northern range [16]. Only the Indian population is recognized as a distinct sub-species, *P. l. persica* [17], whilst North African populations (i.e. the Barbary lion) are mentioned in the context of both the IUCN's Indian and African lion accounts [4].

In the 16–18th centuries many accounts reported lions in the western Maghreb (northern Morocco) near the Atlantic and Mediterranean coasts [7], [18]. Up to the 1830s lions were still seen in these coastal areas, the Rif mountains and the Marmora forest [19], however records remain sparse throughout the 19th century [20]. By 1880 lions had retreated south of the Bou Regreg and Taza passes [19] into the Atlas Mountains and areas bordering the Sahara, where human populations were largely nomadic. Previous commentators suggested Barbary lions were extirpated sometime between 1920 and 1930 [7], [19], [20]. Later sightings, however, have since been documented with the last in the High Atlas in 1942 [18].

In the eastern Maghreb (Algeria and Tunisia) lions frequented coastal-forested regions, the Tell Atlas and the Aurès mountain ranges. By the late 1800s, Tunisian sightings were confined to localities adjacent to the Algerian provinces of Souk Arras and Tebessa. Although no lions were shot in Tunisia after 1891 [7], rumors of their survival persisted in the 1900s in the Khmir Mountains and near Feriana [21]. The population was probably contiguous across the border, corresponding to home range sizes equivalent to those observed in present-day sub-Saharan Africa and India [22], [23], [24]. In Algeria, lions persisted into the 1890s and hunting accounts, the capture of wild cubs and photographs of tame lions were still widely reported [7], [20]. These observations, alongside interview evidence, suggest that a small population survived in Algeria well into the 20th century [25], long after 1893 when Algeria's supposedly last lion was shot [7].

Several reviews have considered the history of sightings [7], [19], [20], [26], [27] and the IUCN recognizes that lions persisted in Morocco into the 1940s [17]. Nevertheless these reviews have inadvertently missed, ignored or have lacked access to many local sources and literature accounts. A systematic evaluation of the last sightings in the Maghreb will provide more complete insights into persistence of remnant lion populations and the resilience of large mammalian carnivores to human pressure.

Several menageries in Europe held Barbary lions in medieval times [9] and they were popular exhibits in public zoological gardens in the 1800s [7], [20]. By the early 1900s zoos and circuses in Europe and North America often promoted their lions as "Barbary" [13], although true representatives were said to be only found in the collection of the Sultan of Morocco, derived from animals caught by local tribes [20]. The significance of this collection was not recognized until the 1970s after the lions were moved from the Royal Palace, Rabat, to a new zoo at Temara when a study identified animals with physical characteristics of the Barbary lion [20]. Despite several attempts, a formal scientific breeding program is yet to be established (Frey pers. comm.). Nevertheless, captive breeding has experienced a recent renaissance [28], [29] and a studbook for these animals (hereafter 'Moroccan Royal lions') has been developed [13].

Debate surrounds the authenticity of Moroccan Royal lions as descendants of wild Barbary lions [13]. One concern is that Moroccan Royal lions hybridized with sub-Saharan African lions potentially introduced to the collection before the 1970s [20]. Definitive genetic matches have not yet been established between Moroccan Royal lions and wild Barbary lions [13]. The few studies of Barbary lions that utilize museum samples [14] are limited by the scarcity of wild-origin reference specimens. Most recent genetic studies [30], [31], [32] rely on Royal lion samples from two zoos, covering at most 4 (possibly only 2) of the 12 maternal bloodlines [13]. In the absence of genetic data, owing to the lack of verified wild specimens, and working on the precautionary presumption that Barbary lions are taxonomically distinct from other lions, assessment of purity relies on 'pedigree information' which at best can be taken from the recently established studbook of lions of known ancestry arising from the Moroccan Royal Collection [13].

The North African-Asian population of lions [14] is only represented by today's Asiatic lion (*c.*350 wild individuals and *c.*100 zoo captives), so the potential significance of captive Moroccan Royal lions (*c.*90 individuals) is not trivial [13]. In the absence of definitive genetic comparisons we examine the generational separation between lions surveyed in Temara Zoo in 1974 and wild-caught Barbary ancestors. This infers the degree of opportunity for hybridization had non-Barbary individuals bred within the Royal collection, and the potential purity and relevance of the extant captive population.

Our study re-examines the historical decline and extirpation of lions in North Africa. We calculate likely extinction dates for different populations and thus patterns of decline. We also examine new estimates of generation time in the captive population originally derived from wild North African ancestors and thus the likelihood of introgression with sub-Saharan individuals.

Materials and Methods

2.1 Sightings Information

Existing literature was reviewed for accounts of sightings and kills by date and location, including documented interviews in addition to recent interviews by the authors (Tables 1–6). The collation includes sightings, photographs, accounts and recollections since 1839. Observations have been taken from recorded accounts from first-hand sources (written recordings of interviews), from direct interviews of eyewitnesses or a first-hand recollection of an eyewitness account; all sightings in the dataset are therefore from second-hand sources or better.

Morocco and the Morocco-Algeria border are considered as the western Maghreb, whilst the eastern Maghreb covers Tunisia and northern Algeria. The meridian bisects sightings west and east, the most proximal being separated by 220 km of desert and temporally by nearly 70 years. The nearest contemporaneous sightings across this divide occurred in 1912 and are separated by a distance of over 450 km. A small central population in the Saharan Atlas (observed in 1898, 1912, 1920 and 1935) possibly traversed both regions but, owing to the size of the dataset, exclusion of these 3 sightings made very little difference.

2.2 Inferring Extinction Dates

While the last sighting of a species is often used as the time of extinction, it rarely corresponds to the true extinction date [33].

Following Solow [33], let sighting times in years be order $t_1<t_2<\ldots<t_n$, where $t_1=0$. Sampling from the uniform distribution, the unbiased estimate of extinction time is

$$\hat{T}_E=\frac{n+1}{n}t_n \tag{1}$$

and the expected year of extinction is $\hat{T}_E$ plus the year of t_1. The upper bound, T_E^u, of 1-α CI for T_E is

$$T_E^u=t_n/\alpha^{1/n}, \tag{2}$$

where α = 0.05 (after [33]). Several other methods of inferring dates of extinction have been developed (see Solow [33] for a review). However, in a study of putatively extinct North American birds, Vogel et al. [34] showed that the simple Poisson process model, used here, had as good or better fit than other more complex models, such as the truncated exponential and Weibull.

Here we calculate the extinction dates for sightings of lions in the Western (Table 1; $n=16$) and Eastern Maghreb (Tables 2–6; $n=66$).

Table 1. The last sightings of lions in the Western Maghreb: Morocco to Western Sahara, 1830s–1940s (Rif Mountains, Anti Atlas, Middle Atlas and High Atlas).

	Location	Season/Year	Observation Type	Solitary (0)/Group (1)	Original source	Ref.
1	Guelaya (the region around Melilla)	Before 1839		0		[19]
2	Cap Spartel-Tanger	Before 1839	Shot, male	0	Maalem Hamed	[51]
3	Mamora forest	1839	Frequent lions	1		[51]
4	Rif region	Summer 1839	Many observed and shot	1	Maalem Hamed	[51]
5	Jebel Kebir (Tanger)	1846	Shot, male	0	Hooker and Ball	[19]
6	Megader pass (at 2300 m)	1864	Camp protected against lion attack		Gerhard Rohlfs	[52]
7	Rif mountains	1895	Last lion killed in the Rif	0	de Planhol, 2004	[53]
8	M'Hamid, south of Zagora (Morocco-Algeria border)	c.1900	Observations close to water points	1	Local inhabitants	[18]
9	Djebel Ebrit (Ain Leuh-Timahdit)	c.1900	Observations	1		[54]
10	Budaa woods (Azrou)	1901	Observations, frequent lions	1	Marquis of Segonzac	[19]
11	Middle Atlas mountains	1911	Observations	1	Engel	[19]
12	Zaián forests, Beni Mgild (Khenifra)	1911	Lions	1		[52]
13	Oued Ifrane	Winter 1917	Observation, male	0		[54]
14	Azrou vicinity	1920	Observation of a single male	0		[20], [55]
15	Middle Atlas mountains	1922	Shot	0		[56]
16	Atlas Mountains (on the Casablanca-Dakar flight)	1925	Male lion seen from the air	0	Photograph Flandrin, M.	[57]
17	Ouiouane area (Ain Leuh)	1930	Observations of few lions, tracks	1		[58], [20]
18	Toubkal massif (now a NP)	Summer 1930	Observations at 3000 m	1	Local residents	[18]
19	Hassi Aggou and Hassi Tighissit (Assa - TanTan)	1935	Observations at water points	1	Bensalem, M.; Ennah, M.	[18]
20	Hassi Aggou vicinity	1939	Observations of 2 lions	1	Hunter told Monteil 1951	[18]
21	Tizi-n-Tichka pass (Marrakesh- Tadderte)	1942	Shot	0	Minet, J.	[18]

NP = one of several National Parks established across the region since the 1940s.

2.3 Testing Behavioral Change

We tested whether lions appear to change their behavior as populations decreased in size to operate singly ('0') rather than in groups ('1') as a binary set of sightings [35].

In a series of N sightings, $X_1, X_2,\ldots, X_N$, each sighting X_i is coded as $X_i = 1$ or $X_i = 0$. Of the N sightings, let m equal the number of sightings of lions in groups and n equal the number of sightings of lions singly. Then

$$m = \sum_{i=1}^{N} X_i \quad (3)$$

and $n = N - m$. The cumulative number of group sightings at each point is then determined. This frequency is

$$S_j = \sum_{i=1}^{i} X_i \quad (4)$$

where j = 1,2,...,N.

The statistic for testing the hypothesis of change is

$$D_{m,n} = \max\left|\frac{N}{mn}\left(S_j - \frac{jm}{N}\right)\right| \quad (5)$$

The expression is evaluated for all values of j from 1 to N-1. $D_{m,n}$ is the largest absolute difference observed in the sequence. We may reject the H_o at 0.05 if $D_{m,n} \geq 1.36\sqrt{N/mn}$

Here we test the null hypothesis that there has been no change in the probability of sightings of lions in groups over time in the Western and Eastern populations.

2.4 Estimation of Generations between Moroccan Royal Lions and Wild-caught Specimens

Hemmer [20] suggested that 20 generations separated Moroccan Royal lions surveyed in 1974 from wild lions captured c.1899, suggesting a 3.75-year generation time and multiple opportunities for hybridization with non-Barbary lions. The average age at which a wild female lion gives birth is 6.5–8 years [11], [36]. In the Moroccan Royal lion studbook, the mean generation time through to today's juveniles is 7.4 years (STD±3.2) in a range of 6.6–10 years [37]. The 30 breeding females since 1974 have a mean age at first surviving litter of 6.3

Table 2. Lion sightings in the eastern Maghreb of Algeria and Tunisia 1830–1850 (Ksour Mountains, Saharan Atlas, Tell Atlas, Ouled-Nail, Aurès Mountains).

	Location	Season/Year	Observation Type	Solitary (0)/Group (1)	Original source	Ref.
22	Oued Khalaad, Tunisia	1832	Observation of 16 lions	1	Greenville	[59]
23	Djebel Guezoul (Tiaret)	1836	Shot (100 shot during his life)	1	Mohamed Ben Esnoussi	[60]
24	Souk Tleta (Chelif region)	Winter 1837	Shot, male, female and 2 cubs	1	Agha Djendel	[61]
25	Ain Sbaa (Bou Saada)	1840	Male, attacking a sleeping man	0		[62]
26	Djebel Mezioun	Spring 1840	Observation of female and cubs	1		[63]
27	Oued Tegedemt (South Tiaret)	Summer 1840/42	Observations 2	1	Colonel Scott	[64]
28	El Diss vicinity	1843	Observed female, Shot male	1		[65]
29	Theniet El Had (now a NP)	Winter 1844	Shot, 3 (14 lions in his life)	1	Mokhtar ben el Arbi	[66]
30	Taza vicinity	1844	Lion attack	0		[66]
31	Miliana vicinity	1844	Lions attack on livestock	1	Bugeaud	[67]
32	Gaada (Djebel Amour)	Before 1845	Refuge place	0		[68]
33	Sidi Bel Abbés vicinity	Winter 1845	Shot, large lion	0		[69]
34	Sbeitla, Tunisia	1845	Observations	1	Bruce	[59]
35	Kef vicinity (Tunisia)	Spring 1845	Skin female	0		[70]
36	Mahouna Gorges (near Guelma)	Summer 1845	Shot, 2	1		[71]
37	Bibans	Summer 1845	Roaring	0		[72]
38	Saida	1846		1	Berbrugger 1846	[73]
39	Staouéli	1846	Attack, male	0		[74]
40	Guelma vicinity	Winter 1847	Shot	0		[71]
41	Theniet El Had (actually NP)	Winter 1847	Shot	0		[66]
42	Djebel Guerioun (Aurès Mounts)	Winter 1848	Fresh faeces and an old male seen	0		[71]
43	Montagne des lions (Oran-Arzew)	Winter 1849	Observations	1		[75]
44	Saf Saf valley (Skikda region)	Winter 1849	Observations	1		[75]
45	Zerazer (Aurès Mounts)	Winter 1850	Shot, 2	1		[71]
46	Tebessa	Spring 1850	Fresh male skin	0		[76]
47	Djebel Bouarif (Aurès Mounts)	Summer 1850	Winter refuge; 30 lions in the area	0		[77]
48	Fedjoudj (Aurès Mounts)	1850	Winter refuge	0		[77]

NP = one of several National Parks established across the region in the 1980s–1990s.

years (STD±2.9). These observations suggest that Moroccan Royal lions have a generation time similar to the birthing age for wild lions, 6.5 years [36]. We calculate the number of generations between the animals moved to Temara Zoo and the likely last wild-caught animals, based on a calculated inferred extinction date. We also calculate the number of generations occurring between sightings of wild Barbary lions as an indicator of persistence.

Results

3.1 Sightings of Lions in North Africa

The list of recent historical lion sightings in North Africa includes data for the western Maghreb from 1840 to the 1940s (Table 1) and the eastern Maghreb from the 1830s to the 1950s (Tables 2–6). Sightings become infrequent in the eastern Maghreb after the 1890s and in the western Maghreb after the 1920s. Our examination of historical accounts reveals that lions occupied the Saharan Atlas (Fig. 1), much further south than previously reported by Schnitzler [27]. The precise location of the aerial photograph on the Casablanca-Dakar route (Fig. 2) is unknown, however our research has identified that it was taken in 1925 when flights commenced [38]. A postcard edition of the image has recently been discovered with the caption "Un lion photographié en avion dans l'Atlas" (Fig. 3). Since this predates discussions on the extermination of lions in the region [19], its significance at that time was then unknown. The photograph by Flandrin is the last known image of a wild Barbary lion.

3.2 Recent Extinction of Lions in North Africa

In recent years it has become widely asserted that the animal shot in 1942 on the Tizi-n-Tichka pass in Morocco's High Atlas Mountains [18] has been considered the last wild Barbary lion [4]. However, our analysis suggests that wild lions actually persisted longer in Algeria until 1958 (T_E), with an upper bound for the 95% CI of $T_E^u = 1962$. This is 10 years after 1948 (T_E), the estimated extinction date of the western (Morocco) population. However, the estimated 95% confidence intervals have a substantial overlap: with upper bounds (T_E) of 1962 for the

Table 3. Lion sightings in the eastern Maghreb of Algeria and Tunisia, 1851–1860 (Ksour Mountains, Saharan Atlas, Tell Atlas, Ouled-Nail, Aurès Mountains).

	Location	Season/Year	Observation Type	Solitary (0)/Group (1)	Original source	Ref.
49	Ourten valley Khenchela (Aurès)	Autumn 1851	Observation, male, female	1		[71]
50	Drean	Spring 1852	Frequent, Shot male	1		[78]
51	Baba Ali (Mitidja)	Summer 1852	Observation, male	0		[79]
52	El Diss vicinity	1852	Shot, female	0		[65]
53	Milianah vicinity	Spring 1853	Observation	0		[80]
54	Khenchela (Aurès Mounts)	Summer 1853	Shot, male (Shot 25 lions during his life)	0		[71]
55	Djebel Onk Jemel (Aurès Mounts)	Autumn 1853	Shot, female	0		[71]
56	Ferdjioua vicinity	Spring 1855	SidiBou Akas	1	Vicomte de Noé	[81]
57	Oued Tafna	1855	Tracks	0	Vicomte de Noé	[81]
58	El Harrouch vicinity	1856	Shot, male, female	1		[65]
59	Beni Salah forest	1856–1857	Shot male & female, (shot 39 lions in his life)	1	Ahmed Ben Amar	[82]
60	Medjerda vicinity, Tunisia	1856–1857	Male, cubs, shot female	1	Ahmed Ben Amar	[82]
61	Ain Taoura (vicinity of Souk Ahras)	1856–1857	2 cubs, shot female	1	Ahmed Ben Amar	[82]
62	El Kala vicinity (now a NP)	1856–1857	4 cubs, shot female	1	Ahmed Ben Amar	[82]
63	Ain Temouchent area	Spring 1858	Observation	1		[83]
64	El Kef region, Tunisia	1858	Numerous lions	1		[84]
65	Azzaba vicinity	1858	Shot, 2	1		[83]
66	Skikda-Azzaba	1858	Observation	0		[83]
67	Berrrahal	Winter 1858	Tracks, roaring, 3	1		[85]
68	Oued Saf Saf - Oued Zergua	Winter 1858	Numerous lions	1		[85]
69	Skikda-Azzaba	Autumn 1859	Observation, tracks	1		[86]
70	Boghar forest	1860	Observations	1		[87]
71	Tiaret	1860		1	Leroux	[73]
72	Mascara	1860		1	Leroux	[73]
73	Tazoult-Hammam Essalihine	Winter 1860	Shot male, (Shot 30 lions during his life)	0	Chassaing, F.	[85]

NP = one of several National Parks established across the region in the 1980s–1990s.

Algerian population and 1965 for the western (Morocco) population.

3.3 Change in Behavior

The frequency of observations, particularly after the 1920s, which involves solitary animals rather than groups (Tables 1–6), may be entirely due to a dwindling population with lower density or a potential change in social behavior (from group to solitary living). Owing to the uncertainty surrounding some sightings, and therefore the order of occurrence of single and group sightings of lions, we analyzed the date at the two extremes; all groups occurring as late as possible and then as early as possible (only relevant to the eastern population). For the western population $D_{m,n} = 0.351$ compared with a critical value of 0.658 ($N = 18$), so that we cannot reject the H_o and suggest that there has been no change in behavior. Likewise for the eastern population no change in behavior was detected even though the inhabited areas are adjacent to human settlements; $D_{m,n}$ ranged from 0.235 to 0.249 compared to a critical value for both of 0.250 ($N = 121$).

3.4 Persistence of Lions

If we consider the gap size between sightings, lions appear able to persist whilst unseen by humans for at least a generation, if not two. A more rigorous analysis of gaps sizes for sightings is not possible due to ambiguity in some sighting dates. In the western Maghreb lions went unseen for 10 years, reducing to 3–5 years towards the end of the record. In the eastern Maghreb the gap sizes between later sightings were generally greater than 5 years, although precision is difficult owing to uncertainty in some dates. In one case the gap could be >10 years (1900s to 1910–1912).

3.5 Generations between Moroccan Royal Lions and Wild-caught Specimens

Current breeding animals in the Moroccan Royal Lion studbook [13] are separated from the Royal Palace animals by just four generations (in one case, three generations) and recent cubs [39] are five generations removed. Hemmer's [20] estimate of generations between wild ancestors and the lions in the Royal collection in the 1970s, would suggest that today's surviving descendants stand 25 generations from wild ancestors (Fig. 4). However, assuming a generation time of 6.5 years [36], our

Table 4. Lion sightings in the eastern Maghreb of Algeria and Tunisia, 1861–1880 (Ksour Mountains, Saharan Atlas, Tell Atlas, Ouled-Nail, Aurès Mountains).

	Location	Season/Year	Observation Type	Solitary (0)/Group (1)	Original source	Ref.
74	Miliana vicinity	Winter 1861	Observations, many lions	1	French man	[88]
75	Djebel Chaambi (now a NP), Tunisia	Spring 1862	Frequent lions	1		[89]
76	Sahel (Algiers vicinity)	Winter 1862	Shot	0	Chassaing, F.	[90]
77	Tebessa mountains, Kasserine, Tunisia	1863	Few lions left on the steppes	1	Citation	[73]
78	Nechmeya vicinity	Autumn 1863	Observation	0		[91]
79	Ourten valley	Autumn 1863	Observations, lion spot	1		[91]
80	Tazoult vicinity	Autumn 1863	Observation, lion killing a cow	1		[91]
81	Mouzaia vicinity (Chréa NP)	1864	Observation, male	0	Le Tell	[92]
82	Tebessa border	1864	Fresh tracks, roaring	0		[91]
83	Beni Salah forest	Spring 1865	Citation	1		[93]
84	Oued Zenati vicinity	Spring 1865	Shot	0		[93]
85	Djebel Riless	Summer 1866	Shot, 3	1		[66]
86	Azzaba vicinity	1866	Shot, male, female	1		[65]
87	Djebel Tangout	1866	Observation, male after burning forest	0		[65]
88	Ras El Ma (Azzaba)	1866	Observation, male	0		[65]
89	Oued Saida	1867	Observations	1		[94]
90	Djurdjura (now a NP)	1867	Observations	1		[95]
91	Fetzara lake	1867	Observations	1		[95]
92	Feriana, Tunisia	1868			Monchicourt	[73]
93	Ouarsensis	1869			Leroux	[73]
94	Djebel Bouarif (Aurès Mounts)	1869	Winter refuge, observed and tracks also seen	1		[74]
95	Souk Ahras vicinity	1873	Shot, 5	1		[54]
96	Djebel Bissa	Winter 1874	Observation	0	El Mobacher	[96]
97	Annaba vicinity	Spring 1875	Observations	1		[59]
98	Ain Mimoun (Aurés Mounts)	Spring 1875	Observations, 2 lions roaring	1		[59]
99	Skikda vicinity	1875	Shot 6 lions	0	Cheret, C.	[97]
100	Souk Ahras vicinity	1877	Shot, 3	1		[54]
101	Souk Ahras vicinity	1878	Shot, 4	1		[54]
102	Souk Ahras vicinity	1879	Shot, 5	1		[54]
103	Tebessa forest	1879	Twenty lions killed/year	1		[98]
104	Souk Ahras vicinity	1880	Shot, 3	1		[54]
105	Northwest Tunisia	1880	Four shot, one photographed	1		[99]

NP = one of several National Parks established across the region in the 1980s–1990s.

analysis places today's Moroccan Royal lions as much closer descendants; with the youngest breeders only 16 generations removed from wild ancestors (Fig. 4).

Discussion

Our analysis suggests that relatively recent sightings of lions (1940s and 1950s) are not exceptional. Whilst we corroborate recognized sightings in Morocco in the 1940s, we also suggest that lions persisted in Algeria into the late 1950s, nearly seventy years later than considered in previous reviews. Barbary lions experienced a combination of factors of decline consistent with the threats to modern-day felids; habitat availability, wild prey availability, livestock husbandry and management, human behaviour, land use and socio-economics [40], [41].

A more recent extinction date for Barbary lions raises several questions: (a) how lions managed to persist in degraded North African ecosystems; (b) whether later persistence in North Africa provides insights for conserving marginal lion populations in West and Central Africa; (c) whether wild Barbary lions were more recently taken into captivity, making current animals closer descendants than previously considered.

Table 5. Lion sightings in the eastern Maghreb of Algeria and Tunisia, 1881–1900 (Ksour Mountains, Saharan Atlas, Tell Atlas, Ouled-Nail, Aurès Mountains).

	Location	Season/Year	Observation Type	Solitary (0)/Group (1)	Original source	Ref.
106	Djurdjura (now a NP)	1880–95	Shot by French man	0		[52]
107	Aurès Mounts	1880–95	Shot	0		[52]
108	El Kala (vicinity NP)	1880–95	Several shot	1		[52]
109	Souk Ahras vicinity	1881	Shot, 1	0		[54]
110	Ain Drahem, Tunisia	1881	Observation	1		[54]
111	Feidja (now a NP), Tunisia	1881	Observation	1		[54]
112	Khang el Melah (Djelfa)	Summer 1881	Citation	1	Guy de Maupassant	[100]
113	Djebel Chelia (Aurès Mounts)	1884	Last lions in cedar forest	1		[52]
114	Zaccar (Djelfa)	1884	Isolated kills	1		[52]
115	Tebessa	1885	Shot, male	0		[101]
116	Ksenna forest	1886	Observations	1		[102]
117	Ghardimaou, Tunisia	1887	Lions in forest	1	Lataste, 1887	[99]
118	Feriana/Medjerda forest, Tunisia	1887		1	Lataste, 1887	[99]
119	Djebel Mssid vicinity	1887		1	Lataste, 1887	[99]
120	Ghardimaou/Souk Ahras (border area)	1888	Shot, male	0		[103]
121	Seraidji (Edough Mount)	Summer 1890	Shot, young female	0		[103]
122	Djedeida, Tunisia	1890	Drowned young lion in dam	0		[52]
123	Babouch (South Tabarka), Tunisia	1891	Shot, male	0		[104]
124	Souk Ahras vicinity	1891	Shot male	0		[104]
125	Edough Mount	1892	Observations, full grown male	0	Werner	[52]
126	Ain Beida	Spring 1893	Shot male	0	Langelet, P.	[105]
127	Batna vicinity	1893	Shot female	0		[104]
128	Oued Sahel valley	1893	Observations	1		[106]
129	Isser Valley	1893	Observations	1		[106]
130	Sebaou valley	1893	Observations	1		[106]
131	Thabourt Bouzgueur	1893	Observations, Winter refuge	1		[106]
132	Aurès Mounts	1894	Lion hunting a horse	0	Germaine Tillon	[107]
133	Djebel Amour	1898	Numerous lions	1		[108]

NP = one of several National Parks established across the region in the 1980s–1990s.

4.1. Recent Persistence of Lions in North Africa

Previous assumptions that lions have been absent from North African ecosystems for most of the past century is challenged by our analysis. We suggest that small populations persisted largely undetected by humans for several generations, indeed many decades. In Morocco, an inhospitable interior always made encounters with lions rare, whilst in Algeria sightings across the population's range occurred on an almost annual basis up to 1894, although thereafter less frequently than every 1 lion generation (6–7 years). Our calculation of the time of extinction (T_E) corresponds with the recent suggestion that the last Barbary lion was probably lost in the destruction of forests north of Setif in 1958 during the French-Algerian War [25].

4.2 Adaptation of Lions in North African Ecosystems

Up to the late 1800s, hunters reported lions traversing from northwest Algeria, westwards into Morocco and from northeastern Algeria eastwards into Tunisia [19]. After the 1880s, the pattern of sightings suggest that lion populations retreated broadly in two directions; in Morocco southwards away from coastal regions through the Rif, Middle and High Atlas Mountains and the Saharan fringes; and in Algeria eastwards into the Tell Atlas and the Aurès Mountains bordering Tunisia. In both regions small populations survived at low densities in remote areas for several generations. Literature and oral accounts suggest that lions persisted through certain behavioral adaptations (hunting domestic livestock, engaging in nocturnal activity, living in small groups or pairs) and shifts in range (leaving deforested localities, moving to outlying areas and higher altitudes, and following water points in arid regions). Many of these particular behavioral adaptations have since been observed in contemporary populations of lions in human-dominated landscapes in sub-Saharan Africa [42], [43] as well as in restricted available habitat in India [11].

Our analysis of historical records in the Maghreb suggests that lions appear not to adapt their social behavior as a response to human habitat pressure. There was no significant change from group to solitary living by lions in the remote western Maghreb regions, whilst importantly even the eastern Maghreb lion

Table 6. Last lion sightings recorded in the eastern Maghreb of Algeria, 1900–1960 (Ksour Mountains, Saharan Atlas, Tell Atlas, Ouled-Nail, Aurès Mountains).

	Location	Season/Year	Observation Type	Solitary (0)/Group (1)	Original source	Citation Text S1
134	Beni Salah (Chréa NP vicinity)	1900's	Observations	1	Inhabitants of Beni Salah	[109]
135	Zaccar-AinTorki	1910–12	Observations	0	MrsDedreuil-Paulet	Difallah, pers. comm.
136	Aurès Mounts	1911	Shot, Male, female	1	Sassorossi family	[107]
137	Bejaia vicinity	1912	Shot	0		[110]
138	Ain Sefra	1912	Shot	0	Khazene, A,	Fellous, pers. comm.
139	Biskra (lion probably from Aurès ranges)	1917	Old lion	0	Seen by the writer	[111]
140	Djebel Tameda, south of Boussemghoun	c.1920	Last lion shot in the Saharan Atlas	0	Gueniche Ahmed	Bahmane L to Fellous
141	Between Ain Talawane and Ain Roua (Setif)	Late 1920s	Observation (in spring)	1	Report by old man	Difallah, pers. comm.
142	Guenzet-Babor Mount (North Setif)	1930	Observation	0	Interview local people	[112]
143	North Setif	1930	Observations	1	Interview local people	[112]
144	DjebelDirah (Sour El Ghozlane)	1930s	Shot	0	Kalem, pers. comm.	Fellous
145	Boussam (Menaceur-Zaccar mounts)	Winter 1935	Male lion shot attacking a cow	0	Bounaceur Farid	Fellous
146	Djebel es Somm (Djebel Amour)	1935	Male lion attacked a camel	0	Hamami Bachir	Fellous
147	Unknown location in Algeria	1943	Shot		Keeling, pers. comm.	[113]
148	North Setif	Late 1940s	Observations	1	Interview local people	Difallah, pers. comm.
149	Beni Ourtilane (North of Setif)	1956	Observation	0	People on a bus	[112]

populations did not exhibit a significant change in group living in the final decades of their existence, despite increasing encroachment by human populations. Although lion population density is typically lower in desert and semi-desert areas, pride size generally remains comparable in desert ecosystems and moister ecosystems with higher lion density [22]. Interestingly, human presence has been shown to provide advantages to lion cub survival by

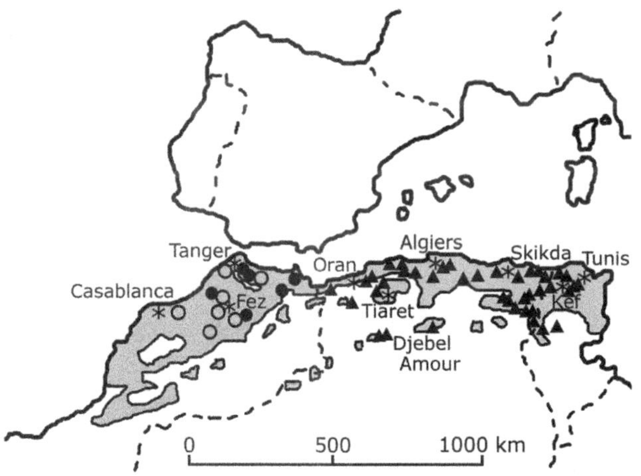

Figure 1. Distribution of historical reports of lions in North Africa (AD 1500–1900). Grey shading indicates Mediterranean scrubland ecosystems [8]. Earliest accounts in the western Maghreb from 16th to the 18th century are indicated as open circles [7], [19]. Documented sightings in known years from 1800 to 1900 are indicated as black circular markers in the western Maghreb (1–7 in Table 1); triangular markers indicate sightings in eastern Maghreb (22–133 in Tables 2–6). Asterisks (*) denote locations of human population centers. Dashed lines indicate national boundaries.

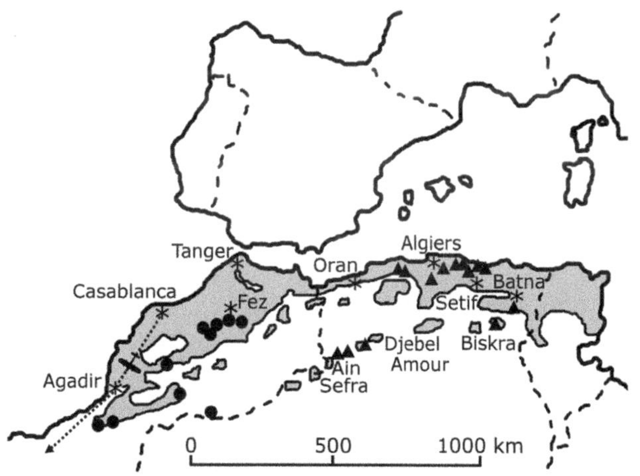

Figure 2. Distribution of recent sightings of lions in North Africa (AD 1900–1960). Grey shading indicates Mediterranean scrubland ecosystems [8]. Circular markers indicate sightings in western Maghreb (8–21 in Table 1); triangular markers indicate sightings in eastern Maghreb (134–149) from incidents described in Table 6. The dotted line indicates the air route across the Atlas Mountains (Casablanca-Agadir-Dakar) during which the last wild lion was photographed. Asterisks (*) denote locations of human population centers. Dashed lines indicate national boundaries.

Figure 3. A lion seen in the Atlas Mountains, during a flight on the Casablanca-Dakar air route. The photograph taken by Marcelin Flandrin in 1925 is the last visual record of a wild 'Barbary' lion of North Africa.

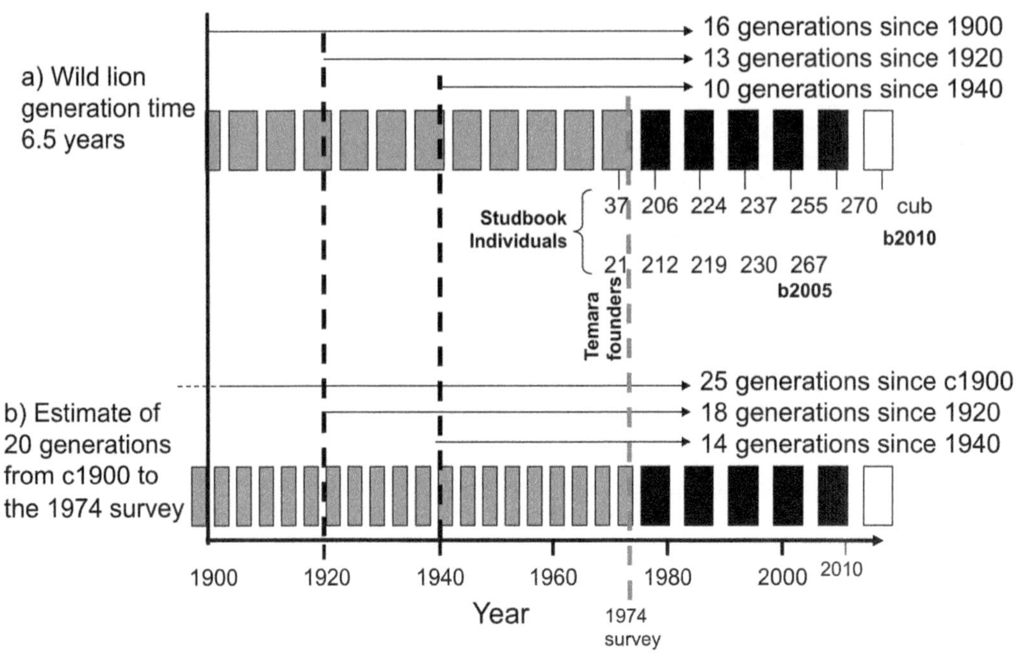

Figure 4. Estimates of captive generations since wild collection in North Africa for current Moroccan Royal lions. Grey boxes indicate estimated lion generations based on suggestions by: (a) Packer et al. [38], and (b) Hemmer [20]. Black boxes are the five known generations in the European studbook [13] since the 1974 survey at Temara Zoo. Generational positions for two studbook maternal lines are illustrated for a female cub (white box) born to studbook female 270, and a young male (267) born to female 230, tracing to founder females 37 and 21 respectively.

removing secondary predators [44], and this may have enabled lion persistence in the Maghreb.

Final extirpation of lions through hunting was a response to livestock predation; a prey dependency probably driven by reduced habitat and fewer wild ungulates [18]. This final collapse may have been exacerbated by the species' behavioral need to remain in social groups with proportionally greater local resource requirements. Group living has been shown to be similarly maintained by lion populations in semi-arid environments such as the Kalahari and Etosha [45] [46]. In contrast, leopards (*Panthera pardus*) still persist at lower population densities in Morocco [1].

4.3 Heritage of Moroccan Royal Lions

Moroccan Royal Lions have been described as "an obvious relic of the original Barbary lion gene pool" [20]. Our analysis suggests fewer post-wild generations have occurred during the period where this population has existed solely in captivity. Today's lions, descended from the Moroccan Royal collection, could be closer relatives to wild ancestors than previously considered. In the absence of definitive nuclear DNA profiles from wild Barbary lion specimens, the precautionary principle suggests that Moroccan Royal lions should be conserved as descendants of the Barbary lion until science can tell us otherwise. The Moroccan Royal lions offer one of the few scenarios in which restoration of lions into regions where the species is long extinct could be envisioned as having useful conservation value [47].

Conclusions

Insights from historical sightings are relevant to current lion conservation. We suggest that wild lions persisted in the Maghreb into the 1950s, much later than previously recognized. The lion is a well-known, visible and potentially threatening species, yet small populations survived in North Africa decades after being generally considered extinct. This persistence reflects the recent rediscovery of a small population of Barbary leopard nearly 20 years after the last previous sighting and a decade since being declared extinct [1] [48]. Careful consideration should be given to mammalian carnivores currently presumed extinct or near-extinct in other regions, coupled with a greater understanding of extinction patterns and the conservation potential in relict populations.

Finally, we suggest caution when considering the current conservation status of lions. Although lions in the Maghreb adapted to reduced population density, prey availability and habitat encroachment, our analysis reveals that lion group-living behavior did not change significantly as human pressures increased. As a pride-forming species [45], *P. leo* populations are prone to collapse, whereas other felids may survive at lower local population densities by not living in social groups [49]. Lions in today's small populations in Central and West Africa persist [16] [50], even if rarely seen, in fragmented remnants, yet clearly exist at the edge of a precipitous drop into extinction. Continued, carefully considered conservation effort remains vitally important.

Acknowledgments

We thank K. Difallah, J. Edwards, A. Harland, W. Frey, L. Bahmane, A. Laabed, A. Djebali, A. and M. Laroui, A. Kalem, A. Khazene, F. Bounacer and B. Hamami for their assistance. The authors also thank Prof Alfred Roca, Dr Andrew Kitchener and two anonymous reviewers for their helpful comments.

Author Contributions

Conceived and designed the experiments: SAB AF NY DLR. Performed the experiments: SAB AF DLR. Analyzed the data: SAB DLR. Contributed reagents/materials/analysis tools: SAB AF DLR. Wrote the paper: SAB AF NY DLR.

References

1. Henschel P, Hunter L, Breitenmoser U, Purchase N, Packer C, et al. (2008) *Panthera pardus*. In IUCN 2010. IUCN Red List of Threatened Species. Version 2010.4. (Accessed 28 February 2011: http://www.iucnredlist.org/details/15954/0).
2. Can OE (2004) Status, conservation and management of large carnivores in Turkey. Convention on the Conservation of European Wildlife and Natural Habitats, Standing Committee Meeting 24, 29th November-3rd December, T-PVS/Inf(2004) 8. Strasbourg: Council of Europe.
3. Karanth KU, Chellum R (2009) Carnivore conservation at the crossroads. Oryx 43: 1–2.
4. IUCN (2012) IUCN Red List of Threatened Species. Version 2012.1. (Accessed 26 June 2012: www.iucnredlist.org).
5. Tumenta PN, Kok JS, van Rijssel JC, Buij R, Croes BM, et al. (2009) Threat of rapid extermination of the lion (*Panthera leo leo*) in Waza National Park, Northern Cameroon. Afr. J. Ecol. 48: 888–894.
6. Burton AC, Sam MK, Kpelle DG, Balangtaa C, Buedi EB, et al. (2011) Evaluating persistence and its predictors in a West African carnivore community. Biol. Conserv. 144: 2344–2353.
7. Guggisberg CAW (1963) Simba: the life of the lion. London: Bailey Bros. and Swinfen.
8. Dobson M (1998) Mammal distributions in the western Mediterranean: the role of human intervention. Mammal Rev 28: 77–88.
9. Barnett R, Yamaguchi N, Shapiro B, Sabin R (2008) Ancient DNA analysis indicates the first English lions originated from North Africa. Contributions to Zoology 77: 7–16.
10. Masseti M (2010) Holocene mammals of Libya: A biogeographical, historical and archaeozoological approach. J Arid Env 74: 794–805.
11. Nowell K, Jackson P (1996) Wild cats, status survey and conservation action plan. Gland: IUCN/SSC Cat Specialist Group.
12. Patterson BD (2007) On the nature and significance of variability in lions (*Panthera leo*). Evol Biol 34: 55–60.
13. Black S, Yamaguchi N, Harland A, Groombridge J (2010) Maintaining the genetic health of putative Barbary lions in captivity: an analysis of Moroccan Royal Lions. Eur J Wildl Res 56: 21–31.
14. Barnett R, Yamaguchi N, Barnes I, Cooper A (2006) The origin, current diversity and future conservation of the modern lion (*Panthera leo*). Proc R Soc Lond B Biol Sci 273: 2119–2125.
15. Mazák JH (2010) Geographical variation and phylogenetics of modern lions based on craniometric data. J Zool 281: 194–209.
16. Bauer H, Nowell K, Packer C (2008) *Panthera leo*. Gland, Switzerland: IUCN Red List of Threatened Species. Version 2010.4 (Accessed 16 November 2010: http://www.iucnredlist.org/apps/redlist/details/15951/).
17. Breitenmoser U, Mallon DP, Ahmad Khan J, Driscoll C (2008) *Panthera leo ssp. persica*. In: IUCN 2012. IUCN Red List of Threatened Species. Version 2012.1. (Accessed 23 August 2012: http://www.iucnredlist.org/details/15952/0).
18. Cuzin F (2003) Les grands mammiferes du marocmeridional (Haut Atlas, Anti Atlas et Sahara): distribution, écologie et conservation. PhD thesis, Université Montpellier.
19. Cabrera A (1932) Los mamíferos de Marruecos. Seria Zoologica. Madrid: Trabajos del Museo Nacional de Ciencias Naturales.
20. Hemmer H (1978) Grundlagen und derzeitiger Stand des Zuchtprogrammes zur Rückerhaltung des Berberlöwen (*Panthera leo leo*). In: Seifurt S, Müller P, editors. Congress Report, 1st International Symposium on the Management and Breeding of the Tiger, 11th and 12th October 1978 in Leipzig, Abb. 1. Zoological Garden. Leipzig: International Tiger Studbook. 65–72.
21. Johnston HH (1911) Tunisia. The Encyclopedia Britannica, 11th Edition, vol. 27. Cambridge: Cambridge University Press.
22. Celesia GC, Peterson AT, Peterhans JCK, Gnoske TP (2009) Climate and landscape correlates of African lion (*Panthera leo*) demography. Afr J Ecol 48: 58–71.
23. Jhala YV, Mukherjee S, Shah N, Chauhan KS, Dave CV, et al. (2009) Home range and habitat preference of female lions (*Panthera leo persica*) in Gir forests, India. Biodivers Conserv 18: 3383–3394.
24. Visser HD, Müller L, Tumenta PN, Buij R, de Iongh HH (2009) Factors affecting lion (*Panthera leo*) home range, movement and diet in Waza National Park, Cameroon. J Zool 300: 131–142.
25. Barnett R, Yamaguchi N, Shapiro B, Nijman V (2007) Using ancient DNA techniques to identify the origin of unprovenanced museum specimens, as illustrated by the identification of a 19th century lion from Amsterdam. Contributions to Zoology 76: 87–94.
26. Bartosiewicz L (2009) A lion's share of attention: archeozoology and the historical record. Acta Archaeologica Academiae Scientiarum Hungaricae. doi 10.1556/AArch.59.2008.2.28

27. Schnitzler AE (2011) Past and present distribution of the North African-Asian lion subgroup: a review. Mammal Rev. 41: 220–243. doi 10.1111/j.1365-2907.2010.00181.x.
28. Bowkett AE (2009) Recent captive-breeding proposals and the return of the ark concept to global species conservation. Conserv Biol 23:773–776. doi 10.1111/j.1523-1739.2008.01157.x.
29. Gippoliti S (2012) *Ex situ* conservation programmes in European zoological gardens: Can we afford to lose them? Biodivers Conserv 21: 1359–1364. doi 10.1007/s10531-012-0256-8.
30. Dubach J, Patterson BD, Briggs MB, Venzke K, Flamand J, et al. (2005) Molecular genetic variation across southern and eastern geographic ranges of the African lion, *Panthera leo*. Conserv Genet 6: 15–24.
31. Antunes A, Troyer JL, Roelke ME, Pecon-Slattery J, Packer C, et al. (2008) The evolutionary dynamics of the lion *Panthera leo* revealed by host and viral population genomics. PLoS Genetics 4: 11: e1000251.
32. Bertola LD, van Hooft WF, Vrieling K, Uit de Weerd DR, York DS, et al. (2011) Genetic diversity, evolutionary history and implications for conservation of the lion (*Panthera leo*) in West and Central Africa. J Biogeogr 38: 1356–1367.
33. Solow AR (2005) Inferring extinction from a sighting record. Math Biosci 195: 47–55.
34. Vogel RM, Hosking JRM, Elphick CS, Roberts DL, Reed JM (2009) Goodness of fit of probability distributions for sightings as species approach extinction. Bull Math Biol 71: 701–719 doi 10.1007/s11538-008-9377-3.
35. Siegel S, Castellan Jr NJ (1988) Nonparametric statistics for the behavioral sciences, 2nd Ed. New York: McGraw-Hill Book Company.
36. Packer C, Pusey AE, Eberly LE (2001) Egalitarianism in female African lions. Science 293: 690–693.
37. Black S, Harland A (2009) Moroccan Royal Lions: justification for a new EEP to support lion sub-species conservation. Paper presented at Felid TAG Meeting. Copenhagen, Denmark: Proceedings of the 26th Annual EAZA Conference, 17th September.
38. Davies RAG (1964) A history of the world's airlines. Oxford: Oxford University Press.
39. Zoo-Olomouc (2010) Přírůstky 6.8.2010 Lev berberský. Czech Republic: Olomouc Zoo (Accessed 21 March 2012: http://zoo-olomouc.cz/app/sekce/23/prirustky?articles[offset] = 40).
40. Bauer H, de Iongh H, Sogbohossou E (2010) Assessment and mitigation of human-lion conflict in West and Central Africa. Mammalia 74: 363–367.
41. Inskip C, Zimmermann A (2009) Human-felid conflict: a review of patterns and priorities worldwide. Oryx 43: 18–34.
42. Mogensen NL, Ogutu' JO, Dabelsteen T (2011) The effects of pastoralism and protection on lion behaviour, demography and space use in the Mara Region of Kenya. African Zool 46: 78–87.
43. Valeix M, Hemson G, Loveridge AJ, Mills G, Macdonald DW (2012) Behavioural adjustments of a large carnivore to access secondary prey in a human-dominated landscape. J App Ecol 49: 73–81.
44. Kissui BM, Mosser A, Packer C (2010) Persistence and local extinction of lion prides in the Ngorongoro Crater, Tanzania. Popul Ecol 52: 103–111.
45. Yamaguchi N, Cooper A, Wedelin L, Macdonald DW (2004) Evolution of the mane and group-living in the lion (*Panthera leo*): a review. J Zool Lond 263: 329–342.
46. Sunquist ME, Sunquist F (2002) Wild cats of the World. Chicago: University of Chicago Press.
47. Hunter TB, White P, Henschel P, Frank L, Burton C, et al. (2013) Walking with lions: why there is no role for captive-origin lions *Panthera leo* in species restoration. Oryx 47: 19–24.
48. Agencia EFE (2010) Hallan una población de leopardos que se daba por extinguida en el Magreb. (Accessed 28 February 2011: http://www.telecinco.es/informativos/cultura/noticia/1274139/1274139).
49. Courchamp F, Clutton-Brock T, Grenfell B (1999) Inverse density dependence and the Allee effect. Trends Ecol Evol 14: 405–410.
50. Henschel P, Azani D, Burton C, Malanda G, Saidu Y, et al. (2010) Lion status updates from five range countries in West and Central Africa. Cat News 52: 34–39.
51. Drummond-Hay JH (1861) Morocco and the Moors. Western: its tribes and savage animals. London: Murray.
52. Guggisberg CAW (1963) Simba: the life of the lion. London: Bailey Bros and Swinfen.
53. Schnitzler AE (2011) Past and present distribution of the North African-Asian lion subgroup: a review. Mammal Rev 41: 220–243.
54. Lavauden L (1932) Les grands félins de l'Afrique du Nord et leur disparition. Le Chéne. Soc. Forestière Mediterranéenne et Coloniale. N°4 Janvier: 208–234.
55. Grzimek B (1975) The lion, In Grzimek, B. (Ed.), Grzimek's Animal Life Encyclopedia, vol. 12, Mammals III.Van Nostrand Reinhold, New York, 353.
56. Yadav PR (2004) Vanishing and endangered species. New Delhi: Discovery Publishing House.
57. Black S (2008) Investigating the feasibility of reintroducing lions (*Panthera leo*) as a flagship for the Moroccan Atlas Mountains. MSc thesis, University of Kent.
58. Panouse JB (1957) Les mammifères du Maroc. Trav Inst Sci Chérif Sér Zool n°5. Rabat. 206 p.
59. Playfair RL (1877) Travels in the footsteps of Bruce in Algeria and Tunis. London: G. Kegan Parl & Co.
60. Daumas E (1855) Moeurs et coutumes de l'Algérie. Paris: Sind bad Bibliothequearabe.
61. Roches L (1904) Dix ans à travers l'Islam, 1834–1844. Paris: Librairie Academique Didier.
62. Carette E (1844) Exploration scientifique de l'Algérie pendant les années 1840–1841–1842. Paris: Imp Roy Tome II: 355.
63. Gerard J (1864) Chasse au lion. Cie Editeurs. Québec: JN Duquest.
64. Scott KC (1842) A journal of residence in the Esmailla of Abd-El-Kader. London: Whittaker & Co.
65. Fenech EV (1867) Récits et chasses d'Algérie. Typographie Denis Aimé.
66. Margueritte A (1869) Chasses de l'Algérie et notes sur les arabes du Sud. 1ère edition. Paris: JouvetEditeurs.
67. Bloch A (2002) Miliana par les textes. Algiers, Editions Zyriab.
68. Daumas E (1845) Le sahara Algérien. Paris: Langlois & Leclercq.
69. Anon. (1845) Journal des haras, des chasses et des courses de chevaux. Imprimerie Parent Tome III. Bruxelles.
70. Kennedy JC (1846) Algeria and Tunis in 1845. Vols I & II. London: H Colburn Publisher.
71. Gerard J (1892) Le tueur de lions. Paris: Paris librairie Hachete & Cie.
72. Saint Marie C (1846) Algeria in 1845. A visit to a French possessions in Africa. London: R Bentley.
73. de Planhol X (2004) Le paysage animal. L'homme et la grande faune: une zoogéographie historique. Paris: Fayard.
74. Palisser G (1982) Les lions et panthères dans le Sahel. L'Algérianiste n°20, Décembre, Narbonne, France.
75. Lestiboudois T (1853) Voyage en Algérie, ou études sur la colonisation de l'Afrique française. Lille: Imp Danel.
76. Saint Arnaud AJL (1864) Lettres du maréchal de saint Arnaud.1832–1854. 3 ème édition. Paris: M Levy frères.
77. Gerard J (1859) La chasse au Lion. Paris: Paris Librairie nouvelle.
78. Dufour L (1856) Notice sur un lion tué en Algérie. Examen necroscopique. Paris: H & C Noblet Imprimerie.
79. D'Esschavannes MJ (1852) Revue de l'Orient, de l'Algérie et des colonies. Paris: Rouvé J Tome XII: 221–222.
80. Bombonnel CL (1893) Bombonnel, le tueur de panthères, ses chasses racontées par lui même. 7eme édition, Paris: Librairie Hachette.
81. Anon. (1860a) The hunting grounds of the Old World. London: Sanders Otley & Co.
82. Gastineau B (1863) Chasses au Lion et à la Panthère en Afrique. Paris: Paris Librairie Hachette.
83. Blakesley JW (1859) Four months in Algeria: with a visit to Carthage. London: MacMillan & Co.
84. Dunant JH (1858) Notice sur la régence de Tunis. Genève: Imprimerie Jules Fick.
85. Anon. (1860b) Journal des chasseurs. Vol. 24, 2eme sem, mai-octobre: 239–248.
86. Thierry-Mieg C (1861) Six semaines en Afrique, souvenirs de voyage. Paris: M Levy frères.
87. Bernard A, Lacroix N (1906) L'évolution du nomadisme en Algérie. Alger: Edition Jourdan.
88. Windham WG (1862) Notes in North Africa: guide to the sportsman and tourist in Algeria and Tunisia. London: Ward & Lock.
89. Guerin V (1862) Voyage archéologique dans la régence de Tunis. Paris: TI Plon.
90. Baroli M (1967) La vie quotidienne des français d'Algérie (1830–1914) Paris: Librairie Hachette.
91. Ormsby J (1864) Autumn rambles in North Africa. London: Longman & Green.
92. Anon. (1936) Revue Chasse et pêche Nord Africaines. Mars n° 48: 7.
93. Carteron C (1866) Voyage en Algérie. Paris: J Hetzel Librairie.
94. Berard V (1867) Indicateur general de l'Algérie. Alger: Typo Bastide, 3eme edit. 600 p.
95. Loche V (1867) Histoire naturelle des mammifères. Paris: Arthis Bertrand Editeur.
96. Tounsi G (1998) Un lion surgit dans la tribu des Sinfita à Tenès. Evocation. Journal Le Jeune Indépendant N°14 du 25 au 31 mars. 24 p.
97. Simmonds PL (1877) Animal Products: their preparation, commercial uses, and value. London: Chapman & Hall. 416 p.
98. Bourde P (1880) A travers l'Algérie. Souvenirs de l'excursion parlementaire, septembre-octobre 1879. Paris: G Charpentier.
99. Gharaibeh BM (1997) Systematics, distribution, and zoogeography of mammals in Tunisia. PhD thesis, Texas Tech University, Lubbock.
100. Emmanuel J (2003) Maupassant, guy, sur les chemins d'Algérie (1881–1890). Paris: Magella & Cie.
101. Lataste F (1885) Étude de la faune des vertébrés de Barbarie (Algérie, Tunisie et Maroc). Actes de la Société Linnéenne de Bordeaux 39: 129–289.
102. Robert G (1891) Voyage à travers l'Algérie: notes et croquis. Paris: Imprimerie de G Rougier.
103. Seurat LG (1924) Zoologie forestière de l'Algérie. Alger: Direction des Forêts.
104. Lavauden L (1926b) Les vertébrés du Sahara. Tunis: Imp A Guénard.
105. Anon. (1893) Rubrique annonce. Le Chenil, n°12 du 23 mars: 137–138.
106. Hanoteau A, Letourneux A (1893) La kabylie et les coutumes kabyles. Tome I. Paris: Imprimerie Nationale.
107. Anon. (1898) La revue hebdomadaire. Paris: Librairie Plon. 378 p.

108. Martin JP (2006) Le Belezma: au forgeron de Batna. Paris: L'Harmattan.
109. Desparmet J (1939) Coutumes, institutions, croyances des indigènes de l'Algérie. Tome I, Imprimerie.Alger, La typo-litho and J Carbonel.
110. De Smet K (1982) La disparition des grands fauves en Algérie. Bull Forest Conserv Nat Alger 1: 23–25.
111. Ossendowski FA (1927) The breath of the desert: the account of a journey through Algeria and Tunisia (translated by Palen LS). London: George Allen and Unwin.
112. Haddadou MA (1994) L'Algérie mystérieuse. Alger.
113. Yamaguchi N, Haddane B (2002) The North African Barbary lion and the Atlas lion project. International Zoo News 49: 321.

15

Influence of Introgression and Geological Processes on Phylogenetic Relationships of Western North American Mountain Suckers (*Pantosteus*, Catostomidae)

Peter J. Unmack[1,2]*, Thomas E. Dowling[3,4], Nina J. Laitinen[1], Carol L. Secor[3], Richard L. Mayden[5], Dennis K. Shiozawa[1], Gerald R. Smith[6]

1 WIDB 401, Department of Biology, Brigham Young University, Provo, Utah, United States of America, 2 National Evolutionary Synthesis Center, Durham, North Carolina, United States of America, 3 School of Life Sciences, Arizona State University, Tempe, Arizona, United States of America, 4 Department of Biological Sciences, Wayne State University, Detroit, Michigan, United States of America, 5 Department of Biology, Saint Louis University, St. Louis, Missouri, United States of America, 6 Museum of Zoology, The University of Michigan, Ann Arbor, Michigan, United States of America

Abstract

Intense geological activity caused major topographic changes in Western North America over the past 15 million years. Major rivers here are composites of different ancient rivers, resulting in isolation and mixing episodes between river basins over time. This history influenced the diversification of most of the aquatic fauna. The genus *Pantosteus* is one of several clades centered in this tectonically active region. The eight recognized *Pantosteus* species are widespread and common across southwestern Canada, western USA and into northern Mexico. They are typically found in medium gradient, middle-elevation reaches of rivers over rocky substrates. This study (1) compares molecular data with morphological and paleontological data for proposed species of *Pantosteus*, (2) tests hypotheses of their monophyly, (3) uses these data for phylogenetic inferences of sister-group relationships, and (4) estimates timing of divergence events of identified lineages. Using 8055 base pairs from mitochondrial DNA protein coding genes, *Pantosteus* and *Catostomus* are reciprocally monophyletic, in contrast with morphological data. The only exception to a monophyletic *Pantosteus* is *P. columbianus* whose mtDNA is closely aligned with *C. tahoensis* because of introgression. Within *Pantosteus*, several species have deep genetic divergences among allopatric sister lineages, several of which are diagnosed and elevated to species, bringing the total diversity in the group to 11 species. Conflicting molecular and morphological data may be resolved when patterns of divergence are shown to be correlated with sympatry and evidence of introgression.

Editor: Laurent Viriot, Team 'Evo-Devo of Vertebrate Dentition', France

Funding: This work was supported in part by funds from the US Bureau of Reclamation to TED and CLS, by the National Evolutionary Synthesis Center (NESCent) and NSF EF-0905606 to PJU and NSF DEB 0240184, EF 0431326, DEB-0817027, DBI-0956370 and DEB-1021840 to RLM. The funders had no role in study design, data collection and analysis, decision to publish, or preparation of the manuscript.

Competing Interests: The authors have declared that no competing interests exist.

* E-mail: peter.pub@unmack.net

Introduction

Western North America experienced intense geological activity resulting in major topographic changes over the past 15 million years, during which time the modern freshwater fish fauna of the region evolved [1,2]. The North American continental crust moved westward over subducting plates, and was extended (stretched) by nearly 100 percent, causing development of about 170 mountain ranges ca 100 km long, separating narrow, north-trending basins (Fig. 1). These roughly parallel basins extend from southern Oregon and Idaho, across Nevada and western Utah into Mexico. The Great Basin is flanked on the west by the Sierra Nevada and on the east by the Colorado Plateau and Rocky Mountains [3,4]. Changes in elevation, caused by major north-south normal faults, resulted in hundreds of short, swift mountain streams often ending in isolated basins. These geological changes caused the diversification of lineages of many aquatic organisms [5,6].

The evolution of basin and range topography caused isolation, which permitted diversification within the aquatic fauna [7,8]. Tectonism and volcanism modified drainage patterns to form several huge, long-lived lakes such as Miocene Chalk Hills Lake [9] and Pliocene Glenns Ferry Lake in southwest Idaho [10], late Miocene Hopi Lake in northwest Arizona [11], Pleistocene Lake Bonneville in Utah [12–14], Pleistocene Lake Lahontan and associated lakes in Nevada [15] and southern Oregon [16], Pleistocene Tulare Lake, California [17], and many other fish-inhabited lakes in the Great Basin [7,18]. These sometimes-isolated basins were the original crucibles of local evolution of subspecies and species of various fishes in the Miocene and Pliocene [6].

Today, the major rivers of western North America are composites of more ancient rivers. The Colorado River flows southwestward across the Colorado Plateau through Grand Canyon to the Grand Wash fault zone, where the river enters the Great Basin and adds to its fish fauna [19,20]. North of the Great Basin, the Snake River flows across southern Idaho to Hells Canyon, where it was captured in the late Pliocene by the Columbia River [21,22]. The Bear River flows north from the

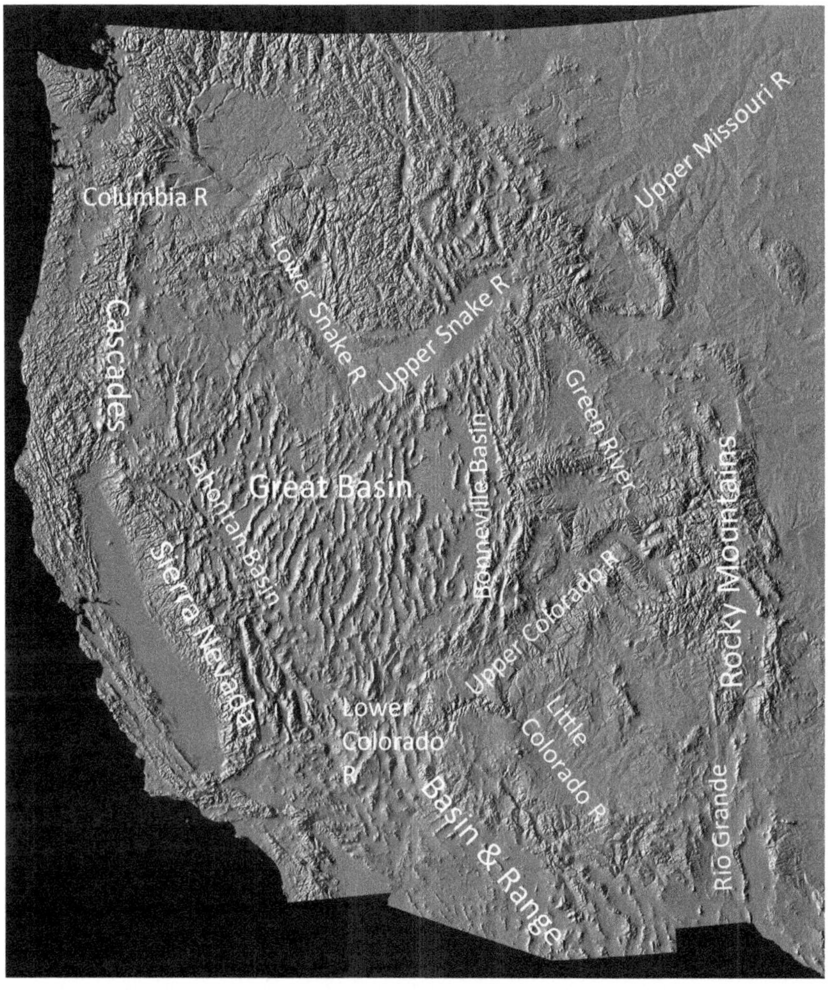

Figure 1. Major Late Cenozoic topographic features, drainage basins, and rivers responsible for isolation and evolution of *Pantosteus*. The hydrographic Great Basin, between the Sierra Nevada and the Colorado Plateau, is the northern, much-extended section of the Basin and Range province, which continues south into Mexico and east to the Rio Grande Rift.

Uinta Mountains to southeast Idaho, where it was diverted south into the Bonneville Basin about 100 Ka [23], causing the Bonneville Basin, Utah, to overflow and become tributary to the Snake and Columbia rivers in Idaho. In the Miocene the Lahontan Basin, Nevada, flowed west to the Pacific Ocean and later overflowed to the Upper Snake River [6,15]. Uplift of the Sierra Nevada and rise of the Cascade Mountains changed climates across the Great Basin [24,25] and influenced drainage connections. Each of these geological and hydrologic events is hypothesized to have influenced evolution of fish populations to be discussed (Fig. 1).

The geological and climatic history of the west also caused high extinction rates, resulting in a depauperate fish fauna relative to other areas of North America [26]. Currently restricted basins have fewer than 25 native fishes and high endemism, typically between 25 and 74% [27]. The Rocky Mountains and associated mountain ranges of the continental divide [2] have largely isolated western rivers from species-rich drainages in the tectonically stable eastern USA. Some western fish clades have Paleogene origins and represent endemic lineages whose evolution is consistent with the active geologic history of this region. Examples include western clades in the Salmonidae, Catostomini [6], Cyprinidae [28,29], Cyprinodontidae [30], Goodeidae [31,32] and *Cottus* [33]. Additionally, evidence has been provided for frequent episodes of introgression of some fishes [34,35], resulting in complete mtDNA replacement [36,37], and the evolution of species of hybrid origin [37,38].

Mountain suckers (*Pantosteus*, sometimes considered a subgenus of *Catostomus*) is one of several clades centered in the tectonically active part of western North America. Species of this lineage occur from southwestern Canada, western USA, and into northern Mexico, inhabiting middle-elevation reaches of rivers of the Basin and Range, Coast Ranges, Cascade Mountains, Rocky Mountains, east to the Black Hills and Sierra Madres [20,34] (Figs. 1, 2). These cool-water, benthic fishes were investigated by early western American ichthyologists, from Cope in 1872 [39] to Snyder in 1924 [40]. Smith [34] presented a classification recognizing six polytypic species of *Pantosteus* (distributions in Fig. 2). The four southern forms included *P. plebeius* in the Rio Grande Basin and its major tributary the Rio Conchos, the Rio Nazas and several Pacific basins in Mexico, *P. santaanae* in the Los Angeles Basin, *P. clarkii* in the Lower Colorado Basin below Grand Canyon, including the Pluvial White, Virgin, Bill Williams and Gila basins, and *P. discobolus* in the Upper Colorado Basin, Upper Snake River and northern Bonneville Basin. The northern group consists of *P. platyrhynchus* in the Great Basin, Columbia-Snake, Fraser, upper

Saskatchewan, Missouri and Green drainages; and *P. columbianus* in the Columbia-Snake drainage. All species except *P. santaanae* exhibit geographic variation, with subgroups that include morphologically differentiated populations in adjacent drainages. Some species contain variants recognized as subspecies: *P. discobolus jarrovii* in the Zuni River, New Mexico, and other drainages in northeast Arizona [41], *P. columbianus hubbsi*, in the Wood River drainage, Idaho [34], and *P. columbianus palouseanus* in the Palouse River, Washington [34]. An additional species, *P. nebuliferus*, from the Nazas Basin, was recognized by [42].

Smith and Koehn [43] conducted a phylogenetic analysis of 16 species of *Catostomus* and *Pantosteus* using both morphological and biochemical data, yielding two general groups in *Catostomus*: (1) a lineage with large, low-elevation, slow-water species in a paraphyletic subgenus *Catostomus*, and (2) a lineage of smaller mountain suckers adapted to cooler and swifter rivers at higher elevations. *Pantosteus columbianus* has traits diagnostic for both *Catostomus* and *Pantosteus*. Smith and Koehn [43] and [20] hypothesized that this taxon is a product of introgressive hybridization. One study has used DNA sequences to examine broader relationships among catostomine species. Doosey et al. [44] examined six species of *Pantosteus* and recovered five of them in a monophyletic group nested within *Catostomus*. The only species of *Pantosteus* excluded from this lineage was *P. columbianus*.

The above studies illustrated how complex distribution of traits can impact the ability to delimit taxa and interpret their evolutionary origins. This is especially true for *Pantosteus* and *Catostomus* as indicated by the proposed paraphyly of some taxa.

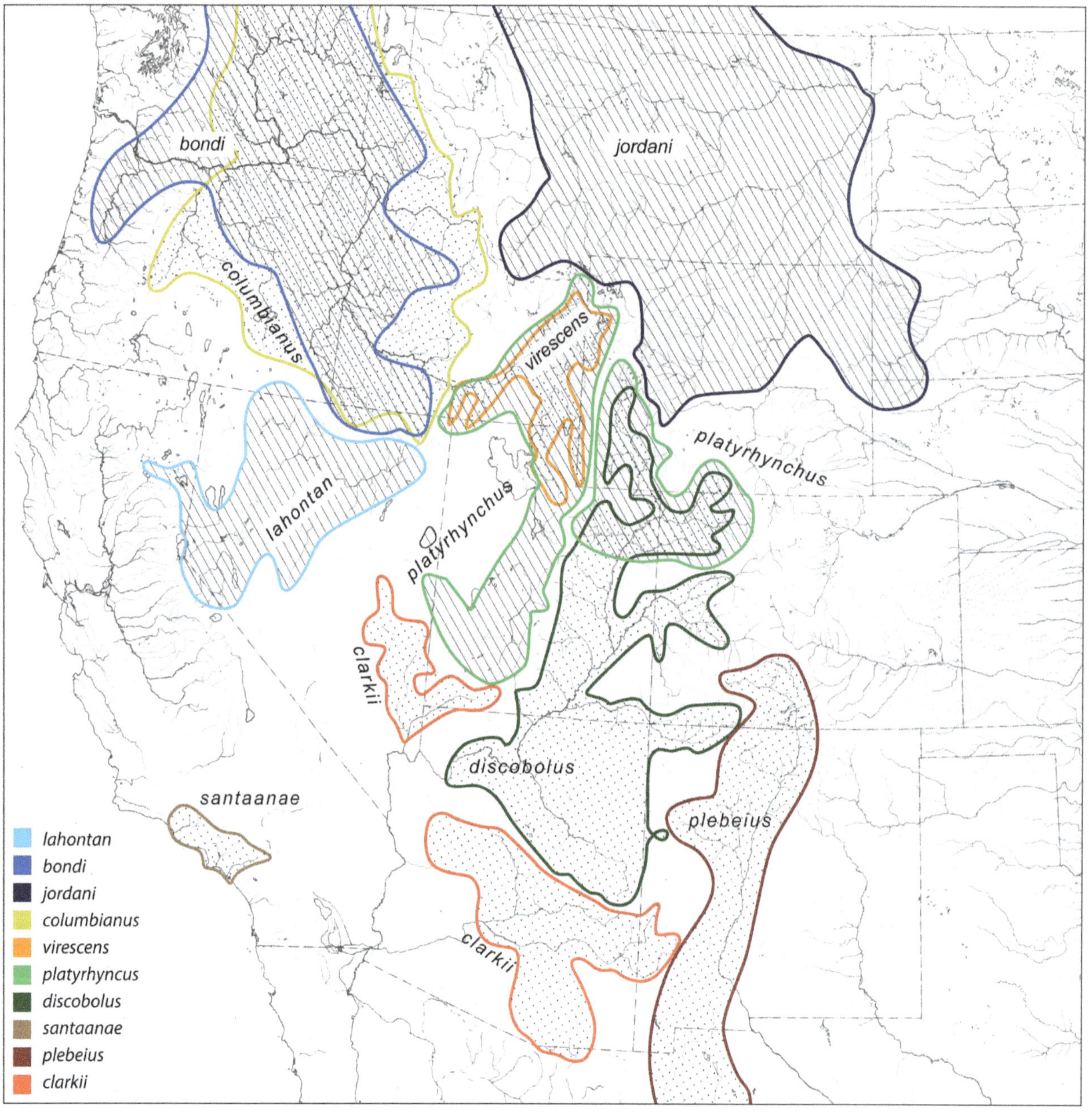

Figure 2. Distribution of the 11 species of *Pantosteus*. Border species are found north and south of the margins of this map.

The goals of the present work are to (1) compare molecular data with morphological and paleontological data for the proposed species of *Pantosteus*, (2) test hypotheses of their monophyly, (3) use these data for phylogenetic inferences of sister-group relationships, and (4) estimate timing of divergence events of identified lineages. With these insights into the reconstructed evolutionary history of the group, we evaluate the variation and distribution of traits found in *Pantosteus* lineages to revise the taxonomy of the genus and consider the impact of geological processes on diversity in this group.

Materials and Methods

Ethics Statement

Permission to undertake field work and collect specimens was obtained under the following permits: Colorado fisheries research permit 01-AQ908, Idaho fisheries research permits F-84-90, F 84-90-10, Nevada fisheries research permit S21662, Utah fisheries research permit 4coll78, Arizona fisheries research permit SP592500, Mexico scientific collection permit issued to Héctor Espinosa, Permiso de Pesca de Fomento 0765. Specimens were obtained under Brigham Young University Institutional Animal Care and Use Committee (IACUC) approval 070403 and 10-1202, Arizona State University IACUC approval 09-1018R, and Saint Louis University IACUC approval 2134.

Study Taxa and Sampling

We conducted range-wide molecular surveys of each species and selected individuals that represented both geographic and genetic diversity within the genus (Table 1). Outgroup taxa to represent the broader diversity within Catostomidae [44,45] include: *C. ardens*, *C. catostomus*, *C. commersonii* (GenBank accession AB127394.1), *C. insignis*, *C. macrocheilus*, *C. tahoensis*, *Cycleptus elongatus* (AB126082.1), *Hypentelium nigricans* (AB242169.1), *Minytrema melanops* (DQ536432.1), *Moxostoma erythrurum* and *M. poecilurum* (AB242167.1) (Table 1).

DNA Isolation, Amplification, and Sequencing

Genomic DNA was extracted from muscle tissue using the DNeasy Tissue Kit (QIAGEN Inc., Chatsworth CA) or by phenol-chloroform extraction as described in [46]. Nine of the 13 mtDNA protein coding genes (ND1, ND2, ND4L, ND4, ND5, ND6, ATPase6/8, cyt*b* and partial sequence from COIII) were amplified, representing approximately half of the mitochondrial genome (8055 bp). We obtained a large proportion of the mitochondrial genome in order to be able to obtain robust phylogenetic relationships as using a small number of genes provided poor resolution [47]. We did not pursue nuclear gene sequencing due to the low resolution of nuclear genes in the same species of catostomids [48], which indicates that a large number of nuclear loci would be required to provide informative phylogenetic results. Single and nested PCR amplification strategies were used to obtain product for different gene combinations. Details of the primers and nesting combinations are in Figure S1. For nested PCR the first reaction size was 10 µL. This first PCR reaction was then diluted to 1:49, and 1 µL of this product was added to a second 25 µL reaction. All other single reactions were 25 µL. Final concentrations for PCR components were as follows: 25 ng template DNA, 0.25 µM of each primer, 0.625 units of Taq DNA polymerase, 0.1 mM of each dNTP, 2.5 µL of 10X reaction buffer and 2.5 mM $MgCl_2$. Amplification parameters were as follows: 94°C for 2 min followed by 35 cycles of 94°C for 30 s, 48°C for 30 s, and 72°C for 60 s (in the first nested reactions this was increased by 1 min per each thousand bp), and 72°C for 7 min. PCR products were examined on a 1% agarose gel using SYBR safe DNA gel stain (Invitrogen, Eugene, OR, USA) and purified using a Montage PCR 96 plate (Millipore, Billerica, MA, USA). Sequences were obtained via cycle sequencing with Big Dye 3.0 dye terminator ready reaction kits using 1/16th reaction size (Applied Biosystems, Foster City, CA, USA). Sequencing reactions were run with an annealing temperature of 52°C (following the ABI manufacturer's protocol), cleaned using Sephadex columns in MultiScreen 96 well assay plates (Millipore, Billerica, MA, USA), and then dried. Most sequences were obtained using an Applied Biosystems 3730 XL automated sequencer at the Brigham Young University DNA Sequencing Center. All sequences obtained in this study were deposited in GenBank, accession numbers KJ441082-KJ441387 and the sequence alignment was deposited in Dryad, doi:10.5061/dryad.51mm0.

Analysis of Sequence Data

Sequences were edited using Chromas Lite 2.0 (Technelysium, Tewantin, Queensland, Australia) and imported into BioEdit 7.0.5.2 [49]. Sequences coding for amino acids were aligned by eye and checked via amino acid coding in MEGA 4.0.2 [50] to test for unexpected frame shift errors or stop codons. Editing resulted in 8055 base pairs representing the complete sequence for the nine genes, plus 22 bp of COIII. Phylogenetic analyses were performed under maximum likelihood (ML) using RAxML 7.2.8 [51,52] by bootstrapping with 1000 replicates with the final best ML tree being calculated using the GTRGAMMA model on the CIPRES cluster at the San Diego Supercomputer Center. Maximum parsimony (MP) analysis was conducted with PAUP* 4.0b10 [53] using a heuristic search with 1000 random additions and TBR branch swapping. Tree lengths reported for MP include both informative and uninformative characters. Robustness of nodes for MP was estimated by bootstrapping with 1000 replicates using a heuristic search with 10 random additions and TBR branch swapping. The ML tree from the DNA analysis presented in this study was deposited in TreeBASE, accession number TB2:S15337, (http://purl.org/phylo/treebase/phylows/study/TB2:S15337). Average between-species genetic distances were calculated based on the proportion of shared differences (p-distance) using MEGA for each lineage within *Pantosteus*, *Catostomus* as a whole, and then the remaining outgroups as a whole.

Morphology and Age Estimates

Eighty-seven morphological characters and eight fossil occurrences were taken from [20]. Calibration of the rate of evolution requires fossils to be assigned to specific branches based on synapomorphies shared by the fossils and recent taxa. This is achieved by cladistic interpretation of apomorphic characters for each fossil. These taxa are then represented as terminal taxa on specific branches. Confidence limits of estimates of times of cladistic branching have two primary sources of error: (1) fossils are unlikely to represent a lineage soon after its divergence from its sister lineage [54], and (2) both sister lineages do not necessarily possess new apomorphies of the lineages immediately after their initial genetic isolation. Fossils of both lineages should ideally contribute to estimation of age of a node, but both are rarely available in the record. For these reasons it is not obvious which end of the stem branch of a crown group should be assigned the age represented by the fossil. We choose to correct the age of the oldest fossil bearing synapomorphies of a lineage in question with the method of [54], which uses the density of the record of other fossils from that lineage to estimate the age of origin of the stem branch. The number of million-year time horizons or independent localities in which fossils of the lineage occur are applied in an

Table 1. Locality data for all *Pantosteus* samples and outgroups.

species	locality	sub-basin	major basin
Pantosteus clarkii	White R	Pluvial White	L Colorado
P. clarkii	Virgin R	Virgin	L Colorado
P. clarkii	Francis Ck	Bill Williams	L Colorado
P. clarkii	Aravaipa Ck	Gila	L Colorado
P. columbianus	Salmon Falls Ck	L Snake	Columbia
P. discobolus	Rifle Ck	U Colorado (CO)	U Colorado
P. discobolus	Havasau Ck at mouth	L Colorado (AZ)	L Colorado
P. discobolus	Tsaile Ck	San Juan	U Colorado
P. discobolus	Wheatsfield Ck	San Juan	U Colorado
P. discobolus	East Clear Ck	Little Colorado	L Colorado
P. discobolus	Upper Little Colorado R	Little Colorado	L Colorado
P. discobolus jarrovii	Rio Nutria	Little Colorado	L Colorado
P. jordani	Whitewood Ck	Missouri (SD)	Mississippi
P. jordani	Lil Popo Agie R	Missouri (WY)	Mississippi
P. lahontan	McDermitt Ck	Quinn	Lahontan
P. lahontan	Lower Truckee R	Truckee	Lahontan
P. lahontan	E Walker R	Walker	Lahontan
P. nebuliferus	Arroyo Penon	Nazas	Nazas
P. platyrhynchus	Blackrock Ck	U Snake	U Snake
P. platyrhynchus	Bear R	Bear	Bonneville
P. platyrhynchus	Weber R	Weber	Bonneville
P. platyrhynchus	Soldier Ck	Spanish Fork	Bonneville
P. platyrhynchus	San Pitch R	Sevier	Bonneville
P. platyrhynchus	Salina Ck	Sevier	Bonneville
P. platyrhynchus	Mammoth Ck	Sevier	Bonneville
P. platyrhynchus	Fremont L	U Green	U Colorado
P. plebeius	S Fork Palomas Ck	Rio Grande	Rio Grande
P. plebeius	R Santa Clara	Santa Clara	Guzmán
P. plebeius	R Escalariado	Escalariado	Guzmán
P. plebeius	Arroyo Ureyna	Conchos	Conchos
P. plebeius	Arroyo Riito	Fuerte	Fuerte
P. plebeius	R Miravalles	Miravalles	Miravalles
P. santaanae	Big Tujunga Ck	Tujunga	Los Angeles Basin
P. santaanae	San Gabriel R	San Gabriel	Los Angeles Basin
P. santaanae	Santa Ana R	Santa Ana	Los Angeles Basin
P. bondi	Similkameen R	Columbia	Columbia
P. bondi	Willamette R	Willamette	Columbia
P. bondi	Salmon Falls Ck	L Snake	Columbia
P. virescens	Twin Ck	U Snake	U Snake
P. virescens	Tin Cup Ck	U Snake	U Snake
P. virescens	Goose Ck	U Snake	U Snake
P. virescens	Weber R	Weber	Bonneville
P. virescens	Echo Ck	Weber	Bonneville
Catostomus ardens	Strawberry Res (introduced)	Duchesne	U Colorado
Catostomus catostomus	Boundary waters	Lake Superior	Great Lakes
Catostomus commersonii	GenBank AB127394.1		
Catostomus insignis	Eagle Ck	Gila	L Colorado
Catostomus macrocheilus	Grimes Ck	L Snake	Columbia
Catostomus macrocheilus	Palouse R	L Snake	Columbia

Table 1. Cont.

species	locality	sub-basin	major basin
Catostomus tahoensis	Walker R	Walker	Lahontan
Cycleptus elongatus	GenBank AB126082.1		
Hypentelium nigricans	GenBank AB242169.1		
Minytrema melanops	GenBank DQ536432.1		
Moxostoma erythrurum	Middle Fork Vermilion R	Mississippi (IL)	Mississippi
Moxostoma poecilurum	GenBank AB242167.1		

equation that uses these data to estimate the probability that the oldest fossil represents the first (unobserved) appearance of the lineage [54]:

$$a=(1-C_1)^{-1/(H-1)}1$$

where a is the confidence interval as a fraction of the total known stratigraphic range, C_1 is the 50% or 95% confidence level, and H is the number of known fossil horizons. The 50% interval is chosen as the mid-point of the distribution for calculation of the node calibration points. These were entered into the BEAST analyses. The ages of the oldest fossils, their identities, apomorphies, calibration points, locations, number of horizons, stratigraphy, and catalog numbers are given below, following best practice recommendations by [55].

An important assumption is that the fish of interest could have lived in a depositional environment below the oldest fossil, so that the absence of fossils possessing synapomorphies of a specific lineage can be interpreted as absence of the lineage at that time. The probability that the earliest fossil occurrence represents the first occurrence following cladogenesis is then used to correct the age estimate. The corrected age is applied to the node joining the taxa bearing the morphological synapomorphies supporting their sister-group relationship (Fig. 3).

BEAST 1.7.1 [56] was used to estimate molecular divergence times of mtDNA lineages based on corrected fossil age estimates. We generated input files using BEAUti 1.7.1. The dataset was trimmed to single representatives per species/lineage because having a mix of within- and between-species data complicates dating owing to different processes for estimating within- versus between-species rates [57] (S. Ho, pers. comm.). The analysis used an uncorrelated lognormal relaxed molecular clock with rate variation following a tree prior using the speciation birth-death process, a GTR+I+G model (identified using the AIC in Modeltest 3.7, [58]). The topology was constrained to match the ML results.

Calibrations were based on eight time horizons [20], with age of occurrence estimated for two nodes with two date estimates in each based on [54]. These two dates, 5.5 Ma and 12.7 Ma were given a log-normal prior with a standard deviation of 1. BEAST analyses were run for 50 million generations, with parameters logged every 10000 generations. Multiple runs were conducted to check for stationarity and to ensure that independent runs were converging on a similar result. The log and tree files from four runs were combined in LogCombiner 1.7.1 with a 10% burn-in. The combined logfile was examined in Tracer 1.5, while the combined treefile was summarized using TreeAnnotator 1.7.1 with the mean values placed on the maximum clade credibility tree.

The diagnoses of recent and fossil forms with specimen disposition for taxa in the morphological phylogenetic analysis are detailed in [20]. Identities, localities, and curatorial data for 347 western catostomin skeletons can be accessed at the University of Michigan Museum of Zoology Fish Division website. The morphological tree was estimated with PAUP, using 87 partly new osteological and morphological traits defined and listed in matrix format in [20]. Inferred instances of introgression are plotted on the tree, based on observed scatter of non-congruent morphological and molecular character states. Alternative hypotheses (such as convergence and retained polymorphisms) were examined by contrasting molecular and morphological trees. High bootstrap support values were used to identify strongly supported nodes in each analysis. Conflict between nodes on these two trees allowed rejection of alternative hypotheses.

Results and Discussion

Phylogenetic Analyses using mtDNA Sequences

Sequence analysis of 58 specimens (Table 1) yielded 4777 invariant characters, 507 variable but parsimony uninformative characters, and 2771 parsimony informative characters. Maximum-likelihood analysis recovered one tree with a likelihood score of −60385.098771 (Fig. 3). Phylogenetic analysis of mtDNA sequences provided resolution among recognized species [20,34] as well as identification of divergent lineages within four of the species. The *Pantosteus* mtDNA lineage was monophyletic, except that *P. columbianus* mtDNA was closely aligned with *C. tahoensis*, although it is morphologically in the *Pantosteus* clade. Both maximum likelihood and maximum parsimony methods recovered the same major lineages within *Pantosteus*, but with some different branching sequences among regional populations. In general, bootstrap analyses provide strong support (>95%) for most deep nodes (Fig. 3). In the following summary we only provide bootstrap values when support was less than strong (<95%). Mean p-distance between these 13 major lineages within *Pantosteus* varied between 2.2% and 10.3% (Table S1), while divergence between sister lineages varied between 2.2% and 8.3%. Outgroup lineages differed from *Pantosteus* by 11.4–19.2% pairwise sequence divergence (Table S1) except for the 2.7% p-distance between *C. tahoensis* and *P. columbianus*.

The earliest separation within the *Pantosteus* mtDNA clade is between the *P. discobolus-plebeius* clade and a clade (formerly part of *P. platyrhynchus*) comprising three groups of populations in the Columbia drainage, Missouri drainage and Lahontan Basin whose relationship is (Missouri (Lahontan, Columbia)). Sister to that lineage is the clade including all other *Pantosteus* species. The morphological *Pantosteus platyrhynchus* group has two forms of mtDNA on the tree–Bonneville Basin and Upper Snake River samples contain *P. discobolus* DNA, raising the possibility that the morphological traits are homoplasious (see below).

The next oldest mtDNA lineage in *Pantosteus* includes samples of southern forms (primarily from Mexico) that extend from the Nazas Basin north into the Rio Grande Basin and its past connectives. In the ML tree, the lineage including *P. nebuliferus* and *P. plebeius* from the Miravalles drainage was sister to a well-supported group of *P. plebeius* (93% bootstrap support) from southern (rios Conchos and Fuerte) and northern (Rio Grande and Guzman basins) lineages. In addition, a *P. plebeius* haplotype from the Rio Grande was found among samples taken from *P. discobolus jarrovii* populations from the upper Little Colorado River (Nutria Creek). This pattern is consistent with their history of introgression [41].

The remaining *Pantosteus* mtDNA lineages form a clade herein identified as the *P. discobolus* group (*sensu stricto*). This group consists of: *P. clarkii* from the Lower Colorado Basin, *P. santaanae* in the Los Angeles area, *P. discobolus* from the Upper Colorado Basin, *P. virescens* from the Weber, Bear, and Upper Snake rivers, and *P. platyrhynchus* from the Bonneville Basin, Upper Snake River, and upper Green River (Fig. 2). The *P. clarkii-P. santaanae* clade is sister to a clade including *P. discobolus*, *P. virescens*, and *P. platyrhynchus*. In this latter group, samples of *P. virescens* from the Weber, Bear, and Upper Snake rivers were sister to *P. discobolus* (Upper Colorado Basin) and *P. platyrhynchus* (Bonneville Basin). The high estimates of sequence divergence within the *P. discobolus* group suggest re-evaluation of the morphology of the Bonneville-Upper Snake populations, and leads to the recognition of these as *P. virescens* in this paper.

Morphological traits (gill-raker counts, vertebral counts, pre-dorsal scale counts, lip papillae, sparse caudal interradial pigment, and deep caudal peduncles) of many individuals in the upper Green River suggest that these populations are similar to *P. platyrhynchus* from the Bonneville and Upper Snake drainages; however, individuals also possess scattered morphological traits of *P. discobolus* consistent with introgression between these species (see Fig. 22 in [20]). The presence of *P. discobolus* mtDNA in *P. platyrhynchus* in the Bonneville Basin renders the mtDNA lineages of these forms paraphyletic (Fig. 3). Samples of *P. platyrhynchus* mtDNA from the Bonneville Basin were subdivided into well-differentiated Sevier River and northern Bonneville-Upper Snake River lineages.

Maximum parsimony analysis of the mtDNA sequence data, with all characters weighted equally, recovered a single most parsimonious tree with a length of 11,151 (CI = 0.398, RI = 0.727). Two major topological differences exist between ML and MP trees; these are highlighted on a reduced MP tree (inset in Fig. 3). Among *P. plebeius* and relatives, the MP tree differs from the ML tree with *P. nebuliferus* and *P. plebeius* from the Rio Miravalles sister to the northern lineage of *P. plebeius* (found in only 63% of MP bootstrap replicates), with a southern group of *P. plebeius* sister to *P. nebuliferus* and Rio Grande-Guzman Basin lineages. The second difference in the MP tree involves species from the Bonneville, Colorado and Los Angeles basins, where *P. clarkii* rather than *P. santaanae* is the sister to the clade that includes all *P. discobolus* plus *P. platyrhynchus* from the Bonneville Basin. This placement of *P. clarkii* is found in 100% of the MP bootstrap replicates.

Contrast of Molecular and Morphological Traits in *Pantosteus*

Pantosteus Cope is a monophyletic group based on osteological and morphological characters (if evidence for introgression in *P. columbianus* and other extreme hybrids are excluded). *Pantosteus* is morphologically diagnosed by: dentaries with their anterior, distal process turned mesially; the maxilla with a low dorsal flange and a prominent anterior flange; the hyomandibula with enlarged anterior and posterior flanges and an extensive fossa and associated processes (absent in *Catostomus*); the dorsolateral ridge of the pterotic is nearly vertical; the preopercle is deep at the center, approaching half-moon shape; the lips are large and fleshy with large notches at the junction of the upper and lower lips, a shallow notch between the lower lips, a prominent cartilaginous ridge on the lower jaw, more numerous gill rakers; and smaller body size than *Catostomus* (*sensu stricto*). *Pantosteus columbianus* is intermediate between *Pantosteus* and *Catostomus* in many of the above characters; *P. plebeius* and *P. nebuliferus* are intermediate in the lip characters [20,34]. Phylogenetic analysis of catostomins with 87 morphological and osteological traits (deposited in Dryad, doi:10.5061/dryad.51mm0) provided a well resolved tree (Fig. 4, modified from [20]), with bootstrap analysis providing strong support (>80%) for many deeper nodes, especially *Pantosteus*. The MP tree from the morphological analysis presented in [20] was deposited in TreeBASE, accession number TB2:S15337, (http://purl.org/phylo/treebase/phylows/study/TB2:S15337).

Phylogenetic analysis of mtDNA variation among taxonomic lineages as previously constituted are not consistent with monophyly of the genera, or monophyly of the *P. platyrhynchus* group, *P. discobolus*, or *P. columbianus*. The length of internodal branches (and concomitant bootstrap values) defining each of the mtDNA lineages that render these taxa as non-monophyletic are sufficiently large that sorting of mtDNA lineages is not a likely explanation for these incongruent results. Therefore, this conflict could be the result of convergence in morphological traits or introgressive transfer of mtDNA [35]. Both processes were hypothesized in previous comparative studies to explain similar patterns of morphological and biochemical characters [34,41,43]. Evaluation of these conflicts involves the use of cladistic analysis (Fig. 4) and fossils [20] to provide historical perspective, comparison with cases of present-day introgressive hybridization and statistical evaluation of phylogenetic trees. The 10 skull, jaw-bone, lip, and gill-raker traits are sufficiently numerous, anatomically diverse, and congruent, leading to high bootstrap values on the morphological tree (Fig. 4). Therefore, morphological homoplasy is considered unlikely. Because conflicting nodes are highly supported in both analyses, the most likely explanation for this discordance is introgression.

Regardless of the reasons for incongruence, some major revisions in the current taxonomy are required by the molecular data. Given the complicated nomenclatorial history of this group, we provide a summary of names we are resurrecting (Table 2) as well as morphological support for these designations. This is presented in the context of mtDNA and morphological phylogenies (Figs. 3, 4), which reveal examples of paraphyly involving *Catostomus* and *Pantosteus* as well as some of the lineages within *Pantosteus*.

Interpretation and Diagnosis of Species

Morphological and molecular results are mostly consistent with previous literature for several species (*P. clarkii*, *P. santaanae*, *P. plebeius* and *P. nebuliferus*). Diagnoses and discussion for each of these species are presented below; see also [20,34]. The remaining lineages, *P. columbianus* and those formerly referred to as *P. discobolus* and *P. platyrhynchus*, exhibit conflict between morphological and molecular traits that are considered in detail, along with hypotheses to explain the complicated polymorphisms and mixtures of traits among populations in the following sections.

Pantosteus clarkii (Baird and Girard) inhabits tributaries to the Lower Colorado Basin, including the Gila, Salt, Bill Williams,

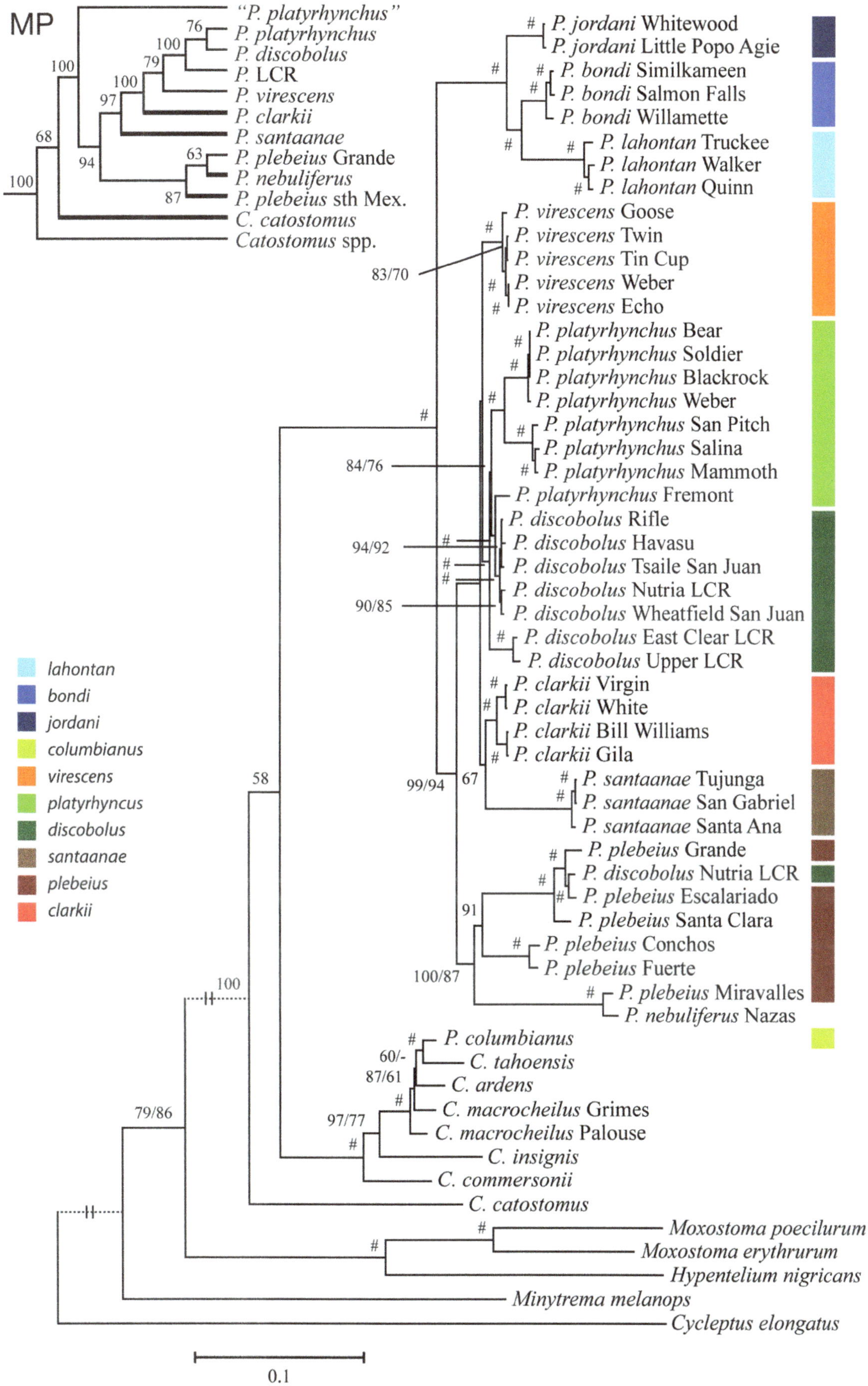
MP
"P. platyrhynchus"
P. platyrhynchus
P. discobolus
P. LCR
P. virescens
P. clarkii
P. santaanae
P. plebeius Grande
P. nebuliferus
P. plebeius sth Mex.
C. catostomus
Catostomus spp.
P. jordani Whitewood
P. jordani Little Popo Agie
P. bondi Similkameen
P. bondi Salmon Falls
P. bondi Willamette
P. lahontan Truckee
P. lahontan Walker
P. lahontan Quinn
P. virescens Goose
P. virescens Twin
P. virescens Tin Cup
P. virescens Weber
P. virescens Echo
P. platyrhynchus Bear
P. platyrhynchus Soldier
P. platyrhynchus Blackrock
P. platyrhynchus Weber
P. platyrhynchus San Pitch
P. platyrhynchus Salina
P. platyrhynchus Mammoth
P. platyrhynchus Fremont
P. discobolus Rifle
P. discobolus Havasu
P. discobolus Tsaile San Juan
P. discobolus Nutria LCR
P. discobolus Wheatfield San Juan
P. discobolus East Clear LCR
P. discobolus Upper LCR
P. clarkii Virgin
P. clarkii White
P. clarkii Bill Williams
P. clarkii Gila
P. santaanae Tujunga
P. santaanae San Gabriel
P. santaanae Santa Ana
P. plebeius Grande
P. discobolus Nutria LCR
P. plebeius Escalariado
P. plebeius Santa Clara
P. plebeius Conchos
P. plebeius Fuerte
P. plebeius Miravalles
P. nebuliferus Nazas
P. columbianus
C. tahoensis
C. ardens
C. macrocheilus Grimes
C. macrocheilus Palouse
C. insignis
C. commersonii
C. catostomus
Moxostoma poecilurum
Moxostoma erythrurum
Hypentelium nigricans
Minytrema melanops
Cycleptus elongatus
lahontan
bondi
jordani
columbianus
virescens
platyrhyncus
discobolus
santaanae
plebeius
clarkii
0.1

Figure 3. Maximum likelihood (ML) tree for *Pantosteus* based on analysis of mitochondrial DNA. A collapsed maximum parsimony (MP) topology is shown in the upper left corner, the thick branches indicate differences in topology to ML. All bootstrap values are based on 1000 pseudoreplicates, a # symbol represents bootstrap values over 95 for both ML/MP. Single bootstrap values are shown for nodes which are different to MP. The tree is rooted with *Cycleptus elongatus*. Locality details are provided in Table 1.

Pluvial White and Virgin rivers, which were connected to the Lower Colorado Basin below Grand Wash, but isolated from the Upper Colorado Basin and *P. discobolus* prior to the connecting flow through Grand Canyon at about 4.8 Ma [59,60]. *Pantosteus clarkii* shares most of its morphological traits and molecular sequences with *P. santaanae*, *P. discobolus*, and *P. virescens*. MtDNA data presented here support the hypothesis that isolation of these populations is relatively recent. Examples of introgression between *P. clarkii* and *P. discobolus* and between *P. clarkii* and *P. platyrhynchus* have been found in samples from the Virgin River and Shoal Creek (Sevier drainage), respectively [34] (and Secor and Dowling, unpub. data).

Pantosteus santaanae Snyder is a generalized, small-bodied mountain sucker found on the Los Angeles Plain, geographically isolated from *P. clarkii* in the Lower Colorado Basin. Osteologically, it shares features with *P. clarkii* consistent with the ML tree; however, these are plesiomorphic, consistent with its position in the MP tree.

Pantosteus plebeius (Baird and Girard) is found in the Rio Grande Basin and associated internal drainages, as well as headwaters of adjacent Pacific coastal drainages in Mexico. It and its sister species, *P. nebuliferus*, comprise the sister group to the *discobolus-virescens-clarkii-santaanae* clade. Analysis of allozymes also identifies another distinct lineage in the Rio Mesquital, Mexico, that was not sampled here; additional sampling and characterization is therefore required [61].

Pantosteus nebuliferus (Garman) is found in the Nazas and Aguanaval basins, although mtDNA of this form is also found in the population in the Rio Miravalles of the Rio Piaxtla drainage. This suggests that *P. nebuliferus* is either more widespread than first thought, or populations in the Rio Piaxtla are introgressed.

Pantosteus discobolus (Cope) was described from the Green River in the Upper Colorado Basin. The divergent mtDNA of *P. virescens* of the Bonneville Basin and Upper Snake River forms, previously included in *P. discobolus*, is consistent with the depth of the caudal peduncle and other unique morphological features. These differences suggest recognition of the name *P. virescens* Cope. Therefore, we follow [40] in separating *P. virescens* (Weber, Bear and Upper Snake rivers) from *P. discobolus* of the Upper Colorado Basin (see below). *Pantosteus discobolus*, *P. virescens*, *P. clarkii*, *P. santaanae* and *P. plebeius* form a clade diagnosed by mtDNA as well as morphological traits [20]. As constituted here, *P. discobolus* is the large mountain sucker (adults 200–350 mm standard length) of the larger rivers in the Upper Colorado Basin. *Pantosteus virescens* (Cope in Cope and Yarrow) was based on a type specimen said to be from the San Juan drainage in Colorado, but [40] concluded that the type locality was mislabeled and *P. virescens* was identical to specimens that he collected from the Weber River, Utah, and the Bear River, Wyoming, both tributary to the northern Bonneville Basin. The distribution of *P. virescens* is consistent with that of other species in the Bonneville Basin and Upper Snake River: multiple connections existed between the Bonneville Basin and the Upper Snake River in the late Pleistocene with several fish species being exchanged at that time [7]. Similarly, stream captures between the upper Green River and the Upper Snake River have been postulated based on geological [62,63] and fish evidence [2], yet the upper Green River populations have mtDNA of *P. discobolus*, despite possible influence from *P. virescens* and *P. platyrhynchus*. *Pantosteus virescens* was formerly abundant as the largest of the mountain suckers in the Weber, Bear, and Upper Snake rivers, but it is rare or absent in most former habitats, and modern specimens are smaller in size, especially in introgressed populations. Populations of *P. virescens* in the Upper Snake River drainage morphologically introgressed into *P. platyrhynchus* (p. 108 and Fig. 22 of [34]). Thus, the small extant *P. virescens* populations present in the Weber and Bear rivers should be provided high conservation significance as they represent the only remaining non-introgressed populations of this dwindling species.

The name *P. discobolus jarrovii* (Cope) has been assigned to a complex of ancient Arizona and New Mexico lineages that share mixed molecular, fin-ray, lip, and jaw characteristics of *P. discobolus* and *P. plebeius*. They are found in headwaters of the Little Colorado River in northeastern Arizona and northwestern New Mexico. The populations possess molecular, morphological and allozyme traits that originated in two separate but adjacent drainages, but which have been variously transferred in the Pleistocene via stream captures [34,41,64]. The populations are diverse and threatened with extinction. Their complex patterns of introgressed molecular, morphological and osteological characters are being described elsewhere to deal with controversial but critical conservation issues (Dowling et al., unpublished data).

Pantosteus platyrhynchus (Cope) was previously recognized as widely distributed throughout the Lahontan and Bonneville basins, Columbia and Snake rivers, and the headwaters of the Missouri drainage and Upper Colorado Basin; however, high mtDNA divergence among several lineages from these regions suggests that the taxonomy requires revision to more accurately reflect the apparent diversity. Former populations of *P. platyrhynchus* in the Missouri drainage are here referred to *P. jordani* Evermann; those from the Lahontan Basin are recognized as *P. lahontan* Rutter; and those from the Columbia-Lower Snake rivers were described as *P. bondi* [20].

Pantosteus platyrhynchus, as here diagnosed, occupies the Bonneville drainage and its former connective, the Upper Snake river, as well as small headwater streams in the Green River. The mtDNA of *P. platyrhynchus* in the Bonneville Basin is derived from *P. discobolus* (Fig. 3). This reflects the non-congruence of morphological traits, which are otherwise allied to northern populations in the Columbia River, Lahontan Basin, and Missouri drainage rather than *P. discobolus*. In addition, *P. platyrhynchus* contains two forms of mtDNA haplotypes diverged after acquisition of *P. discobolus* mtDNA with consistent slight morphological differences. One form occurs in the Sevier River (southern Bonneville Basin) and the other in the northern Bonneville Basin and Upper Snake River.

There is evidence for introgression between *P. platyrhynchus* and sympatric *Pantosteus* species. Populations in the Upper Snake and Green rivers are included in *P. platyrhynchus* based on the majority of their morphological traits; however, they have *P. virescens* and *P. discobolus* mtDNA, respectively, and mixtures of morphological traits of those species [34]. Broader sampling across this region is required to determine whether mtDNA introgression is partial or complete. The geographic pattern of variation in the number of predorsal scales suggests introgression between populations of *P. platyrhynchus* in Shoal Creek (Bonneville Basin) and *P. clarkii*

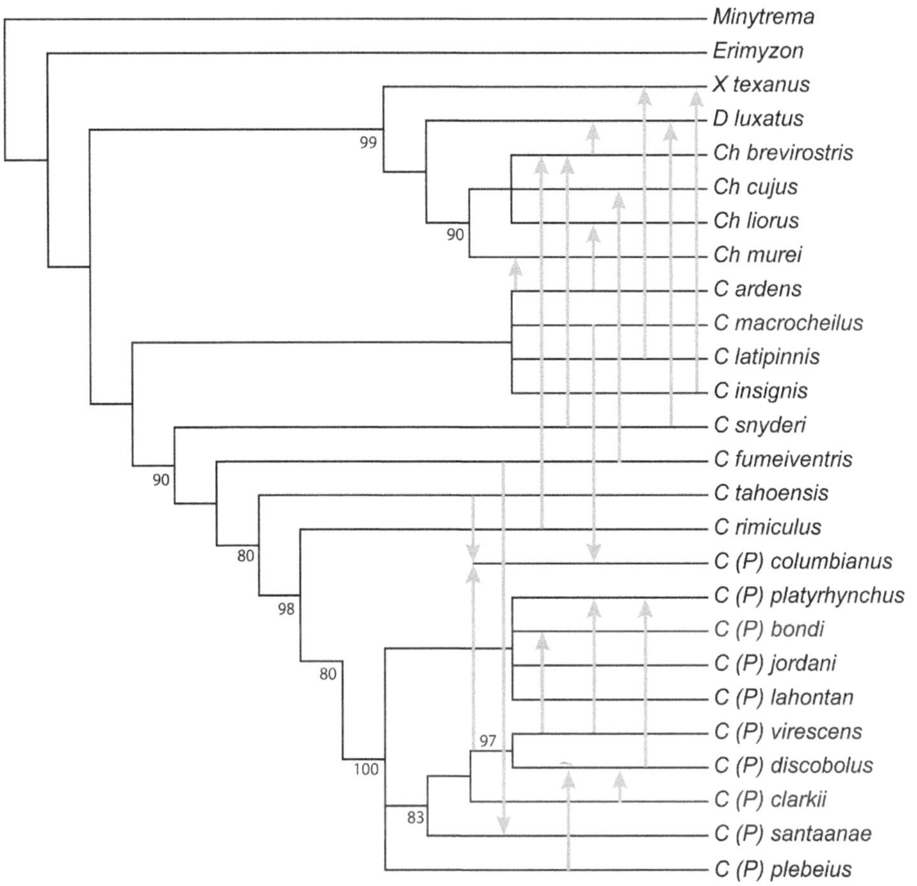

Figure 4. Morphology-based phylogenetic tree based on PAUP analysis of 87 characters of *Pantosteus* and other catostomin species from the genera *Xyrauchen, Deltistes, Chasmistes,* and *Castostomus.* Arrows indicate introgression of mtDNA and morphological traits (modified from [20]). Topology is based on the strict consensus of 63 equally shortest maximum parsimony trees (length = 252, CI = 0.53, RI = 0.83). Bootstrap values of >80 and above shown at nodes (from 1000 replicates).

introduced from the Virgin River (Secor and Dowling, unpublished data).

Pantosteus jordani Evermann is a small-bodied species from the upper Missouri drainage from the Black Hills of South Dakota to western Wyoming, Montana and Alberta, with Pleistocene fossils from western Kansas [20]. It is similar to *P. platyrhynchus*, with which it was formerly combined.

Table 2. Proposed changes to the taxonomy of *Pantosteus* species.

original taxonomy	changed taxonomy	distribution
Pantosteus clarkii	unchanged	Lower Colorado River Basin
P. columbianus	unchanged	Columbia River Basin
P. discobolus	*P. discobolus*	Colorado River Basin
P. discobolus	*P. virescens*	northern Bonneville and upper Snake River basins
P. discobolus jarrovii	unchanged	upper Little Colorado River
P. nebuliferus	unchanged	Nazas and Aguanaval basins
P. platyrhynchus	*P. bondi*	Columbia River Basin (minus upper Snake)
P. platyrhynchus	*P. jordani*	Missouri River Basin
P. platyrhynchus	*P. lahontan*	Lahontan Basin
P. platyrhynchus	*P. platyrhynchus*	Bonneville, upper Snake and Green River basins
P. plebeius	unchanged	Grande, Guzman, Conchos, and Yaqui basins
P. santaanae	unchanged	Los Angeles region

Pantosteus lahontan Rutter occupies the Lahontan Basin of Nevada, Oregon and California. It is similar to *P. platyrhynchus*, but has a slenderer hyomandibula, 30% as broad as long, with a cup-like post-dorsal process surrounding the post-dorsal fossa of the hyomandibula. It has 42–47 predorsal scales, usually 39–42 post-Weberian vertebrae, and 25–35 gill rakers on the outer row of the first arch [20].

Pantosteus bondi (Smith, Stewart, and Carpenter) occurs in the Columbia drainage, based upon samples from the Willamette, Similkameen, Boise and Salmon Falls drainages. This species has formerly been referred to as either *P. jordani* or *P. platyrhynchus* (see synonymy in [34]). It is diagnosed from *P. platyrhynchus* by its reciprocally monophyletic mtDNA and by its combination of a continuous but slightly emarginate ridge anterior to the opercular condyle of the hyomandibula, prominent post-dorsal fossae and process at the post-dorsal tip of the pterotic process of the hyomandibula, prominent post-mesial flange of the hyomandibula overlapping the antero-dorsal tip of the preopercle, medium width of the hyomandibula (about 34% as wide as long) and 29–37 gill rakers in the external row on the first gill arch [20].

Pantosteus columbianus Eigenmann and Eigenmann has a mixture of morphological traits of *P. virescens* and a species of *Catostomus*. Its mtDNA is most similar to that of *C. tahoensis* of the Lahontan Basin, suggesting hybrid origin involving that species. This interpretation requires immigration of *C. tahoensis* into the Snake River drainage, possibly at a time of Plio-Pleistocene spillovers from the Lahontan Basin to the Lower Snake River documented by [65], where it was apparently introgressed and absorbed into the populations of *P. virescens* that formerly occupied the Lower Snake River (according to fossil evidence, described below). Morphologically, *P. columbianus* is intermediate between those species in its lip shape and in many osteological traits, but shares scale counts and shapes of the jaws and suspensorium with *P. virescens* rather than *C. tahoensis* [20,34].

Dating mtDNA Lineages

Fossils provided bases for calibration points for two nodes in the tree, 12.7 Ma for the node at the base of *P. nebuliferus, P. plebeius, P. virescens, P. discobolus, P. clarkii*, and *P. santaanae*, and 5.5 Ma for the node ancestral to the clade of *P. virescens, P. discobolus, P. clarkii*, and *P. santaanae*. These are calibration points based on corrections of fossil ages [54]. The pertinent data on the fossils are as follows. The 12.7 calibration point is based on an oldest fossil age of 11.6 Ma–a synapomorphic postero-lateral ridge of a hyomandibula, which is shared with the clade including *P. discobolus, P. virescens, P. clarkii, P. santannae, and P. plebeius* (Figs. 3, 4). The key specimen (among three, 8.5–11.6 Ma) is the holotype hyomandibula of *P. hyomyzon* (UO F-55749) from UO locality 2337 in the Juntura Formation, Malheur County, Oregon, constrained by the overlying ash dated 11.5 Ma [20]. The number of horizons is seven.

The 5.5 Ma calibration point is based on a 4.5 Ma synapomorphic, broad and obtusely-angled shape of the fourth Weberian rib of *P. oromyzon*, this apomorphy is shared with the clade including *P. discobolus, P. virescens, P. clarkii, and P. santannae* (Figs. 3, 4). The key specimen is the holotype (one of two specimens of) *P. oromyzon* (UMMP 42356) from the basal Glenns Ferry Formation, Owyhee County, Idaho, constrained by the overlying Glenns Ferry oolite estimated to be 4.5 Ma [20]. The number of horizons is three. The fossil is approximately the same age as *P. asitus* of this clade, a dentary and other *Pantosteus* bones, from White Narrows, Clark County, Nevada [20]. We do not include fossil *P. columbianus* in the calibration of molecular rates because the hybrid origin of this taxon causes discordance between morphological and mtDNA topologies. Other populations have not been introgressed to this extent and estimated divergence times are offered as hypotheses with some support in the form of geological and hydrologic correlations.

We explored a variety of subsets of our sequence dataset to examine the effects of sample selection on divergence estimates, primarily to determine the effect of including outgroup samples from *Catostomus* on our estimates within *Pantosteus*. In all cases the results were similar (within 10%, results not shown). Sequences from sixteen individuals were selected to represent the phylogenetic diversity found within *Pantosteus* along with seven appropriate outgroup samples within *Catostomus*. The BEAST analysis was constrained to fit the ML topology (Fig. 3), with dates based on their mean and 95% highest posterior density (HPD) presented in Fig. 5; all parameter estimates had effective sample sizes greater than 2600. Each of the major nodes and associated geological events are discussed in detail below. These estimates of divergence time are based on mtDNA; therefore, they may be influenced by introgression and must be interpreted with this caveat in mind. In cases where extensive introgression of morphology and mtDNA have occurred (e.g., *P. discobolus jarrovii, C. columbianus*), this method provides an estimate of the date at which this mtDNA lineage was transferred into that form, allowing for inferences of the age of hybridization events.

The estimated mean time of divergence between *Pantosteus* and *Catostomus* (exclusive of *C. catostomus*) was 25.1 Ma (95% HPD of 35.3–16.6 Ma) [4]. This corresponds to a period of high mountains throughout the area that was in the early process of becoming the Basin and Range Province. The location of the earliest fossils in the proto-Columbia Basin, which corresponds with both high fossil and extant catostomid diversity in the West, suggests that the northern Great Basin is a plausible region in which initial divergence took place. The middle Miocene Nevada-Washington rift and the development of the Oregon-Idaho Graben in southern Oregon [66] represent major tectonic structures and events possibly related to this time and place of early catostomin evolution.

The three oldest fossils of *Pantosteus*, 11.6, 10.3 and 8.5 Ma, are from separate drainages in central Washington and southeast Oregon [20], at the northern edge of the Basin and Range Province [3]. The estimated age of first occurrence of the clade bearing synapomorphies of these fossils (Fig. 5, mean 14.5 Ma, 95% HPD of 17.3–12.8 Ma) coincides with the extensional opening up of the Basin and Range topography. Because nearly all of the taxa in *Pantosteus* are in or adjacent to either the northern or southern ends of the Great Basin (Fig. 2, 6), it is probable that the distribution and differentiation of these fishes owes its history to habitats and barriers created in this geological province. *Pantosteus* inhabit medium-gradient streams of the size that flow from the hundreds of north-south trending mountain ranges of the Basin and Range province (Fig. 1). These mountains also form barriers that isolate populations and initiate differentiation, but this differentiation is periodically interrupted by moist-dry climatic cycles that allow dispersal through renewed aquatic connections [26].

The relationship between geography and phylogenetic relationships based on mtDNA is presented in Fig. 6. The oldest branches of *Pantosteus*, based on molecular analyses, are (1) the broad *P. plebeius-P. discobolus* clade, distributed south through the Great Basin to Mexico, including the upper Rio Grande Basin in the Rio Grande rift, and onto the Colorado Plateau, and (2) the *P. platyrhynchus* group, distributed in the north from the Columbia drainage and northern Great Basin to the upper Missouri drainage. Our analyses suggest that the first great pathway of dispersal and divergent evolution, following the initial split, was south through the Great Basin.

In the Basin and Range province, the *P. plebeius* and *P. discobolus* clades began diverging around 10.5 Ma (Fig. 5, 95% HPD of 13.9–6.6 Ma), a time of tectonic, volcanic, and climatic changes in the northern Great Basin [3,4,66]. The *P. discobolus* clade differentiated into its five modern species in the northern Great Basin and Colorado Plateau about 5.4 Ma (Fig. 5, 95% HPD of 8.1–3.1 Ma). The Upper and Lower Colorado basins achieved their connection through Grand Canyon at 4.8 Ma [59,60], so it is likely that differentiation was occurring in pre-existing drainages in the Grand Wash area [19,67,68] between 10 and 5 Ma. At this time, the Colorado Plateau was elevating but probably had less *Pantosteus* habitat than the Great Basin [2,11,69]. Four of the five species in the *P. discobolus* group dispersed through and differentiated in the Great Basin west of the emerging Colorado Plateau (Figs. 1, 6).

The Bonneville and Lahontan basins, in the eastern and western halves of the Great Basin, were extending in the late Miocene, causing numerous north-south drainage connections. The Colorado River was in a stage preceding the modern Grand Canyon at this time, although older canyons were present [68,70,71]. *Pantosteus* (and *Catostomus*) must have expanded their range through the Basin and Range area where the Colorado Plateau and the northern and southern portions of the Great Basin meet (Figs. 1, 6), to establish the southern clades in Mexico (see below) during this period.

The problem of the relationships of *P. santaanae* to either *P. clarkii* or to the remainder of this clade is not resolved by the phylogenetic analysis of mtDNA as indicated by conflict between ML and MP analyses. Morphological data are also ambiguous. An earlier date of divergence is plausible if the ancestors date back to the dispersal southward to Mexico. These estimates might be impacted, however, by their higher rates of sequence evolution (as evidenced by their long branch, Fig. 3) induced by warm temperatures and small body size [72].

The map and trees (Figs. 2, 3, 4, 5, 6) suggest that the *P. platyrhynchus* group originated 14.5 Ma (Fig. 5, 95% HPD of 17.3–12.8 Ma) and diversified over the last 6.9–4.8 Ma (Fig. 5, 95% HPD of 10.5–2.0 Ma) into four species: *P. bondi* in the Columbia drainage, *P. jordani* in the upper Missouri drainage, *P. lahontan* in the Lahontan Basin, and *P. platyrhynchus* in the Bonneville Basin, Upper Snake River and the Green River. This diversification corresponds to a time of major volcanic and tectonic activity in the Columbia region [73]. These lineages probably began in three pre-Columbia River drainages of the northern Basin and Range Province, where the three oldest *Pantosteus* fossils were recovered. All except one differentiated in the Northern Great Basin; that one exception, *P. jordani*, evolved initially in the Yellowstone region, now in the headwaters of the Missouri River, which at that time drained to Hudson Bay [74]. Later, in the Pleistocene, there were exchanges among the Bear, Snake and Green rivers east of the Snake River Plain (Figs. 1, 6) in southwest Wyoming [2,6,62], which were possibly responsible for the transfer of individuals of *P. platyrhynchus* from west to east (see below).

The northern Bonneville/Upper Snake and southern Bonneville populations of *P. platyrhynchus* last shared a common ancestor 2.3 Ma (Fig. 5, 95% HPD of 3.7–1.0 Ma). This separation between the southern and northern Bonneville Basin is a common pattern shared by two other fishes: *Lepidomeda copei* [75] and *Rhinichthys osculus* [76]. The Sevier Basin was isolated from the Salt Lake Basin by a barrier higher than the present, from at least 3 Ma until the early Pleistocene [77]. The recent connection of the southern and northern halves of the Bonneville Basin via high stands of Lake Bonneville appears to have not facilitated movement between them, which may be due to these species being stream specialists that are not typically found in lakes.

The divergence of Mexican and Rio Grande Basin *Pantosteus*, between 8.4 and 1.1 Ma (Fig. 5, 95% HPD of 11.8–0.3 Ma), suggest dispersal from Idaho southward through Nevada and Utah prior to this time (see [70]). The drainage changes responsible for divergence of *P. plebeius* and *P. nebuliferus* in central Mexico are not known.

Importance of Introgressive Hybridization in *Pantosteus*

Comparison of mtDNA distributions among taxa with morphological data, biogeography and paleontology reveals noncongruence that suggests either extensive morphological convergence or introgressive transfer of mtDNA among lineages in the ancient past (Fig. 4). Several considerations suggest resolution in favor of the introgressive genetic transfer hypothesis. (1) Paleontological, morphological and biogeographical interpretations are often consistent with each other, but not with the mtDNA phylogeny. (2) The discrepancies in the mtDNA phylogeny involve taxa known to have experienced introgressive hybridization in recent times [34], thus allowing for the possibility of introgression occurring in the past. (3) Some examples of introgression among fishes of western North America have been observed in the fossil record [35] or have been verified by identification of geological evidence of stream capture consistent with the transfer leading to the introgression [41]. (4) Lastly, the broad scatter of morphological traits within individuals of an introgressed population show individually unique patterns of variation, suggesting different degrees of penetration of different alleles, epigenetic interactions and selective regimes. That is, there are unusually variable and unique patterns of assorted polymorphisms among individuals [34,35,78–80]. The principal examples of introgressive hybridization in *Pantosteus* are the Zuni Sucker (*P. discobolus jarrovii*), the Bonneville Basin, Snake River and Green River *P. platyrhynchus*, and *P. columbianus*, all of which demonstrate geographically patchy and non-concordant distributions of characters and alleles typical of introgression [37,81,82]. We now discuss each example of introgression below.

Pantosteus plebeius introgression with *P. discobolus* was documented with morphological, biogeographic and geologic data by [34], and the hypothesis tested and supported with biochemical data [41,64]. The taxon, *P. discobolus jarrovii* Cope was based on the downstream cline of mixed and intermediate characters of the introgressed population, inferred to have resulted from a stream capture and fish transfer in the Pleistocene, with an estimated mean age of 1.1 Ma (Fig. 5, 95% HPD of 2.1–0.3 Ma). Because of multiple haplotypes, morphologies and conservation problems, these unusually complex populations are under separate investigation with more detailed sampling (Dowling at al., unpublished).

Pantosteus discobolus mtDNA has replaced that of *P. platyrhynchus* throughout its range as exhibited by the discordance of morphological and mtDNA characters. Morphological synapomorphies detailed by [34] and discussed above support the placement of *P. platyrhynchus* in the clade of populations, all formerly classified as *P. platyrhynchus*, distributed in the Columbia-Snake drainage, Missouri drainage, Green River and northern Great Basin. These populations are well known to hybridize with sympatric catostomins elsewhere [34,79], so it is no surprise to find mtDNA of the *P. discobolus* group in *P. platyrhynchus* in the Upper Snake River and Green River. It was unexpected, however, to find all sampled individuals of Bonneville Basin and Upper Snake River *P. platyrhynchus* bearing *P. discobolus* mtDNA, dating from 3.4 Ma (Fig. 5, 95% HPD of 5.2–1.8 Ma). Introgression between

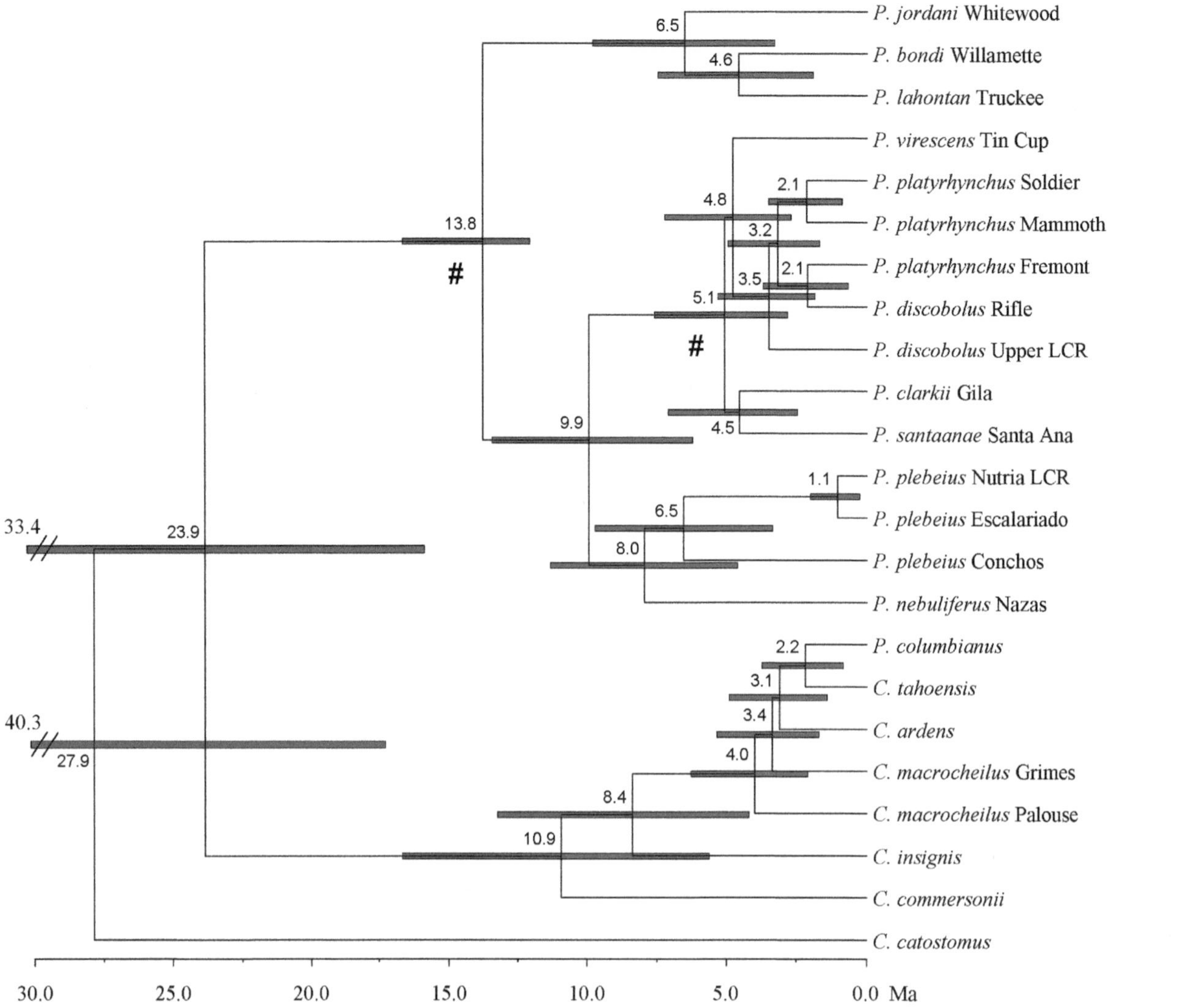

Figure 5. Bayesian molecular clock estimates for *Pantosteus* based on analysis of mitochondrial DNA. Horizontal bars represent the 95% highest posterior density ranges. Locality details are provided in Table 1, OTU order and labels match those in Fig. 3.

these species in the Green River, and between *P. platyrhynchus* and *P. virescens* in the Upper Snake River, were described by [34]. However, no morphological traits of *P. discobolus* were observed in *P. platyrhynchus* populations in the Bonneville Basin. The presence of *P. discobolus* mtDNA in Bonneville Basin and Upper Snake River *P. platyrhynchus* populations is especially puzzling because *P. discobolus* is allopatric to Bonneville Basin *P. platyrhynchus*; *P. virescens* is the sympatric congener (Fig. 3). *Pantosteus discobolus* and *P. virescens* are larger fish, preferring larger, downstream reaches; *P. platyrhynchus* is a headwater species, but where stream habitats overlap they are sympatric. Additional sampling is required across the range of *P. platyrhynchus* to determine if patterns of introgression are ubiquitous.

The hypothesis that these populations of *P. platyrhynchus* are actually genetic *P. discobolus* that have evolved gill-raker, scale, fin, pigment, and osteological traits that converge on *P. platyrhynchus*, is rejected based on the diversity, congruence, and independence of the morphological evidence. This assumption of independence of diverse morphological traits should be tested, however, with studies of nuclear genes responsible for morphology [83].

Geomorphic evidence of stream capture, possibly responsible for the transfer of fish between the Snake and Green rivers in southwest Wyoming, is noted above. Preliminary sampling indicates that the introgressed traits, which seemed more or less stabilized when observed in the Snake and Green rivers by ([34], see graph, p. 109), had become more prevalent in samples examined for the present study. *Pantosteus platyrhynchus* samples in the Upper Snake River exhibit *P. virescens* morphological traits and mtDNA (Fig. 3) to varying degrees. *Pantosteus discobolus* mtDNA (Fig. 3) and morphological traits were scattered in *P. platyrhynchus* specimens in the Green River.

In particular, the *P. platyrhynchus* in Fremont Lake, in the upper Green River of Wyoming, has mtDNA that differs by a p-distance of 1.1% from *P. discobolus* collected in the upper Grand River at Rifle, Colorado (Fig. 3). The *P. discobolus* mtDNA introgressed into *P. platyrhynchus* is estimating the differentiation of *P. discobolus* in the Green River from populations in the upper Grand River (2.3 Ma, 95% HPD of 3.9–0.7 Ma, Fig. 5), not *P. platyrhynchus*.

Pantosteus columbianus appears to have a complex history of introgression between *C. tahoensis* and *P. virescens* ([20] and this paper) and a later history of introgression with *C. macrocheilus* [79].

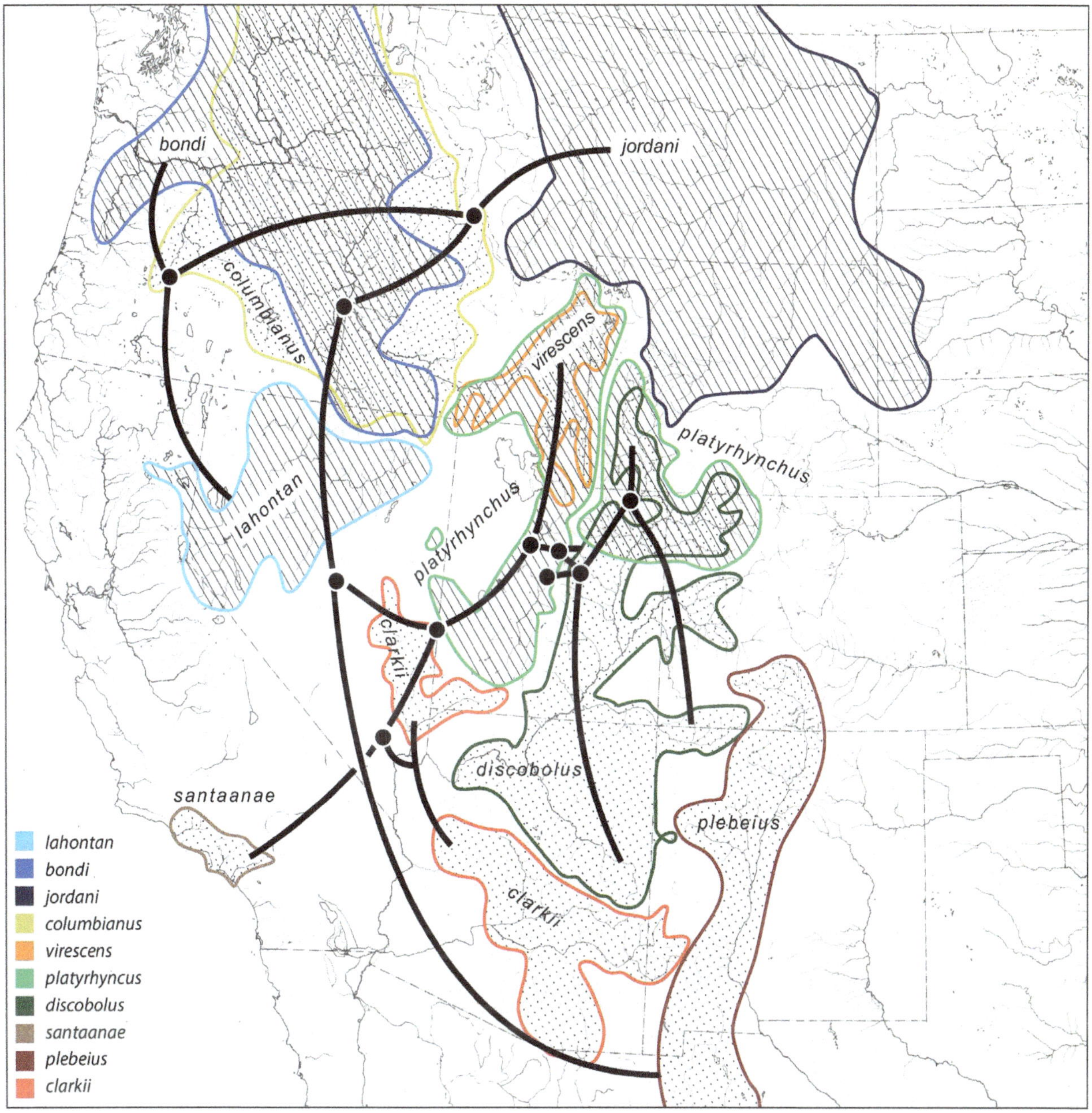

Figure 6. ***Pantosteus*** **species distribution patterns with overlay of the species relationships from Fig. 5.** Estimated ages of branching events are in Fig. 5.

Pantosteus columbianus was described originally in *Pantosteus* and later in *Catostomus* because of its mixture of diagnostic traits [20]. Its lips are intermediate, most of its osteological traits are diagnostic of *Pantosteus*, and its mtDNA is most similar to that of *C. tahoensis*. The genetic divergence between the mtDNA of *P. columbianus* and *C. tahoensis* implies a mean age of introgression of 2.3 Ma (Fig. 5, 95% HPD of 4.0–0.9 Ma). Fossils of *P. columbianus* in the western Snake River plain between 3 and 2 Ma help constrain the time of introgression that formed *P. columbianus*. The absence of *P. virescens* below the falls of the Snake River, notwithstanding fossil evidence of its occurrence there at 4.5 Ma, implies that the immigration of *C. tahoensis* during spillover from the Lahontan Basin to the Snake River drainage in the Pleistocene [65], resulted in complete genetic absorption of the two parental populations into a species of introgressed origin, *P. columbianus*. Further tests of this hypothesis with nuclear genes should be instructive. Sub-populations of *P. columbianus* are isolated in the headwaters of the Palouse River, Washington, the Wood River, Idaho, and the Deschutes and Malheur rivers, Oregon. Their morphological differences from the central populations suggest an early widespread form of the species that was more *Pantosteus*-like, especially in bones and lip structure. These indicate significant morphological change during their existence [34].

Conclusions

The Basin and Range Province is a unique setting in which to study the effects of insular isolation on diversification of plants and of aquatic and terrestrial animals [65,84,85]. The fragmentation of a late Mesozoic mountain belt along the southwest margin of North America by crustal thinning and extension created hundreds of narrow valleys separated by hundreds of elongate mountainous remnants (Fig. 1). The valleys were frequently isolated except when filled and externally drained [65] and the mountains were isolated except when cool climates forced transitional and boreal life, including mountain suckers, down and out across valley floors. During late Neogene glacial/interglacial stages, fluctuations in barriers and connections led to about 20 cycles of isolation, differentiation, and repeated secondary genetic contact among populations of organisms. Only the most recent of these glacial and post-glacial stages is well studied.

The distribution of mtDNA among populations of *Pantosteus* reflects this cyclic history. Many instances of secondary contact resulted in hybridization, and some populations experienced introgression, or even amalgamation (Dowling et al., 1997 [86]), resulting in conflicting patterns in morphological and DNA interpretations of relationships. Most species of western catostomids [79] and cyprinids [78] show evidence of hybridization and all 11 species of *Pantosteus* have hybridized with sympatric relatives [34,79]. Morphological and molecular evidence presented here suggests that introgression was ancient (e.g., pre-human influence) in *P. tahoensis*×*P. virescens* (= *P. columbianus*), *P. platyrhynchus*×*P. discobolus*, *P. platyrhynchus*×*P. virescens*, *P. discobolus*×*P. clarkii*, *P. discobolus*×*P. plebeius* (= *P. d. jarrovii*) and possibly *P. plebeius*×*P. nebuliferus*. Many other hybrid individuals, e.g., *P. platyrhynchus*×*P. clarkii* and many hybrids involving *Catostomus* [79], show distributions and characters that suggest recent hybridization in the presence of environmental disturbance and inter-basin transfer by humans. Environmental disturbance in the form of compromised spawning sites were presumably responsible for many ancient introgressions.

Ancient introgression causes the production of morphotype/haplotype combinations that yield misleading phylogenetic reconstructions [35] and misidentification of specimens by scientists and managers. Changes in biodiversity, negative or positive, are of considerable interest [86–88]. Loss of species is a possibility, as in the case of dwindling populations of *P. virescens* in the Upper Snake, Bear, and Weber rivers. Replacement of species, such as *C. tahoensis* and *P. virescens* by *P. columbianus* in the Columbia drainage is an interesting example. Post-glacial expansion of *P. columbianus* north to 55 degrees in Canada is evidence of an adaptable genome. Whether the added genetic variability in introgressed *P. platyrhynchus* in the Green River is adaptive, neutral, or harmful in its changing environment is not known, but the introgressed alleles have increased between 1963 [34] and the present study.

Demonstration of introgression has taxonomic as well as evolutionary implications. Smith et al. [20] noted that introgressive homoplasy renders *Pantosteus* and other catostomine genera polyphyletic and subsumed them in the genus *Catostomus* to maintain monophyly in the group. In the present paper, well-established names of genera are conserved in the interest of stability in the classification system and uniqueness of the clades. Comparative studies of population genetics, genomics and impact of diverse sources of genetic variation in populations of *P. discobolus jarrovii*, *P. platyrhynchus*, *P. lahontan*, *P. jordani*, and other catostomins will contribute to an understanding of the remarkable failure of selection to reduce heterospecific matings and introgression in this group.

Supporting Information

Figure S1 All primers used to generate catostomid sequences.

Table S1 Mean between species p-distances as percentages (lower left) for each *Pantosteus* lineage, *Catostomus* as a whole, and the remaining outgroups as a whole.

Acknowledgments

We thank the many people who assisted with all aspects of fieldwork, sample curatorial work, and for providing samples and other assistance including K. Bertrand, P. Cavalli, C. Dillman, H. Espinosa, S. Gregory, G. Gunnell, D. Markle, D. Nelson, L. Schultz, E. Taylor and A. Webber. J. Megahan helped with the figures and N. Carpenter provided dates for the fossils. Thanks to the various government agencies which provided permission to sample and obtain specimens.

Author Contributions

Conceived and designed the experiments: PJU TED NJL CLS RLM DKS GRS. Performed the experiments: PJU TED NJL CLS RLM GRS. Analyzed the data: PJU TED GRS. Contributed reagents/materials/analysis tools: PJU TED NJL CLS RLM DKS GRS. Wrote the paper: PJU TED NJL CLS RLM DKS GRS.

References

1. Cavender TM (1986) Review of the fossil history of North American freshwater fishes. In: Hocutt CH, Wiley EO, editors. The Zoogeography of North American Freshwater Fishes. New York, NY: John Wiley and Sons. pp. 699–724.
2. Spencer JE, Smith GR, Dowling TE (2008) Middle to late Cenozoic geology, hydrography, and fish evolution in the American Southwest. In: Reheis MC, Hershler R, Miller DM, editors. Late Cenozoic Drainage History of the Southwestern Great Basin and Lower Colorado River Region: Geologic and Biotic Perspectives: Geological Society of America Special Paper 439: 279–299.
3. Faulds JE, Varga RJ (1998) The role of accommodation zones and transfer zones in the regional segmentation of extended terranes. In: Faulds JE, Stewart JH, editors. Accomodation Zones and Transfer Zones: the Regional Segmentation of the Basin and Range Province, Geological Society of America, Special Paper 323: 1–46.
4. Chamberlain CP, Mix HT, Mulch A, Hren MT, Kent-Corson ML, et al. (2012) The Cenozoic climatic and topographic evolution of the western North American Cordillera. Am J Sci 312: 213–262.
5. Hershler R, Sada D (2002) Biogeography of Great Basin aquatic snails of the genus *Pyrgulopsis*. In: Hershler R, Madsen DB, Currey DR, editors. Great Basin Aquatic Systems History: Smithsonian Contributions to the Earth Sciences 33: 255–276.
6. Smith G, Dowling T, Gobalet K, Lugaski T, Shiozawa D, et al. (2002) Biogeography and timing of evolutionary events among Great Basin fishes. In: Hershler R, Madsen D, Curry D, editors. The Great Basin: Cenozoic Geology and Biogeography, Great Basin Aquatic Systems History: Smithsonian Contributions to Earth Sciences 33: 175–234.
7. Hubbs C, Miller R (1948) Correlation between fish distribution and hydrographic history in the desert basins of southwestern United States. Bulletin of the University of Utah Biological Series 38: 18–166.
8. Taylor D (1985) Evolution of freshwater drainages and molluscs in western North America. In: Smiley CJ, editor. Late Cenozoic History of the Pacific Northwest. San Francisco, CA: American Association for the Advancement of Science. pp. 265–321.
9. Kimmel PG (1975) Fishes of the Miocene-Pliocene Deer Butte Formation, southeast Oregon. University of Michigan Papers on Paleontology 14: 69–87.
10. Smith GR, Swirydczuk K, Kimmel PG, Wilkinson BH (1982) Fish biostratigraphy of late Miocene to Pleistocene sediments of the western Snake River Plain, Idaho. In: Bonnichsen B, Breckenridge RM, editors. Cenozoic Geology of Idaho, Idaho Bureau of Mines and Geology, Bulletin 26: 519–542.
11. Dallegge T, Ort M, McIntosh W, Perkins M (2001) Age and depositional basin morphology of the Bidahochi Formation and implications for the ancestral upper Colorado River. In: Young RA, Spamer EE, editors. Colorado River:

Origin and Evolution; Proceedings of a Symposium Held at Grand Canyon National Park, June, 2000 Grand Canyon Association Monograph. pp. 47–52.
12. Oviatt C (2002) Bonneville basin lacustrine history: the contributions of G.K. Gilbert and Ernst Antevs. In: Hershler R, Madsen DB, Currey DR, editors. Great Basin Aquatic Systems History: Smithsonian Contributions to the Earth Sciences 33: 121–128.
13. Sack D (2002) Fluvial linkages in Lake Bonneville subbasin integration. In: Hershler R, Madsen DB, Currey DR, editors. Great Basin Aquatic Systems History: Smithsonian Contributions to the Earth Sciences 33: 129–144.
14. Currey DR (1990) Quaternary palaeolakes in the evolution of semidesert basins, with special emphasis on Lake Bonneville and the Great Basin, USA. Palaeogeogr, Palaeoclimatol, Palaeoecol 76: 189–214.
15. Reheis M, Sarna-Wojcicki A, Reynolds R, Repenning C, Mifflin M (2002) Pliocene to middle Pleistocene lakes in the western Great Basin: ages and connections. In: Hershler R, Madsen DB, Curry DR, editors. Great Basin Aquatic Systems History: Smithsonian Contributions to Earth Sciences 33: 53–108.
16. Negrini RM (2002) Pluvial lake sizes in the northwestern Great Basin throughout the Quaternary Period. In: Hershler R, Madsen DB, Currey DR, editors. Great Basin Aquatic Systems History: Smithsonian Contributions to the Earth Sciences 33: 11–52.
17. Davis OK (1999) Pollen analysis of Tulare Lake, California: Great Basin-like vegetation in Central California during the full-glacial and early Holocene. Rev Palaeobot Palynol 107: 249–257.
18. Benson L (2004) Western lakes. In: Gillespie AR, Porter SC, Atwater BF, editors. The Quaternary Period in the United States, Developments in Quaternary Science: Elsevier Science. pp. 185–199.
19. Howard KA, Lundstrom SC, Malmon DV, Hook SJ (2008) Age, distribution, and formation of late Cenozoic paleovalleys of the lower Colorado River and their relation to river aggradation and degradation. In: Reheis MC, Hershler R, Miller DM, editors. Late Cenozoic Drainage History of the Southwestern Great Basin and Lower Colorado River Region: Geologic and Biotic Perspectives: Geological Society of America Special Paper 439: 391–410.
20. Smith GR, Stewart JD, Carpenter NE (2013) Fossil and recent mountain suckers *Pantosteus*, and significance of introgression in catostomin fishes of Western United States. Occas Pap Mus Zool Univ Mich 724: 1–59.
21. Wheeler HE, Cook EF (1954) Structural and stratigraphic significance of the Snake River capture, Idaho-Oregon. J Geol 62: 525–536.
22. Smith GR, Morgan N, Gustafson E (2000) Fishes of the Mio-Pliocene Ringold Formation, Washington: Pliocene capture of the Snake River by the Columbia River. University of Michigan Papers on Paleontology 32: 1–47.
23. Bouchard DP, Kaufman DS, Hochberg A, Quade J (1998) Quaternary history of the Thatcher Basin, Idaho, reconstructed from the $^{87}Sr/^{86}Sr$ and amino acid composition of lacustrine fossils: implications for the diversion of the Bear River into the Bonneville Basin. Palaeogeogr, Palaeoclimatol, Palaeoecol 141: 95–114.
24. Retallack GJ (2004) Late Miocene climate and life on land in Oregon within a context of Neogene global change. Palaeogeogr, Palaeoclimatol, Palaeoecol 214: 97–123.
25. Kohn MJ, Fremd TJ (2008) Miocene tectonics and climate forcing of biodiversity, western United States. Geology 36: 783–786.
26. Smith GR, Badgley C, Eiting TP, Larson PS (2010) Species diversity gradients in relation to geological history in North American freshwater fishes. Evol Ecol Res 12: 693–726.
27. Abell R, Thieme ML, Revenga C, Bryer M, Kottelat M, et al. (2008) Freshwater ecoregions of the world: a new map of biogeographic units for freshwater biodiversity conservation. Bioscience 58: 403–414.
28. Dowling TE, Tibbets CA, Minckley W, Smith GR, McEachran J (2002) Evolutionary relationships of the plagopterins (Teleostei: Cyprinidae) from cytochrome *b* sequences. Copeia 2002: 665–678.
29. Bufalino AP, Mayden RL (2010) Molecular phylogenetics of North American phoxinins (Actinopterygii: Cypriniformes: Leuciscidae) based on RAG1 and S7 nuclear DNA sequence data. Mol Phylogen Evol 55: 274–283.
30. Echelle AA, Carson EW, Echelle AF, Van Den Bussche R, Dowling TE, et al. (2005) Historical biogeography of the new-world pupfish genus *Cyprinodon* (Teleostei: Cyprinodontidae). Copeia 2005: 320–339.
31. Parenti LR (1981) A phylogenetic and biogeographic analysis of cyprinodontiform fishes (Teleostei, Atherinomorpha). Bull Am Mus Nat Hist 168: 335–557.
32. Webb SA, Graves JA, Macias-Garcia C, Magurran AE, Foighil DO, et al. (2004) Molecular phylogeny of the livebearing Goodeidae (Cyprinodontiformes). Mol Phylogen Evol 30: 527–544.
33. Kinziger AP, Wood RM, Neely DA (2005) Molecular systematics of the genus *Cottus* (Scorpaeniformes: Cottidae). Copeia 2005: 303–311.
34. Smith GR (1966) Distribution and evolution of the North American catostomid fishes of the subgenus *Pantosteus*, genus *Catostomus*. Misc Publ Mus Zool Univ Mich 129: 1–132.
35. Smith GR (1992) Introgression in fishes: significance for paleontology, cladistics, and evolutionary rates. Syst Biol 41: 41–57.
36. Dowling TE, DeMarais BD (1993) Evolutionary significance of introgressive hybridization in cyprinid fishes. Nature 362: 444–446.
37. Gerber AS, Tibbets CA, Dowling TE (2001) The role of introgressive hybridization in the evolution of the *Gila robusta* complex (Teleostei: Cyprinidae). Evolution 55: 2028–2039.
38. Demarais BD, Dowling TE, Douglas ME, Minckley W, Marsh PC (1992) Origin of *Gila seminuda* (Teleostei: Cyprinidae) through introgressive hybridization: implications for evolution and conservation. Proc Natl Acad Sci USA 89: 2747–2751.
39. Cope ED (1872) Recent reptiles and fishes. Report on the reptiles and fishes obtained by the naturalists on the expedition. U.S. Geological Survey of Wyoming & Contiguous Territories (1871). U.S. Geological Survey of Wyoming & Contiguous Territories Part 4 (art. 8): 432–442.
40. Snyder JO (1924) Notes on certain catostomids of the Bonneville system, including the type of *Pantosteus virescens* Cope. Proceedings of the United States National Museum 64: 1–6.
41. Smith G, Hall JG, Koehn RK, Innes DJ (1983) Taxonomic relationships of the Zuni mountain sucker, *Catostomus discobolus yarrowi*. Copeia 1983: 37–48.
42. Miller RR, Minckley WL, Norris SM (2005) Freshwater fishes of Mexico. Chicago, IL: The University of Chicago Press.
43. Smith GR, Koehn RK (1971) Phenetic and cladistic studies of biochemical and morphological characteristics of *Catostomus*. Syst Biol 20: 282–297.
44. Doosey MH, Bart Jr HL, Saitoh K, Miya M (2010) Phylogenetic relationships of catostomid fishes (Actinopterygii: Cypriniformes) based on mitochondrial ND4/ND5 gene sequences. Mol Phylogen Evol 54: 1028–1034.
45. Smith G (1992) Phylogeny and biogeography of the Catostomidae, freshwater fishes of North America and Asia. In: Mayden RL, editor. Systematics, historical ecology, and North American freshwater fishes. Stanford, CA: Stanford University Press. pp. 778–826.
46. Tibbets CA, Dowling TE (1996) Effects of intrinsic and extrinsic factors on population fragmentation in three species of North American minnows (Teleostei: Cyprinidae). Evolution 50: 1280–1292.
47. Duchêne S, Archer FI, Vilstrup J, Caballero S, Morin PA (2011) Mitogenome phylogenetics: the impact of using single regions and partitioning schemes on topology, substitution rate and divergence time estimation. PLoS ONE 6: e27138.
48. Bart HL, Reneau PC, Doosey MH, Bell CD (2010) Evolutionary divergence of duplicate copies of the growth hormone gene in suckers (Actinopterygii: Catostomidae). International Journal of Molecular Sciences 11: 1090–1102.
49. Hall TA (1999) BioEdit: a user-friendly biological sequence alignment editor and analysis program for Windows 95/98/NT. Nucleic Acids Symp Ser 41: 95–98.
50. Tamura K, Dudley J, Nei M, Kumar S (2007) MEGA4: molecular evolutionary genetics analysis (MEGA) software version 4.0. Mol Biol Evol 24: 1596–1599.
51. Stamatakis A (2006) RAxML-VI-HPC: maximum likelihood-based phylogenetic analyses with thousands of taxa and mixed models. Bioinformatics 22: 2688–2690.
52. Stamatakis A, Hoover P, Rougemont J (2008) A rapid bootstrap algorithm for the RAxML web servers. Syst Biol 57: 758–771.
53. Swofford DL (2002) PAUP*. Phylogenetic Analysis Using Parsimony (*and Other Methods). Version 4. Sunderland, Massachusetts: Sinauer Associates.
54. Marshall CR (1990) Confidence intervals on stratigraphic ranges. Paleobiology 16: 1–10.
55. Parham JF, Donoghue PCJ, Bell CJ, Calway TD, Head JJ, et al. (2012) Best practices for justifying fossil calibrations. Syst Biol 61: 346–359.
56. Drummond AJ, Rambaut A (2007) BEAST: Bayesian evolutionary analysis by sampling trees. BMC Evol Biol 7: 214–214.
57. Ho SY, Larson G, Edwards CJ, Heupink TH, Lakin KE, et al. (2008) Correlating Bayesian date estimates with climatic events and domestication using a bovine case study. Biol Lett 4: 370–374.
58. Posada D, Crandall KA (1998) MODELTEST: testing the model of DNA substitution. Bioinformatics 14: 817–818.
59. Spencer JE, Patchett PJ, Pearthree PA, House PK, Sarna-Wojcicki AM, et al. (2013) Review and analysis of the age and origin of the Pliocene Bouse Formation, lower Colorado River Valley, southwestern USA. Geosphere 9: 444–459.
60. Roskowski JA, Patchett PJ, Spencer JE, Pearthree PA, Dettman DL, et al. (2010) A late Miocene–early Pliocene chain of lakes fed by the Colorado River: evidence from Sr, C, and O isotopes of the Bouse Formation and related units between Grand Canyon and the Gulf of California. Geol Soc Am Bull 122: 1625–1636.
61. Ferris SD, Buth DG, Whitt GS (1982) Substantial genetic differentiation among populations of *Catostomus plebeius*. Copeia 1982: 444–449.
62. Hansen WR (1985) Drainage development of the Green River basin in southwestern Wyoming and its bearing on fish biogeography, neotectonics, and paleoclimates. The Mountain Geologist 22: 192–204.
63. Beranek LP, Link PK, Fanning CM (2006) Miocene to Holocene landscape evolution of the western Snake River Plain region, Idaho: Using the SHRIMP detrital zircon provenance record to track eastward migration of the Yellowstone hotspot. Geol Soc Am Bull 118: 1027–1050.
64. Crabtree CB, Buth DG (1987) Biochemical systematics of the catostomid genus *Catostomus*: assessment of *C. clarki*, *C. plebeius* and *C. discobolus* including the Zuni sucker, *C. d. yarrowi*. Copeia 1987: 843–854.
65. Reheis MC, Hershler R, Miller DM (2008) Late Cenozoic Drainage History of the Southwestern Great Basin and Lower Colorado River Region: Geologic and Biotic Perspectives: Geological Society of America Special Paper 439. 432 p.
66. Cummings ML, Evans JG, Ferns ML, Lees KR (2000) Stratigraphic and structural evolution of the middle Miocene synvolcanic Oregon-Idaho graben. Geol Soc Am Bull 112: 668–682.
67. Hunt CB (1960) Cenozoic geology of the Colorado Plateau. U S Geol Surv Prof Pap 279: 1–99.

68. Wernicke B (2011) The California River and its role in carving Grand Canyon. Geol Soc Am Bull 123: 1288–1316.
69. Blakey RC, Ranney W (2008) Ancient landscapes of the Colorado Plateau. Grand Canyon, AZ: Grand Canyon Association. 175 p.
70. Lucchitta I (2011) A Miocene river in northern Arizona and its implications for the Colorado River and Grand Canyon. Geological Society of America Today 21: 4–10.
71. Lucchitta I (1972) Early history of the Colorado River in the Basin and Range Province. Geol Soc Am Bull 83: 1933–1948.
72. Estabrook G, Smith G, Dowling T (2007) Body mass and temperature influence rates of mitochondrial DNA evolution in North American cyprinid fish. Evolution 61: 1176–1187.
73. Tolan TL, Martin BS, Reidel SP, Kauffman JD, Garwood DL, et al. (2009) Stratigraphy and tectonics of the central and eastern portions of the Columbia River Flood-Basalt Province: An overview of our current state of knowledge. In: O'Connor JE, Dorsey RJ, Madin IP, editors. Volcanoes to Vineyards: Geologic Field Trips Through the Dynamic Landscape of the Pacific Northwest, Geological Society of America Field Guide 15. pp. 645–672.
74. Lemke RW, Laird WM, Tipton M, Lindvall R (1965) Quaternary geology of northern Great Plains. In: Wright HE, Frey DG, editors. The Quaternary of the United States. Princeton, NJ: Princeton University Press. pp. 15–27.
75. Johnson JB, Dowling TE, Belk MC (2004) Neglected taxonomy of rare desert fishes: congruent evidence for two species of leatherside chub. Syst Biol 53: 841–855.
76. Billman EJ, Lee JB, Young DO, McKell MD, Evans RP, et al. (2010) Phylogenetic divergence in a desert fish: differentiation of speckled dace within the Bonneville, Lahontan, and upper Snake River basins. West N Am Nat 70: 39–47.
77. Oviatt CG (1994) Review of the Quaternary Geology of the Sevier and Black Rock deserts. Utah Geol Assoc Publ 23: 97–103.
78. Hubbs CL (1955) Hybridization between fish species in nature. Syst Zool 4: 1–20.
79. Hubbs CL, Hubbs LC, Johnson RE (1943) Hybridization in nature between species of catostomid fishes. Contribution from the Laboratory of Vertebrate Biology, University of Michigan 22: 1–76.
80. Neff NA, Smith GR (1979) Multivariate analysis of hybrid fishes. Syst Biol 28: 176–196.
81. Dowling TE, Moore WS (1984) Level of reproductive isolation between two cyprinid fishes, *Notropis cornutus* and *N. chrysocephalus*. Copeia 1984: 617–628.
82. Dowling TE, Smith GR, Brown WM (1989) Reproductive isolation and introgression between *Notropis cornutus* and *Notropis chrysocephalus* (family Cyprinidae): comparison of morphology, allozymes, and mitochondrial DNA. Evolution 43: 620–634.
83. Colosimo PF, Hosemann KE, Balabhadra S, Villarreal Jr G, Dickson M, et al. (2005) Widespread parallel evolution in sticklebacks by repeated fixation of ectodysplasin alleles. Science 307: 1928–1933.
84. Blackwelder E, Hubbs C, Antevs E (1948) The Great Basin, with emphasis on glacial and post-glacial times. Bulletin of the University of Utah Biological Series 38: 1–191.
85. Harper KT, Reveal JL (1978) The Great Basin: a symposium. Great Basin Nat Mem 2.
86. Dowling TE, Secor CL (1997) The role of hybridization and introgression in the diversification of animals. Annu Rev Ecol Syst 28: 593–619.
87. Dowling TE, DeMarais BD, Minckley W, Douglas ME, Marsh PC (1992) Use of genetic characters in conservation biology. Conserv Biol 6: 7–8.
88. Dowling TE, Minckley W, Douglas ME, Marsh PC, Demarais BD (1992) Use of molecular characters in conservation biology: implications for management of the red wolf. Conserv Biol 6: 600–603.

16

Characterizing Inland Pacific Northwest American Viticultural Areas with Geospatial Data

Ian-Huei Yau[1]*, Joan R. Davenport[1], Richard A. Rupp[2]

1 Crop and Soil Sciences, Washington State University, Prosser, Washington, United States of America, **2** Crop and Soil Sciences, Washington State University, Pullman, Washington, United States of America

Abstract

American Viticultural Areas are officially recognized appellations for wine grapes (*Vitis vinifera* L.). They represent not only geographic identification for growers, but also economic significance through price premiums for grapes from desirable appellations and wines sourcing grapes from such appellations. Petitions for establishment and official descriptions of American Viticultural Areas in the inland Pacific Northwest have traditionally relied on general descriptions of physical attributes and data from point measurements, namely weather stations. Examination of spatial datasets in a geographic information system provides a more holistic means of assessing viticultural areas and a spatially continuous representation of an area. Comparison of spatial datasets to official appellation descriptions largely corroborate petitioners' claims, often with greater detail, but also highlight some shortcomings of official appellation descriptions. By focusing on spatial data representing environmental factors most important to wine grape production, viticultural areas can be described more thoroughly and accurately and appellations may be more appropriately delineated. We examined inland Pacific Northwest American Viticultural Areas with a geographic information system approach, illustrating the utility of spatial datasets in characterization and delineation of American Viticultural Areas.

Editor: Ben Bond-Lamberty, DOE Pacific Northwest National Laboratory, United States of America

Funding: Funding from: Northwest Center for Small Fruits Research (http://www.nwsmallfruits.org/); Washington State Grape and Wine Research Group (http://www.washingtonwine.org/about.php). The funders had no role in study design, data collection and analysis, decision to publish, or preparation of the manuscript.

Competing Interests: The authors have declared that no competing interests exist.

* E-mail: yau.ianhuei@gmail.com

Introduction

The inland Pacific Northwest (IPNW), the area east of the Cascade Mountains in Washington state, Oregon and Idaho, has developed into a world-class wine grape (*Vitis vinifera* L.) producing region. The vast majority of wine grape production occurs in Washington state, which currently grows nearly 18,000 ha of wine grape vineyards, a 395% increase over the last 18 years [1]. Washington state is second only to California in wine grape production in the United States [2].

The IPNW hosts thirteen American Viticultural Areas (AVAs) acknowledged by the Alcohol and Tobacco Tax and Trade Bureau (TTB) on the basis of national or local name recognition, usage and distinguishing features (e.g., climate, geology, soils, physical features and elevation). Three AVAs, Columbia Valley, Columbia Gorge and Walla Walla Valley, share area between Washington and Oregon. One more appellation, Snake River Valley AVA, straddles the border of Oregon and Idaho. These thirteen AVAs comprise the major grape growing regions of the IPNW (Fig. 1). Burgeoning viticulture in the Snake River Valley is receiving proactive research in the form of cultivar trials [3].

The concept of *terroir* is embedded within the nature of appellations despite the widely variable sites found in many AVAs, especially massive ones like Columbia Valley AVA. The distinction between what an actual vineyard bestows upon a grape and the perception of fruit characteristics from a given AVA often comes down to the physical features of a vineyard and its viticultural practices versus the marketing surrounding an AVA or the notable exploits of outstanding vineyards within the area. Another highly variable aspect of *terroir* is winemaking practice. Unlike Old World appellations, AVAs face fewer enological restrictions and AVA boundaries arguably reflect more physical characteristics of vineyard sites and less consistent anthropological influence. Wine prices were strongly determined by appellation designation, but not by specific site attributes in the Willamette Valley [4]. Delineation of AVAs thus represents a serious economic concern for grape growers and winemakers.

Traditional approaches to appellation delineation rely on point measurements of climatic attributes, a general sense of physical attributes, soil survey and topographic maps. Particularly with climatic data, this leaves much to be desired as mesoclimate (sub-regional to vineyard scale) can vary widely over relatively short distances [5]. Geomorphologic features may be described vaguely or broadly to encompass large areas. Contour lines on topographic maps are often used to delineate portions of appellation boundaries with inaccurate descriptions of maximum and minimum elevations within the proposed area.

Climate is perhaps the ultimate limiting factor in *V. vinifera* production. Growing degree-day (GDD) accumulation is one common method of measuring relative heat accumulation within the growing season. Growing degree-day accumulation for *V. vinifera* is calculated as the summation of average temperatures less a threshold of 10°C between 1 April and 31 October. Amerine and Winkler [6] developed five categories of general wine styles

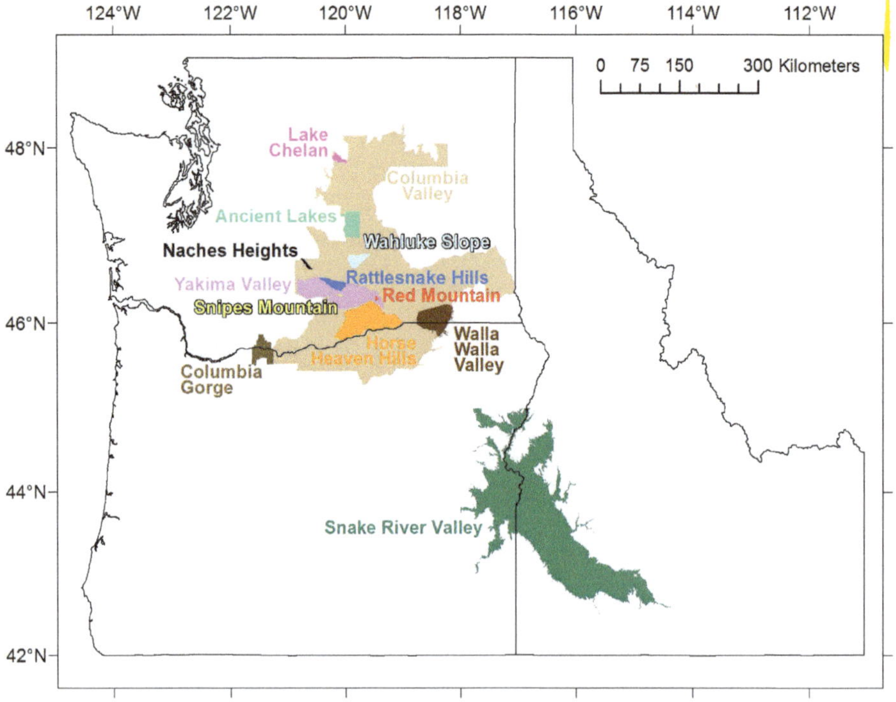

Figure 1. American Viticultural Areas of the inland Pacific Northwest.

from *V. vinifera* grapes in California based on this index of heat accumulation.

Frost-free days (FFD), the period between the last spring and first autumn frosts (0°C), is frequently examined in determining the suitability of an area for grape production [7]. Frost-free days indicate growing season length and serve as a proxy of the period over which *V. vinifera* can develop and ripen. Precipitation during the growing season is linked to primary infection and subsequent favorable microclimate for grape powdery mildew (*Uncinula necator* (Schwein.) Burrill [syn. *Erysiphe necator* Schwein.]), the most common fungal disease of grapes in the IPNW [8]. High humidity accompanied by moderate temperature provides favorable conditions for powdery mildew development [9,10].

Topographic suitability relates to the physical ability to manage a vineyard (i.e. ability for machinery to safely operate on a site) and influence over mesoclimatic (sub-regional to vineyard scale) conditions. Modern vineyard machinery readily operates on slopes of up to 15% if vine rows are parallel to the slope [11] and high-end, self-leveling equipment can handle side slopes of 30 to 35% [12]. Southern-facing slopes are often given greatest preference in the Northern Hemisphere due to higher levels of solar insolation, and consequently heat accumulation. Moderate slopes (5 to 15%) are considered the best sites for grape production [13]. Katabatic cold air flow and subsequent lower ambient temperatures can be severely limiting to shoot development, inflorescence development, pollen formation and flower fertilization. Sloped sites can reduce cold air pooling as they promote air drainage to alternate locations. Sites located above potential cold air pools may also benefit from additional elevation through lower daytime temperatures which can promote fruit quality in hot regions [14]. Unfortunately, slope alone cannot predict mesoclimate conditions and sites must be considered within the greater context of surrounding topography, obstructions to air flow and prevailing winds [15]. Ideal elevations are region-specific and vary with geography and macroclimate. Relative elevations (i.e. height above a valley floor) are more important than absolute elevation above mean sea level [14,16].

V. vinifera tolerates a wide range of soil conditions and no one soil type can be proven superior. Nutrient availability and moisture management are the two major areas of importance governed by soil characteristics [13]. The IPNW hosts a predominance of vineyard sites on Quaternary sediments overlying Miocene basalt. These are commonly Aridisols, the order of soils formed under arid or semi-arid climates, with upper horizons composed of loess (aeolian silt) or sand sheets and lower horizons consisting of stratified silty to gravelly flood sediments [17].

Soil drainage is a critical factor in grape production [13,18]. Waterlogged soils retard vine growth, hinder mechanical operations in the vineyard and favor the development of several root diseases and chlorosis in calcareous soils [19]. Free-draining soils maintain oxygen concentrations near roots and facilitate moderate water stress with proper irrigation management [20].

Unrestricted soil drainage to a depth >2 to 3 m is recommended for vineyards in most situations [18,19]. Failla et al. [21] found grapevine roots at depths >3 m during soil surveys and vines may grow roots to depths >30 m if no impenetrable barriers are present [20]. Only under severe water stress will *V.* vinifera access substantial water from >2 m. Vines draw most of their nourishment from a depth of 50 to 100 cm and readily transpire water from depths up to 2 m, but root concentrations may shift to greater depths under conditions such as competition with vineyard floor cover crops [20,22]. Shallow soils where root penetration is problematic are considered unsuitable for grape production and increase the likelihood of waterlogging [19]. Good soil drainage, along with greater depth, increases grapevines' incentive to grow robust, perennial root structures. Roots act as storage organs for carbohydrates, water and other nutrients, which are important for initial growth of shoots and roots when vines break dormancy [14,20,23].

Under ubiquitous irrigation, IPNW vineyards offer the most flexibility to consistently ripen quality crops when the available water-holding capacity (AWC) of soils is relatively low to moderate. Moisture needs of grapevines vary throughout the growing season and the ability of soils to maintain plant available levels of water without constant irrigation, as well as induce some level of water stress within relatively short time frames, is desirable [18,24].

A readily measured soil characteristic determinant of potential nutrient availability is pH [13,19]. Absorption of many nutrients is optimal in *V. vinifera* at approximately neutral to slightly acidic soil pH (6.6 to 7.2) [25]. Overly alkaline soils lead to deficiencies of phosphorus, iron, manganese, boron and zinc [18]. Acidic soils can generate toxic levels of aluminum, copper and manganese, induce phosphorus deficiency, restrict root growth and lead to grapevine nutrient and soil microbial imbalances [18,26].

Using available spatial datasets and creating datasets from available data to assess AVAs provides a spatially continuous, more accurate means of describing an appellation. By examining IPNW AVAs with a geographic information system (GIS) approach, we will illustrate the utility of spatial datasets in characterization and delineation of AVAs.

Materials and Methods

Geographic information system software packages used were ArcGIS 9.3.1 and 10×(Esri, Redlands, CA). Thematic maps were created from Soil Survey Geographic (SSURGO) databases using Soil Data Viewer 5.2 (NRCS, Lincoln, NE). Tabular data were spatially represented using geographic coordinates provided by responsible organizations. Spatial data were projected to UTM Zone 11 North, North American Datum, 1983 using bilinear sampling for continuous data where necessary. AVA boundaries were mapped following final rule descriptions in the Federal Register, the corresponding, georeferenced U.S. Geological Survey topographic quadrangle maps and topographic lines extracted from the National Elevation Dataset (NED) 1/3 arc-second digital elevation model (DEM) [27,28].

Slope, solar insolation and aspect were calculated from the 1/3 arc-second NED DEM. Slope calculations were made in units of percent rise. Aspect calculations were made in degree units; areas with zero slope were classified as flat.

Solar insolation calculations were performed using the mean latitude of each DEM to 10^{-12} decimal degrees, a sky size of 40,000 cells, 14-day and two-hour intervals. Calculation guidelines recommend the extent of input DEMs be less than one degree latitude [29], but calculations of adjacent surfaces with combined extents of much less than one degree latitude showed notable differences in continuity. Most DEMs masked by county were too large for a single calculation of solar insolation. Digital elevation models were divided vertically whenever possible to maintain relatively constant mean latitude for each county (±0.0001°). Vertical divisions were made with overlap to account for effects of adjacent topography and improved continuity. Divided surfaces were later mosaicked using a blend of overlapping areas. Calculations were made from ordinal day 91 to 304 (1 April to 31 October). Thirty-two azimuth directions were used to calculate viewshed. Eight zenith and azimuth divisions were used to calculate the sky map. A uniform sky diffuse radiation model was used. A diffuse proportion of 0.3 and transmissivity of 0.5 were used, presumptive of generally clear sky conditions [29,30]. Solar insolation surfaces do not represent actual solar accumulation over a climatically relevant period of record and are intended to represent sites' topographic exposure. Output surface units are Watt hours per square meter (Wh/m2).

Thematic maps of soil characteristics were created for drainage class, depth to any restrictive layer, available water-holding capacity (0 to 50 cm) and pH (0 to 50 cm). Dominant component or condition was used as the aggregation method. Vector data was converted to raster data with a resolution of 10 m to adequately represent soil map unit boundaries and match DEM resolution.

Heat accumulation was calculated using Parameter-elevation Regressions on Independent Slopes Model (PRISM) monthly normals, 1971–2000 [31]. This model produces official climate surfaces used by the U.S. Department of Agriculture. Monthly normal maximum and minimum temperatures were averaged for April through October. From these monthly average temperatures, a base of 10°C was subtracted, each surface was multiplied by days in the month and the resulting surfaces were summed. Additionally, a biologically effective degree-day (BEDD) surface was calculated limiting daily heat accumulation for each month to 9C° after Gladstones who observed phenological development in *V. vinifera* to be most active from 10°C to 19°C [18]. A surface of median FFD from 1971–2000 was calculated by subtracting PRISM surfaces of median last spring frost date from median first fall frost date. Monthly PRISM precipitation normals from 1971–2000 were summed for April through October. This growing season precipitation serves as a proxy for *U. necator* risk.

Principal components analysis (PCA) was performed for each IPNW AVA using the ArcGIS Spatial Analyst. Surfaces included in PCA were DEM, soil AWC, soil pH, depth to restrictive layer, GDD, FFD and growing season precipitation. Slope, aspect and solar insolation were derived from the DEM, so including these surfaces in PCA would introduce redundancy. Soil drainage is classified into qualitative categories and was omitted from PCA. Growing degree-day and BEDD surfaces were both derived from the same PRISM datasets and GDD surfaces were included in PCA because the limitation on heat accumulation in BEDD reduced observable variability. All datasets included in PCA were standardized to z-scores by subtracting the mean and dividing by the standard deviation [32].

Results and Discussion

Geospatial analysis of American Viticultural Areas

Table 1 shows means of site characteristics examined by AVA. Table 2 summarizes soil drainage class, a qualitative characteristic, with proportional representation of each class within each AVA. Figure 2 shows proportional distribution of aspect by AVA. These summaries do not represent realistically plantable area within each appellation, but merely the entire area within boundaries excluding water features from the National Hydrography Dataset. The availability of water for irrigation and the associated water rights are often a greater limitation than other facets of site suitability [17,33]. Table 3 shows planted area of the three most common red and white *V. vinifera* varieties and other notable varieties for each available AVA based on a 2011 U.S. Department of Agriculture survey [1]. Discussion of AVAs proceeds in chronological order of establishment.

The approximately 2,900 km^2 Yakima Valley AVA became the first appellation in Washington state on 4 May 1983. The Federal Register describes the western portion of the AVA as a vast, flat expanse and the eastern portion as gently sloping north of the Yakima River [34]. It is the second flattest of IPNW appellations. Spanning from 46°08'N to 46°31'N latitude, the area registers the fourth highest solar exposure.

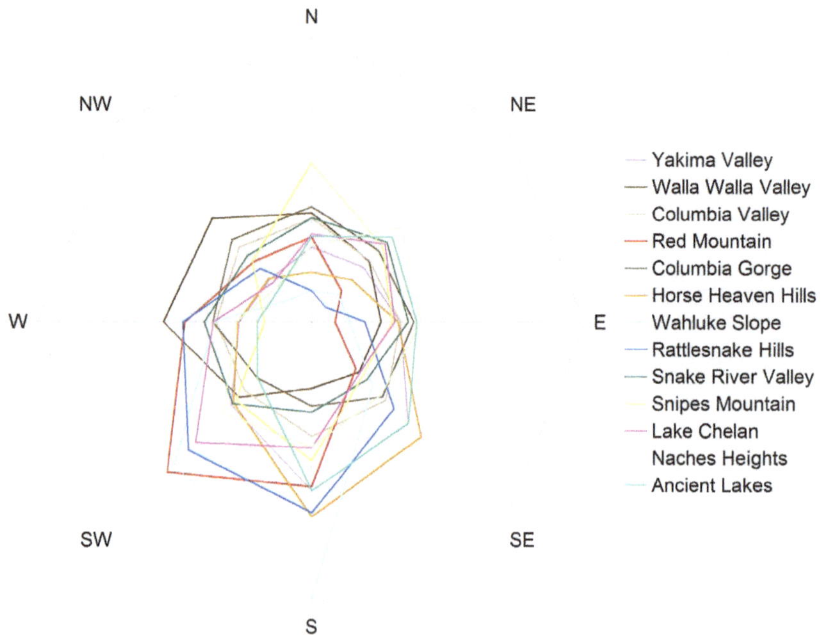

Figure 2. Aspect distribution by American Viticultural Area.

When the petition for establishment of Yakima Valley AVA was filed, it was stated by the petitioner that most vineyards were planted on two soil associations, Warden-Shano and Scootenay-Starbuck silt loams with varying depths over basalt and/or gravel [34]. The area is dominated by well-drained soils with a small component of somewhat poorly drained soils located primarily southwest of the Yakima River in the western portion of the AVA. It has the seventh deepest soils.

Climatic distinction for Yakima Valley AVA was established as warmer than mountains to the west and cooler than areas to the north and east [34]. Heat accumulation does decline precipitously within 24 km to the west of the AVA boundary as elevations climb into the Cascade mountain range (data not shown) and Walla Walla Valley, Red Mountain and Wahluke Slope AVAs, all lying north or east of Yakima Valley, record more GDD than Yakima Valley AVA. Mean FFD was fifth lowest amongst IPNW AVAs potentially due to the flat terrain and associated cold air pooling. Red varieties are dominated by the Bordeaux varieties Merlot and Cabernet Sauvignon and white varieties by the Burgundy variety Chardonnay and Rhine variety Riesling. There are also contin-

Table 1. Mean site characteristics by inland Pacific Northwest American Viticultural Areas sorted by chronological establishment.

	Topographic parameters			Soil parameters			Climate parameters			
	Elevation (m)	Slope (%)	Solar insolation (kWh/m2)	AWC (cm/cm)	Depth (cm)	pH	GDD (C°)	BEDD (C°)	FFD	PPT (mm)
Yakima Valley	331	5.2	1031.2	0.169	165	7.62	1371	1204	162	89
Walla Walla Valley	315	10.6	1007.2	0.186	190	7.36	1487	1269	200	197
Columbia Valley	462	11.3	1022.1	0.163	141	7.28	1278	1110	163	125
Red Mountain	227	7.7	1016.7	0.161	186	7.40	1501	1341	176	89
Columbia Gorge	405	22.8	983.5	0.163	158	6.48	927	908	150	200
Horse Heaven Hills	323	6.6	1031.7	0.169	179	7.52	1435	1236	174	98
Wahluke Slope	251	3.0	1027.9	0.120	174	7.73	1512	1281	173	73
Rattlesnake Hills	433	11.5	1031.6	0.176	108	7.67	1214	1119	148	105
Snake River Valley	874	10.8	1111.9	0.153	120	7.59	1339	1135	139	136
Snipes Mountain	279	17.4	994.4	0.189	166	7.63	1447	1284	167	80
Lake Chelan	488	24.5	951.6	0.143	170	7.43	1198	1055	168	112
Naches Heights	542	10.6	1026.2	0.169	142	7.10	1061	937	119	125
Ancient Lakes	386	4.3	1026.3	0.143	157	7.55	1380	1210	167	83

Abbreviations: AWC (available water-holding capacity), GDD (growing degree-days), BEDD (biologically effective degree-days), FFD (frost-free days), PPT (growing season precipitation).

Table 2. Summary of drainage classes by American Viticultural Area from Soil Survey Geographic database, masked to exclude water features, showing a predominance of soils classed as well drained.

	ED	SED	WD	MWD	SPD	PD	VPD
Yakima Valley	1.9%	5.6%	78.4%	0.6%	13.2%	0.3%	-
Walla Walla Valley	-	3.9%	86.8%	4.9%	4.2%	0.2%	-
Columbia Valley	6.6%	6.9%	83.0%	0.9%	2.0%	0.5%	0.1%
Red Mountain	0.6%	14.9%	84.5%	-	-	-	-
Columbia Gorge	-	0.2%	96.7%	-	2.9%	0.3%	-
Horse Heaven Hills	9.6%	9.4%	80.6%	-	0.3%	0.1%	-
Wahluke Slope	34.7%	33.7%	31.5%	-	<0.01%	-	-
Rattlesnake Hills	-	<0.01%	99.4%	-	0.6%	-	-
Snake River Valley	1.8%	4.6%	86.7%	0.8%	4.7%	1.2%	0.1%
Snipes Mountain	-	0.3%	99.2%	-	0.4%	-	-
Lake Chelan	-	0.9%	90.0%	9.1%	-	-	-
Naches Heights	-	<0.01%	98.3%	-	1.7%	-	-
Ancient Lakes	5.0%	27.5%	67.2%	0.1%	-	0.2%	-
Average	8.6%	8.3%	83.3%	2.8%	3.0%	0.4%	0.1%

Abbreviations: ED (excessively drained), SED (somewhat excessively drained), WD (well drained), MWD (moderately well drained), SPD (somewhat poorly drained), PD (poorly drained), VPD (very poorly drained).

gents of the cooler varieties Gewürztraminer and the Burgundy variety Pinot Noir.

Walla Walla Valley AVA, established 7 March 1984 [35], originally consisted of approximately 1,050 km^2 and was expanded on 27 April 2001 to approximately 1,066 km^2 [36]. The original appellation was bound by the confluence of the north and south forks of the Walla Walla River, the 610 m contour to its south and east, an approximation of the divide between the Walla Walla and Touchet River drainages to the north and the confluence of the Walla Walla and Columbia Rivers to the west. The expanded appellation includes area in the Touchet River drainage [36,37].

Distinctive features listed in the Federal Register were limited both for the original and expanded appellation. According to the NED DEM, the range of elevations, 122 to 697 m, within Walla Walla Valley AVA is greater than the stated 250 to 600 m [36]. It also has the greatest proportion of west and northwest-facing slopes, aspects associated with lower heat accumulation and less solar insolation. The loess derived soils were predominately classed as well drained. The AVA has the second highest mean AWC and the highest mean depth to restrictive layer.

The original and expanded petitions both note relatively high precipitation [35,36]. The AVA has the second highest mean growing season precipitation, just 3 mm fewer than the Columbia Gorge AVA which lays nearly 300 km to the west. PRISM 1971–2000 normals show mean annual precipitation at 44.2 cm and a maximum of 96.1 cm within the AVA (data not shown). This is greater than the mean of 31.8 cm and maximum of 50.8 cm indicated in either final rule description from undisclosed sources [35,36]. The longest mean growing season is also found here at 200 days, which coincides with the description which states 190 to 220 FFD [35]. However, FFD from PRISM 1971–2000 normals also indicate a minimum growing season of 138 (data not shown). This AVA also shows the second-greatest difference between GDD and BEDD. As expected, the greatest differences between mean GDD and BEDD are found in the warmest AVAs and vice versa. However, the third-warmest AVA, Red Mountain, records only 14C° greater GDD than Walla Walla AVA, but Walla Walla AVA records a 58C° greater difference between GDD and BEDD than Red Mountain AVA. This indicates greater heat accumulation during the peak of the growing season with potentially lower temperatures during budbreak and flowering and after veraison. Walla Walla Valley AVA is mostly planted to red varieties which are dominated by Cabernet Sauvignon with notable area planted with the later-ripening Bordeaux variety Petit Verdot, possibly owing to the long growing season.

The immense Columbia Valley AVA, established 13 December 1984, encompasses approximately 46,000 km^2 [38]. It is a diverse region and many smaller appellations have been carved from within it since its establishment. Yakima Valley AVA is included within the Columbia Valley AVA, as is Walla Walla Valley AVA following a boundary adjustment [36]. Since many smaller AVAs have been distinguished by more distinctive characteristics, this discussion will not explore Columbia Valley AVA.

Established 11 June 2001, Red Mountain AVA covers approximately 1,650 ha and is contained entirely within Yakima Valley AVA [39]. The Saddle Mountains Formation of the Columbia River Basalt Group underlies the appellation upon which glacial flood sediments have been topped with loess and dune sands [40].

Elevations within Red Mountain AVA range from 167 to 428 m according to the NED DEM, a greater range than the stated 183 to 305 m [39]. Although largely comprised of southwest-facing slopes, the appellation records the fifth lowest mean solar insolation However this is comparable to many other AVAs. This may be partially due to relatively gentle slopes, fourth lowest amongst IPNW AVAs.

Soils of the Warden-Shano association dominate Red Mountain AVA, including Warden silt loam, Hezel loamy fine sand, Scooteney silt loam and Kiona very stony silt loam [39]. Again, this is an area dominated by well drained soils with a component of somewhat excessively drained soils. Soil depths are second only to Walla Walla Valley AVA.

The petitioner stated heat accumulation on Red Mountain AVA exceeds that of other areas in the Yakima Valley [39], a claim confirmed by PRISM GDD. Although Red Mountain AVA is the second-warmest AVA, it has the fifth-least difference between GDD and BEDD indicating more even heat accumulation over the growing season than Walla Walla Valley AVA. The growing season length in the appellation is second only to Walla Walla Valley AVA. Elevation, southwest slopes, continental air mass movement through the gap between Red Mountain and Rattlesnake Ridge to the west and the Yakima River flowing to the west and around the northern boundary of Red Mountain AVA all likely contribute to this long growing season [39]. Like Walla Walla Valley AVA, vineyards are dominated by red varieties, once again Cabernet Sauvignon and notably the Rhône variety Mourvèdre.

Columbia Gorge AVA, like Columbia Valley and Walla Walla Valley AVAs, is the third to share area in both Washington and Oregon. Established 9 July 2004, this roughly 77,500 ha appellation straddles the Columbia River [41]. Slopes are steep in this appellation with the second highest mean. Elevations range from 23 to 839 m with areas well above the 610 m contour defining a portion of the southern boundary [41]. Columbia Gorge AVA records the second lowest mean solar insolation with the greatest standard deviation (data not shown). This is likely due to a greater proportion of north-facing slopes on the Oregon side of the AVA and hillshade effects due to steep slopes.

Table 3. Planted area (ha) of three most common red and white *Vitis vinifera* varieties and other notable varieties in Washington American Viticultural Areas (AVAs).

	Red varieties					White varieties				
	Cabernet Sauvignon	Merlot	Syrah	Other	Total red	Chardonnay	Riesling	Pinot Gris	Other	Total white
Yakima Valley	679	930	295	80 Cabernet Franc, 32 Pinot Noir	2178	1278	1157	346	170 Sauvignon Blanc, 157 Gewürztraminer	3266
Walla Walla Valley	212	81	67	31 Cabernet Franc, 13 Petit Verdot	433	11	-	-	8 Viognier, 8 Sémillon	95
Columbia Valley	708	586	258	70 Cabernet Franc, 24 Malbec	1745	539	461	100	69 Sauvignon Blanc, 39 Gewürztraminer	1278
Red Mountain	272	74	70	17 Cabernet Franc, 8 Mourvèdre	478	-	-	-	8 Sémillon, 7 Sauvignon Blanc	37
Columbia Gorge	-	8	-	17 Pinot Noir, 2 Tempranillo	58	33	9	28	30 Gewürztraminer	102
Horse Heaven Hills	1344	828	306	107 Cabernet Franc, 51 Grenache	2824	717	354	56	199 Sauvignon Blanc, 36 Viognier	1459
Wahluke Slope	769	623	173	56 Cabernet Franc, 43 Malbec	1792	398	344	-	49 Gewürztraminer, 17 Sauvignon Blanc	898
Rattlesnake Hills	118	149	31	26 Cabernet Franc, 8 Petit Verdot	361	48	183	19	11 Viognier, 3 Chenin Blanc	286
Snipes Mountain	57	45	32	-	154	60	28	-	-	132
Lake Chelan	3	-	15	17 Pinot Noir, 2 Sangiovese	51	5	17	5	11 Gewürztraminer, 6 Viognier	49

Data for Snake River Valley, Naches Heights and Ancient Lakes AVAs were unavailable.
Adapted from [1].

Columbia Gorge AVA is also dominated by well drained soils and has the fifth lowest mean AWC. This contrasts somewhat with final rule descriptions of slow to moderate permeability and high AWC, although the latter is given no quantitative context [41]. The mean depth coincides well with descriptions from the petitioner [41]. The appellation records the lowest mean pH.

The petitioner states that due to relatively cool conditions, *V. vinifera* grown in the AVA are generally early varietals like Pinot Noir and Gewürztraminer [41]. There is also some area planted to Pinot Gris as well as the generally warmer Burgundy variety Chardonnay. Heat accumulation surfaces place the Columbia Gorge as the coolest IPNW AVA, which also unsurprisingly show the least difference between GDD and BEDD. Growing season precipitation is greatest in this AVA with a large degree of longitudinal variability. Annual precipitation ranges over the strong precipitation gradient agree with the 46 cm on the eastern end and 76 cm on the western end (data not shown) stated by the petitioner [41].

Lying to the south of Yakima Valley and Red Mountain AVAs and north of the Columbia River, the approximately 2,300 km^2 Horse Heaven Hills AVA was established on 1 August 2005 [42]. It is bound on the west by a 518 m contour and Pine Creek and by the ridge of the Horse Heaven Hills to the north. As seen previously, elevation ranges from the NED DEM (77 to 670 m) are greater than those stated in the final rule description (61 to 549 m) [42], but in this case, the minimum observed elevation in the DEM is greater than the stated minimum. Slopes are comparable to those in Yakima Valley and Red Mountain AVAs, largely face to the south and southeast and solar insolation is second only to the Snake River Valley, which is at a much lower latitude.

Like many IPNW viticultural areas, Horse Heaven Hills soils are composed of aeolian sand and silt, glacial and slackwater sediments, gravel alluvium and Columbia River basalt colluvium, but the proportion and arrangement of these features distinguish appellations from one another [42]. The Horse Heaven Hills are dominated by soils classed as well drained, but also have one of the greatest proportions of somewhat excessively and excessively drained soils. The third deepest soils are also found here.

Ten-year averages of heat accumulation provided in the petition list, from warmest to coolest, Red Mountain, Walla Walla Valley, Horse Heaven Hills and Yakima Valley AVAs [42]. This corresponds to rankings from GDD and BEDD surfaces. Horse Heaven Hills AVA also ranks fourth in difference between GDD and BEDD surfaces again indicating greater heat accumulation at the peak of the growing season. Red varieties planted are again dominated by Cabernet Sauvignon and Merlot with a notable area of the late-ripening Rhône variety Grenache. White varieties include Chardonnay, Riesling, the Bordeaux variety Sauvignon Blanc and the Rhône variety Viognier.

Sitting atop a 24 km long mega alluvial fan deposited by the Missoula Floods, the approximately 33,000 ha Wahluke Slope AVA was established on 6 January 2006 [43]. Noted topographic features are gently south-facing slopes which decline to the east, south and west, providing for good air drainage, while the 451 m contour of the Saddle Mountains provide the northern boundary [43]. Wahluke Slope registers the lowest mean slope, has by far the greatest proportion of south-facing slopes, fifth highest mean solar insolation and the fourth longest growing season. More continuously similar aspects and low slopes reduce hillshade effects and contribute to the levels of solar insolation.

Deep aeolian sands and silts have the fourth deepest soils with relatively large areas of excessively and somewhat excessively drained soils. Drainage classes are split almost evenly between these two classes and well drained soils. These soils also have the lowest AWC and highest pH.

Based on the state of Washington's Public Agricultural Weather System data from 1994–2003, the petitioner states Wahluke Slope AVA is the driest in the region and recorded 1,674 GDD [43]. It does have the lowest mean growing season precipitation and mean annual precipitation of 192 mm based on PRISM 1971–2000 normals (data not shown). The GDD surface indeed rates it as the hottest of all IPNW AVAs, but BEDD only ranks it as the third hottest. Wahluke Slope AVA also shows the greatest difference between GDD and BEDD surfaces. Again, red varieties are dominated by Cabernet Sauvignon and Merlot and white varieties by Chardonnay and Riesling. Other notable varieties are the Bordeaux variety Malbec and surprisingly Gewürztraminer, typically grown in cooler climates.

Fully within Yakima Valley AVA, the nearly 28,000 ha Rattlesnake Hills AVA was established on 20 March 2006. It is comprised of a portion of east-west hills between the Yakima and Moxee River Valleys and lying between the Hanford Reservation and Union Gap [44].

The petitioner states the appellation contains diverse topography along the southern face of the Rattlesnake Hills, much more variability than the rest of the Yakima Valley and most vineyards have been located on southern ridges and terraces with good air drainage [44]. The fourth steepest mean slope and third highest mean solar insolation are found here with greater standard deviations than Yakima Valley AVA (data not shown). Aspects are dominated by south and southwest-facing slopes.

Soils on the Rattlesnake Hills are dominated by those classed as well drained. They contrast with the more coarsely textured sandy soils of Red Mountain and Horse Heaven Hills AVAs, silty soils found elsewhere in the Yakima Valley and are shallow, especially above 335 m where soil formation was above the depositional influence of the Missoula Floods [44]. These finer textured soils have the third highest AWC and are the shallowest of all IPNW AVAs.

The petitioner states that the Umptanum and Yakima Ridges to the northeast shield the Rattlesnake Hills AVA from Arctic fronts, which are funneled toward Red Mountain and Walla Walla Valley AVAs [44]. This contradicts FFD data, which indicates Rattlesnake Hills AVA has the third shortest growing season at a mean of 148 FFD, 28 days shorter than the mean of Red Mountain AVA and 52 shorter than that of Walla Walla Valley AVA. Heat accumulation claims are more accurate; Rattlesnake Hills AVA is substantially cooler than Red Mountain AVA and the area between the two AVAs is cooler than Rattlesnake Hills AVA (data not shown). The appellation also records the second-least difference between GDD and BEDD reflecting both the relatively low heat heat accumulation and indicating even heat accumulation over the growing season. The most common red varieties are again Merlot and Cabernet Sauvignon with a small area of Petit Verdot despite the relatively short growing season. White varieties are largely comprised of Riesling with small areas planted to Viognier and the Loire variety Chenin Blanc.

Straddling the border between Oregon and Idaho and situated in the rift basin that once contained ancient Lake Idaho, the approximately 21,400 km^2 Snake River Valley AVA was established on 9 April 2007 [45]. Although a 1,040 m contour is stated as its maximum elevation and comprises nearly its entire boundary [45], Snake River Valley AVA contains over 47,000 ha exceeding this elevation according to the NED DEM. A maximum elevation of 1,504 m is located in the western portion of the appellation and many hills and ridges between the 1,040 m contour and drainages exceed the stated maximum elevation. Although the central Snake

River Plain is relatively flat as stated in the petition [45], many steeper drainages in northern, western and southern portions of the appellation (data not shown) push mean slope up to sixth steepest. Snake River Valley AVA records the highest solar insolation of IPNW AVAs due to its lower latitude (42°29'N to 45°N). Aspects are fairly evenly distributed in all directions.

The large, edaphically diverse area is also dominated by well drained soils with small components of both somewhat excessively drained and somewhat poorly drained soils. The final rule description provides little in terms of distinctive soil features, but shallow soils are noted and corroborated by SSURGO data with a mean depth greater only than Rattlesnake Hills AVA.

Mean GDD over the AVA were 165C° fewer than the average of four National Climatic Data Center weather station normals from 1971–2000, the same period of record used to compile the GDD surface [45]. Snake River Valley AVA recorded 149 fewer GDD than Walla Walla Valley AVA. While showing less difference than stated by the petitioners assessing AVAs using point measurements, this does generally agree with their claims [45]. The appellation also recorded the second fewest FFD, but exhibits a range second only to the massive Columbia Valley AVA and a maximum of 188 FFD (data not shown). Finding sites with an adequate growing season and heat accumulation are likely the greatest challenges to grape production within Snake River Valley AVA and large areas included in the boundary may never be suitable for *V. vinifera* production.

A third sub-appellation located entirely within Yakima Valley AVA, Snipes Mountain AVA, was established on 20 February 2009. It is the smallest IPNW AVA at less than 1,700 ha. Lying north of the Yakima River, this appellation contains the landform of its namesake as well as 53 ha on Harrison Hill, east of Snipes Mountain, which is stated to have similar soils and contiguous topography with Snipes Mountain [46].

Although Yakima Valley AVA encompasses a much greater elevation range and maximum elevation than Snipes Mountain, including Rattlesnake Hills AVA to the north, Snipes Mountain AVA is substantially higher than its immediate vicinity, particularly to the south, as indicated by the petitioner. Additionally, steep slopes on the southern face of the landforms make Snipes Mountain the third steepest IPNW AVA. However, these steep slopes may lead to shading from adjacent hills. This, in conjunction with gentle northern slopes and the greatest proportion of northern aspects of all IPNW AVAs face, gives Snipes Mountain the third lowest solar insolation.

The appellation is comprised almost entirely of soils classed as well drained. Although the petitioner states that nearly all soils, lacustrine or alluvial, are now generally dry, classified as Aridisols and contain more rock fragments than elsewhere in the Yakima Valley, Snipes Mountain AVA has the highest AWC of all IPNW AVAs. This would be expected to be lower than that of Yakima Valley AVA, which the petitioner states is composed of 43% Mollisols [46].

Climatically, little is stated in the final rule establishing Snipes Mountain AVA aside from steep southern slopes of Snipes Mountain and Harrison Hill draining cold air freely into the Yakima Valley below [46]. It recorded a mean of 6.4 greater FFD than Yakima Valley AVA. It is also relatively warm and dry during the growing season, recording the fourth highest GDD and the second lowest growing season precipitation. Red varieties include Cabernet Sauvignon, Merlot and the Rhone variety Syrah. White varieties again include Chardonnay and Riesling.

The farthest north sub-appellation (47°52'N) within the Columbia Valley AVA flanks the southeastern 19 km of the 88-km-long Lake Chelan, the third deepest lake in the United States from which the appellation is named. Alpine glaciers descended from the Cascade Mountains during the last ice age, 14,000 to 18,000 y.b.p., carving the Lake Chelan Valley, including the lake. Lake Chelan AVA was established on 29 May 2009 and is roughly 9,700 ha [46].

Lake Chelan AVA has the third highest mean elevation of IPNW AVAs and the NED DEM shows greater maximum elevation within the boundary (1,148 m) than the 999 m unnamed peak in the northwest portion of the appellation mentioned by the petitioner [47]. It contains the steepest mean slope with the steepest occurring at higher elevations along the western bank of Lake Chelan and along the northeastern portions of the appellation (data not shown). Along portions of the northern, southern and eastern banks of Lake Chelan, slopes are more gentle as stated by the petitioner [47]. High latitude and steep slopes at the foot of the Cascade Mountains give Lake Chelan AVA the lowest solar insolation.

Distinctions in soil characteristics given by the petitioner focus on differences between the Lake Chelan Valley and the greater Columbia Plateau, primarily greater proportions of volcanic pumice and ash from the eruption of Glacier Peak approximately 12,000 y.b.p. and lower proportions of loess in the top 50 cm of soils. Most notably, soils with large proportions of volcanic ash, pumice or clays weathered from glass have unusually high AWC [47]. This is not corroborated by SSURGO data which ranks Lake Chelan AVA with the second lowest mean AWC. The appellation is dominated by well drained and moderately well drained soils.

Lake Chelan's massive volume and moderating effect are emphasized by the petitioner as defining the appellation's distinctive climate [47]. Despite its high latitude and elevation, the AVA records the fifth highest FFD. Heat accumulation is also moderated by the lake and Lake Chelan is the third coolest IPNW AVA according to GDD and BEDD. As stated by the petitioner, growing season length and heat accumulation drop precipitously as elevations increase to the north and west and, to a lesser extent, to the south and east across the Columbia River and away from the moderating influence of Lake Chelan (data not shown). Red varieties include Pinot Noir, Syrah and the Italian variety Sangiovese. White varieties include Riesling, and Gewürztraminer.

Naches Heights AVA is completely contained by the Columbia Valley AVA and lies to the northwest of Yakima Valley AVA. Bound by the Tieton River to the west, the Naches River to the north and east and Cowiche Creek and the Congdon (Schuler) Canal to the south, the appellation is defined by its elevated andesite plateau. The boundary surrounds approximately 5,300 ha which the petitioner states is the home of 42 ha of grapes [48].

The plateau is largely covered with moderate slopes, but the inclusion of cliffs along the northern, eastern and southern boundaries gives it the eighth highest mean slope. Approaching the Cascade foothills, mean elevation is second only to Snake River Valley AVA. According to the petitioner, the appellation is dominated by Tieton loam and Ritzville silt loam, which are deep soils with adequate drainage [48]. While 98% of the soils in the area are classed as well drained, Naches Heights has the fourth lowest mean depth, which is not excessively shallow at a mean of 142 cm. AWC is fairly high with a mean of 0.169 cm/cm.

Growing season length does decrease to the southwest, across the Tieton River, as claimed by the petitioner [48], but actually increases to the northeast, across the Naches River (data not shown). It is a very cool area with the shortest growing season

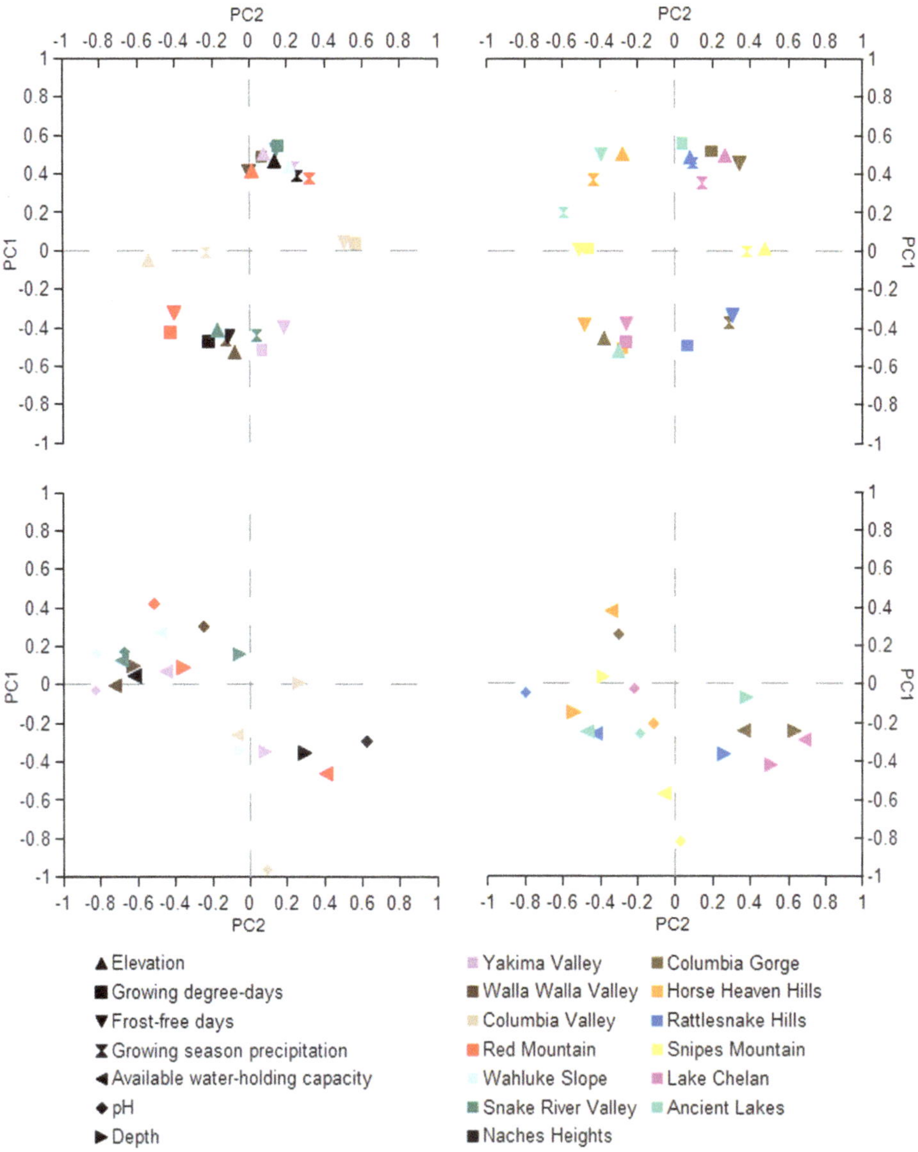

Figure 3. Eigenvectors of first two principal components of seven environmental parameters for 13 inland Pacific Northwest American Viticultural Areas.

amongst other IPNW AVAs and second lowest mean GDD and BEDD.

The newest IPNW AVA, Ancient Lakes of Columbia Valley, was established 19 November 2012 and is also completely contained within Columbia Valley AVA. The petition for the establishment of Ancient Lakes AVA utilized GIS data to describe topographic and edaphic features, but relied on individual weather stations for climatic features. As stated, it is a flat appellation with the second-lowest mean slope [49]. Ancient Lakes AVA also has one of the greatest proportion of east-facing slopes. These features may act in opposition in some ways; gentle slopes tend to increase cold air pooling while east-facing slopes are heated earlier in the day while avoiding more intense afternoon sun when temperatures are higher. These factors, a greater elevation and the AVA's higher latitude (46°58'N to 47°16'N) contribute to its moderate heat accumulation compared to nearby Wahluke Slope AVA.

Ancient Lakes AVA ties for second-lowest mean AWC with Lake Chelan AVA. While warmer than Lake Chelan AVA, Ancient Lakes AVA has soil drainage properties similar to those of Wahluke Slope AVA with proportionally 6% and 30% less area classed as somewhat excessively and excessively drained, respectively. It also has the third-lowest growing season precipitation. This indicates a moderate potential for drought stress when comparing these three appellations.

Principal components analysis

Figure 3 shows the PCA results of the first and second principal components of each IPNW AVA with eigenvectors indicating relative loading of each parameter. At least 60% of variance was explained with two principal components and 85% of variance was explained with four principal components for all AVAs. The first principal component for Snipes Mountain AVA explained the greatest amount of variance (80.6%) while the first principal

Table 4. Key differences between American Viticultural Area (AVA) descriptions in Federal Register and observations in modeled geospatial datasets.

Characteristic	Difference(s)	AVA(s) in which difference(s) were observed
Elevation	wider modeled range than stated	Walla Walla Valley, Red Mountain
	higher modeled maximum than stated contour line	Columbia Gorge
	greater modeled maximum than stated	Horse Heaven Hills, Snake River Valley, Lake Chelan
Slope	steeper than stated	Snake River Valley
Frost-free days	modeled minimum much less than stated range	Walla Walla Valley
	far fewer modeled frost-free days compared to Red Mountain and Walla Walla	Rattlesnake Hills
	modeled growing season decreases to the northeast, contrary to description	Naches Heights
Growing degree-days	cooler over the region than mean of four National Climatic Data Center stations	Snake River Valley
Precipitation	greater modeled than stated	Walla Walla Valley
Drainage	well-drained compared to stated 'slow to moderate permeability'	Columbia Gorge

components for Red Mountain AVA and Ancient Lakes AVA explained the least (34.4% and 36.1%, respectively).

A few patterns emerge from this multivariate analysis. With regards to first principal components, Yakima Valley, Horse Heaven Hills, Wahluke Slope, Rattlesnake Hills, Lake Chelan and Naches Heights AVAs showed similar results. These AVAs have large positive eigenvectors for elevation and growing season precipitation and relatively large negative eigenvectors for GDD and FFD. An inverse of this pattern is seen in Walla Walla Valley, Columbia Gorge, Snake River Valley and Ancient Lakes AVAs with relatively large negative eigenvectors for elevation and relatively large positive eigenvectors for GDD and FFD. Walla Walla Valley, Columbia Gorge and Snake River Valley AVAs also show a relatively large negative eigenvector for growing season precipitation, but Ancient Lakes AVA does not. The more moist Columbia Gorge AVA also shows a relatively large positive eigenvector for soil pH. These relationships reflect the interactions between elevation, GDD, FDD and precipitation. While the negative correlation between elevation and both GDD and precipitation are generally well understood, the inclusion of FFD in this interaction is less expected because this parameter tends to rely more on local topography rather than absolute elevations.

The first principal components of Snipes Mountain and Columbia Valley AVAs return the largest eigenvectors for soil parameters. Both have large negative eigenvectors for soil pH and Snipes Mountain AVA also returns a large negative eigenvector for AWC. These two AVAs also returned nearly zero eigenvectors for elevation, GDD, FFD, growing season precipitation and soil depth. The nearly zero eigenvector for precipitation for Snipes Mountain AVA is interesting since it also has an average growing season precipitation of less than 100 mm, but lack of loading implies less influence of this factor than of soil components.

Those AVAs that returned the greatest loadings for elevation, GDD, FFD and growing season precipitation tended to return larger loadings for soil factors in their second principal components and vice versa. Red Mountain, Columbia Gorge, Horse Heaven Hills and Ancient Lakes AVAs. continued to return relatively large eigenvectors for GDD, FFD and PPT. This indicates climatic variables contribute substantially to variance in these appellations.

Conclusion

Examining existing IPNW AVAs reveals some inaccuracies in their final rulings and quantitative detail in their descriptions not found in their original petitions. Table 4 summarizes key differences found in this study between AVA descriptions in the *Federal Register* and in modeled, geospatial data. Topographic and soil spatial data are easily and freely obtained and should be used more regularly in appellation delineation and characterization. Climatic data is the most difficult to obtain and analyze, but this analysis offers some headway. Indication of data sources when describing climatic characteristics of AVAs, the most temporally variable of distinguishing features, is critical for the relevance of their inclusions. The SSURGO database also includes many other datasets that may be useful in appellation characterization.

Principal components analyses suggest most AVAs were more strongly influenced by a combination of elevation, GDD, FDD and, in most cases, precipitation. A smaller group showed more influence by soil parameters. However, the PCA did not necessarily coincide with how the AVAs ranked in these components.

Delineation and petitioning for new AVAs in the IPNW and elsewhere should rely more on GIS and spatial datasets in the future. Accurate representation of physical characteristics of AVAs is important in comparisons and understanding of wine regions of the United States. As indicated by the predominance of the most popular varieties grown in the region, planting decisions are likely largely market driven. Often more insight can be gained from less common varieties that are planted and for what wine styles they are intended. Further explanation of vineyard areas by variety and more detailed reporting of varieties planted would help refine descriptions of AVAs. As the IPNW continues to develop its wine grape industry and more distinct AVAs are established, spatially comprehensive descriptions of distinguishing features will help contribute to the understanding of unifying characteristics of the region as a whole as well as the variability within and the breadth of its potential.

Author Contributions

Conceived and designed the experiments: IHY JRD. Performed the experiments: IHY. Analyzed the data: IHY. Contributed reagents/materials/analysis tools: RAR. Wrote the paper: IHY JRD.

References

1. United States Department of Agriculture-National Agricultural Statistics Service. (2011) Washington vineyard acreage report 2011. http://1.usa.gov/sYOar2. Accessed 17 February 2013.
2. United States Department of Agriculture-National Agricultural Statistics Service. (2011) Grape release. Available: http://1.usa.gov/taiR8H. Accessed 17 February 2013.
3. Shellie KC (2007) Viticultural performance of red and white wine grape cultivars in southwestern Idaho. HortTechnology 17: 595–603.
4. Cross R, Plantinga AJ, Stavins RN (2011) What is the value of terroir? Am Econ Rev 101: 152–156.
5. Tarara J (2005) A climatology refresher for vineyard managers and winemakers. WSU Wine and Grape Research Extension Newsletter 15(3): 2–5. Prosser, WA.
6. Amerine MA, Winkler AJ (1944) Composition and quality of musts and wines of California grapes. Hilgardia 15: 493–675.
7. Jackson DI, Cherry NJ (1988) Prediction of a district's grape-ripening capacity using a latitude-temperature index (LTI). Am J Enol Vitic 39: 19–28.
8. Grove GG (2004) Perennation of Uncinula necator in vineyards of Eastern Washington. Plant Dis 88: 242–247.
9. Delp CJ (1954) Effect of temperature and humidity on the grape powdery mildew fungus. Phytopathology 44: 615–626.
10. Carroll JE, Wilcox WF (2003) Effects of humidity on the development of grapevine powdery mildew. Phytopathology 93: 1137–1144.
11. Bergmeier E, Striegler RK (2011) Preparations for successful vineyard mechanization. WG3001. Univ Missouri Ext, Columbia.
12. Greenspan M (2008) Vineyard design for mechanical harvesting. Wine Business Monthly. 15 Nov. Available at http://bit.ly/SiFYQU. Accessed 17 February 2013.
13. Jones GV, Snead N, Nelson P (2004) Geology and wine 8. Modeling viticultural landscapes: a GIS analysis of terroir potential in the Umpqua Valley of Oregon. Geoscience Can 31: 167–178.
14. Wolf TK, Boyer JD (2003) Vineyard site selection. Virginia Tech Publication Number 463–020.
15. Jackson D, Schuster D (2001) The production of grapes and wine in cool climates. Gypsum Press, Wellington, NZ.
16. Smith L (2002) Site selection for establishment and management of vineyards. In: Whigham PA, editor. Proc 14th Annu Colloquium Spatial Information Res Cent. 3–5 December. Dunedin, New Zealand. Univ. of Otago, pp. 33–45.
17. Busacca AJ, Meinert LD (2003) Wine and geology – the terroir of Washington State. In: Swanson TW, editor. Western Cordillera and Adjacent Areas. Geological Society of America Field Guide 4. Boulder, CO: GSA. pp. 69–85.
18. Gladstones J (1992) Viticulture and environment. Winetitles, Adelaide, Australia.
19. Jackson RS (2008) Wine science principles and applications. 3rd ed. Academic Press, Burlington, MA.
20. Keller M (2010) The science of grapevines: Anatomy and physiology. Elsevier, Inc.
21. Failla O, Mariani L, Brancadoro L, Minelli R, Scienza A, et al. (2004) Spatial distribution of solar radiation and its effects on vine phenology and grape ripening in an alpine environment. Am J Enol Vitic 55: 128–138.
22. Huggett JM (2006) Geology and wine: a review. Proc Geol Assoc 117: 239–247.
23. Tesic D, Woolley DJ, Hewett EW, Martin DJ (2002) Environmental effects on cv Cabernet Sauvignon (Vitis vinifera L.) grown in Hawke's Bay, New Zealand. 1. Phenology and characterisation of viticultural environments. Aust J Grape Wine Res 8: 15–26.
24. Malheiro AC, Santos JA, Fraga H, Pinto JG (2010) Climate change scenarios applied to viticultural zoning in Europe. Clim Res 43: 163–177.
25. Meinert L, Curtin T (2005) Terroir of the Finger Lakes of New York. In: Bettison-Varga L, et al., editors. 18th Keck Symposium. The Colorado College, Colorado Springs. Keck Geology Consortium, pp. 34–40.
26. Bargmann CJ (2003) Geology and wine 7. Geology and wine production in the coastal region, western Cape Province, South Africa. Geoscience Can 30: 161–182.
27. Gesch D, Oimoen M, Greenlee S, Nelson C, Steuck M, et al. (2002) The national elevation dataset. Photogramm Eng Remote Sensing 68: 5–11.
28. Gesch DB (2007) The national elevation dataset. In: Maune D, editor. Digital elevation model technologies and applications: the DEM users manual, 2nd edition: Am Soc Photogramm Remote Sensing. Bethesda, MD. pp. 99–118.
29. Esri (2010) ArcGIS 10 help. Esri, Redlands, CA.
30. Fu P, Rich PM (2002) A geometric solar radiation model with applications in agriculture and forestry. Comput Electron Agric 37: 25–35.
31. Daly C, Halbleib M, Smith JI, Gibson WP, Doggett MK, et al. (2008) Physiographically sensitive mapping of climatological temperature and precipitation across the conterminous United States. Int J Climatology 28: 2031–2064.
32. Demšar U, Harris P, Brunsdon C, Fotheringham AS, McLoone S (2012) Principal component analysis on spatial data: an overview. Ann Assoc Am Geogr 103: 106–128.
33. Norman DK, Busacca AJ, Teissere R (2004) Geology of the Yakima Valley wine country – a geologic field trip guide from Stevenson to Zillah, Washington. Olympia: Washington Department of Natural Resources. 17 p.
34. Bureau of Alcohol, Tobacco Firearmsand Explosives. (1983) Yakima Valley viticultural area, final rule. Federal Register 48, 4 April: 14374.
35. Bureau of Alcohol, Tobacco Firearmsand Explosives. (1984) Establishment of the Walla Walla Valley viticultural area, final rule. Federal Register 49, 6 February: 4374.
36. Bureau of Alcohol, Tobacco Firearmsand Explosives. (2001) Realignment of the boundary of the Walla Walla Valley viticultural area and the eastern boundary of the Columbia Valley viticultural area, final rule. Federal Register 66, 26 February: 11540–11542.
37. Meinert LD, Busacca AJ (2000) Geology and wine 3: terroirs of the Walla Walla Valley appellation, southeastern Washington State, USA. Geoscience Canada 27: 149–171.
38. Bureau of Alcohol, Tobacco Firearmsand Explosives. (1984) Columbia Valley viticultural area, final rule. Federal Register 49, 13 November: 44895.
39. Alcohol and Tobacco Tax and Trade Bureau. (2001) Red Mountain viticultural area, final rule. Federal Register 66(69), 10 April: 18543–18546.
40. Meinert LD, Busacca AJ (2002) Geology and wine 6. Terroir of the Red Mountain appellation, central Washington state, U.S.A. Geoscience Canada 29: 149–168.
41. Alcohol and Tobacco Tax and Trade Bureau. (2004) Columbia Gorge viticultural area, final rule. Federal Register 69(90), 10 May: 25831–25835.
42. Alcohol and Tobacco Tax and Trade Bureau. (2005) Horse Heaven Hills viticultural area, final rule. Federal Register 70(126), 1 July: 38004–38009.
43. Alcohol and Tobacco Tax and Trade Bureau. (2005) Establishment of the Wahluke Slope viticultural area, final rule. Federal Register 70(234), 7 December: 72707–72710.
44. Alcohol and Tobacco Tax and Trade Bureau. (2006) Establishment of the Rattlesnake Hills viticultural area, final rule. Federal Register 71(32), 16 February: 8206–8211.
45. Alcohol and Tobacco Tax and Trade Bureau. (2007) Establishment of the Snake River Valley viticultural area, final rule. Federal Register 72(46), 9 March: 10598–10603.
46. Alcohol and Tobacco Tax and Trade Bureau. (2009) Establishment of the Snipes Mountain viticultural area, final rule. Federal Register 74(12), 21 January: 3422–3425.
47. Alcohol and Tobacco Tax and Trade Bureau. (2009) Establishment of the Lake Chelan viticultural area, final rule. Federal Register 74(81), 29 April: 19409–19416.
48. Alcohol and Tobacco Tax and Trade Bureau. (2011) Establishment of the Naches Heights viticultural area, final rule. Federal Register 76(240), 14 December: 77696–77699.
49. Alcohol and Tobacco Tax and Trade Bureau. (2012) Establishment of the Ancient Lakes of Columbia Valley viticultural area, final rule. Federal Register 77(202), 18 October: 64033–64036.

Current and Potential Tree Locations in Tree Line Ecotone of Changbai Mountains, Northeast China: The Controlling Effects of Topography

Shengwei Zong[1], Zhengfang Wu[1]*, Jiawei Xu[1], Ming Li[2], Xiaofeng Gao[3], Hongshi He[1,4], Haibo Du[1], Lei Wang[1]

1 School of geographical sciences, Northeast Normal University, Changchun, China, **2** College of Urban and Environment Sciences, Shanxi Normal University, Shanxi, China, **3** School of geography, Beijing Normal University, Beijing, China, **4** School of Natural Resources, University of Missouri-Colombia, Columbia, Missouri, United States of America

Abstract

Tree line ecotone in the Changbai Mountains has undergone large changes in the past decades. Tree locations show variations on the four sides of the mountains, especially on the northern and western sides, which has not been fully explained. Previous studies attributed such variations to the variations in temperature. However, in this study, we hypothesized that topographic controls were responsible for causing the variations in the tree locations in tree line ecotone of the Changbai Mountains. To test the hypothesis, we used IKONOS images and WorldView-1 image to identify the tree locations and developed a logistic regression model using topographical variables to identify the dominant controls of the tree locations. The results showed that aspect, wetness, and slope were dominant controls for tree locations on western side of the mountains, whereas altitude, SPI, and aspect were the dominant factors on northern side. The upmost altitude a tree can currently reach was 2140 m asl on the northern side and 2060 m asl on western side. The model predicted results showed that habitats above the current tree line on the both sides were available for trees. Tree recruitments under the current tree line may take advantage of the available habitats at higher elevations based on the current tree location. Our research confirmed the controlling effects of topography on the tree locations in the tree line ecotone of Changbai Mountains and suggested that it was essential to assess the tree response to topography in the research of tree line ecotone.

Editor: Eryuan Liang, Chinese Academy of Sciences, China

Funding: The study was supported by the Doctorial innovation fund of Northeast Normal University (No. 10SSXT133, http://www.nenu.edu.cn/), National Natural Science Foundation of China (No. 41171038, 41171072 and 41101523, http://www.nsfc.gov.cn/), Doctoral Fund of Ministry of Education of China (No. 20120043110014, http://www.moe.gov.cn/), and the Special Fund of National Seismological Bureau, China (No. 201208005, http://www.cea.gov.cn/). The funders had no role in study design, data collection and analysis, decision to publish, or preparation of the manuscript.

Competing Interests: The authors have declared that no competing interests exist.

* Email: wuzf@nenu.edu.cn

Introduction

Tree line is considered as an important indicator of terrestrial ecosystems to climate change because it is sensitive to environmental change. It has been proved that tree line species have advanced in the past decades, regardless they are in arctic or in the alpine regions [1,2]. Numerous studies show that the temperature is responsible for causing the upward shift of trees as average temperatures have raised globally over the past century [3,4,5]. Increasing temperature may promote the growth of plant species at the tree line. However, the responses of tree species around tree line to the climate warming differ in different regions. Some researches show that tree line species are particularly responsive to changes in summer temperature [6], whereas others suggest that winter temperature may have important effects [7,8]. Variations in average and seasonal temperature changes may influence the establishment of tree recruitments [9]. However, air temperature may not be the only dominant factor controlling tree growth at tree lines. According to Liang et al. (2011), Smith fir (*Abies georgei* var. *smithii*) does not show a significant upward movement, despite warming of the Tibetan Plateau. The warming in the past 200 year is already having a significant impact on the population density of the trees, but not on the position of the Smith fir tree line [10]. Furthermore, soil temperature has greater influences than air temperature on tree growth at tree line position [11]. The effects of temperature on the tree locations around tree line cannot be ignored. However, in mountainous regions, topography may control local environments including microclimates, soil properties, and disturbances [12] and exerts a strong influence on vegetation distribution and succession in tree line ecotone [13,14]. On one hand, topographic shelters may be important controls in determining where pioneer tree species can establish and persist in

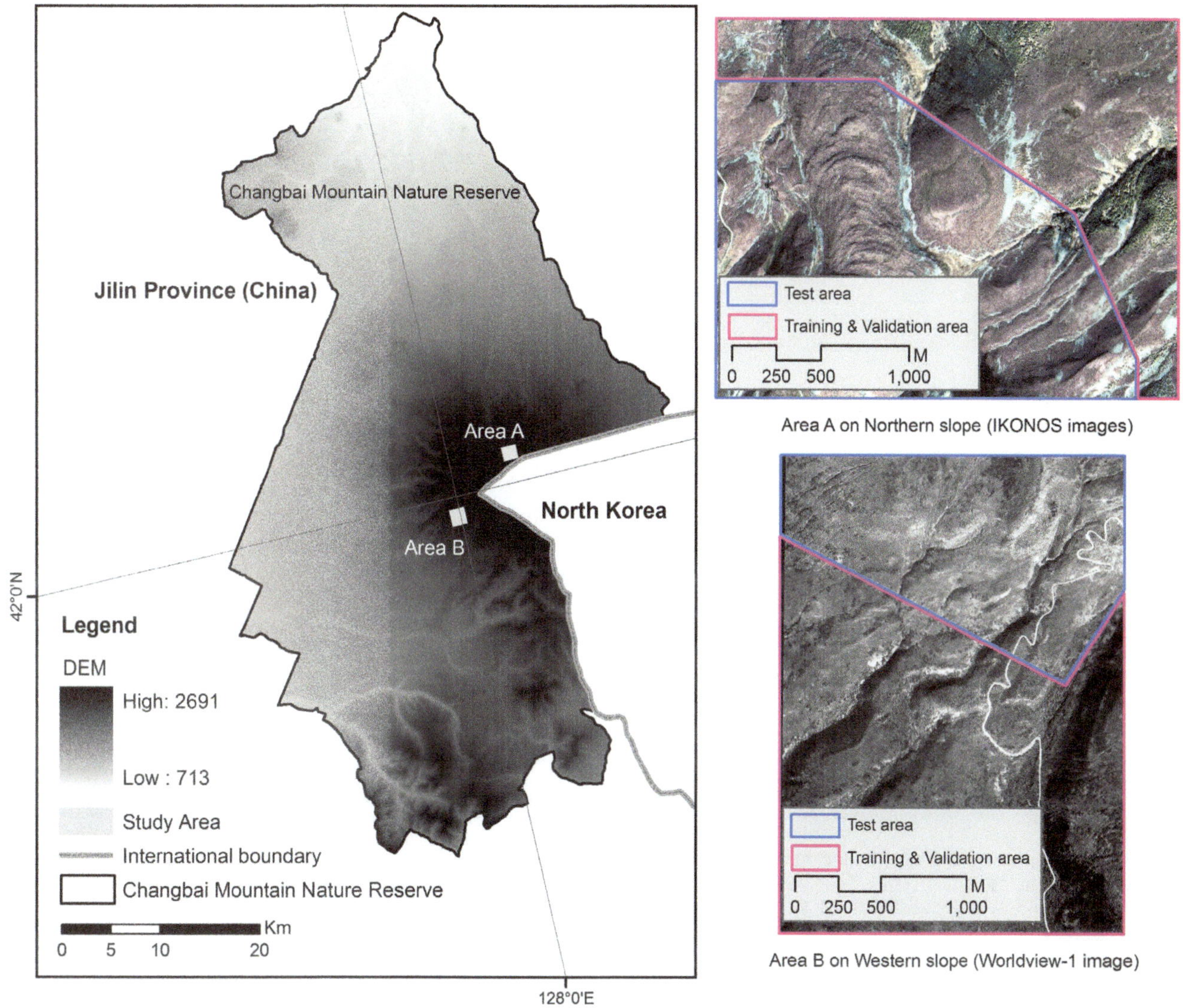

Figure 1. Location of the study areas in Changbai Mountains Nature Reserve in Jilin province, Northeast China. DEM (Digital elevation model) value ranges from 713 to 2681. Area A shown by the IKONOS image (False color image created by combining the blue band, green band, and red band) is on the northern slope. Area B shown by the WorldView-1 image is on the western slope. Red area represented the training and validation area. Blue area represented the test area which is used for the prediction of tree locations.

the initial stages from upslope tree migration to alpine tundra [15]. On the other hand, the effect of varying topography on site conditions combined with natural disturbances may override the effects of temperature [16]. Therefore, it is essential to assess the response of tree species to topography in tree line ecotone.

Tree line ecotone in Changbai Mountains has undergone large changes over the past decades, in particular, on the northern and western sides of the mountains. Previous researches have mainly focused on tree line changes on the northern side. Most of these studies investigated the effects of climate warming on trees and attributed the main effects to the temperature change [17]. According to Wang and Liu (2011), trees cannot reach its potential altitude owing to the asynchronous changes in heat and water at high elevations, which also confirms the effects of climate change [18]. Physiological drought in *Betula. ermanii* during its growing season, caused by variation in water and heat along the altitudinal gradient, restricted the distribution of the *B. ermanii* to higher elevations in the Changbai Mountains [19]. Apart from climate change, some experts have also expressed concerns about the effects of other factors including the shape of forest boundaries (concave and convex boundary) and soil properties [20,21]. However, limited researches have been conducted on the topography of the tree line ecotone in Changbai Mountains. It was notable that the important climatic factors for forests at the tree line ecotone are temperature and solar radiation, which are exactly influenced by the topography [1]. Therefore, the examination of tree locations in tree line ecotone by topographic variables could help explain the variations in tree line ecotone of Changbai Mountains.

Vegetation on the western side of the mountains has also experienced large changes during the past decades. A large windthrow occurred in the tree line ecotone in 1986, which almost destroyed the *B. ermanii* forests [22]. In addition, Herbaceous species from *B. ermanii* forest, represented by *Deyeuxia angustifolia*, have invaded the alpine tundra and led to severe impacts on the alpine tundra ecosystem [23]. Invasion of *D.*

Figure 2. Aerial photograph of the volcanic cone in the Changbai Mountains. The photo was taken from the northern side on October 2010. Region A is on the northern slope. It can be seen that trees in the northern side gradually move upward to the high elevations. Region B is on the western side. It can be seen that tree line position in western side is relatively stable.

angustifolia represented the invasion of the Mountain birch forest [23]. However, it is still not clear whether trees on the western side would move up to high elevations. Therefore, there is an urgent need to study the current state of tree locations on the western side to understand its future changes. The results could help to facilitate the interpretation of changes in alpine tree line ecotone.

In this study, we hypothesize: 1) topographic differences are responsible for the differences in tree locations on northern and western sides of the mountains because the two sides have distinct topographic characteristics; and 2) habitats above the current tree line are available for tree species to establish. Our specific objectives are to 1) interpret the current tree locations; 2) evaluate the controlling effects of topography on the tree locations; and 3) predict the potential habitat of trees locations in tree line ecotone on the northern and western side of the Changbai Mountains.

Study area

The study area (41°58′–42°02′N, 127°58′–128°05′E) is located in the Changbai Mountains National Nature Reserve (CMNNR) in Jilin province, Northeast China (Fig. 1). The CMNNR has vertical forest zones along the altitudinal gradient, in particular, on the northern side of the mountains. The tree line surrounds the volcanic cone at different altitudes on the four sides. The main tree species in tree line ecotone include *B. ermanii*, *Larix olgensis*, and *Alnus tinctoria*. On the northern side, *B. ermanii* has invaded into the alpine tundra zone, which led to the expansion of the tree line over the last few decades [17]. In contrast, there has been no expansion of the trees on western side (Fig. 2). The tree line position on the western side has been steady over the past decades. The study area is not affected by anthropogenic activity. Therefore, it is an ideal area for studying the tree locations, in particular, in the temperate zone of Northeast China. We defined the tree line ecotone as an area of transition spanning from close canopy forests to open area where trees sparsely distributed. We used tree line ecotone to simply refer the area where we made predictions of current and potential tree locations. The definition of tree line ecotone itself did not influence in any shape or form on our results.

The climate of the study area is characterized low temperature, heavy precipitation and a short growing season. Annual mean temperatures in the growing season (June to September) range from 3.37 to 8.82°C (mean temperature is 5.87°C). Annual average precipitation ranges from 700 to 1,400 mm [24]. The mean annual radiation is 506.6 $J.cm^{-2}.a^{-1}$. The mean annual sunlight duration is 2295 h. Harsh environment is harmful to plant organism that limits the photosynthesis capability and vegetative growth [25]. Therefore, the growth rate of alpine trees is very slow. The thin soil (only about 10–25 cm depth) in study area is the alpine tundra meadow soil.

The two areas chosen for this study (Fig. 1) are typical and representative areas on the two sides of the mountains, which is conducive for the comparative analysis of the differences in tree

Table 1. Topographical indices used in this study.

Variable	Description	Calculation	Ecological meaning
Altitude	Altitude above sea level (m)		Temperature, moisture
Slope	Slope angle (degrees)		Solar radiation, soil erosion, moisture
Aspect	Aspect angle (0° to 360°)		Temperature, wind
Eastness	Aspect east-west (1 to −1)	sin(aspect)	Morning/afternoon solar radiation
Northness	Aspect north-south (1 to −1)	cos(aspect)	Summer/winter solar radiation
Plan curvature	Curvature perpendicular to slope	−1 = concave, 1 = convex	Solar radiation, wind, moisture
Profile curvature	Curvature in slope direction	−1 = convex, 1 = concave	Moisture
Hillshade	Hillshade surface	hillshade with sun elevation angle of 45° and sun azimuth angle of 315°	Solar radiation
PRR	Potential relative radiation	sums daily values over the growing season [31]	Solar radiation potential
SPI	Snow potential index	Snow accumulation in topography [12]	Temperature, moisture
Surface roughness	Roughness of regional area	1/cos(slope×π/180) [46]	wind, moisture
Surface undulation	Undulation of regional area	Max(regional altitude)− Min(regional altitude) [46]	wind, moisture
Wetness	Topographic wetness index	$\ln(A_s/\tan\beta)$ [32]	moisture
Windthrow index	Windthrown area by typhoon	0 = non-windthrow, 1 = windthrow [45]	Typhoon disturbance
LST	Land surface temperature	Derived from Landsat TM/ETM+ images [33]	Temperature

locations between the northern and western side. The area on the northern side covers approximately 5.36 km^2 (Fig. 1). The western area covers approximately 5.33 km^2. The red area in Figure 1 represents the training and validation area with 20% of this area used specifically for training and the left used as the validation area. Detailed information can be found in the description of the sampling in the Materials and Methods section. The blue area represents the test area, which is used for the prediction of the potential habitat of trees.

Materials and Methods

The field work assisted by the Scientific Research Academy of the Changbai Mountains (http://cbs.jl.gov.cn/ResearchWeb/main.aspx) was permitted by Administration Commission of Changbai Mountains Nature Reserve (http://www.cbs.jl.gov.cn/weball/main.aspx#). The field studies conducted on the west side (127°58′56.827″E–128°0′22.129″E, 41°58′10.543″E–41°59′36.287″) and north side (128°4′2.67″E–128°5′54.832″E, 42°1′15.245″E–42°2′21.036″E) of the mountains did not involve endangered or protected species.

Data

Studies of individual tree location require high spatial resolution satellite images. Tree lines in Changbai Mountains have a diffused form, which indicates that many tree locations would not be detected if low resolution images were used. Therefore we use the high-resolution images to get the accurate tree location data.

The satellite image data used in this study include IKONOS data for the northern side and WorldView-1 for the western side. The IKONOS data consist of one panchromatic image at 1-m resolution and four multispectral images at 4-m resolution, which are acquired on 2002-09-20. The panchromatic image covers a spectral range from the visible red to the near-infrared region (450–900 nm). The multispectral images covers four bands including the visible blue (445–516 nm), green (506–595 nm), red (632–698 nm), and near-infrared (770–888 nm) bands, respectively. The WorldView-1 data consists of one panchromatic image at 0.5-m resolution with a spectral range of 400–900 nm, which is acquired on 2009-10-09. We consider that the differences in acquired time of the two image data would not affect the comparative results of the two sides because there was little change of tree locations during the period of 2002–2009. The two image data are preprocessed using standardized procedures of radiometrically correction, sensor correction, and geometrically correction.

The DEM (Digital elevation model) data are derived from the 1: 50 000 topographic maps (Data source: National Administration of Surveying, Mapping and Geoinformation (http://chj.jl.gov.cn/article/hdjl/lyzx/?siteid=jilin). The cellsize is confined to 5 m. The DEM error is defined as the elevation difference between the constructed DEM and the corresponding true surface [26]. To eliminate the DEM error, we collected 205 GPS positions as ground truth points. We used the Real-Time Kinematic (RTK) GPS equipment, which can deliver almost instantaneous point coordinates with centimetre-scale accuracy [27]. A reference station of RTK GPS was set up at the Tianchi weather station (128°04′02″E, 42°01′40″N) in the Changbai Mountains with known coordinates. Then a receiver was used to record the ground truth points. After the field survey, the RTK GPS data were imported into ArcGIS for DEM quality assessment and correction. Based on this correction, the height error was <0.25 m. However, the DEM still has some negative values and contained terraces. We therefore used ArcGIS functions to remove these values and fill the sinks [28]. Finally, a high quality DEM was prepared to derive the topographic variables (Table 1). Approximately 14 topographic variables (14 for the northern side and 15 for the

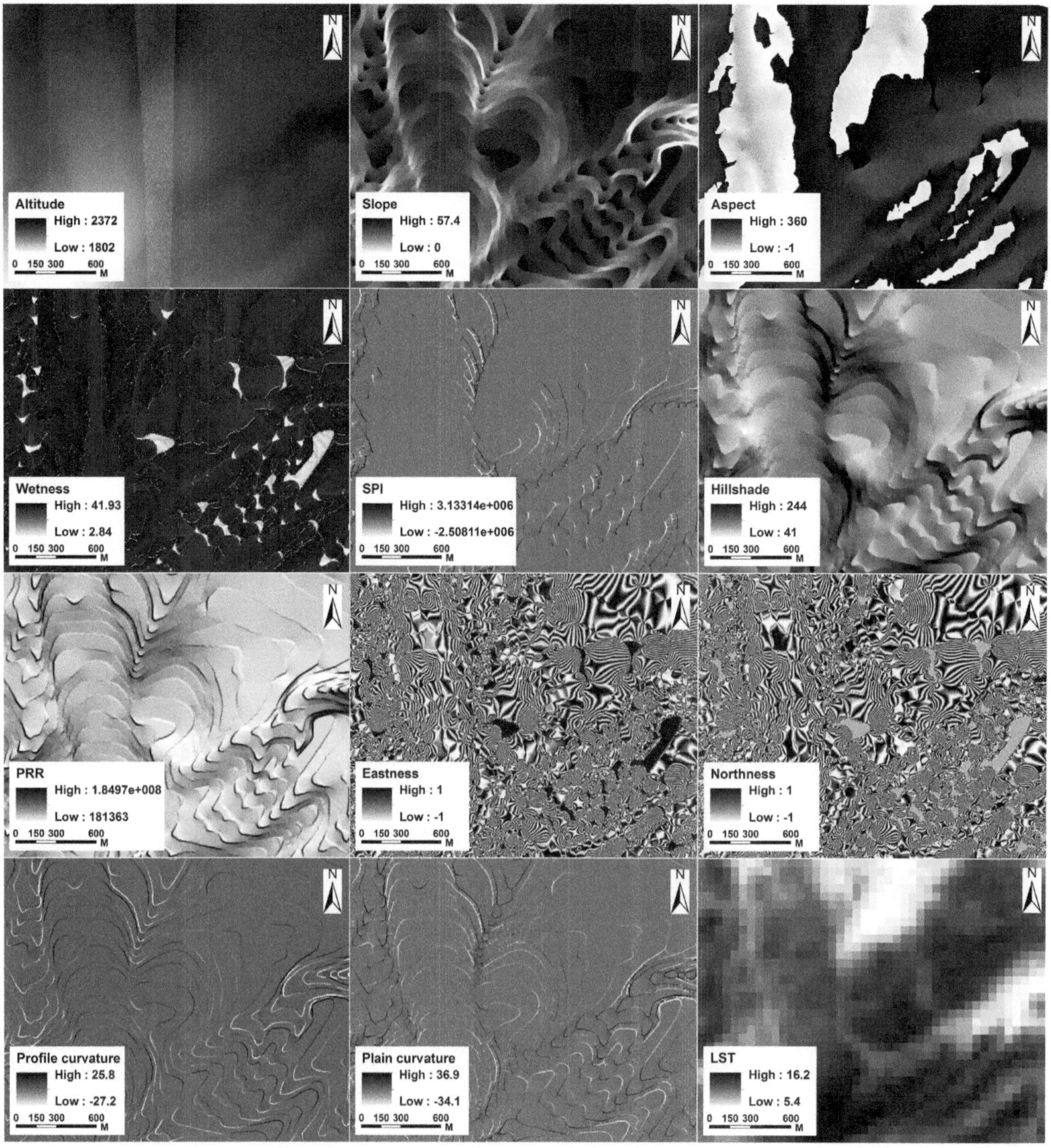

Figure 3. Topographic variables of the northern side derived from the digital elevation model. PRR represents potential relative radiation which sums daily values over the growing season. SPI represents snow potential index which indicates the snow accumulation in topography. LST represents the land surface temperature.

western side) were selected based on our hypothesis that the tree locations may be determined by a variety of topographic factors. These topographic indices, such as the Wetness Index, have significant ecological implications for explaining certain ecological processes. The topographic variables (Figs. 3 and 4) used in this study have been previously applied in several studies [12,29,30]. All the indices were derived from the DEM data. The wetness index is based on the assumption that topography controls the movement of water in terrains with slopes, and thus, it regulates the spatial pattern of soil moisture. High values of the wetness index are found in converging, flat terrain. Low values are typically found in steep, diverging areas. PRR (Potential relative radiation) sums hourly estimates of clear-sky radiation over a day and then sums daily values over the growing season. Each point estimate accounts for topographic shading by surrounding landscape features. Hillshade index reflects the sunlight conditions

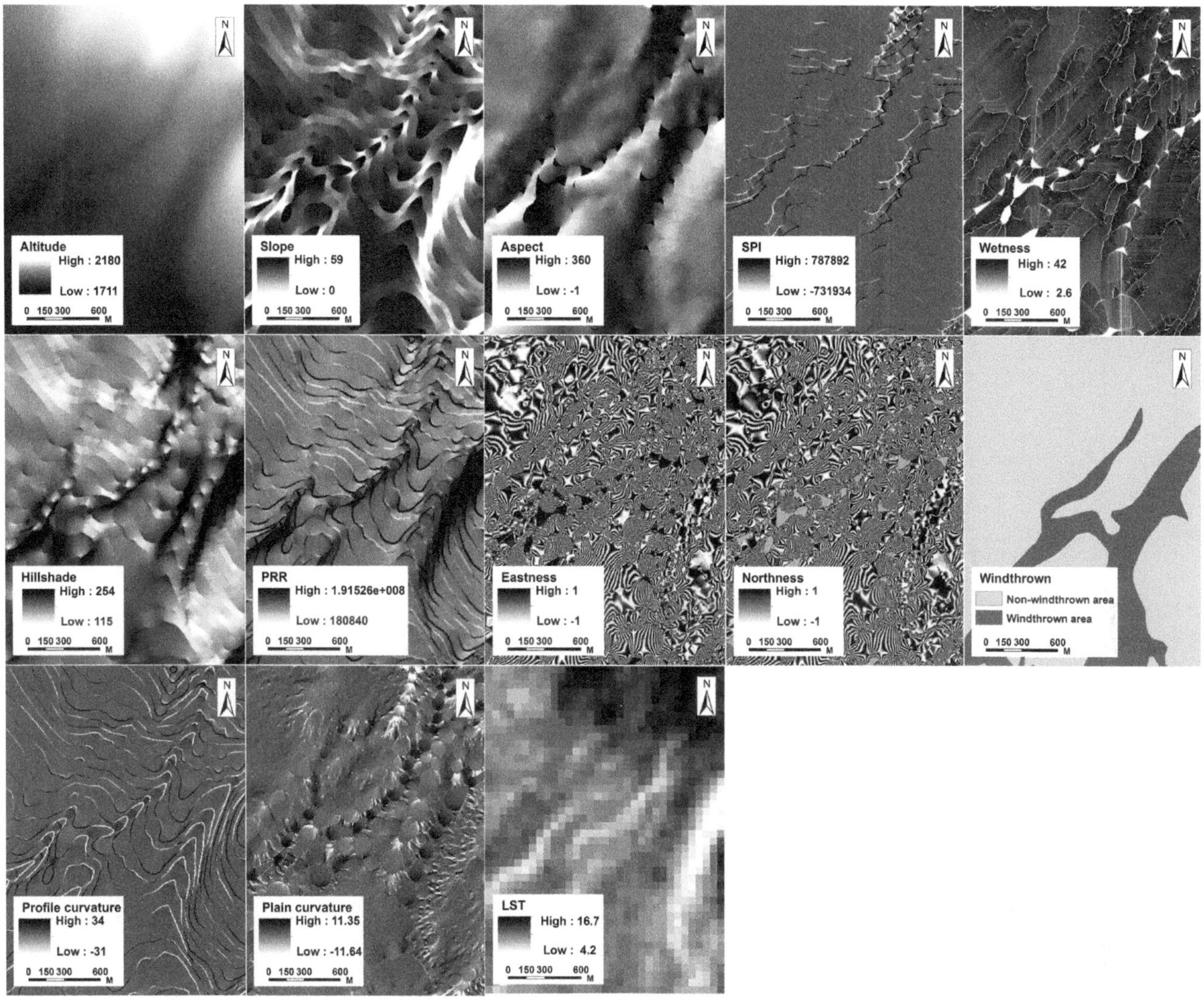

Figure 4. Topographic variables of the western side derived from the digital elevation model. PRR represents potential relative radiation which sums daily values over the growing season. SPI represents snow potential index which indicates the snow accumulation in topography. LST represents the land surface temperature.

in the mountainous area, which is created from DEM raster by considering the illumination angle and shadows. High value represents the shadow areas whereas low value means the bright area. The SPI (snow potential index) represents the potential for snow accumulation at specific sites. Higher values of the index were obtained on slopes that were leeward (i.e. closer to northeast exposure), concave, and at higher elevations. In addition, it was evident that vegetation on the western side had experienced a severe typhoon disturbance in 1986, which resulted in wide-spread windthrown areas. Therefore, a raster of the windthrown areas was produced as the natural disturbance index (Fig. 4) for the western side. Land Surface Temperature (LST) data which represents the temperature condition were derived from Landsat ETM+ data using the method from Li et al (2004) [31]. Corresponding to the acquire time of the IKONOS (September 20, 2002) and Worldview-1 (October 9, 2009) data, we collected two scenes of cloud and snow free Landsat ETM+ images on August 25, 2002 and September 29, 2009. We assumed that there was little change of temperature distribution pattern along altitudinal gradient on the alpine tundra at the end of growing season (late August to early October). Therefore, the LST data could be used as a variable concerning temperature in the process of constructing the model. All the Landsat ETM+ data were preprocessed using standardized procedures of radiometrically correction, sensor correction, and geometrically correction.

Data preparation

For IKONOS images, a high quality one-meter resolution pan-sharpened image was first produced by Gram-Schmidt spectral sharpening. Then, a false color composite image was created by combining the blue band, green band, and red band. Afterwards, tree locations were identified manually with interactive interpretation procedures, supplemented by different image transformations, such as vegetation indices which can minimize shadow and atmospheric condition effects [32]. The tree locations of the WorldView-1 image were identified using unsupervised classification with the ISODATA (Iterative Self-Organizing Data Analysis Technique) method. The results were manually modified with ground truth data. The ground truth data were divided into two parts. One section was used to generate the ROI (region of

Table 2. Correlation matrix of Topographic indices of western northern side.

	western side														
	Surface undulation	**Profile curvature**	**Plan curvature**	**Windthrow index**	**Northness**	**Wetness**	**Eastness**	**Altitude**	**SPI**	**PRR**	**Surface roughness**	**Slope**	**Aspect**	**Hillshade**	**LST**
Surface undulation	1	0.0043	0.0332	0.0668	0.0306	0.3104	0.0598	0.3468	0.0247	0.3707	**0.8774**	**0.9240**	0.3511	0.3656	−0.1607
Profile curvature		1	0.3876	0.0492	0.0126	0.0165	0.0058	0.0346	0.5442	0.2021	0.0007	0.0027	0.0062	0.0163	−0.0658
Plan curvature			1	0.1275	0.0059	0.0694	0.0033	0.1167	0.3930	0.0045	0.0407	0.0254	0.0086	0.0242	−0.0048
Windthrow index				1	0.0017	0.1101	0.0084	0.2936	0.0396	0.1875	0.0712	0.0296	0.0679	0.2557	−0.1115
Northness					1	0.1116	0.0228	0.0147	0.0070	0.0122	0.0315	0.0472	0.0726	0.0082	0.1414
Wetness						1	0.1706	0.1695	0.0247	0.0227	0.2197	0.3912	0.5477	0.0881	0.2397
Eastness							1	0.0187	0.0091	0.0037	0.0483	0.0787	0.1300	0.0260	0.2224
Altitude								1	0.0286	0.0749	0.2339	0.3151	0.0856	0.2057	0.4870
SPI									1	0.1079	0.0112	0.0106	0.0046	0.0073	0.2516
PRR										1	0.4302	0.3197	0.3249	0.5123	0.1714
Surface roughness											1	**0.9179**	0.3246	0.5136	−0.1299
Slope												1	0.3766	0.3050	−0.1020
Aspect													1	0.0317	−0.4717
Hillshade														1	0.1737
LST															1
	northern side														
	Surface undulation	**Aspect**	**Slope**	**Surface roughness**	**PRR**	**SPI**	**Eastness**	**Hillshade**	**Northness**	**Wetness**	**Altitude**	**Plan curvature**	**Profile curvature**	**LST**	
Surface undulation	1	0.2757	**0.8889**	**0.8627**	0.3019	0.0100	0.0276	0.3470	0.0210	0.2995	0.2660	0.0015	0.0080	−0.0131	
Aspect		1	0.3086	0.2419	0.2015	0.0009	0.0310	0.0264	0.0095	0.1850	0.0332	0.0026	0.0045	−0.3451	
Slope			1	**0.8483**	0.3441	0.0027	0.0503	0.5696	0.0328	0.4164	0.2241	0.0090	0.0043	-	
Surface roughness				1	0.2669	0.0031	0.0202	0.3282	0.0128	0.2343	0.1763	0.0116	0.0027	−0.0158	
PRR					1	0.0708	0.0151	0.4923	0.0009	0.1222	0.0430	0.0015	0.0581	0.4012	
SPI						1	0.0053	0.0101	0.0007	0.0450	0.0459	0.4962	0.6272	−0.2430	
Eastness							1	0.0333	0.0163	0.1581	0.0067	0.0004	0.0017	0.0221	
Hillshade								1	0.0191	0.2863	0.0966	0.0088	0.0091	0.3524	
Northness									1	0.1009	0.0126	0.0032	0.0000	0.0028	
Wetness										1	0.1253	0.0771	0.0279	0.0201	
Altitude											1	0.0547	0.0513	−0.1574	
Plan curvature												1	0.4293	−0.0256	
Profile curvature													1	−0.0205	
LST														1	

PRR represents potential relative radiation which sums daily values over the growing season. SPI represents snow potential index which indicates the snow accumulation in topography. LST represents the land surface temperature.

Table 3. Classification accuracy of tree locations of IKONOS and WorldView-1 images.

		Tree class		Non-tree class		Overall accuracy (%)
Data type	Kappa statistics	User accuracy (%)	Producer accuracy (%)	User accuracy (%)	Producer accuracy (%)	
IKONOS	0.9275	97.56	95.86	98.02	97.78	96.5933
Worldview-1	0.9649	99.77	98.66	99.46	98.45	99.4425

interest) data. The other was used to examine the accuracy of the classification result. The tree species examined in the study area included *B. ermanii* and *L. olgensis*. On the northern side, there were few *A. tinctoria* distributed in the tree line ecotone. The final classification map was a binary (0 = tree absence; 1 = tree presence). GCPs (Ground control points) were important for accuracy assessment of classification map, which were also obtained using GPS during August 2012. We obtained 183 GCPs of tree sites on the northern side, and 273 GCPs on the western side. Collection of GCPs for assessing the accuracy of the image classification and DEM was conducted simultaneously. The classification map of the IKONOS image and the WorldView-1 image were resampled at a resolution of 5 m using bilinear interpolation to match the resolution of the DEM and the derived variables. All image data were converted from disparate sources to a common format defined in Arc/Info grid with Gauss-Kruger projection and Krasovsky 1940 spheroid.

Sampling method

Species distributional data often display spatial autocorrelation [33]. Consequently, in the modeling process, the spatial autocorrelation may bias parameter estimates. To avoid the autocorrelation effect, a useful solution is to increase the sample distance and decrease the correlation between points [12]. We used the "Sample" tool of the ArcGIS toolbox to obtain the sample data used for model construction. Sample distance ranged from 1 to 20 pixels. Then, the sample data was imported into SPSS (SPSS Inc., Chicago, IL, USA) to check the spatial autocorrelation using the method of "Time Series Analysis". We found that the autocorrelation coefficient was below 0.530 after the sample distance was increased to 4 pixels apart. Therefore, the optimal sample distance was determined as 4 pixels.

Logistic regression model

In this study, logistic regression model was applied to predict the tree/non-tree locations based on the topographic variables:

$$P_{(Tree)} = \frac{e^{(a+b_1x_1+b_2x_2+\ldots+b_nx_n)}}{1+e^{(a+b_1x_1+b_2x_2+\ldots+b_nx_n)}} \quad (1)$$

In the equation, a was the constant; $b_{1\text{-}n}$ were the coefficients of the variables ($x_{1\text{-}n}$). In this study, the included variables were shown in Table 1. P was a probability value from 0 to 1.0, which was the probability value of the tree locations.

Before importing the topographic variables into the logistic regression procedure, the multicollinearity of all the explanatory variables which would bias the model estimate was checked. The linear correlations of the variables were tested among the variables. The correlation matrix (Table 2) among the variables showed that, on the western side, the surface undulation was highly correlated with the surface roughness (0.8774) and slope (0.9240), and that the surface roughness was highly correlated with slope (0.9179). On the northern side, the surface undulation was highly correlated with the surface roughness (0.8627) and slope (0.8889), and the surface roughness was highly correlated with slope (0.8483). Therefore, surface roughness and surface undulation were excluded. For the regression, 'binary logistic regression' was used with the forward stepwise algorithm based on a maximum-likelihood ratio test. We ran the model with all the cut-off points between 0 and 1.0 with intervals of 0.1, and obtained correct classification rates for presences, absences and both. The cut-off value of P was finally set to 0.04 for the western side and 0.06 for the northern side. Values above this threshold were considered as tree presences, while values below the threshold were considered as tree absences.

Results

Tree locations in the tree line ecotone

The overall accuracies of the tree locations classification with the IKONOS and WorldView-1 images were 96.59% and 99.44%, respectively (Table 3). The identification ability of non-tree location in the IKONOS image was better than that of tree location, whereas in the WorldView-1 image, the identification ability of the tree location was better than that of the non-tree location. This may be because trees at the upper boundary on the western side are easy to identify because of their large-diameter size. By contrast, small trees at the upper boundary of the tree line ecotone cannot be easily identified on the northern side.

On the northern side, there is a significant upward shift trend of Mountain birch. Trees with a height ≤0.5 m at the upper boundary cannot be detected. On the western side, there are no trees showing an upward shifting trend. Compared with field observation data, trees with a height ≤0.3 m at the upper boundary cannot be detected. However, this limitation would not influence the classification results because the number of the short trees is fairly little. The upmost altitude a tree can currently reach is 2140 m asl on the northern side and 2060 m asl on western side. The upward shift of *B. ermanii* appears to promote the tree locations to high elevations on the northern side.

Logistic regression model

The variables in the logistic regression model for the northern side of the mountains were altitude, aspect, wetness, PRR (Potential relative radiation), hillshade, SPI (Snow potential index), and slope (Table 4). Among them, altitude, SPI, and aspect contributed significantly to the regression model, which indicates that they are driving factors in determining tree locations on the northern side. The excluded variables (LST, eastness, northness, plain curvature, and profile curvature) suggested that the stand orientation and soil erosion may not influence the tree locations. It

Table 4. Logistic regression models including the model parameters of the northern and the western sides.

	Northern side					Western side			
	Coefficient	**S.E.**	**Wald**	**Sig.**		**Coefficient**	**S.E.**	**Wald**	**Sig.**
Altitude	0.002587	0.000164	**468.651**	.000	**Aspect**	0.003056	0.000495	266.497	.000
SPI	0.002798	0.000272	**204.687**	.000	**Wetness**	0.040304	0.000874	155.904	.000
Aspect	0.05056	0.000025	**169.335**	.000	**Slope**	0.011182	0.000957	79.585	.000
Slope	0.003115	0.000302	55.742	.000	Altitude	−0.002898	0.001062	42.879	.000
Hillshade	0.002969	0.000000	20.684	.001	PRR	0.001758	0.000001	20.441	.000
PRR	0.000464	0.001705	10.050	.002	LST	−0.002466	0.000258	16.967	.000
Wetness	0.000121	0.000002	6.669	.036	Plain curvature	0.050949	0.027044	6.305	.005
Constant	16.842537	0.333254	56.852	.001	Constant	3.000672	0.854591	1.393	.118
Classification accuracy		Observed	Predicted	Correctly predicted	Classification accuracy		Observed	Predicted	Correctly predicted
	Non-tree	11088	9782	85.1%		Non-tree	23521	20986	89.2%
	Tree	4492	4218	89.7%		Tree	4364	3704	84.9%
	Overall accuracy			83.6%		Overall accuracy			87.1%

Classification accuracy was shown with the numbers of pixels that were observed and predicted in the two classes of tree and non-tree.
The variables that have considerable effects on the models are marked in bold. Significance of all variables: $P < 0.001$.
PRR represents potential relative radiation which sums daily values over the growing season.
SPI represents snow potential index which indicates the snow accumulation in topography. LST represents land surface temperature.

Table 5. Classification tables showing the predictive accuracies of the complete model.

Area		Northern side			Area		Western side		
		Observed	Predicted	Correctly predicted			Observed	Predicted	Correctly predicted
Validation	Non-tree	44349	39201	85%	Validation	Non-tree	94081	72564	74%
	Tree	17967	15334	79%		Tree	17455	13015	72%
	Overall accuracy			82%		Overall accuracy			73%
Test	Non-tree	134893	113057	81%	Test	Non-tree	70305	55679	76%
	Tree	1863	1165	69%		Tree	1612	1326	65%
	Overall accuracy			75%		Overall accuracy			71%

Shown are the numbers of pixels that were observed and predicted in the two classes of tree and non-tree.

was notable that LST was excluded which may because LST has negative relationship with most of the important variables as shown in Table 2. The overall predictive accuracy of the model was 83.6% with tree predictive accuracy of 89.7% and non-tree predictive accuracy of 85.1%. For the validation area, the overall predictive accuracy was 82% with tree predictive accuracy of 79% and non-tree predictive accuracy of 85% (Table 5). The percentage of correctly predicted tree pixels was relatively low in the validation area compared with that in the training area.

The variables in the logistic regression model for the western side were aspect, wetness, slope, altitude, PRR, LST, and plain curvature (Table 4). Aspect, wetness, and slope contributed significantly to the model, which indicates that they are important factors in determining tree locations on the western side. The excluded variables (eastness, profile curvature, hillshade, SPI, northness, and windthrow index) suggest that the stand orientation, soil erosion, snow cover, and wind disturbance may not influence the tree locations. Although the current tree locations are greatly affected by wind disturbance, the windthrow effect is considered only as a temporary disturbance which would not influence the tree locations. The overall predictive accuracy of the model was 87.1% with tree predictive accuracy of 84.9% and non-tree predictive accuracy of 89.2% (Table 4). In the validation area, the overall predictive accuracy was 73% with tree predictive accuracy of 72% and non-tree predictive accuracy of 74% (Table 5). Because of the wind disturbance, the tree locations scatter within the tree line ecotone, which results in low predictive accuracy in the validation area.

Prediction of potential habitats above the tree line

We applied the logistic regression model to predict the potential habitats at high elevations. In the test area on the northern side, the overall predictive accuracy was 75% with tree predictive accuracy of 69% and non-tree predictive accuracy of 81% (Table 5). The model predictive accuracy was subjected to the scattered distribution of trees. The invasion of *B. ermanii* is widely accepted on the northern side. Our results indicate that there are habitats available for tree recruitments above the current tree line. Trees may potentially move to higher elevations as they have done over the past years on the northern side (Fig. 5). The upmost altitude a tree may reach is approximately 2160 m asl based on the current tree distribution features.

In the test area on the western side, the overall predictive accuracy was 71% with tree predictive accuracy of 65% and non-tree predictive accuracy of 76% (Table 5). The scattered tree location greatly affected the predictive accuracy. The predicted tree location appeared to cluster in patches (Fig. 6). There are habitats available for tree recruitments above the current tree line, which indicate that trees may potentially move to higher elevations. However, the upward shifting trend of trees was not obvious. The upmost altitude a tree may reach is slightly higher than 2060 m asl, which is still lower than that of northern side.

Discussion

Comparison of tree locations between the northern and western sides

There are great differences in the characteristics of tree locations between the northern and western sides of the mountains. The current tree locations on the western side are subject to the windthrow effect. However, our model results show that windthrow is not the decisive factor. There is no evidence that windthrow affected tree line advance based on our field observation. Aspect, wetness, and slope had a significant influence

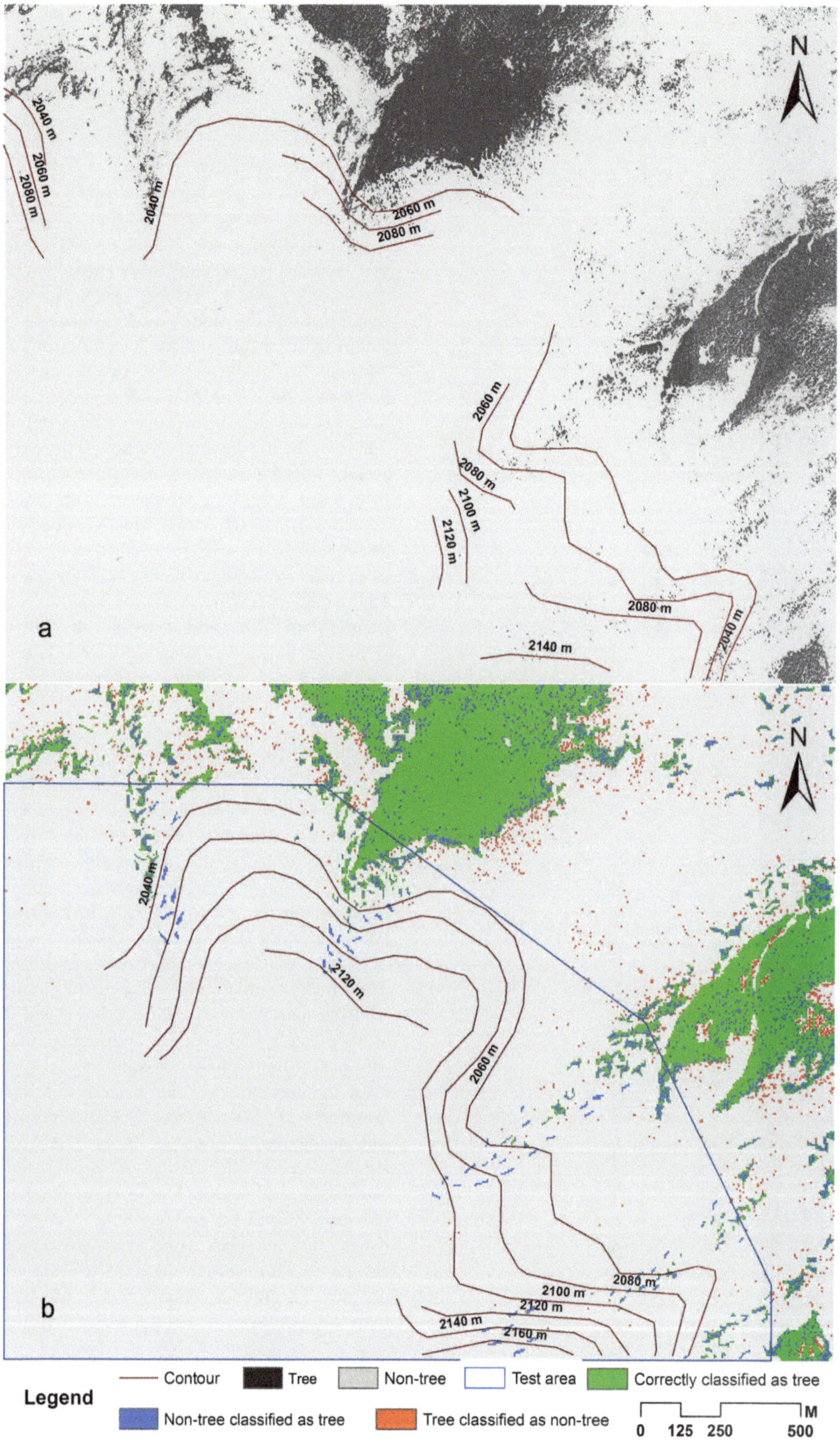

Figure 5. Classification and prediction maps of tree locations in the northern side. (a) Classification result of IKONOS images in the northern side. Tree classification was shown in black. Non-tree classification was shown in grey. (b) Predicted tree locations in the training area, validation area, and the test area. Green indicates correctly predicted trees. Blue indicates areas where tree is predicted where in fact tree was not present; red indicates areas where no tree is predicted where in fact tree was present.

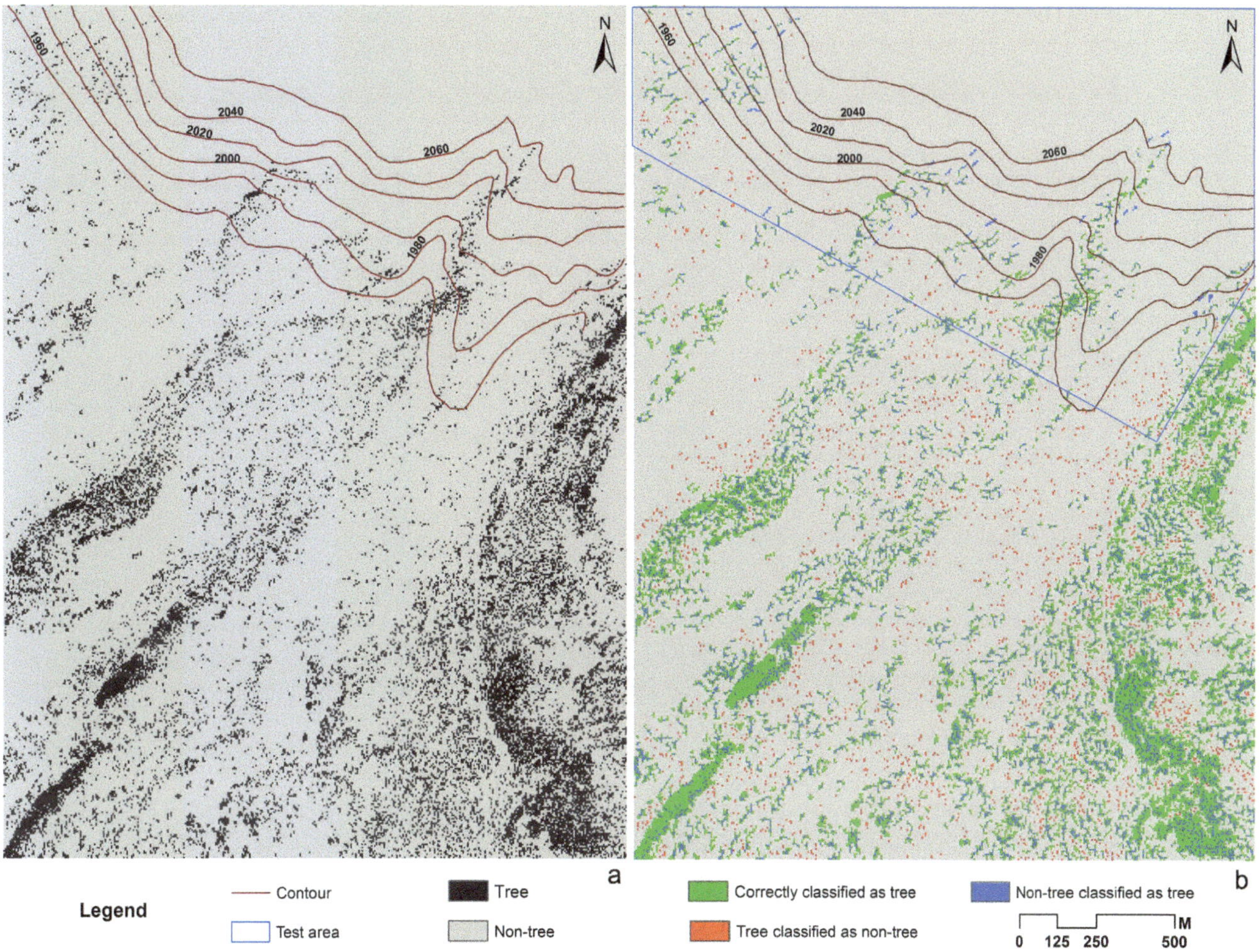

Figure 6. Classification and prediction maps of tree locations in the western side. (a) Classification result of WorldView-1 image in the western side. Tree classification was shown in black. Non-tree classification was shown in grey. (b) Predicted tree locations in the training area, validation area, and the test area. Green indicates correctly predicted trees. Blue indicates areas where tree is predicted where in fact tree was not present; red indicates areas where no tree is predicted where in fact tree was present.

on the tree locations. Conversely, altitude, SPI, and aspect are the dominant factors on the northern side. Environmental conditions between the northern and western sides differ mainly in the wind and precipitation regimes. Unlike the northern side, the western side is on the windward side and westerly winds prevail during the growing season [34]. Differences in the precipitation regimes may have caused differences in air and soil moisture, which contributed to the differences in tree distribution patterns. On the western side, wetness in topography affects the tree locations, which confirms our hypothesis that topographic differences are responsible for the differences in tree locations on northern and western sides of the mountains. In previous researches, scientists has attributed the variations of tree locations between the two sides to climate change and volcano activity [35,36]. Shi and Li (2000) pointed out that the form of the tree line boundary in the Changbai Mountains was affected by the heterogeneity of micro-topography [21]. These studies confirmed the effect of topography on the local environment. However, as an important factor in the study of tree locations in tree line ecotone, temperature represented by LST data was excluded in the process of constructing the model on the northern side of the mountains, whereas it was marginally important compared to aspect, wetness, and slope on the western side of the mountains (Table 4), suggesting the effects of temperature were not significant in affecting tree location. Although temperature has been always emphasized as a major controlling factor, our study concluded that topography should be considered as an important factor in explaining the variations in tree locations in the Changbai Mountains.

Potential tree locations

The current upmost altitude of tree location (2140 m asl) on the northern side is higher than that (2060 m asl) on the western side. We hypothesized that habitats above the current tree line are available for trees. The model prediction results confirmed our hypothesis, indicating that no matter on the northern or western side of the mountains, pioneer tree species like *B. ermanii* may potentially move to these habitats in the future. The tree lines in the Changbai Mountains have a diffused form, which is more likely to advance than those with an abrupt or krummholz form [2]. Previous studies have demonstrated that *B. ermanii* could expand its distribution range to higher elevations on the northern side, as it has expanded in the past decades [17]. Our analysis showed that *B. ermanii* could also move up on the western side, which would elevate the tree line position. However, it is noteworthy that the current position has been stable for many years on the western side of the mountains. This may be because the 1986 typhoon (~10, 000 ha) have largely destroyed the vegetation in the tree line ecotone, which has delayed the advancement of *B. ermanii* [22]. In addition, soils in the western side of the mountains are not as developed as in the northern sides of the mountains, which could limit tree growth in the western side of the mountains [36]. Some important tree growth parameters such as biomass, age, and size, and future climate change predictions were not included in our models, which could limit our model prediction. Nevertheless, our research may help explain variations of tree locations and reveal topographic controls within the tree line ecotone in this temperate alpine region.

Application of the high resolution image

The accuracy of the detection of tree locations depends on the spatial resolution of the available data as well as the subsequent methods [37]. Low-resolution satellite images, such as Landsat TM images, have been widely used in the detection of forest-tundra boundaries [38]. However, such macro-scale analysis often produced coarse scale results. In addition, the analysis required numerous inputs of field validation data to ensure the accuracy of the classification [39]. By contrast, high-resolution satellite images are more suitable for the detection of tree locations in the tree line ecotone [40,41]. Unfortunately, they are difficult to acquire because of the relatively high cost compared with other commercial sensors. The IKONOS images benefit from multi-spectral bands and perform well in the assessment of tree locations [42]. To our knowledge, the WorldView-1 image has been rarely used in the study of tree locations because of the lack of spectral information. By contrast, the WorldView-2 (2 m spatial resolution) images with eight multispectral bands is commonly used in the extraction of tree information [43,44]. In our study, a cloud-free WorldView-1 image could effectively identify the tree locations in the tree line ecotone. Misclassification errors in classification could be easily eliminated using few field validation data. When applying classification results to modeling analysis, particular attention should be given to eliminate the spatial autocorrelation. Otherwise, the predictive accuracy of the model will be considerably low.

In conclusion, we investigated the controlling effects of topography on the tree locations in tree line ecotone of the Changbai Mountains. Our results show that topography should not be ignored because it plays an important role in determining the tree locations. Tree locations in the tree line ecotone on the four sides (northern, southern, eastern and western) of the mountains were different. Such variations could be explained by the differences in topography. To better clarify the features of the tree locations in the Changbai Mountains, we will continue our research on the eastern and southern sides of the mountains.

Acknowledgments

I would like to thank Xiangjun Meng, Jie Liu, and Shaoping Li for their assistance in the field work.

Author Contributions

Conceived and designed the experiments: ZFW JWX. Performed the experiments: ML XFG. Analyzed the data: HBD LW. Contributed reagents/materials/analysis tools: SWZ HSH. Wrote the paper: SWZ HSH.

References

1. Körner C (1998) A re-assessment of high elevation treeline positions and their explanation. Oecologia 115: 445–459.
2. Harsch MA, Hulme PE, McGlone MS, Duncan RP (2009) Are treelines advancing? A global meta-analysis of treeline response to climate warming. Ecology letters 12: 1040–1049.
3. Grace J, Berninger F, Nagy L (2002) Impacts of climate change on the tree line. Annals of Botany 90: 537–544.
4. Dullinger S, Dirnböck T, Grabherr G (2004) Modelling climate change-driven treeline shifts: relative effects of temperature increase, dispersal and invasibility. Journal of Ecology 92: 241–252.
5. Lloyd AH, Fastie CL (2002) Spatial and temporal variability in the growth and climate response of treeline trees in Alaska. Climatic Change 52: 481–509.
6. Körner C, Paulsen J (2004) A world-wide study of high altitude treeline temperatures. Journal of Biogeography 31: 713–732.
7. Kullman L (2007) Tree line population monitoring of *Pinus sylvestris* in the Swedish Scandes, 1973–2005: implications for tree line theory and climate change ecology. Journal of Ecology 95: 41–52.
8. Rickebusch S, Lischke H, Bugmann H, Guisan A, Zimmermann NE (2007) Understanding the low-temperature limitations to forest growth through calibration of a forest dynamics model with tree-ring data. Forest Ecology and Management 246: 251–263.
9. Elliott GP, Baker WL (2004) Quaking aspen (*Populus tremuloides* Michx.) at treeline: a century of change in the San Juan Mountains, Colorado, USA. Journal of Biogeography 31: 733–745.
10. Liang E, Wang Y, Eckstein D, Luo T (2011) Little change in the fir tree-line position on the southeastern Tibetan Plateau after 200 years of warming. New Phytologist 190: 760–769.
11. Körner C, Hoch G (2006) A Test of Treeline Theory on a Montane Permafrost Island. Arctic, Antarctic, and Alpine Research 38: 113–119.
12. Brown DG (1994) Predicting vegetation types at treeline using topography and biophysical disturbance variables. Journal of Vegetation Science 5: 641–656.
13. Van Cleve K, Viereck L, Dyrness C (1996) State factor control of soils and forest succession along the Tanana River in interior Alaska, USA. Arctic and Alpine Research 28: 388–400.
14. Malanson GP, Cairns DM (1997) Effects of dispersal, population delays, and forest fragmentation on tree migration rates. Plant Ecology 131: 67–79.
15. Resler LM, Butler DR, Malanson GP (2005) Topographic shelter and conifer establishment and mortality in an alpine environment, Glacier National Park, Montana. Physical Geography 26: 112–125.
16. Holtmeier F-K, Broll G (2007) Treeline advance–driving processes and adverse factors. Landscape Online 1: 1–33.
17. Yu DP, Tang LN, Wang SX, Wang SZ, Dai LM (2007) Quantitative methodologies for ecotone determination on north slope of Changbai Mountains. Chin J Appl Ecol 15: 1760–1764.
18. Wang XD, Liu HQ (2011) Water and Heat Changes of *Betula ermanii* Treeline on Northern Slope of Changbai Mountains. Progress in geography 30: 313–318.
19. Wang QW, Qi L, Tian J, Zhou WM, Ding H, et al. (2011) Effects of altitudinal gradient on water use efficiency of Betula ermanii on the northern slope of Changbai Mountains, Northeast China. Chinese Journal of Applied Ecology 22: 2227–2232.
20. Wang XD, Liu HQ (2011) Population dynamics of Betula ermanii associated with soil change on treeline of northern slope of Changbai Mountains. Geographical research 30: 531–539.
21. Shi PL, Li WH (2000) Boundary from effects of timberline ecotone on colonization of woody plants and timberline dynamics in Changbai Mountain. Acta ecologica sinica 20: 573–580.
22. Niu LJ, Liang Y, Wang SX, Xu JW, Zhang R, et al. (2010) Forest recovery state in wind disaster area of Changbai Mountains, Northeast China. Chinese Journal of Applied Ecology 21: 1381–1388.
23. Zong SW, Xu JW, Wu ZF (2013) Investigation and Mechanism Analysis on the Invasion of Deyeuxia.angustifolia to Tundra Zone in Western Slope of Changbai Mountain. Journal of Mountain Science 31: 448–455.
24. Zheng D, Wallin DO, Hao Z (1997) Rates and patterns of landscape change between 1972 and 1988 in the Changbai Mountain area of China and North Korea. Landscape Ecology 12: 241–254.
25. Körner C (2003) Alpine Plant Life: Functional Plant Ecology of High Mountain Ecosystems. New York: Springer Berlin Heidelberg press.
26. Fisher PF (1993) Algorithm and implementation uncertainty in viewshed analysis. International Journal of Geographical Information Science 7: 331–347.
27. Chang HC, Ge L, Rizos C (2004) Assessment of digital elevation models using RTK GPS. Journal of Geospatial Engineering 6: 13–21.

28. Hutchinson M (1989) A new procedure for gridding elevation and stream line data with automatic removal of spurious pits. journal of Hydrology 106: 211–232.
29. Pierce Jr KB, Lookingbill T, Urban D (2005) A simple method for estimating potential relative radiation (PRR) for landscape-scale vegetation analysis. Landscape Ecology 20: 137–147.
30. Schmidt F, Persson A (2003) Comparison of DEM data capture and topographic wetness indices. Precision Agriculture 4: 179–192.
31. Li F, Jackson TJ, Kustas WP, Schmugge TJ, French AN, et al. (2004) Deriving land surface temperature from Landsat 5 and 7 during SMEX02/SMACEX. Remote Sensing of Environment 92: 521–534.
32. Katoh M (2004) Classifying tree species in a northern mixed forest using high-resolution IKONOS data. Journal of Forest Research 9: 7–14.
33. Dormann CF, McPherson JM, Araújo MB, Bivand R, Bolliger J, et al. (2007) Methods to account for spatial autocorrelation in the analysis of species distributional data: a review. Ecography 30: 609–628.
34. Liu XG, Zhou H (2009) Analysis on the regional precipitation characteristics of western slope in Changbai Mountain. Jilin meteorology 3: 25–26.
35. Wang XD, Liu HQ (2012) The Dynamics Response of *Betula ermanii* Population and Climate Change on Different Slopes Aspect of North Slope, Changbai Mountains. Scientia geographica sinica 32: 199–201.
36. Jin YH, Xu JW, Liang Y, Zong SW (2013) Effects of Volcanic Interference on the Vegetation Distribution of Changbai Mountain. Scientia geographica sinica 33: 203–208.
37. Püssa K, Liira J, Peterson U (2005) Transitional and abrupt edges of forest to forest to non-forest bounaries on medium resolution Landsat TM winter images. ForestSat 2005. Boras. pp. 79.
38. Calef MP, David McGuire A, Epstein HE, Scott Rupp T, Shugart HH (2005) Analysis of vegetation distribution in Interior Alaska and sensitivity to climate change using a logistic regression approach. Journal of Biogeography 32: 863–878.
39. Rees G, Brown I, Mikkola K, Virtanen T, Werkman B (2002) How can the dynamics of the tundra-taiga boundary be remotely monitored? Ambio: 56–62.
40. Zhang YJ, Xu M, Adams J, Wang X (2009) Can Landsat imagery detect tree line dynamics? International Journal of Remote Sensing 30: 1327–1340.
41. Wulder M, White J, Niemann K, Nelson T (2004) Comparison of airborne and satellite high spatial resolution data for the identification of individual trees with local maxima filtering. International Journal of Remote Sensing 25: 2225–2232.
42. Radoux J, Defourny P (2007) A quantitative assessment of boundaries in automated forest stand delineation using very high resolution imagery. Remote Sensing of Environment 110: 468–475.
43. Pu R, Landry S (2012) A comparative analysis of high spatial resolution IKONOS and WorldView-2 imagery for mapping urban tree species. Remote Sensing of Environment 124: 516–533.
44. Huang HG, Cao B. Experiment on extracting forest canopy height from Worldview-2. Fuzzy Systems and Knowledge Discovery; 2011; Shanghai. IEEE. pp. 2614–2617.
45. Shao G, Shugart H H, Zhao G, et al. (1996) Forest cover types derived from Landsat Thematic Mapper imagery for Changbai Mountains area of China. Canadian Journal of Forest Research 26: 206–216.
46. Liu HJ, Hu YM, Bu RC, Liu JT, Leng WF (2009) Impacts of climate changes on the landscape patterns of potential mire distributions in northern Great Khing'an mountains. Advances in water science 20: 105–110.

18

Analysis of Feature Intervisibility and Cumulative Visibility Using GIS, Bayesian and Spatial Statistics: A Study from the Mandara Mountains, Northern Cameroon

David K. Wright[1]*, Scott MacEachern[2], Jaeyong Lee[3]

1 Department of Archaeology and Art History, Seoul National University, Seoul, Republic of Korea, **2** Department of Anthropology, Bowdoin College, Brunswick, Maine, United States of America, **3** Department of Statistics, Seoul National University, Seoul, Republic of Korea

Abstract

The locations of *diy-geδ-bay* (DGB) sites in the Mandara Mountains, northern Cameroon are hypothesized to occur as a function of their ability to see and be seen from points on the surrounding landscape. A series of geostatistical, two-way and Bayesian logistic regression analyses were performed to test two hypotheses related to the intervisibility of the sites to one another and their visual prominence on the landscape. We determine that the intervisibility of the sites to one another is highly statistically significant when compared to 10 stratified-random permutations of DGB sites. Bayesian logistic regression additionally demonstrates that the visibility of the sites to points on the surrounding landscape is statistically significant. The location of sites appears to have also been selected on the basis of lower slope than random permutations of sites. Using statistical measures, many of which are not commonly employed in archaeological research, to evaluate aspects of visibility on the landscape, we conclude that the placement of DGB sites improved their conspicuousness for enhanced ritual, social cooperation and/or competition purposes.

Editor: Lalit Kumar, University of New England, Australia

Funding: This work was supported by an international faculty research grant from Seoul National University, the National Geographic Society's Committee for Research and Exploration (grant # 9494-13), and the National Research Foundation of Korea Grant #2013S1A5B6043901 funded by the Korean government. The funders had no role in study design, data collection and analysis, decision to publish, or preparation of the manuscript.

Competing Interests: The authors have declared that no competing interests exist.

* Email: msafiri@snu.ac.kr

Introduction

How and why people settle onto landscapes have been focal research questions in archaeology since the advent of the discipline (e.g., [1–3]). Post-processual archaeology challenged the assumption that the motivations guiding human behavior can be analytically reduced to testable sets of assumptions [4]. However, it is now widely accepted that a middle ground exists between generating absolute laws of governing human behavior and human agents behaving independent of systems or repeatable principles [5–7]. The use of scientific mapping tools can be used to pattern human settlements, the construction of archaeological features and provide insight into how human agency expresses itself onto a landscape.

Monumental architecture is commonly seen as a means for solidifying political hegemony [8–10] or promoting social cohesion [11–13] within or between communities. In Africa, the explanations for the construction of walled settlements and monumental architecture are as diverse as the continent itself [14–17]. The diversity of human populations on the African continent gave rise to different forms of political complexity that exhibit unique characteristics worthy of situating in the global discourse of social ascendancy [18].

In the mid-1990s, a series of tell mound sites at the edge of the Mandara Mountains were excavated in order to study the evolution of social and political systems over the last three millennia (Fig. 1) [19]. The culmination of these cultural developments resulted in the rise of predatory, slave-raiding plains states. More recent research at a set of monumental stone sites (the *diy-geδ-bay* or DGB sites) in the northwestern Mandara massif indicates a parallel history of variable political complexity in the mountains over the last 800 years at least, with some evidence for occupation at least a century or two before that [20]. Archaeological and historical data strongly indicate that the relationship between mountains and plains was extremely complex, and that both regions offered resources and challenges that mutually influenced political and cultural developments in the other [20–22].

The motivations behind the construction of DGB sites remain a source of speculation, with one theory being that they were initially constructed as a response to profound droughts in the 15th and 16th centuries CE [19]. This seems to have been a particularly intense period of occupation at the two largest sites, DGB-1 and DGB-2, but excavations on the former has established that initial stages of construction two place at least two centuries before the period of drought [19]. Based on examination of site locations on 1:50,000 topographic maps, Nicolas David [23] has hypothesized

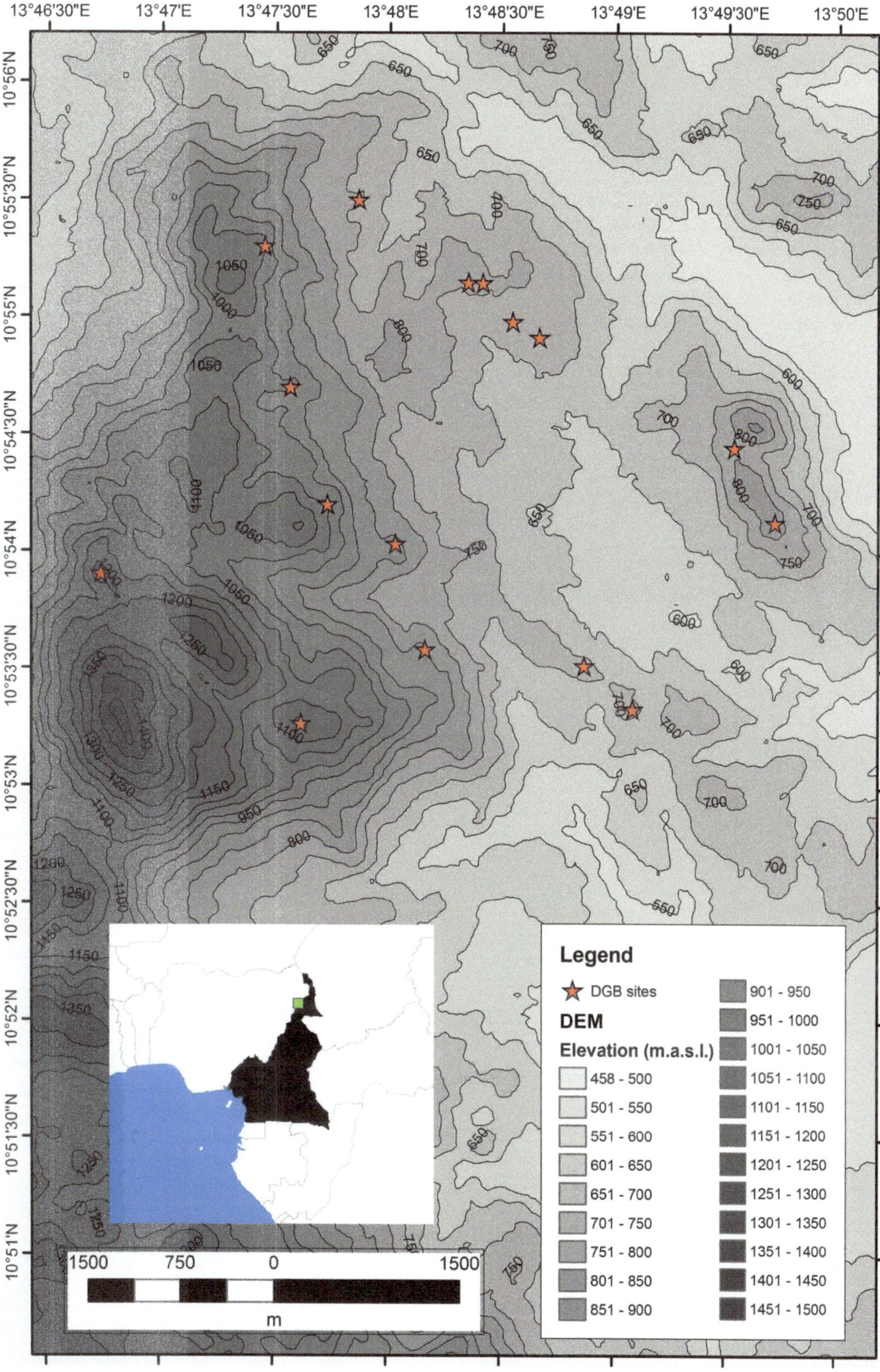

Figure 1. Location of the project area and DGB sites.

that the sites were situated non-randomly on high ground in order to maximize their visibility over surrounding terrain and their intervisibility, possibly in the course of ritual and/or political competition between local communities and their leaders. However, this hypothesis of non-random location in the service of intervisibility has heretofore remained untested using modern computing methods.

In this paper, we describe novel ways of determining relative intervisibility of archaeological sites as well as the cumulative viewshed of the surrounding landscape based on a Geographical Information Systems (GIS) analysis. Using DGB sites as our case study, we statistically analyze whether the locations of sites were randomly located above an elevational threshold of 700 m.a.s.l. or were selected in order to maximize their viewshed. The results of our statistical tests reject the null hypothesis that DGB site locations were randomly selected and we argue that although other causative site selection factors cannot be ruled out, the viewscape of the DGB sites was certainly a factor in where sites were constructed.

Background to the research area: DGB sites

The landscape of the northern Mandara Mountains in northern Cameroon and Nigeria is one of the most densely populated in Central Africa, with densities of up to 220 people/km^2 in a dry Sudanian environment. It is thus by necessity an almost entirely domesticated landscape, with all of the arable land under agriculture and with virtually every tree and shrub exploited for human needs. The region is extremely diverse ethnically, linguistically and culturally, with more than 20 different languages spoken in an area of about 2000 km^2, and with a complex sociopolitical history to match.

The 6.35-×-6.5 km area covered in this study includes 16 DGB sites that have been previously identified by archaeologists [20,23]. These sites are scattered across volcanic inselbergs of the Mandara Mountains, which comprise the northeastern segment of the Cameroon Volcanic Line [24]. The Mandara Mountains are mapped as trachyte and rhyolitic plugs [25]. The topography is highly irregular with steep peaks and dissected valleys. The native vegetation is comprised of mosaics of *Pennistum* sp. grasslands and *Acacia* sp. woodlands and allies [26], but has been heavily anthropogenically modified into an agricultural landscape during the Late Holocene with intensive human settlement [20].

The 16 DGB sites are prehistoric monumental architectural complexes made up of varying combinations of rubble-filled platforms and terraces, and faced with distinctive and carefully arranged dry-stone facades. A number of them, and all so far excavated, contain a variety of interior passages, staircases, courtyards and chambers. The two largest, DGB-1 and DGB-2, are situated just over 100 m apart and are at least in part contemporary; they thus form a single monumental site complex over an area of about 2.5 ha (Fig. 2.). The other DGB sites are significantly smaller. Seventeen dates from DGB-1 establish occupation during the 13th to 17th centuries AD, while the one reliable date from DGB-2 is from the 15th century [19]. Two dates from DGB-1, the only other excavated DGB site, are of the late 14th–15th century AD [23]. It is quite possible that further excavations would increase the period of site usage at these latter sites.

The functions of the DGB sites are not completely understood. They do not appear to have defensive functions, given their

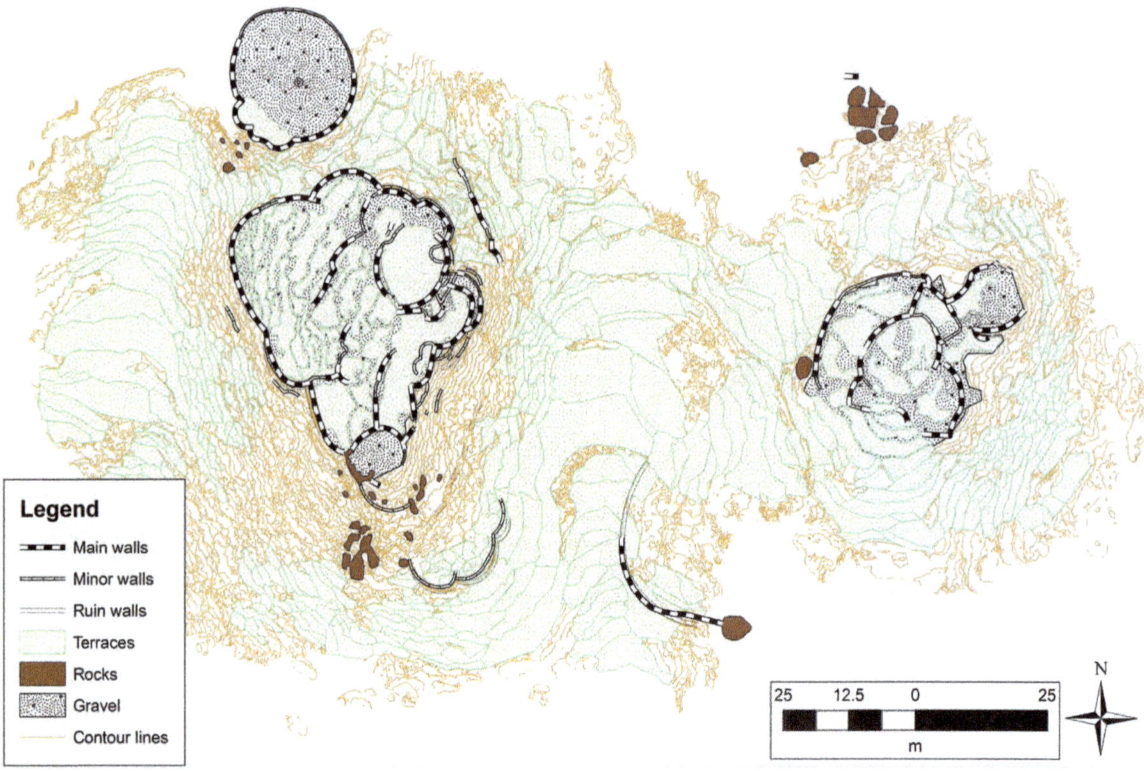

Figure 2. Laser phase scans using a Leica HDS of the DGB-1 and DGB-2 complexes (complements of Heinz Rüther and the Zamani Project).

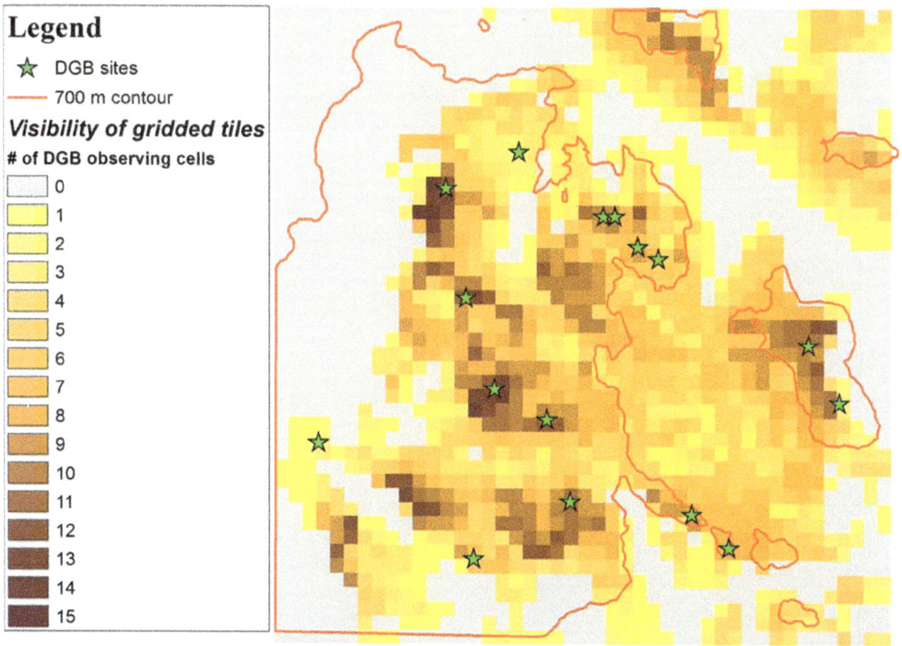

Figure 3. Observation point analysis of 2025 tiles showing the number of Actual DGB sites able to see each cell.

positioning, unfortified wall features and lack of close access to water. There is some evidence for contemporary domestic occupation on DGB-1 and in the area between DGB-1 and DGB-2, but they do not seem to have been primarily habitation sites. They provide a striking impression of being in some sense platforms for display, almost for theater: their raised open platforms would certainly afford easy visibility for public activities, while sunken courtyards, subterranean passages and chambers would accommodate private ceremonialism, with stairways linking the two worlds.

David [23] has located the functioning of the DGB sites in the context of modern Mandara montagnard ritual practice, in ceremonialism involving water and fecundity, while MacEachern [27] emphasizes their potential role in the increasingly complex political world of the southern Lake Chad Basin in the mid-second millennium AD, when predatory states and slave-raiding were becoming important elements in the plains north of the Mandara Mountains. The contemporary capital of one of these states, Wandala, was located only 30 km north of the DGB sites, at Keroua/Kirawa on the modern Cameroon-Nigeria border. It is very likely that the DGB sites fulfilled both of these roles simultaneously. Their size and imposing architectural features, along with the labor that would have been invested in their construction, makes it almost certain that they would have played an important role in sociopolitical relationships in the region during their period of use. Given that fact, it will be extremely useful to know if they were situated to maximize their viewsheds and their intervisibility. This would imply that the 16 sites were in fact of an interacting regional cultural system, and might increase the likelihood that some element of comparison, cooperation and/or competition was involved in their use. There are precedents for such regional ceremonialism in the area [28], particularly with the *maray* ceremony held at varying intervals among mountain communities. The 'passing' of the ritual between communities might be relevant to questions of intervisibility among ceremonial sites.

The locations of the sites have been previously hypothesized to be situated according to principles of intervisibility [23], potentially for reasons of interregional cooperation and/or competition. Herein, we articulate two sets of testable hypotheses:

(1a) H1: The locations of DGB sites are situated to maximize intervisibility of the sites to one another.

(1b) H0(null): The locations of DGB sites are randomly situated above 700 m.a.s.l. and are not situated to maximize intervisibility to one another.

(2a) H2: The locations of DGB sites are situated to maximize the cumulative viewshed of the surrounding valleys.

(2b) H0: The locations of DGB sites are randomly situated above 700 m.a.s.l. and are not situated to maximize the cumulative viewshed of the surrounding valleys.

By comparing the locations of actual DGB sites (hereafter, Actual DGB) to permutations of stratified random "DGB sites" (hereafter, Random DGB-1, 2, 3…), we intend to accept or reject the hypotheses based on statistical measures of randomness. We are not attempting to eliminate other potential factors for site selection (geology, aspect, soil fertility of adjacent land, visibility for astronomical purposes, etc.), merely testing an earlier hypothesis using the best fit of data generated with modern computing techniques.

Visibility studies in archaeology

Landscape visibility and feature intervisibility are related concepts in GIS data processing. The landscape visibility is concerned with determining whether features are located in order to see positions at lower elevations, and potentially vice-versa [29]. Methods analyzing feature intervisibility typically determine whether features at relatively similar elevations might have been constructed in their given locations in order to facilitate their ability to see one another. In both cases, the "viewshed" is considered a tangible cultural asset, worthy of enhancing or restricting for some socially important reason. While determining whether or not sites were constructed in order to enhance visibility

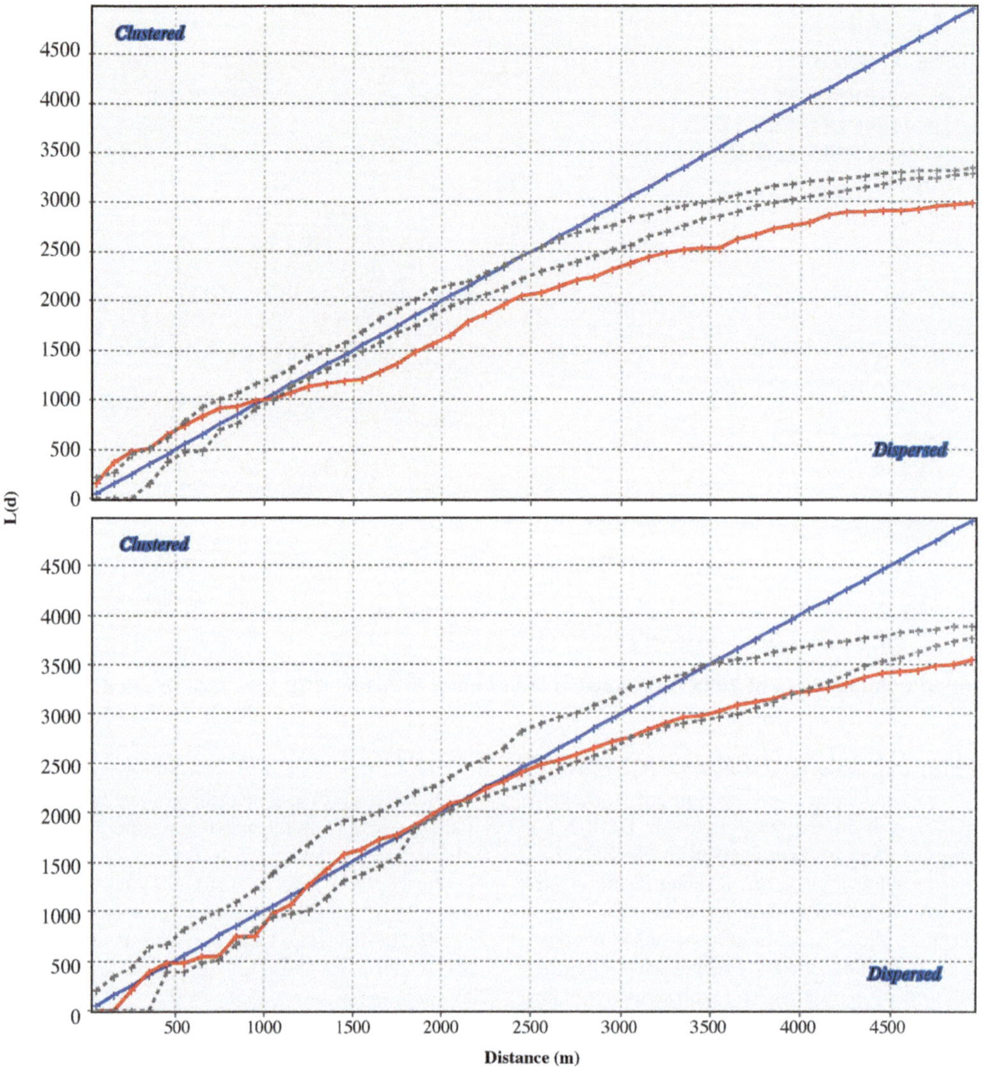

Figure 4. Ripley's K analysis of clustering of Actual DGB sites (top) and Random DGB-1 (bottom). The blue line represents the predicted values, the red line represents the actual values analyzed and the dotted lines represent the 2-σ confidence envelope.

has become relatively simple, getting at the emic cultural dimensions of feature construction has become no less simple than in the days preceding automated spatial computational techniques.

The intervisibility of archaeological sites or features has been a topic of discussion since the advent of GIS. An early pioneer, Kvamme [30], advocates testing samples of sites against a "background standard" in order to identify random vs. non-random patterns of site placement. Wheatley [31] analyzes the placement Neolithic funerary monuments using a Kolmorgov-Smirnoff (K–S) test of cumulative frequency distributions to argue that the patterns show intentional intervisibility of "long barrow" sites in Stonehenge, but the degree to which the sites are visible to the landscape in general is variable according to region (see also [32] for a similar study on rural settlement in Greece). K–S tests are used to determine whether samples exceed the statistical probability of being randomly distributed. In Kvamme's [30] analysis, "the sampling frame of...cell values...serves as the background referent distribution". However, Fisher [33] argues that such studies are most effective if there is a control set of randomized points to compare in order to test random vs. non-random effects. Additionally, in recent years, K–S has been determined to be a less effective test of normality in datasets compared to other non-parametric tests such as Shapiro-Wilk, Lilliefors and Anderson-Darling [34].

A second type of visibility analysis includes the ability of observers to see points on a landscape from archaeological features [33,35,36]. This has been called "cumulative viewshed" analysis and has been employed in numerous archaeological studies [30–33,37,38]. Good visibility of a landscape to and from an archaeological site or feature has been argued to have military applications of site defensibility [39–42], resource acquisition [43–45] or to solidify some aspect of social power/cohesion through a feature's visibility [46–50]. Regardless of the purported reasons for doing so, it is clear that prehistoric people frequently made informed choices about how to situate archaeological sites to either enhance or reduce visibility relative to points on a landscape.

Early pioneering work on landscape visibility was very labor intensive and lacked statistical rigor because of a lack of data processing tools (e.g., [37,51]), but Geographic Information

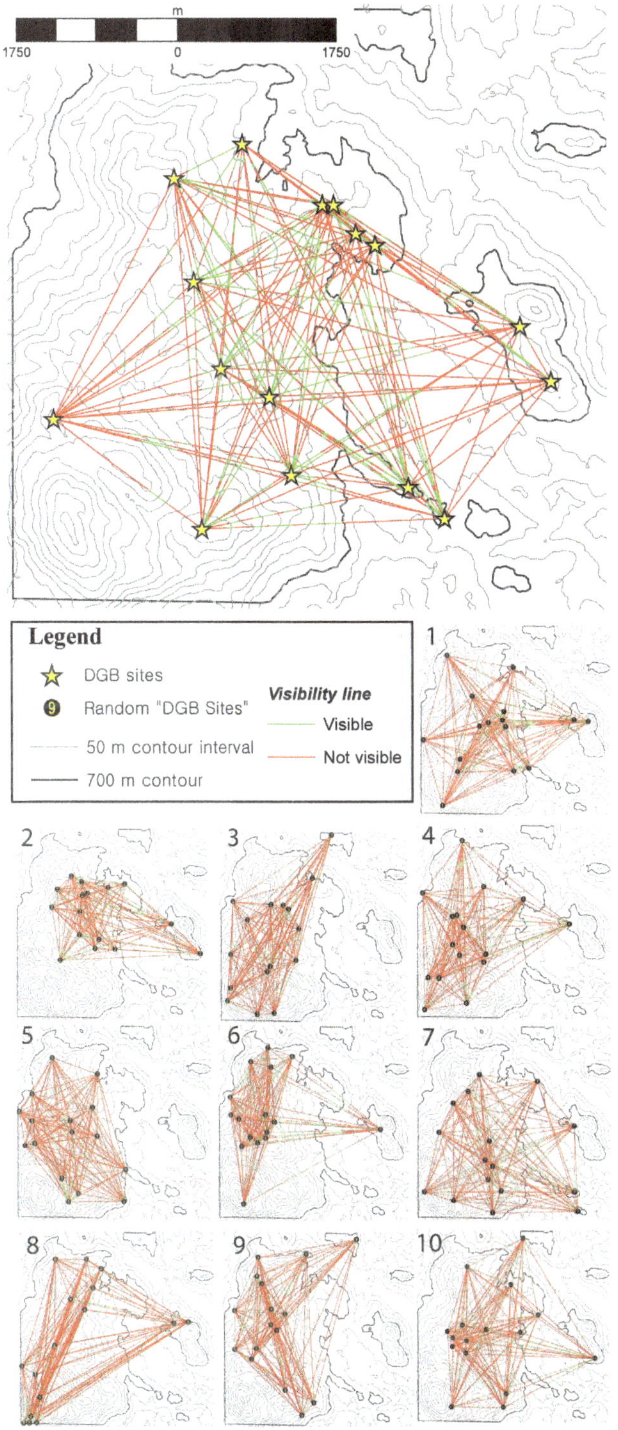

Figure 5. Intervisibility lines generated using the "Line of Sight" tool in ArcGIS10.1 Spatial Analyst (vectors created using v.net.visibility in GRASS6.4) of Actual DGB and 10 permutations of Random DGB sites.

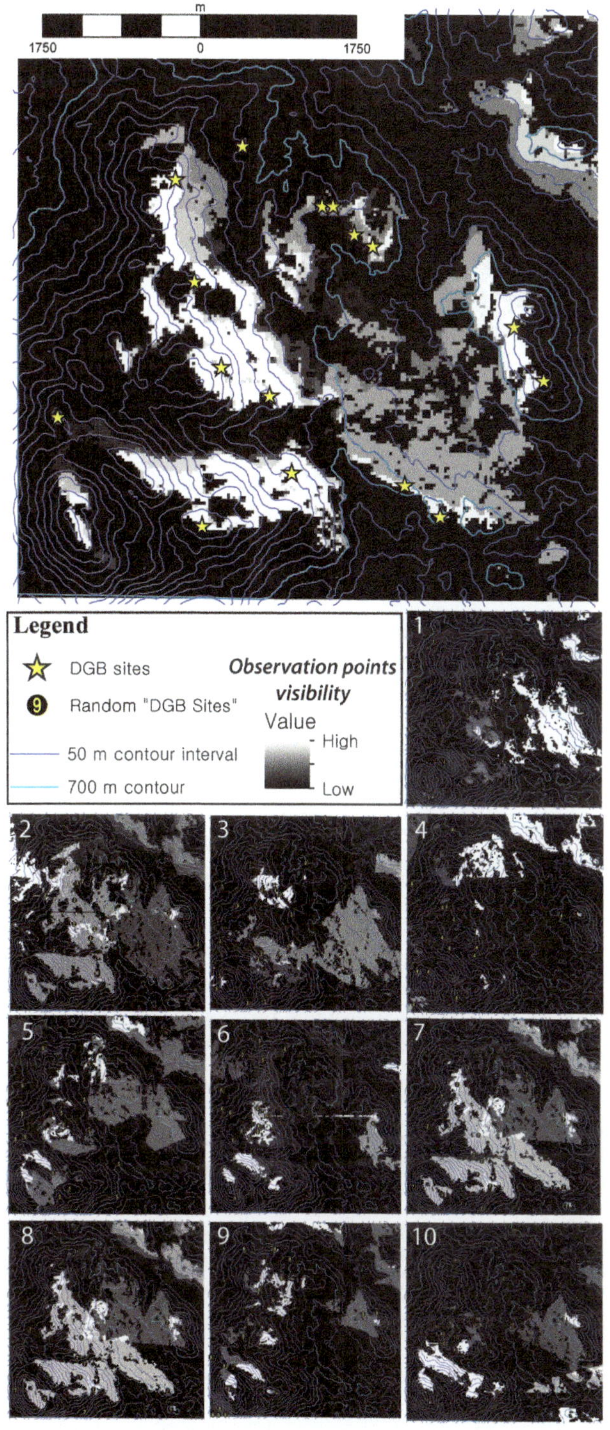

Figure 6. Viewshed analysis in ArcGIS 10.1 of Actual DGB and 10 permutations of Random DGB sites.

Systems (GIS) now provide ample computing power and statistical tools for determining visible and invisible aspects of a landscape from an observer point. The use of Monte-Carlo simulations to create randomized sets of data has become increasingly popular to create control samples of data populations [33,52–54]. The advantage of such simulations is that they remove any possible subjectivity in the analysis, while the disadvantage is that they are difficult to accommodate using stratified random sampling methods.

The fundamental principle of visibility and intervisibility in GIS is to generate lines of sight between given points on a landscape. In the case of two features (Point A, Point B), a line of sight is created as a straight vector between A and B and the topography which

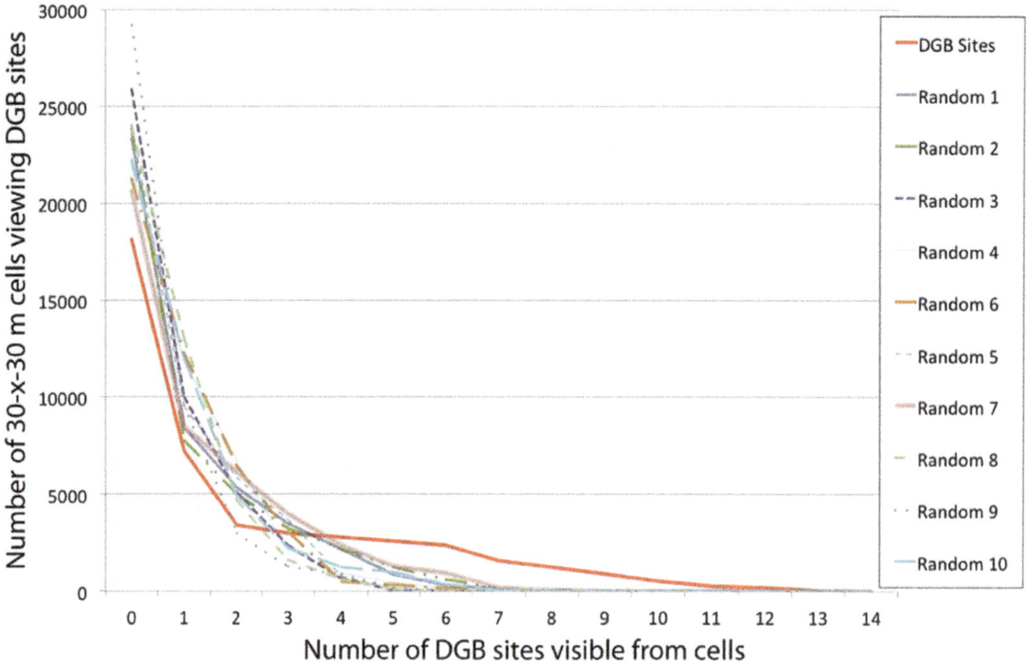

Figure 7. Comparison of the number of DGB sites (Actual, Random-1– Random-10) visible from 30-×-30-m grid cells within the study area.

would prohibit or facilitate intervisibility is generated from a Digital Elevation Model (DEM) of the terrain. Today, DEMs are commonly available to resolutions of up to 10 m, but most portions of Africa (the subject of this study) are covered by 30-×-30 m grids by United States Geological Survey (USGS) satellites (e.g., http://glovis.usgs.gov). Whether or not Point A is visible to Point B is determined on the basis of analyzing the presence/absence of topographic features that might interfere with the vector connecting the two nodes. Although visibility itself is hindered by physical distance relating to the curvature of the earth and recognizability of objects to the human eye over long distances [35], we presume that the DGB sites were theoretically intervisible on the basis that no two intervisible sites were more than 5 km from each other (which is roughly the distance at which the earth curves out of sight to the naked eye on a continually flat surface).

Methods

Spatial organization of the data using GIS

Previous studies of intervisibility have advocated one-sample tests comparing a site sample against a hypothetical "background" sample (the strongest advocate and most commonly cited of these studies is [30]). However, in applying the K–S test, background samples have typically been truly randomized to check distributions of actual vs. hypothetical data to delineate statistical outliers. In the present analysis of DGB sites, we apply a stratified random sampling method for the distribution of hypothetical DGB sites (> 700 m.a.s.l.), because a truly randomized distribution of hypothetical sites across the project area would not reflect the true conditions of site selection.

After a DEM of the project area was generated, a spider diagram of DGB sites was created in ArcGIS10.1 as potential sight lines between the facilities. In order to test whether the distribution of DGB sites was situated for intervisibility, random points were generated in a stratified fashion within a polygon shapefile created from a contour above 700 m.a.s.l. This corresponds to the approximate minimum elevation on which DGB sites were located. Therefore, we make the assumption that for any myriad of reasons (e.g., defense, drainage, visibility), there was a cultural proclivity to select locations above 700 m.a.s.l. Ten sets of 16 random points were generated using the "Create Random Points" tool in ArcGIS10.1 with the constraining feature being the elevationally-derived contour line shapefile. A minimum distance of 121 m was assigned between the random points based on the distance between DGB-1 and DGB-2. Lines of sight vectors were constructed using the"v.net.visibility" tool in GRASS6.4. The result was the construction of 120 vectors, which is the total number of lines that interconnect all 16 points (points do not connect to themselves and connections are not made twice). Intervisibility lines were created using the 3D Analyst toolbox ("Line of Sight" tool) from the lines of sight vectors. The DEM was the input surface, and line features were the sight lines created.

Slope of Actual and Random DGB sites were calculated from the raster DEM, then aggregated and averaged to control whether site construction was a function of factors other than intervisibility. Slope values were generated automatically in ArcGIS10.1 by extracting the elevation of plotted DGB points from the Mandara Mountain DEM compared to the eight surrounding raster cells then calculating a % of elevation change. The data were tabulated to show the variance between the slope of Actual and Random DGB sites.

A separate raster map of the observability of DGB sites was created to assess whether site locations were situated so that they were more visible from different landscape points. Based on the resolution of the DEM, the raster cell value was set to include a 30-×-30 m area. Observer points were set as the Actual DGB sites or Random DGB sites according to the point features created randomly. A viewshed analysis was executed according to a simple yes/no (binary) function to determine whether a raster cell has visibility of at least one observer point factoring in potential

Table 1. Comparative analysis of numbers of DGB sites (Actual, Random-1-10) visible from 30-×-30-m grid cells within the study area.

Number of Sites Visible	DGB Actual	Random-1	Random-2	Random-3	Random-4	Random-5	Random-6	Random-7	Random-8	Random-9	Random-10	mean (ALL)	1-σ	mean (Random-1-10)
0	18132	23372	23846	25894	21873	24090	21274	20721	23996	29189	22240	23147.9	2884.3	23649.5
1	7226	8427	7782	10028	11853	9532	12254	8468	13043	9609	12017	10021.7	1991.1	10301.3
2	3384	5329	5029	5111	6716	5863	6434	6136	4678	2985	4945	5146.4	1170.5	5322.6
3	2961	3474	3218	2363	2135	3691	3069	3985	1612	1253	2209	2724.5	874.6	2700.9
4	2762	2145	2178	635	917	780	488	2359	605	865	1240	1361.3	830.5	1221.2
5	2545	845	1215	47	434	102	312	1272	117	183	991	733.0	755.2	551.8
6	2362	313	606	6	150	24	126	916	30	0	340	443.0	698.6	251.1
7	1568	132	163	0	6	2	87	172	3	0	79	201.1	458.4	64.4
8	1227	38	44	0	0	0	40	46	0	0	21	128.7	364.8	18.9
9	906	9	3	0	0	0	0	9	0	0	2	84.5	272.5	2.3
10	499	0	0	0	0	0	0	0	0	0	0	45.4	150.5	0
11	250	0	0	0	0	0	0	0	0	0	0	22.7	75.4	0
12	185	0	0	0	0	0	0	0	0	0	0	16.8	55.8	0
13	60	0	0	0	0	0	0	0	0	0	0	5.5	18.1	0
14	17	0	0	0	0	0	0	0	0	0	0	1.5	5.1	0
MEDIAN	1568	132	163	0	6	2	87	172	3	0	79	201.1	458.4	64.4
1-σ	4601.7	6173.7	6221.5	6936.5	6211.0	6493.1	6120.0	5561.9	6767.7	7681.3	6218.9	6229.3	807.0	6438.6
invis. km^2	16.3	21.0	21.5	23.3	19.7	21.7	19.1	18.6	21.6	26.3	20.0	20.8	2.6	21.3
visible km^2	23.4	18.6	18.2	16.4	20.0	18.0	20.5	21.0	18.1	13.4	19.7	18.8	2.6	18.4
weighted avg.	**2.3**	**1.0**	**1.1**	**0.7**	**0.9**	**0.8**	**0.9**	**1.3**	**0.7**	**0.5**	**0.9**	*1.0*	*0.5*	0.9

Invisible vs. visible km^2 represent the total area that can see ≥1 DGB site. Weighted average multiplies the number of sites visible by the number of cells visible to and from the sites divided by the total number of cells in the project area. Note: Total cells analyzed are 44,084; each cell is 900 m^2.

Table 2. Spearman's-ρ ranked analysis comparing the average cumulative viewshed of cells from DGB Random-1–10 against individual values.

Average DGB Random 1–10 vs.	Spearman's R	Spearman's t	p-level
Actual DGB	0.9820	18.7350	0.0000
Random1	1.0000	0.0000	1.0000
Random2	1.0000	0.0000	1.0000
Random3	0.9389	9.8330	0.0000
Random4	0.9661	13.4907	0.0000
Random5	0.9661	13.4907	0.0000
Random6	0.9860	21.3307	0.0000
Random7	1.0000	0.0000	1.0000
Random8	0.9661	13.4907	0.0000
Random9	0.9027	7.5631	0.0000
Random10	1.0000	0.0000	1.0000
Average*	**0.9734**	**13.9906**	

*Average of Spearman's t includes only those with positive values.

topographic barriers. The "Viewshed" tool in ArcGIS10.1 determines the number of observer points visible from a given raster cell on the landscape. The values are then classified from high to low with "high" values being a computer-determined value of numerous sites visible from an observer point, and "low" values are those with no visibility to an observer point. The data are then be displayed graphically or exported into table format for statistical analysis.

Cumulative viewshed analysis

Once stratified random sample sets were created and exported into tabular format, statistical means for testing the distribution of actual vs. random sites were generated to test the hypotheses. Weighted means of cumulative viewsheds [33] were generated from the number of DGB sites (Actual DGB, Random DGB) visible from 30-×-30 m cells within the project area ($n = 44084$). Spearman's-ρ correlations were performed on comparing individual DGB sites (Actual and Random) to the computed average of the stratified Random DGB (1–10) sites using R statistical software. Spearman's-ρ is a measure of concordance between two related, but independent sets of variables, taking into account the related degree of monotonicity of the datasets [55]. Spearman's-ρ values vary between +1 and −1 depending on the degree of correlation of ranked data. A perfect positive correlation of variables is a +1, a perfect negative correlation is a −1, and a non-correlation value is 0.

Two-way statistical tests

Additionally, binary sets of intervisible DGB sites were created from lines of sight generated from both Actual DGB and Random DGB points. A Cochran's Q test was performed on the intervisible points of the 16 DGB sites compared to the 10 permutations of stratified random DGB sites located above 700 m.a.s.l. Cochran's Q is designed to analyze statistical significance of differences in binary populations [56]. For the purpose of the present study, intervisible sites were classified as "1" and non-intervisible sites were classified as "0". For example, DGB sites 3 and 4 were intervisible, so they were assigned a 1, while DGB sites 3 and 14 were not intervisible, so they were assigned a 0. McNemar's contingency tests were performed comparing the randomness of Actual DGB vs. Random DGB-1, 2, 3... as well as Random DGB sites to one another. The McNemar's contingency tests determine whether two sets of binary data are normally distributed [57,58]. Such methods have not been applied to spatial data in archaeology, but are common in medical (e.g., [59]) and ecology-related (e.g., [60]) fields.

Bayesian logistic regression

We postulate a Bayesian logistic regression model tests the hypotheses $H_{0(null)}$ and H_1, i.e. whether the locations of DGB sites are situated to maximize intervisibility of the sites to one another. The test model follows several steps beginning with a spatial organization of the data and concluding with a probability-based calculation. First, we retained the same rectangular-shaped region as previous analyses containing 16 DGB sites and subdivided this region by segmenting the west-to-east line and the south-to-north line into 45 equal length intervals, respectively, resulting in 2025 tiles (Fig. 3). The decision was made to analyze ~2000 tiles in order to capture enough geographical resolution to have meaningful insight into apparent visibility or non-visibility of sites, but not too much data that would make the analysis overly complex for the 16 DGB sites. The area captured within each tile approximated nine DEM cells plus perimeter values of an additional 16 cells extracted by a rectangular mask in ArcGIS10.1 to estimate the elevation of each 20,375 m^2 tile.

Among the 2025 tiles within the overall project area, we take $n = 1236$ tiles located ≥700 m.a.s.l. for the logistical regression analysis. For tile $i = 1,2,\ldots,1236$, we created two variables, y_i and x_i. We let $y_i = 1$ if a DGB site was built in tile i and $y_i = 0$ otherwise, and let x_i be the number of DGB sites that are visible from tile i excluding the DGB site on tile i. If the DGB sites are situated to maximize intervisibility of the sites to one another, the chance of $y_i = 1$ would increase, as x_i gets larger.

Second, we postulate the logistic regression model assuming that $i = 1,2,\ldots, n$, y_i follows Bernoulli distribution with probability $p(x_i)$ which is a function of the number of DGB sites intervisible from tile i. The function $p(x)$ is the probability that a DGB site is built in a tile when there are x DGB sites visible from the tile. The functional form that relates to the probability and the number of visible DGB sites is the logit function; in particular, $\text{logit}(p(x)) = \log$

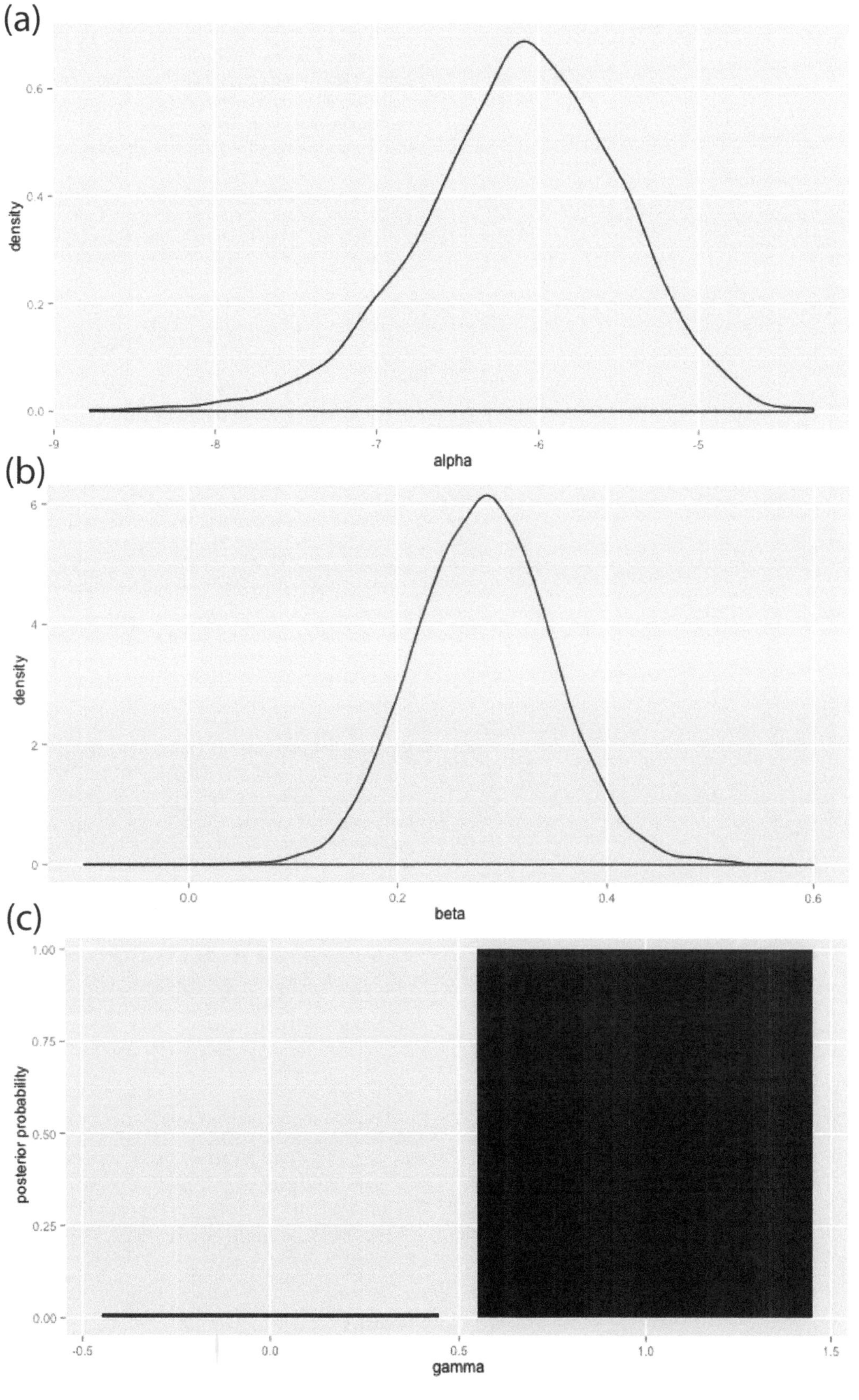

Figure 8. Posterior means and standard deviations of α, β, and γ using a Bayesian logistical regression model to test the intervisibility DGB sites across 2025 tiles, each of which measures 20,375 m^2.

Table 3. Average and 1-σ calculations of elevation (m.a.s.l.) and slope (%) of Actual and Random DGB sites.

	DGB	DGB	DGB	DGB	DGB	DGB	DGB	DGB	DGB	DGB	DGB	mean		mean
	Actual	Random-1	Random-2	Random-3	Random-4	Random-5	Random-6	Random-7	Random-8	Random-9	Random-10	(ALL)	1-σ	(Random-1-10)
Average elevation	866.4	890.1	846.5	952.5	967.3	913.5	946.2	917.1	971.6	911.1	918.6	918.3	39.9	923.4
1-σ elevation	157.3	193.4	107.8	187.3	191.9	172.4	170.1	163.5	215.4	122.3	169.3	168.2	31.2	169.3
Average slope	10.9	18.8	20.3	19.9	18.6	16.4	15.0	17.3	20.5	16.0	14.6	17.1	2.9	17.7
1-σ slope	9.2	7.7	9.0	8.9	6.7	9.2	6.2	11.4	10.7	9.6	7.6	8.7	1.6	8.7

$[p(x)/(1-p(x))] = \alpha+\beta x$. The parameter α is logit (p) when $x=0$, i.e., there are no visible DGB sites from the tile. The parameter β is amount of increase in probability that a DGB site is built in the logit scale as x increases. If $\beta=0$, the number of visible DGB sites does not affect the probability of building DGB site; if $\beta>0$, the probability of building DGB sites gets larger as x gets larger; if $\beta<0$, the probability of building DGB sites gets smaller as x gets larger.

Since we are interested in whether the number of intervisible DGB sites affect the probability of building DGB sites, we formulate the question in the following statistical hypothesis testing problem:

$$H_{0(null)}: \beta=0 \text{ vs } H_1: \beta^{1} 0$$

For the prior of α and β, we take $\alpha \sim N(0, 10000)$ whose variance is very large representing vague prior knowledge on α, and $\beta \sim 0.5\ \delta_0 + 0.5\ N(0, 10000)$, where δ_0 is the probability distribution which has probability 1 at 0. The latter prior represents that β is 0 with probability 0.5 and follows N (0, 10000) with probability 0.5. We are effectively putting equal prior probabilities to H_0 and H_1, respectively. In the actual implementation of the prior of β, we introduce another random variable γ, which follows Bernoulli distribution with probability 0.5. In particular, $\gamma=1$ if $\beta \neq 0$ and $\gamma=0$ if $\beta=0$. The samples from the posterior distribution of α, β, and γ are obtained by Markov chain Monte Carlo; in particular we use the R2jags package of R statistical language to generate the posterior sample from the posterior.

Ripley's K function

Ripley's K function (multidistance spatial cluster analysis) was run on the projected data points to determine the degree of clustering or dispersion at different distances. The Ripley's K analysis determines whether the number of neighbors for a set of points at a given distance is greater or less than that of a hypothetical distribution of points [61]. A total of 50 distance bands were generated with a computed confidence envelope of nine permutations per analytical distance. The outer boundary was simulated using the ArcGIS10.1 "Outer Boundary Simulator," which simulates points outside the study area in order to minimize edge effects. The beginning distance band was 50 m, increasing by 100-m increments in each iteration of the analysis. We applied the "Simulate Outer Boundary Values" using an automatically-generated, minimum-enclosing rectangle for the project area as the feature class of the study area.

The data are not perfectly suited for Ripley's K analysis because the randomized site locations were stratified with the user-defined minimum elevation of site locations being ≥700 m.a.s.l. However, because the distribution of land ≥700 m.a.s.l. is relatively well distributed across the study area, Ripley's K patterns were deemed as potentially instructive in showing a site selection bias in the distribution of Actual DGB sites when compared to their stratified random counterparts [53,54].

Results

Ripley's K

The various measures of spatial analysis demonstrate that the placement of DGB sites on the landscape for intervisibility concerns was not random. Ripley's K analysis can be highly sensitive to the size of the project area being evaluated [62,63], and the results of our analysis bear this out. The Ripley's K analysis demonstrates clustering of Actual DGB sites at the scale of ≤300 m (exceeding the 95% confidence envelope), a moderate degree of clustering ≤1000 m (within the 95% confidence envelope), but dispersal at >1000 m (Fig. 4). In contrast, the randomly placed points simulating the potential DGB sites (Random DGB 1 was selected for the figure, but all results are similarly distributed) show a random distribution ≤2500 m with dispersion thereafter (Fig. 4). Dispersion measured >2500 m is statistically significant, but analytically insignificant as the project area itself is only 39.7 km^2. Therefore, dispersion measured above the 2500-m scale in Random DGB sites reflects the effects of the boundaries in the simulation, as the computations include hypothetical neighbors outside the project area itself. When compared to the 10 permutations of stratified Random DGB sites, the clustering of the Actual DGB sites ≤300 m and moderate clustering between 300 and 1000 m is provisionally interpreted as a function of a necessary principle of intervisibility-the sites need to be within view of each other and dispersion of the sites would hamper observers' abilities to see other sites. The clustered pattern of Actual DGB sites ≤1000-m scale is interpreted to reflect the apparent preference to locate DGB sites in locations in which other DGB sites are visible. However, the 10 permutations of Random DGB sites do not reflect this bias, so, predictably, their distribution is not patterned to accommodate this concern.

Cochran's Q and McNemar's contingency tests

In order to further test this assumption, a strict measure of the degree to which the placement of sites was related to their ability to see one another was generated. The comparison of the intervisibility of Actual DGB sites compared to the 10 permutations of random sites was performed using the Cochran's Q test (Fig. 5; Table S1). The test statistics of the 11 sites ($n = 240$) yielded a Cochran's Q value of 444.502 ($df = 10$, asymptotic significance = 0.000). With only two degrees of freedom (comparing Actual DGB sites with Random DGB-1 and Random DGB-2), the Cochran's Q value was 98.709 (asymp. sig. = 0.000). When removing the Actual DGB sites from the Cochran's Q, the value was 59.621 ($df = 9$, asymp. sig. = 0.000); while with two degrees of freedom (comparing Random DGB-1, −2 and −3), the value was 15.955 (asymp. sig. = 0.000). We interpret the asymptotic significance of Cochran's Q among the Random DGB sites as reflecting the bimodal nature of the intervisibility of the stratified random sample with Random DGB-1, −2, −6, −7 and −8 having 10 or more intervisible sites and Random DGB-3, −4, −5, −9 and −10 having six or less intervisible sites. The stratified random sampling of the test DGB sites skewed intervisibility of sites higher than a truly random distribution, but not as high as the actual distribution of DGB sites.

Therefore, in order to test whether there was a systematic bias in selecting stratified random DGB sites, a McNemar contingency test was performed comparing all stratified random samples (Random DGB-1– Random DGB-10) dichotomously with Actual DGB sites. The results are shown in Table S2 and demonstrate that the differences between the actual distribution of intervisible DGB sites and stratified random samples are statistically significant above the 99% confidence envelope. Furthermore, five permutations of McNemar's test were performed on the samples. The samples selected were non-biased (1 vs. 2, 3 vs. 4, 5 vs. 6, 7 vs. 8, 9 vs. 10) and demonstrate no statistical difference in the intervisibility of sites at the 95% confidence interval (Table S2). The results clearly demonstrate that the stratified random selection of hypothetical DGB sites did not inherently bias the intervisibility of sites against the actual distribution of sites.

Spearman's-ρ test

Analyzing landscape visibility (a.k.a. cumulative viewshed) of Actual DGB sites vs. Random DGB-1– Random DGB-10, shows that Actual DGB sites command a higher visibility of the surrounding landscape compared to the test sites (Fig. 6; Fig. 7; Table 1). The Spearman's-ρ analysis shows almost no differences in the ranks of data between Actual and Random DGB sites and an average of the stratified random DGB sites. A Spearman's-ρ analysis of ranks of the number of sites visible from individual 30-×-30 m grid cells from the Actual and Random DGB sites vs. the average ranks of Random DGB-1– Random DGB-10 (Table 2) show strong positive relationships between the ranks of data. All R values were >0.9 with Random DGB-9 having the weakest correlation coefficient and Random 1, 2, 7 and 10 showing perfect (+1) positive correlations. Actual DGB sites had R-values of 0.98 and τ-values of 18.74, which wasn't significantly different than the average values. We interpret the strong positive relationships between these data as reflecting the relatively good aspect of visibility of sites >700 m.a.s.l. selected in the stratified random sampling method undertaken in this study.

However, a more categorical approach to the data shows that subtle visibility conditions may have played a bigger role than the scaled data analysis suggests. Visible portions of the landscape from Actual DGB sites (23.4 km^2) was almost 2-σ above the mean distribution compared to Random DGB-1– Random DGB-10. Additionally, while there were no instances of portions of the landscape where ≥10 Random DGB sites were visible, there were 1011 cells totaling 0.9 km^2 where ≥10 Actual DGB sites were visible. The weighted average of visible cells on the landscape of Actual DGB sites (2.3 sites per cell) was >2-σ above the next highest permutation of Random DGB-2 (1.1 sites per cell). The results demonstrate that not only were intervisible locations factored into the construction of Actual DGB sites, but the ability of DGB sites to view the surrounding landscape (and probably vice versa) appears to have been a consideration.

Testing with Bayesian logistic regression model

The posterior means and standard deviations of α, β, and γ calculated from the posterior samples are −6.1014, 0.2828, 0.9990 and 0.61727, 0.07143, 0.03161, respectively (Fig. 8). The posterior probability of $\beta \neq 0$ or equivalently $\gamma = 1$ is 0.0001. Since we used the equal prior probability for $H_{0(null)}$ and H_1, the ratio of the posterior probabilities of $H_{0(null)}$ and H_1 is the same as the Bayes factor B_{10} and we obtained $\log B_{10} = 6.90$. According to Kass and Raftery's [64] interpretation of the Bayes factor, this is decisive evidence of H_1; furthermore, the posterior probability of $\beta > 0$ is 0.999. On that basis, we conclude that rejecting $H_{0(null)}$ indicates that the probability of building a DGB site is larger for the tiles with a greater number of intervisible DGB sites.

Secondary factors (elevation and slope)

Secondary factors of site location were also evaluated pertaining to elevation and slope. Actual DGB sites are ≤1.3-σ below the average elevation of all DGB sites (Actual and Random DGB-1– Random DGB-10; Table 3). However, the slope of Actual DGB sites are 2.1-σ less than average DGB sites (Actual and Random DGB-1– Random DGB-10), indicating that there is a much stronger preference for constructing Actual DGB sites on lower slope than lower elevation. Maintaining a constant elevation of DGB sites is not interpreted as having played a significant role in site selection, as Actual DGB sites are within 1-σ of the mean standard deviation of all DGB sites analyzed in this study. Therefore, principles of intervisibility of DGB sites to each other, the visibility of DGB sites to places on the landscape (and vice versa) as well as the secondary factor of low-slope topography (which may have facilitated and enhanced the first two factors) are interpreted as prime factors in determining the locations of Actual DGB sites in Mandara Mountains of northern Cameroon.

Discussion

Based on the GIS and related statistical analyses of the locations of DGB sites, the null hypotheses (H_0–the locations of DGB sites are randomly located above 700 m.a.s.l.) can be rejected. The statistical evidence demonstrates that there is >99% probability that good intervisibility of DGB sites and/or the surrounding landscape was considered when deciding where to construct sites. The GIS analysis confirms previous research hypotheses [23] that the inferred intervisibility of DGB sites was significant, possibly for maintaining relations of cooperation and/or competition among the communities farming the adjacent hill-slopes. DGB sites retain community ritual and patrimonial functions today [65], and the archaeological evidence suggests that their earliest functions were related at least in part to ritual use [19].

As cited above, intervisibility studies of point features on archaeological landscapes are increasingly common because of their relative ease using GIS computational techniques. We employ relatively novel statistical measures for archaeological research (Cochran's Q, McNemar's, Bayesian logistic regression) to crosscheck the effectiveness of our stratified randomization of potential DGB sites within a circumscribed study area. The advantage of two-way tests over K–S tests is that they are designed to compare binary tables rather than means of distribution. Since the analytical design of the intervisibility study was to compare identical distributions of points, means testing are not suitable to test the hypotheses.

In this study, we reject the null hypothesis that the Actual DGB sites are randomly located, but based on the Cochran's Q analysis, we must also reject the null hypothesis for the Random DGB sites. However, we argue that the virtue of the stratified random sample augmented the inherent intervisibility of the sites in the sample area. For this analysis, we stress that the actual value of the Cochran's Q test provides the significance of the difference between the Actual and Random DGB sites. In this case, given the relative degrees of freedom analyzed, inclusion of the Actual DGB sites yielded Cochran's Q values 5–7× higher than when only Random DGB-1– Random DGB-10 were analyzed. McNemar's contingency tests individually comparing the Actual DGB sites to each of the Random DGB sites confirms that the actual distribution of sites is significantly different than random sites, while the same test performed on Random DGB sites to one another did not show significant differences. These results are further supported with Bayesian logistic regression, which tests whether there is a mathematical preference for site construction within highly visible portions of the landscape.

Weighted means and Spearman's-ρ analysis of the cumulative viewshed shows that DGB sites were either positioned to see or be seen from the surrounding landscape. Stratified random permutations of DGB sites above 700 m.a.s.l. have significantly smaller cumulative viewsheds than the Actual DGB sites. The Ripley's K analysis shows clustering of DGB sites ≤1000 m, and the clusters appear to have been oriented to maximize the cumulative as well as intervisible viewshed.

Possible consideration for a high viewshed is the fact that sites are located on low slope benches of the Mandara massif, which also coincidentally commanded better views of the surrounding areas than any of the Random DGB sites. Site slope has been previously determined to be a high predictor of archaeological settlement, because people do not like to construct settlements in high-slope environments [66–68]. Although this explanation suffices for explaining cumulative viewsheds, it does not explain the intervisibility of the features to one another.

These analyses do not exclude other selective factors, such as geology, soil fertility, or social boundaries, but they do demonstrate that seeing and/or being seen by other DGB sites and from points on the surrounding landscape featured prominently in the choice of locations for these sites. As already noted, there are a number of modern ceremonial cycles still practiced in the region that encompass multiple communities, and where intervisibility could well have played a part in some aspects of the ritual, especially in determining when the beginning of the ceremony was 'passed' from one community to another [28]. While the details of, for example, the modern *maray* ceremonial cycle do not fit with the architectural details of the DGB sites, *maray* at least provides a conceptual framework for understanding interacting, community-based rituals in the area. In addition, it is quite likely that at least some of the sites, and especially the DGB-1/DGB-2 complex, played a sociopolitical role involving conspicuous display – and conspicuousness is, of course, closely related to visibility and intervisibility.

Conclusions

Cumulative viewshed and two-way analyses of archaeological sites from northern Cameroon compared to 10 permutations of stratified random sites above 700 m.a.s.l. within a restricted study area concludes that placement of the sites was non-randomly situated to enhance intervisibility between sites, and visibility between the sites and the surrounding landscape. A Bayesian logical regression model also rejects the null hypothesis that site construction location was random according to principles of enhanced visibility. Although other contingent settlement factors cannot be excluded, the locations of the sites were ideally located for maximizing the viewshed of DGB sites, and this likely relates to the interpreted social functions of the sites.

Ripley's K analysis shows a clustering ≤200 m (above the 95% confidence envelope) and moderate to weak degree of clustering 200–1000 m (within the 95% confidence envelope) reflecting the placement of the sites to maximize the view of intervening areas within the Mandara Mountains. Spearman's-ρ analyses and weighted means of 30-×-30-m grid cells to Actual DGB sites is ~2-σ above the mean for all 11 permutations of cumulative visibility analyzed. Furthermore, intervisibility of the sites was demonstrated using binary statistical techniques directly comparing the Actual DGB sites vs. permutations of Random DGB sites. Both Cochran's Q and McNemar contingency tests indicate that Actual DGB sites have significantly better intervisibility than Random DGB sites in all of the test cases, which is further supported using a logistic regression model.

It is not clear how many of the DGB sites were contiguously utilized, although the available radiocarbon dates indicate some degree of contemporaneity. Intervisibility of some of the features does seem to have affected their placement within the Mandara Mountains along with placement of the sites on lower slope portions of the inselbergs compared to Random DGB sites. This information will be important in continuing analysis of the functioning of the DGB sites. Available archaeological data indicate that that functioning was complex, probably involving both ritual and sociopolitical roles, and the sites probably acted within both local and regional systems of interaction and competition. The placement of the sites in locations that increase visibility and intervisibility certainly reinforces our appreciation of their significance in cultural systems in the Mandara Mountains in the mid-second millennium AD.

Supporting Information

Table S1 Results of Cochran's Q tests. (a) Actual DGB and Random DGB-1–10, (b) Actual DGB and Random-1–2, (c) Random DGB-1–10, (d) Random DGB-1–3. Cochran's Q tests are binary with the number of non-intervisible sites coded as 0, while the number of intervisible sites are coded as 1.

Table S2 McNemar's test comparing (a) Actual DGB sites vs. 10 independent permutations of DGB Random sites and (b) permutations of Random DGB sites against each other.

Acknowledgments

We thank the people of the Mandara Mountains who have provided generous insights into their unique lives to SM over the years. GIS laboratory facilities were provided by the Department of Archaeology and Art History in the College of Humanities at Seoul National University. We thank reviewer Mal Ridges and the editors for consideration of the manuscript.

Author Contributions

Conceived and designed the experiments: DKW JL. Performed the experiments: DKW JL. Analyzed the data: DKW SM JL. Contributed reagents/materials/analysis tools: DKW JL. Contributed to the writing of the manuscript: DKW SM JL. Designed and conceived the geospatial experiments and GIS methods: DKW. Conducted fieldwork identifying sites used in this manuscript: SM. Designed and conceived logistic regression experiment: JL.

References

1. Parsons JR (1972) Archaeological Settlement Patterns. Annual Review of Anthropology 1: 127–150.
2. Binford LR (1980) Willow Smoke and Dogs' Tails: Hunter-Gatherer Settlement Systems and Archaeological Site Formation. American Antiquity 45: 4–20.
3. Hassan FA (1981) Demographic archaeology. New York: Academic Press.
4. Hodder I (1991) Post-modernism, post-structuralism and post-processual archaeology. In: Hodder I, editor. The meanings of things: material culture and symbolic expression. London: Unwin Hyman. pp. 64–78.
5. Earle TK, Preucel RW, Brumfiel EM, Carr C, Limp WF, et al. (1987) Processual Archaeology and the Radical Critique [and Comments and Reply]. Current Anthropology 28: 501–538.
6. O'Brien M, Lyman RL (2004) History and Explanation in Archaeology. Anthropological theory 4: 173–197.
7. Pauketat TR (2001) Practice and history in archaeology. Anthropological theory 1: 73–98.
8. Kirch PV (1990) Monumental architecture and power in Polynesian chiefdoms: A comparison of Tonga and Hawaii. World Archaeology 22: 206–222.
9. Sinopoli CM (1994) The Archaeology of Empires. Annual Review of Anthropology 23: 159–180.
10. Brück J (2001) Monuments, power and personhood in the British Neolithic. Journal of the Royal Anthropological Institute 7: 649–667.
11. Bonanno A, Gouder T, Malone C, Stoddart S (1990) Monuments in an island society: The Maltese context. World Archaeology 22: 190–205.
12. Dillehay TD (2004) Social landscape and ritual pause: uncertainty and integration in Formative Peru. Journal of Social Archaeology 4: 239–268.
13. Letesson Q, Vansteenhuyse K (2006) Towards an Archaeology of Perception: 'Looking' at the Minoan Palaces. Journal of Mediterranean Archaeology 19: 91–119.
14. Garlake P (2002) Early Art and Architecture of Africa. Oxford: Oxford University Press.
15. Shaw T, Sinclair P, Andah BW, Okpoko AI (1993) Archaeology of Africa: foods, metals and towns. London: Routledge.
16. Connah G (1987) African Civilizations Precolonial Cities and States in Tropical Africa: An Archaeological Perspective. Cambridge: Cambridge University Press.
17. Hildebrand EA (2013) Is monumentality in the eye of the beholder? Lessons from constructed spaces in Africa. Azania: Archaeological Research in Africa 48: 155–172.
18. McIntosh SK, editor (1999) Beyond chiefdoms: pathways to complexity in Africa. Cambridge: Cambridge University Press.
19. MacEachern S, David N (2013) Monumental architecture in mountain landscapes: the diy-geδ-bay sites of northern Cameroon. Azania: Archaeological Research in Africa 48: 241–262.
20. MacEachern S (2012) Wandala and the DGB sites: political centralisation and its alternatives north of the Mandara Mountains, Cameroon. Azania: Archaeological Research in Africa.
21. MacEachern S (1993) Selling the iron for their shackles: Wandala-montagnard interactions in northern Cameroon. Journal of African History 34: 247–270.
22. MacEachern S (2002) Beyond the belly of the house: space and power in the Mandara Mountains. Journal of Social Archaeology 2: 179–219.
23. David N (2008) Performance and agency: the DGB sites of Northern Cameroon. Oxford: British Archaeological Reports.
24. Fitton JG, Dunlop HM (1985) The Cameroon line, West Africa, and its bearing on the origin of oceanic and continental alkali basalt. Earth and Planetary Science Letters 72: 23–38.
25. Elangwe HN (1979) Carte Géologique de la République Unie du Cameroun. Yaounde: Minister of Mines and Power.
26. Letouzey R, Fotius G (1985) Carte Phytogéographique du Cameroun. Toulouse: Institut de la Carte Internationale de la Végétation.
27. MacEachern S (2012) The Holocene history of the southern Lake Chad Basin: archaeological, linguistic and genetic evidence. African Archaeological Review: 1–19.
28. von Graffenried C (1984) Das Jahr des Stieres, ein Opferrituel der Zulgo und Gemjek in Nordkamerun. Fribourg: Universitätsverlag.
29. Conolly J, Lake M (2006) Geographic Information Systems in Archaeology. Cambridge: Cambridge University Press.
30. Kvamme KL (1990) One-Sample Tests in Regional Archaeological Analysis: New Possibilities through Computer Technology. American Antiquity 55: 367–381.
31. Wheatley D (1995) Cumulative viewshed analysis: a GIS-based method for investigating intervisibility, and its archaeological application. In: Lock GR, Stančič Z, editors. Archaeology And Geographic Information Systems: A European Perspective. London: Taylor and Francis. pp. 171–186.
32. Sevenant M, Antrop M (2007) Settlement models, land use and visibility in rural landscapes: Two case studies in Greece. Landscape and Urban Planning 80: 362–374.
33. Fisher P, Farrelly C, Maddocks A, Ruggles C (1997) Spatial Analysis of Visible Areas from the Bronze Age Cairns of Mull. Journal of Archaeological Science 24: 581–592.
34. Razali NM, Wah YB (2011) Power comparisons of Shapiro-Wilk, Kolmogorov-Smirnov, Lilliefors and Anderson-Darling tests. Journal of Statistical Modeling and Analytics 2: 21–33.
35. Ogburn DE (2006) Assessing the level of visibility of cultural objects in past landscapes. Journal of Archaeological Science 33: 405–413.
36. Llobera M (2001) Building Past Landscape Perception With GIS: Understanding Topographic Prominence. Journal of Archaeological Science 28: 1005–1014.
37. Lake MW, Woodman PE, Mithen SJ (1998) Tailoring GIS Software for Archaeological Applications: An Example Concerning Viewshed Analysis. Journal of Archaeological Science 25: 27–38.
38. O'Sullivan D, Turner A (2001) Visibility graphs and landscape visibility analysis. International Journal of Geographical Information Science 15: 221–237.
39. Jones EE (2006) Using Viewshed Analysis to Explore Settlement Choice: A Case Study of the Onondaga Iroquois. American Antiquity 71: 523–538.
40. Jones EE (2010) An analysis of factors influencing sixteenth and seventeenth century Haudenosaunee (Iroquois) settlement locations. Journal of Anthropological Archaeology 29: 1–14.
41. Sakaguchi T, Morin J, Dickie R (2010) Defensibility of large prehistoric sites in the Mid-Fraser region on the Canadian Plateau. Journal of Archaeological Science 37: 1171–1185.
42. Kay S, Sly T (2001) An application of Cumulative Viewshed Analysis to a medieval archaeological study: the beacon system of the Isle of Wight, United Kingdom. Archeologia e Calcolatori 12: 167–179.
43. Llobera M, Fábrega-Álvarez P, Parcero-Oubiña C (2011) Order in movement: a GIS approach to accessibility. Journal of Archaeological Science 38: 843–851.
44. Bauer A, Nicoll K, Park L, Matney T (2004) Archaeological site distribution by geomorphic setting in the southern lower Cuyahoga River Valley, northeastern Ohio: Initial observations from a GIS database. Geoarchaeology 19: 711–729.
45. Krist FJ, Brown DG (1994) GIS modeling of Paleo-Indian period caribou migrations and viewsheds in northeastern Lower Michigan. Photogrammetric Engineering and Remote Sensing 60: 1129–1137.
46. Chapman HP (2003) Rudston 'Cursus A'– Engaging with a Neolithic Monument in Its Landscape Setting Using GIS. Oxford Journal of Archaeology 22: 345–356.
47. Bradley R, Harding J, Mathews M (1993) The siting of prehistoric rock art in Galloway, south-west Scotland. Proceedings of the Prehistoric Society 59: 129–145.

48. Howey MCL (2007) Using multi-criteria cost surface analysis to explore past regional landscapes: a case study of ritual activity and social interaction in Michigan, AD 1200–1600. Journal of Archaeological Science 34: 1830–1846.
49. Lambers K, Sauerbier M (2006) GIS-based visibility studies of the Nasca geoglyphs at Palpa, Peru. In: Baltsavias M, Gruen A, van Gool L, Paterak M, editors. Recording, Modeling and Visualization of Cultural Heritage: Proceedings of the International Workshop, Centro Stefano Franscini, Monte Verita, Ascona, Switzerland, May 22–27, 2005. London: Taylor & Francis. pp. 249–261.
50. Bongers J, Arkush E, Harrower M (2012) Landscapes of death: GIS-based analyses of chullpas in the western Lake Titicaca basin. Journal of Archaeological Science 39: 1687–1693.
51. Fraser D (1983) Land and Society in Neolithic Orkney. Oxford: British Archaeological Reports 117.
52. Vanacker V, Govers G, van Peer P, Verbeek C, Desmet J Jr, et al. (2001) Using Monte Carlo Simulation for the Environmental Analysis of Small Archaeologic Datasets, with the Mesolithic in Northeast Belgium as a Case Study. Journal of Archaeological Science 28: 661–669.
53. Crema ER, Bevan A, Lake MW (2010) A probabilistic framework for assessing spatio-temporal point patterns in the archaeological record. Journal of Archaeological Science 37: 1118–1130.
54. Bevan A, Conolly J (2006) Multiscalar Approaches to Settlement Pattern Analysis. In: Lock G, Molyneaux B, editors. Confronting Scale in Archaeology: Springer US. pp. 217–234.
55. Scarsini M (1984) On measures of concordance. Stochastica 8: 201–218.
56. Patil KD (1975) Cochran's Q Test: Exact Distribution. Journal of the American Statistical Association 70: 186–189.
57. McNemar Q (1947) Note on the sampling error of the difference between correlated proportions or percentages. Psychometrika 12: 153–157.
58. Cox DR (1970) Analysis of Binary Data. London: Methuen.
59. Chan IS (2003) Proving non-inferiority or equivalence of two treatments with dichotomous endpoints using exact methods. Statistical Methods in Medical Research 12: 37–58.
60. Guastaferro F, Castrignanò A, Benedetto D, Sollitto D, Troccoli A, et al. (2010) A comparison of different algorithms for the delineation of management zones. Precision Agriculture 11: 600–620.
61. Dixon PM (2006) Ripley's K Function. In: El-Shaarawi AH, Piegorsch WW, editors. Encyclopedia of Environmetrics. New York: John Wiley & Sons. 1796–1803.
62. Goreaud F, Pélissier R (1999) On explicit formulas of edge effect correction for Ripley's K-function. Journal of Vegetation Science 10: 433–438.
63. Tang M, Tang S, Lei X, Zhang H, Hong L, et al. (2003) Edge correction of Ripley's K(d) function on population spatial pattern analysis. Acta Ecologica Sinica 23: 1533–1538.
64. Kass RE, Raftery AE (1995) Bayes Factors. Journal of the American Statistical Association 90: 773–795.
65. Datouang Djoussou JM (2011) Archaeology and cultural heritage in Cameroon: the case of the DGB sites. Antiquity.
66. Warren RE, Asch DL (2000) A predictive model of archoaelogical site location in the eastern Prairie Peninsula. In: Wescott K, Brandon RJ, editors. Practical Applications of GIS for Archaeologists: A Predictive Modelling Toolkit. London: Taylor & Francis. pp. 6–36.
67. Kohler TA, Parker S (1986) Predictive models for archaeological resource location. In: Schiffer M, editor. Advances in Archaeological Method and Theory, Vol 9. New York: Academic Press. pp. 397–452.
68. Kvamme KL (1990) The fundamental principles and practice of predictive archaeological modeling. In: Voorrips A, editor. Mathematicsand Information Science in Archaeology: A Flexible Framework. Bonn: Studies in Modern Archaeology, vol. 3. Holos-Verlag. pp. 257–295.

19

Elevational Distribution and Ecology of Small Mammals on Africa's Highest Mountain

William T. Stanley[1]*, Mary Anne Rogers[1], Philip M. Kihaule[2], Maiko J. Munissi[3]

1 The Field Museum of Natural History, Department of Science and Education, Chicago, Illinois, United States of America, 2 University of Dar es Salaam, Department of Zoology, Dar es Salaam, Tanzania, 3 Southern Highlands Conservation Programme, Mbeya, Tanzania

Abstract

Mt Kilimanjaro is Africa's highest mountain, and an icon for a country famous for its mammalian fauna. The distribution and abundance of small mammals on the mountain are poorly known. Here we document the distribution of shrews and rodents along an elevational gradient on the southeastern versant of Kilimanjaro. Five sites were sampled with elevational center points of 2000, 2500, 3000, 3500 and 4000 m, using a systematic methodology of standard traps and pitfall lines, to inventory the shrews and rodents of the slope. Sixteen species of mammal were recorded, including 6 shrew and 10 rodent species, and the greatest diversity of both was found at 3000 m, the elevational midpoint of the transect. No species previously unrecorded on Kilimanjaro were observed. Two genera of rodents that occur in nearby mountains (*Hylomyscus* and *Beamys*) were not recorded. *Myosorex zinki*, the only mammal endemic to Mt. Kilimanjaro, which previously was known by only a few specimens collected in the ericaceous or moorland habitat, was found in all but one (the lowest) of the sites sampled, and was one of the most widespread species of small mammal along the gradient. Two shrews (*Crocidura allex* and *Sylvisorex granti*) and one rodent (*Dendromus insignis*) were found throughout the entire transect, with *Dendromus* being observed at our highest trap point (4240 m). As in similar faunal surveys on other mountains of Tanzania, rainfall influenced the sample success of shrews, but not rodents. Trap success for rodents at 3500 m was notably low. This study contributes further justification for the conservation of the forest habitat of Mt. Kilimanjaro.

Editor: Robert Guralnick, University of Colorado, United States of America

Funding: Funding for this study came from the Field Museum of Natural History, including awards from the Marshall Field, Barbara Brown and Ellen Thorne Smith Funds of the Field Museum. The funders had no role in study design, data collection and analysis, decision to publish, or preparation of the manuscript.

Competing Interests: The authors have declared that no competing interests exist.

* Email: bstanley@fieldmuseum.org

Introduction

Knowing the distribution of organisms along an elevational gradient is critical to understanding the evolution and ecology of montane biotic systems, and to designing conservation strategies to maintain them. These reasons have motivated elevational surveys of small mammals in various areas of the world including Chile [1], Costa Rica [2], Malaysia [3], Philippines [4], [5], [6], Taiwan [7], and Tanzania [8]. Goodman, Ganzhorn & Rakotondravony [9] summarize some of the important biotic inventories along elevational gradients in Madagascar. Each of these studies elucidate both specific and broadly general patterns that help explain the mechanisms influencing the distributions of mammals along such gradients with significant implications for biogeographic analysis and conservation priorities [10]. Indeed, such surveys have served as vital baselines for comparison to subsequent inventories in testing the influence of climatic vicissitudes or habitat alteration. For example, range shifts in various mammalian species were documented in Yosemite Valley, California, with two similar surveys separated by almost a century [11].

Knowledge of the ecology and behavior of the targeted faunas help frame considerations of the results of systematic sampling along gradients. For example, Stanley & Hutterer [8] documented patterns of distribution along an altitudinal gradient in the Udzungwa Mountains of Tanzania that differed between shrews and rodents, and suggested that the amount of coincident rainfall influenced shrew, but not rodent, capture rates. Such observations must be factored into deciphering the results of systematic sampling along elevational gradients, and surveys using identical methodologies on other mountains should help to reveal whether such observations are unique to particular sites or more common across multiple gradients.

Mt. Kilimanjaro is the highest mountain in Africa and an icon for a region renowned for its unique mammalian fauna. Ironically, the mammals that inhabit the habitats of this volcano are relatively unknown, with most historical attention focused on larger species, leading to calls for complete inventories of the fauna of the mountain [12]. To date, the most comprehensive summary of our overall understanding of the mammalian fauna of Kilimanjaro remains that presented by Grimshaw, Cordeiro & Foley [13], who provided a faunal list of the mountain, and described past studies of Kilimanjaro's mammalian fauna. Few studies employing systematic sampling have taken place on Kilimanjaro [14] and only one [15] used a systematic survey to document the presence and distribution of small rodents and shrews along elevational gradients on the mountain. The lack of detailed biotic vertebrate surveys, such as those of small mammals, hampers efforts to monitor ecological change over time on the mountain. Thompson

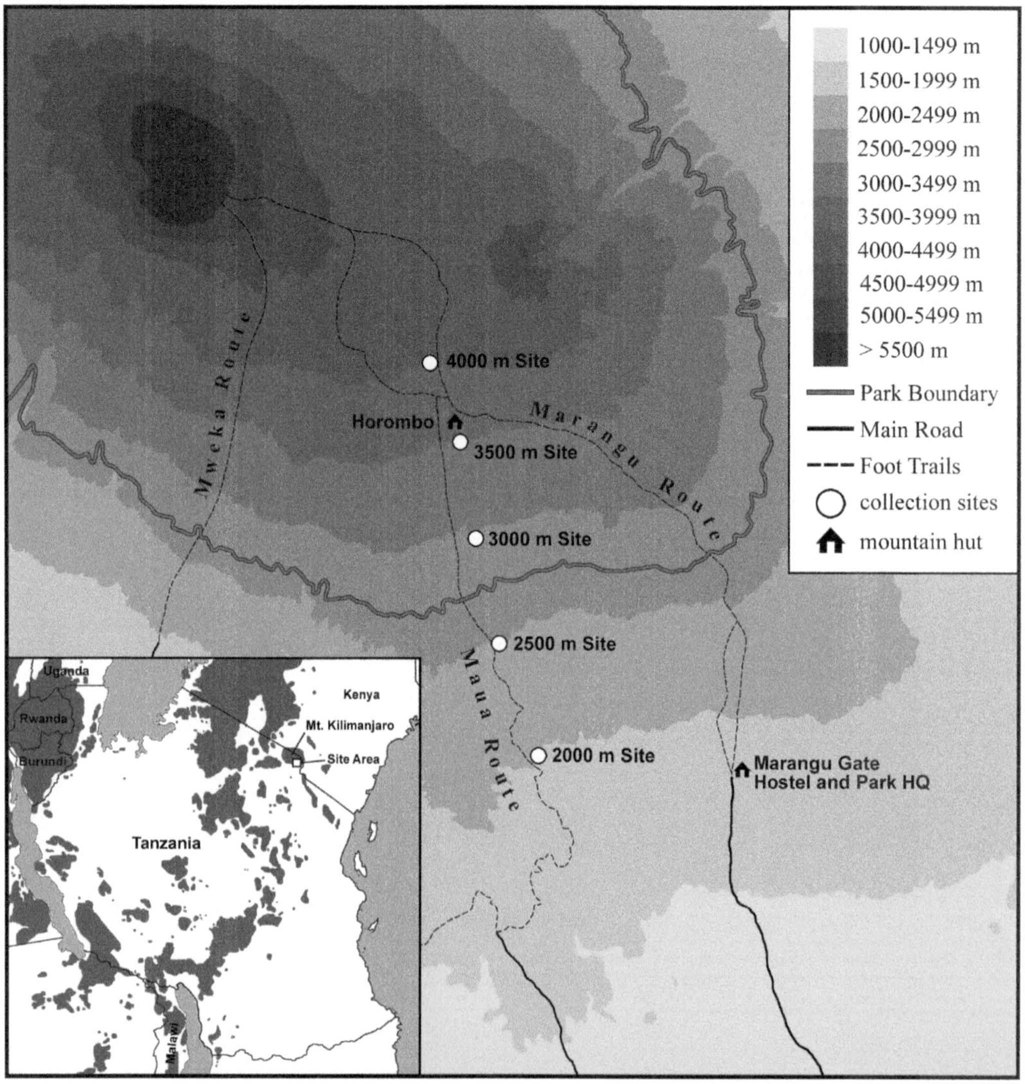

Figure 1. Map of Mt. Kilimanjaro showing routes, elevational contours and study sites.

et al. [16] suggest that climate change is affecting the habitat and ecology of Kilimanjaro, and baseline data for the distribution and abundance of various plants and animals are needed to judge the effect of such changes, as has been done elsewhere [11].

Using a standardized sampling regime that has been utilized in several other montane sites of Tanzania over the past two decades [8], [17], [18], [19] we surveyed the small mammals (shrews and rodents) at five different elevations and habitats along the southeastern versant of Mt. Kilimanjaro. Our study had three principal goals: 1) to initiate intensive surveys of the elevational distribution and abundance of small mammals along the transect sampled; 2) to test for differences between rodents and shrews in their relationship to elevation and response to different trapping methodologies; and 3) to compare the generated results to similar studies on Kilimanjaro and other mountains of Tanzania.

Materials and Methods

Study Site

Mt. Kilimanjaro is in northeastern Tanzania and reaches an elevation of 5895 m. An extinct volcano, the mountain is the conglomeration of three volcanoes: Kibo (the highest, most prominent and familiar), Mawenzi (the second peak of the mountain), and Shira (a plateau) [20]. Because the mountain is a popular destination for climbers, there are numerous paths that originate in the lowlands and run up the side of the mountain [21]. Two such routes that are on the southeastern (and wettest) versant are "Marangu" and "Mweka". Between these two is the "Maua" path which is currently closed to tourists, and is used by Kilimanjaro National Park (KINAPA) staff to access and maintain facilities within the park. Between 17 July and 31 August 2002, we sampled the small mammals (shrews and rodents) at five different elevations, ranging from roughly 2000 to 4000 m, along the "Maua" route on the southeastern slope of Mt. Kilimanjaro (Figure 1).

The specific localities, elevations, habitats (sensu Mwasaga [22]) and dates of sampling are listed below. The elevations given for each site are centered at the associated camp and sampling efforts spanned roughly 100–200 m above and below the camp. For this reason, we labeled each camp at the closest 500 m interval (2043 = 2000 m; 2470 = 2500 m, etc.). Temperature and rainfall for each site (measured at camp) are listed in Table 1:

Table 1. Climatic data for each of the sites sampled on Mt. Kilimanjaro in July-August, 2002.

Elevation (m)	Daily Minimum Temperature (°C)	Daily Maximum Temperature (°C)	Daily rainfall (mm)
2000	8.5°±1.7	14.1°±2.1	4.2±3.0
	5–10°	12–18°	0–9
	N=9	N=8	N=9 (8)
2500	3.6°±2.2	11.6°±1.9	1.6±0.3
	1–6.5°	9–15°	0–1.8
	N=9	N=9	N=8 (2)
3000	2.2°±2.5	9.3°±2.5	1.5±3.1
	−2–5°	6–12°	0–9.5
	N=9	N=8	N=9 (6)
3500	−0.9°±1.2	12.3°±3.5	0.1±0.1
	−3–1°	7.5–17°	0–0.2
	N=9	N=8	N=8 (3)
4000	−6.8°±3.6	20.5°±5.9	1.5±3.4
	−12–−1°	11–25.5°	0–10.2
	N=9	N=8	N=9 (3)

Totals given as mean ± standard deviation, range and sample size (number of days measured). Sample size for rainfall is given as number of days monitored and (number of days with rain).

Site 1–2000 m). 4 km N, 1.5 km W Maua, 3°14.404' S, 37°27.502' E, 2043 m; lower montane forest; 23–30 August 2002.

Site 2–2500 m). 7 km N, 2.5 km W Maua, 3°12.459' S, 37°26.818' E, 2470 m; upper montane forest; 17–25 July 2002.

Site 3–3000 m). 10.5 km N, 3.5 km W Maua, 3°10.627' S, 37°26.413' E, 2897 m; ecotone between montane forest and ericaceous zone; 26 July-03 August 2002.

Site 4–3500 m). 13.5 km N, 4 km W Maua, 3°08.941' S, 37°26.133' E, 3477 m; ericaceous zone; 4–12 August 2002.

Site 5–4000 m). 16 km N, 4.5 km W Maua, 3°07.566' S, 37°25.600' E, 3995 m; ecotone between ericaceous and alpine zones; 13–21 August 2002.

Trapping Procedure

We used identical sampling techniques to those employed in similar small mammal surveys in other Tanzanian forests [8], [19], [23], [24], [25], [26], [27], [28]. Pitfall and trap lines were set in different microhabitats at each site, to sample shrews and small rodents (<200 g). Each pitfall line consisted of 11, 15 l buckets spaced 5 m apart, and placed so the upper rim was flush with the ground level. A 50 cm high vertical plastic fence was placed over the buckets, bisecting the openings. Most shrews and very small rodents were captured with this technique. Trap lines were installed using three types of traps: Museum Special traps, 14×7 cm; Victor rat traps, 17.5×8.5 cm (both manufactured by Woodstream Corporation, Lititz, Pennsylvania, USA); and medium-sized Sherman live traps, 23×9.5×8 cm (H.B. Sherman Traps Inc., Tallahassee, Florida, USA). Each line was composed of between 20 to 70 traps, with the Museum Special and Victor traps making up approximately 85% of each line. Traps were baited with pieces of freshly fried coconut coated in peanut butter, which was renewed each afternoon. Further details are outlined in Stanley, Goodman & Newmark [29].

All traps and buckets were checked once in the early morning and again in the late afternoon. Not all traps or buckets were employed for equal amounts of time (some trap lines were set the first day of the survey, others were installed on the second), so we use the measures "trap-night" and "bucket-night" (one trap or bucket in operation for one 24 hr period-0700 to 0700 hrs) to quantify sampling effort. We refer to the success rate of each method as either "trap success" or "bucket success", and calculate these values by dividing the number of individuals captured by the number of trap-nights or bucket-nights and multiplying by 100. In discussions involving the two trapping methodologies combined, the term "sampling-night" refers to either one trap-night or one bucket-night, and "sample success" refers to the success rate for the two methodologies combined. The latter is calculated by dividing the number of individuals captured by the number of sampling-nights and multiplying by 100.

Standard external measurements and reproductive status were recorded for each specimen, which was then either prepared as a study skin and skeleton or preserved in 10% formalin, and later transferred to 70% EtOH. Specimens are deposited in the Field Museum of Natural History (FMNH) with a portion to be returned to Museum of Zoology, University of Dar es Salaam (UDSM). We follow the taxonomy of Carleton & Stanley [30], Holden [31], Hutterer [32], and Musser & Carleton [33].

Ethics Statement

Permits for the collection and export of specimens were provided by the Tanzania Commission for Science and Technology (Ref# 2002-232-ER-90-172), the Tanzania Ministry of Natural Resources and Tourism (Wildlife Division; Ref# GD/R.40/1/22), and the Tanzania National Parks (Ref # TNP A44). Import of specimens into USA was approved by the US Fish and Wildlife Service (3177-W10214-9/18/02). Shrews and rodents were euthanized following the protocol approved by the American Society of Mammalogists [34], and the study was approved by the Field Museum of Natural History.

Results

During the survey, we accumulated 11,562 sample-nights (8361 trap-nights and 3201 bucket-nights) and trapped 612 small

Table 2. Trapping totals for rodents and shrews by trap technique on the southeastern slope of Mt. Kilimanjaro in July-August, 2002.

Elevation	2000 m	2500 m	3000 m	3500 m	4000 m	Totals
BUCKETS						
# bucket-nights	616	649	649	638	649	3201
# individuals	84	75	86	51	33	329
(% bucket success)	(13.6)	(11.5)	(13.2)	(8.0)	(5.1)	(10.3)
# species	10	5	7	3	5	13
# shrews	68	74	79	48	30	299
(% bucket success)	(11.0)	(11.4)	(12.2)	(7.5)	(4.6)	(9.3)
# shrew species	5	4	4	2	3	6
# rodents	16	1	7	3	3	30
(% bucket success)	(2.6)	(0.1)	(1.1)	(0.5)	(0.5)	(0.9)
# rodent species	5	1	3	1	2	7
TRAPS						
# trap-nights	1600	1776	1785	1600	1600	8361
# individuals	67	68	57	3	88	283
(% trap success)	(4.2)	(3.8)	(3.2)	(0.2)	(5.5)	(3.4)
# species	5	9	11	2	4	12
# rodents	65	63	48	3	84	263
(% trap success)	(4.1)	(3.5)	(2.7)	(0.2)	(5.2)	(3.1)
# rodent species	4	7	8	2	3	8
# shrews	2	5	9	0	4	20
(% bucket success)	(0.1)	(0.3)	(0.5)		(0.2)	(0.2)
# shrew species	1	2	3	0	1	4
TOTAL						
# sample-nights	2216	2425	2434	2238	2249	11562
# individuals	151	143	143	54	121	612
(% sample success)	(6.8)	(5.9)	(5.9)	(2.4)	(5.4)	(5.3)
# species	11	13	14	4	6	16

Table 3. Elevational distribution of Soricomorpha species along the southeastern slope of Mt. Kilimanjaro in July-August, 2002.

Elevation	2000 m	2500 m	3000 m	3500 m	4000 m	Totals
Species						
Crocidura allex	24	19	40	45	30	158
Crocidura hildegardeae	7	0	0	0	0	7
Crocidura monax	21	29	26	0	0	76
Crocidura olivieri	2	2	0	0	0	4
Myosorex zinki	0	3	4	3	3	13
Sylvisorex granti	16	26	18	0[a]	1	61
Total # individuals	70	79	88	48	34	319
Total # species	5	5	4	2+1[a]	3	6
Total # sample-nights	2216	2425	2434	2238	2249	11562
Sample success (%)	3.1	3.2	3.6	2.1	1.5	2.7
Total # caught in buckets	68	74	79	48	30	299
Total # bucket-nights	616	649	649	638	649	3201
Bucket success (%) for pitfall lines	11.0	11.4	12.2	7.5	4.6	9.3

Only specimens caught in traps or buckets are included in totals.
[a]presence inferred from occurrence at lower and higher sites.

Table 4. Elevational distribution of rodent species along the southeastern slope of Mt. Kilimanjaro in July-August, 2002.

Elevation	2000 m	2500 m	3000 m	3500 m	4000 m	Totals
Species						
Otomys angoniensis	0	1	1	0	0	2
Otomys tropicalis	0	4	1	0[a]	7	12
Dendromus insignis	4	1	5	5	21	36
Dendromus melanotis	5	1	4	0	0	10
Grammomys dolichurus	3	6	6	0	0	15
Lophuromys aquilus	23	25	17	0	0	65
Praomys taitae	37	25	3	0	0	65
Rhabdomys dilectus	0	0	11	1	59	71
Graphiurus murinus	9	1	5	0	0	15
Tachyoryctes daemon	0	0	2	0	0	2
Total # individuals	81	64	55	6	87	293
Total # species	6	8	10	2+1[a]	3	10
Total # sample-nights	2216	2425	2434	2238	2249	11562
Sample success (%)	3.6	2.6	2.2	0.3	3.9	2.5
Total # caught in traps	65	64	48	3	84	264
Total # trap-nights	1600	1776	1785	1600	1600	8361
Trap success (%)	4.1	3.6	2.7	0.2	5.2	3.1

Only specimens caught in traps or buckets are included in totals.
[a]presence inferred from occurrence at lower and higher sites.

mammals, including 319 shrews representing 6 species, and 293 rodents representing 10 species (Tables 2, 3, 4). Sampling success for shrews was significantly greater in buckets than in traps ($X^2 = 695.2$, $P<0.05$), and significantly more rodents were caught in traps than in buckets ($X^2 = 44.8$, $P<0.05$), a pattern observed in past studies on small mammals of Tanzania [8], [23], [24]. In 8361 trap-nights, 283 mammals were captured for an overall trap success of 3.4%. Of the mammals caught in traps, 263 were rodents (3.1% trap success for rodents) and 20 were shrews (0.2% trap success). In the 3201 bucket-nights, 329 mammals were captured for a total bucket success of 10.3%. Of these, 299 were shrews (9.3% success) and 30 were rodents (0.9% success). This striking pattern was evident not only across the entire survey, but also at each of the five sites sampled (Table 2). Shrew species caught in traps included *Crocidura allex*, *C. monax*, *C. olivieri*, and *Myosorex zinki* (weighing between 3.6–51.0 g). While most of the rodents caught in buckets were relatively small (i.e. *Dendromus insignis*; 7–20 g), both specimens of *Tachyoryctes daemon* (240–290 g) were captured in buckets. Other rodent species captured in buckets included *Grammomys dolichurus*, *Graphiurus murinus*, *Praomys taitae*, and *Rhabdomys dilectus*.

The number of captures (and overall sample success) at each elevational site ranged from 54 [2.4%] at 3500 m to 151 [6.8%] at 2000 m (Table 2). For shrews alone, the lowest values were observed at the 4000 m site (34 [1.5%]) and the highest values at the 3000 m site (88 [3.6%]; Tables 2, 3). For rodents, the lowest (6 [0.3%]) and highest (87 [3.9%]) values were observed at the 3500 m and 4000 m sites, respectively (Tables 2, 4). The cumulative number of species trapped reached an asymptote at all sites except 2500 m site (Figure 2), where *Dendromus insignis* and *Otomys angoniensis* were captured on the last day of trapping.

We examined the relationship of four daily capture parameters (number of individuals, number of species, number of new species [i.e. previously unsampled at a given site], and cumulative number of species) with cumulative sample-nights for both type of trapping methodology (Table 5) and mammalian order (Table 6). We chose cumulative sample-nights instead of day of sampling period because of the differences in sampling effort among sites (Table 2). Based on correlation analysis, there was a significant positive correlation between cumulative sample-nights and cumulative species across all sites for trap lines, bucket lines and both sampling methods combined (Table 5). The same pattern was generally evident at each site, although, in some cases, correlation values were high, but not significant. The correlation of the cumulative number of shrew species caught in buckets with cumulative number of bucket nights was significant at the 2500 m site, and high at all other sites. For both shrews and rodents, there was generally a negative correlation between cumulative sampling effort and new species captured. Notable exceptions include trap lines at the 3000 and 3500 m sites. There was no notable correlation between the number of species and cumulative sampling effort across the entire transect or at each site, with the exception of shrew species captured in buckets at the 2000 m site. The correlation between number of individuals and sampling effort varied among sites. There was a significantly negative pattern exhibited by pitfall lines and both trap and pitfall lines combined at the 3500 m site, but no such relationship exhibited at the 3000 and 4000 m sites. Table 6 presents the same analyses as Table 5, but is focused on the taxonomic groups sampled, and the patterns are similar.

The effect of rainfall on captures is presented in Table 7. Generally, there was a stronger and more positive correlation between rainfall and daily captures of shrews, than there was for rodents. Over the entire transect, the capture of individual shrews in both buckets and traps was significantly correlated with the amount of rainfall each day, but the capture of individual rodents was not. A graphic representation of the differences between shrew and rodent captures with respect to rainfall amount is presented in

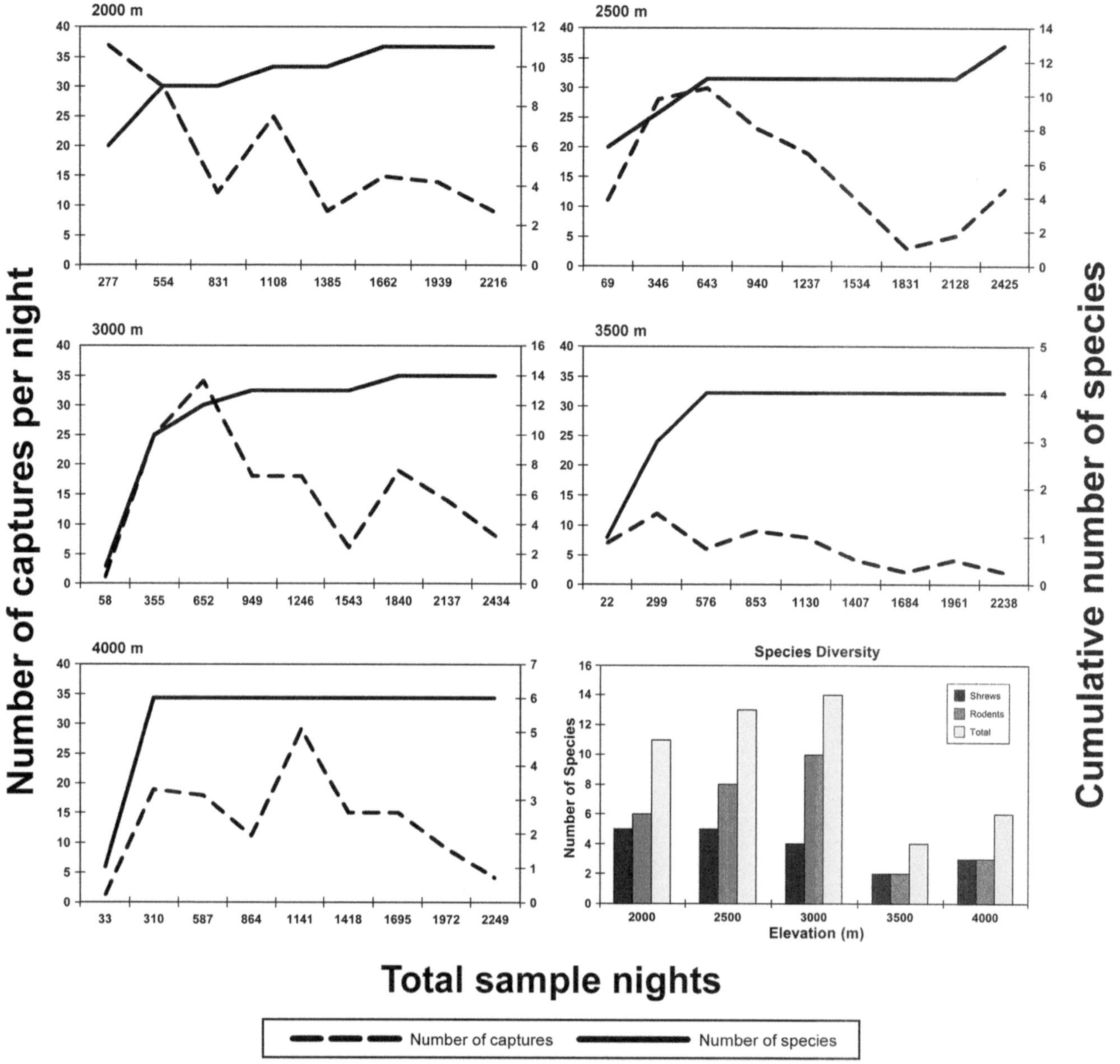

Figure 2. Species accumulation curves (for both pitfall and trap lines combined) for each site. The dashed lines represent the number of captures each day; the solid lines represent the cumulative number of new species for the site observed each day. The graph at the lower right shows the number of specimens of shrew, rodent and mammal captured at each site.

Figure 3. The overall relationship between rainfall and captures of shrews was not as strong as in other elevational surveys of mammals in Tanzania [8].

There was a significantly negative relationship between elevation and the total number of shrew species collected (Table 8). Additionally, elevation was negatively correlated with total number of individual shrews collected and sample success for shrews, with *r* values high, but not significant. Rodents showed no such notable pattern. The relationship between elevation and total sample success, number of individual mammals, and number of species collected for shrews and rodents combined was generally negative, but not significant. The least number of mammals, and species collected was at the 3500 m site. The greatest number of individuals noted was at the lowest site (2000 m), and the highest species diversity was observed at the 3000 m site. In most cases, the forested sites showed greater abundance and species diversity than the habitats above tree line (Tables 2, 3, 4).

Captures in any individual trap or bucket were rare events. Although there was a 10.3% bucket success for all mammals captured, and 329 animals (299 shrews and 30 rodents) were collected in 385 buckets (77 buckets installed at each of five sites), most buckets captured no animals. Over the entire survey, 203 buckets caught nothing, 100 took one animal, 43 trapped two, 26 captured three, 7 caught four animals, 3 collected five animals, 2 trapped six animals, and ten animals were found in one bucket. Traps showed a similar pattern with 3.4% trap success in 1040 individual traps, and 283 captures (263 rodents and 20 shrews), but 834 traps caught nothing, 148 one, 42 two, 13 three and 3

Table 5. Product-moment correlation coefficients (*r*) of cumulative sample-nights with four parameters of trap/bucket captures.

Daily cumulative sample-nights correlated with (across)	Number of individuals	Number of Species	New Species added	Cumulative species
Total				
traps (rodents only)	0.099	−0.131	−0.428**	0.713**
traps (all captures)	0.068	−0.168	−0.491**	0.778**
buckets (shrews only)	(−0.249)	−0.085	−0.161	0.673**
buckets (all captures)	−0.149	0.168	(−0.217)	0.875**
traps and buckets combined (all captures)	−0.049	−0.100	−0.404**	0.798**
2000 m				
traps (rodents only)	(−0.698)	0.231	(−0.540)	0.865**
traps (all captures)	−0.746*	−0.066	(−0.605)	(0.577)
buckets (shrews only)	−0.800*	(−0.668)	(−0.684)	(0.577)
buckets (all captures)	−0.698	0.126	−0.882**	0.900**
traps and buckets combined (all captures)	−0.786*	−0.425	−0.752*	0.883**
2500 m				
traps (rodents only)	(−0.541)	−0.353	−0.376	0.727*
traps (all captures)	−0.455	−0.262	(−0.515)	0.776*
buckets (shrews only)	(−0.660)	−0.374	(−0.542)	1.000**
buckets (all captures)	(−0.657)	−0.429	(−0.611)	0.722*
traps and buckets combined (all captures)	(−0.604)	−0.442	(−0.591)	0.819**
3000 m				
traps (rodents only)	−0.424	−0.287	−0.452	0.809**
traps (all captures)	−0.342	−0.177	−0.401	0.826**
buckets (shrews only)	−0.020	0.164	(−0.585)	(0.548)
buckets (all captures)	0.060	−0.137	−0.666*	0.730*
traps and buckets combined (all captures)	−0.213	−0.220	(−0.539)	0.747*
3500 m				
traps (rodents only)	0.169	0.169	−0.126	0.907**
traps (all captures)	0.169	0.169	0.247	0.907**
buckets (shrews only)	0.809**	−0.365	−0.725*	(0.548)
buckets (all captures)	−0.760*	0.000	(−0.645)	(0.548)
traps and buckets combined (all captures)	−0.776*	−0.274	−0.754*	0.675*
4000 m				
traps (rodents only)	−0.310	−0.252	(−0.577)	1.000**
traps (all captures)	−0.309	−0.314	(−0.577)	1.000**
buckets (shrews only)	−0.465	−0.438	(−0.645)	(0.548)
buckets (all captures)	−0.439	−0.259	(−0.628)	0.903**
traps and buckets combined (all captures)	−0.108	−0.030	(−0.523)	(0.548)

Results are given for each sampling method for both targeted groups and everything captured. Values in parentheses represent strong but not significant correlations.
* $=P\leq0.05$; ** $=P\leq0.01$.

four. To test for "trap competition" and to determine if captures were independent with respect to each other, we compared the observed distribution of captures by bucket and by trap to the Poisson distribution. Neither captures by buckets or traps followed the Poisson distribution (G-test for goodness of fit = 84.0 for buckets, 10.0 for traps; $p<0.01$) suggesting a lack of trap or bucket independence. Significantly fewer traps or buckets caught one individual than would have been expected based on the assumption that the frequency of captures follows a Poisson distribution, and significantly more caught 2, or more, than expected [7].

Discussion

Sixteen species of mammal (6 shrews and 10 rodents) were recorded along an elevational transect from roughly 2000 to 4000 m on the southeastern slope of Mt. Kilimanjaro (images of select taxa are presented in Figure 4). Only one of these (*Myosorex zinki*) is endemic to the massif, and none were introduced taxa. The other species have broader distributions, to varying degrees. For example, among the soricomorphs, *Crocidura monax* has been recorded in neighboring mountains within the Eastern Arc Mountains to the southeast of Kilimanjaro, including the North

Table 6. Product-moment correlation coefficients (*r*) of shrew and rodent captures with four parameters of trap success.

Shrew and rodent captures correlated with (across)	Number of individuals	Number of Species	New species added	Cumulative species
Total, shrews	(−0.273)	−0.187	(−0.268)	0.726**
Total, rodents	0.074	−0.158	−0.496**	0.730**
2000 m, shrews	(−0.821)	(−0.668)	(−0.684)	(0.577)
2000 m, rodents	(−0.620)	0.063	0.724*	0.924**
2500 m, shrews	(−0.615)	−0.407	(−0.611)	0.722*
2500 m, rodents	(−0.539)	−0.412	−0.466	0.804**
3000 m, shrews	0.064	0.246	(−0.585)	(0.548)
3000 m, rodents	−0.440	−0.355	(−0.488)	0.794*
3500 m, shrews	−0.803**	−0.365	−0.725*	(0.548)
3500 m, rodents	0.452	0.000	(−0.518)	0.710*
4000 m, shrews	−0.398	−0.438	(−0.645)	(0.548)
4000 m, rodents	0.081	0.405	−0.411	(0.548)

Values in parentheses represent strong but not significant correlations. * $=P \leq 0.05$; ** $=P \leq 0.01$.

Pare and West Usambara Mountains. *Crocidura allex* is known from other mountains of the northern highlands of Tanzania (Meru, Ngorongoro) and the highlands of Kenya (Kenya, Aberdares). *Crocidura hildegardeae* and *Sylvisorex granti* are distributed across Kenya and the montane habitats of the Albertine Rift. Finally, *Crocidura olivieri* is broadly distributed across much of the African continent [32]. Among the ten species of rodents recorded, most are variably distributed across eastern Africa, and some range over larger regions of Africa. For example, *Tachyoryctes daemon* is restricted to northern Tanzania, but murines such as *Grammomys dolichurus* and *Rhabdomys dilectus* range across much of eastern and southern Africa, as does the dormouse, *Graphiurus murinus* [31], [33]. However, many taxonomists have cautioned that some of these soricomorph and rodent taxa are almost certainly species complexes, and work in progress may alter our taxonomic understanding of these groups [31], [32], [33].

We found no species not previously documented on the mountain. There are several published faunal lists for Kilimanjaro, the most complete being that of Grimshaw *et al.* [13], who evaluated the accuracy of previous published records, and developed a working list of likely residents of the mountain. While this list includes every species we documented, there are other small mammals listed by Grimshaw *et al.* [13] that we did not document. For example, among shrews, we have no record of *Crocidura luna*. This species was listed by Grimshaw *et al.* [13] based on voucher specimens at the FMNH, collected at 1400 m, an elevation below our lowest sampling site (2000 m). Similarly, many rodent species listed by Grimshaw *et al.* [13] occur at elevations lower than the range of this study. Examples include genera such as *Aethomys*, *Arvicanthis*, *Lemniscomys*, *Mastomys*, *Pelomys* and *Tatera*. One rodent historically recorded in our elevation sampling range but absent from animals we captured is *Otomys typus* (= *O. orestes zinki*; [35]). The holotype of *O. zinki* Bohmann 1943 was collected at Horombo Hut [36], [37], but the only two species of *Otomys* we documented were *O. angoniensis* and *O. tropicalis*. Two other rodents are notably absent from our inventory: *Beamys hindei* and *Hylomyscus arcimontensis*. Both are residents in forests of the Eastern Arc (including the North Pare Mountains roughly 50 km SE of Kilimanjaro) and Southern Highlands [38], [39], but no voucher specimen is known for either species from Kilimanjaro, or other northern highland sites. The type locality of *Beamys hindei* is Taveta, Kenya [40], and Dieterlen [41] identified a skull collected by C.G. Schillings in 1903 at Moshi as *Beamys*. Both localities are at, or near the base of Mt. Kilimanjaro. While this is not the most common species recorded in recent surveys of montane habitats of Tanzania across the elevational range from 600 to 2000 m [8], [19], given the number of trap nights expended during this survey, we anticipate the capture of *Beamys* if it occurs in the forests of southeastern Kilimanjaro. Records of *Hylomyscus* on Meru [42] and Ngorongoro [43] are now attributed to *Praomys taitae* [30], [38]. Recent surveys of both Meru and Ngorongoro using techniques identical to this study did not record either *Beamys* or *Hylomyscus* (Stanley, unpubl. data). All of this leads us to the conclusion that neither *Beamys* nor *Hylomyscus* currently occur in the forests of Mt. Kilimanjaro.

The trap success for rodents was highest at 4000 m and lowest at the 3500 m site (Table 4). The very low number of rodent captures at 3500 m (three in 1600 trap nights; Table 2) was striking, and is the lowest trap success recorded in similar surveys in montane habitats of Tanzania [8], [19], [24], [39]. Shore & Garbett [14] trapped at 3500 m, roughly the same elevation as our fourth site, but on the Shira Plateau on the western slopes of Kilimanjaro. The species they documented (*Crocidura allex*, *Myosorex blarina zinki* [= *M. zinki*], *Dendromus mesomelas kilimandjari* [= *D. insignis*] and *Rhabdomys pumilio diminutus* [= *R. dilectus*] were the same as in our study at the 3500 m site. One species (*Otomys tropicalis*) recorded by Shore & Garbett at 3500 m was not trapped by us at that elevation, but was collected at sites both lower and higher than 3500 m. Notably, their trap success for small mammals (2.7%; 81 captures in 2995 trap nights) was much higher that of the 3500 m site in this study (0.2%; 3 captures in 1600 trap nights), although the period of the surveys (mid-July to early August) was similar in both studies and trapping extended over several days at each site. The bait used by Shore & Garbett [14] included fried coconut and peanut butter (as in our procedure) but also fish, nuts and oats. However, the 3500 m site in our study was in stark contrast to lower and higher sites along the same transect, leading us to hypothesize that different bait is not the explanation for the lower rodent trap success at 3500 m in this study compared to the patterns documented by Shore &

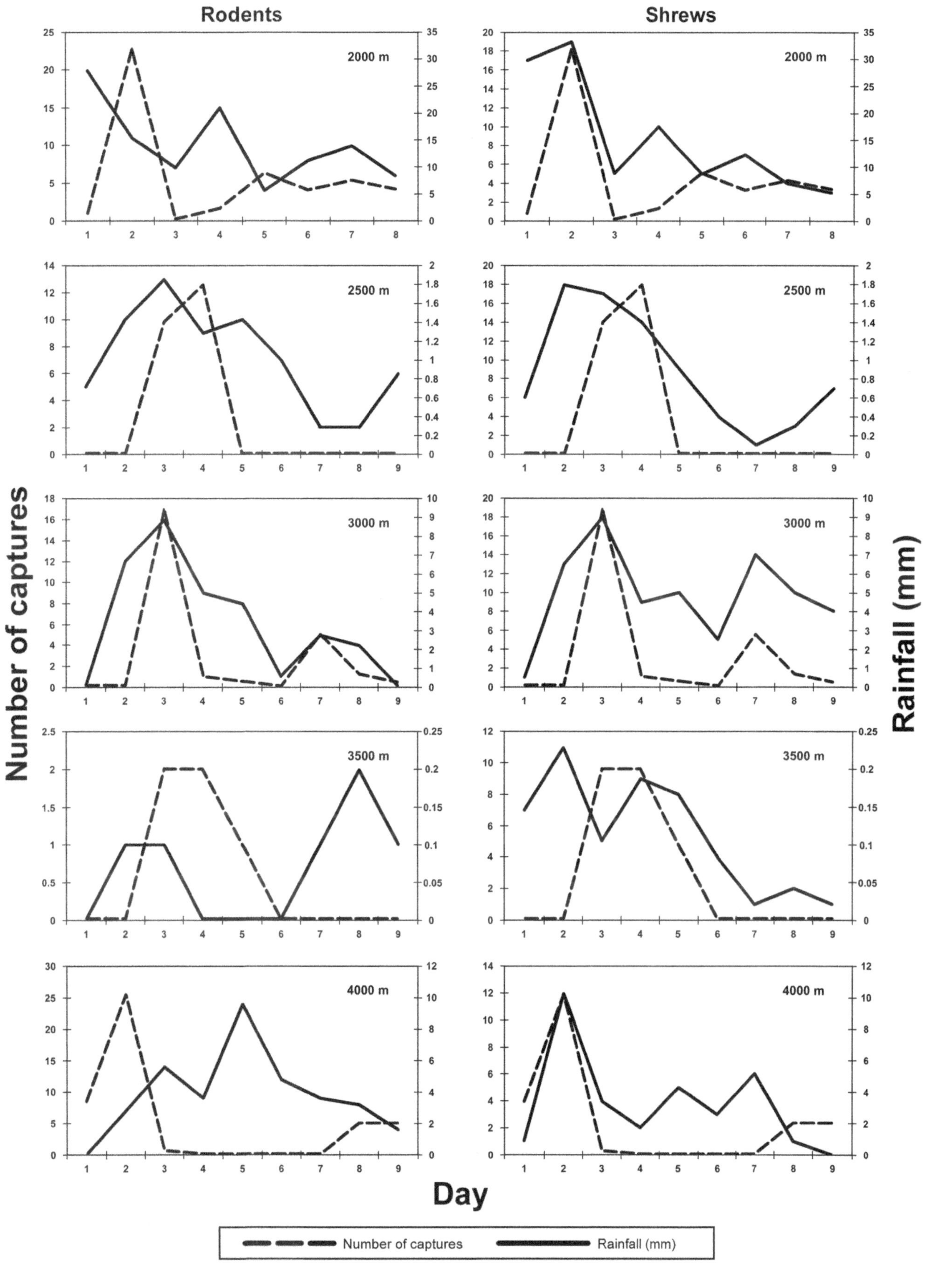
Rodents
Shrews
2000 m
2500 m
3000 m
3500 m
4000 m
Number of captures
Rainfall (mm)
Day
Number of captures
Rainfall (mm)

Figure 3. The relationship between numbers of individuals captured each day of the sampling period, and rainfall, at each site. Rodentia are on the left and Soricomorpha are on the right.

Garbett [14]. One potential explanation might be the proximity of Horombo Hut (3°8'20"S, 37°26'18"E) which was approximately 600 m from our trap lines (no other sites in this study were close to human habitation). The buildings and discarded flour and other foodstuffs generated by people occupying this touristic camp provide shelter and food for rodents. Indeed, while visiting Horombo on 8 and10 August, we saw many *Rhabdomys* moving between buildings. Both repeating our sampling methodology at our site, and sampling with the same techniques at, and at increasing distances from Horombo would be illustrative of the influence of human habitation on the abundance of native rodents in the environs of Kilimanjaro.

Mulungu *et al.* [15] published the results of two elevational transects of Kilimanjaro (along the Shira and Marangu routes) that focused on shrews and rodents. While total trapping effort was less than half of the current study (3600 vs 8361 trap nights), the recorded trap success was much higher (up to 36%). We attribute the higher success of Mulungu *et al.* [15] to the fact that traps were in place for only two nights at each site, and thus a reduction in capture rates typical of longer periods of time was not observed. The soricomorph and rodent species documented by Mulungu *et al.* [15] were identical to this study, with the exception of *Mus triton*, which they recorded at 2300 and 3270 m. *Lophuromys aquilus* was recorded at 3200 and 3590 m by Mulungu *et al.* [15] but our study did not record this rodent above 3000 m. However, one *Lophuromys* was brought to us by a Tanzanian National Parks employee who captured it at Horombo Hut (3760 m). The presence of other species at various elevations documented by Mulungu *et al.* [15] mirrored patterns observed during this study.

As in past studies within Tanzania, the combination of traps and pitfall lines were effective in sampling non-volant small mammal communities at different elevations on Mt. Kilimanjaro [8]. In general, species accumulation curves reached a plateau at each site, with the exception of the 2500 site where we captured *Dendromus insignis* and *Otomys angoniensis* for the first time during the last 24-hours of trapping. Notwithstanding the 2500 m pattern, we are confident that we documented almost all of the species of shrews and small rodents occurring at each site, and thus feel justified in comparing results among different elevational sites of this transect, as well as to results of similar surveys within Tanzania [8].

There was a significantly negative correlation between elevation and shrew species diversity at each site, and while not significant, there was generally lower abundance (as measured by sample success) for shrews as elevation increased (Table 8). However, rodents showed no notable correlation with elevation, either in diversity or abundance. This is in contrast to the patterns observed in the Udzungwa Mountains [8], where diversity and abundance of rodents were positively and significantly correlated with elevation. The same trends were not observed for shrews in the Udzungwas. Another difference between the Udzungwa and Kilimanjaro studies was seen in the overall measures of sample success in relation to elevation. Stanley & Hutterer [8] found either significant or high positive correlations between elevation and total sample success, number of individual mammals, and number of species collected for shrews and rodents combined. Such a relationship on Mt. Kilimanjaro was negative but not statistically significant. Whereas there was no mid-elevational peak (sensu McCain [10]) in the Udzungwa study, the greatest diversity of shrews and rodents on Kilimanjaro was at 3000 m, in the middle of our transect. Indeed, this site was situated at the ecotone between forest and heathland, and species typical of both habitats were present. For example, this was the highest, and lowest site where *Praomys taitae* and *Rhabdomys dilectus*, respectively, were documented and the only site where the two species were found together. In general, there was more species diversity for both shrews and rodents in forest habitats than above treeline (Tables 2,3,4).

Rainfall generally influenced the capture of shrews, but not rodents as was observed in the faunal inventories in the Udzungwa Mountains [8]. Thus, rainfall amounts while sampling shrew diversity or abundance should be considered. In addition, there

Table 7. Product-moment correlation coefficients (*r*) of amount of daily rainfall with four parameters of shrew and rodent daily captures.

Rainfall amount correlated with (across)	Number of individuals	Number of Species	New species added	Cumulative species
Total, shrews (buckets and traps)	0.385*	0.422**	0.086	(0.277)
Total, rodents (buckets and traps)	0.190	(0.280)	(0.230)	0.053
2000 m, shrews	(0.523)	(0.655)	−0.025	0.262
2000 m, rodents	−0.108	(0.523)	0.424	−0.008
2500 m, shrews	(0.592)	(0.502)	−0.050	0.283
2500 m, rodents	(0.544)	0.201	−0.217	0.096
3000 m, shrews	0.719*	0.199	−0.246	0.187
3000 m, rodents	(0.644)	(0.487)	0.139	0.122
3500 m, shrews	0.363	0.378	−0.357	0.236
3500 m, rodents	−0.267	−0.060	0.286	0.334
4000 m, shrews	(0.629)	(0.652)	0.950**	−0.160
4000 m, rodents	−0.411	−0.086	0.927**	−0.160

All captures (both traps and pitfalls) of each group are included. * $=P\leq0.05$; ** $=P\leq0.01$.

Table 8. Product-moment correlation coefficients (*r*) between elevation and trap success.

Elevation correlated with	(*r*)	P
Total number of individual mammals collected	−0.59	>0.05
Total trap success	−0.59	>0.05
Total number of species collected	−0.68	>0.05
Total number of shrews collected	(−0.73)	>0.05
Shrew trap success	(−0.79)	>0.05
Total number of shrew species collected	**−0.95**	**<0.05**
Total number of rodents collected	−0.23	>0.05
Rodent trap success	−0.21	>0.05
Total number of rodent species collected	−0.56	>0.05

Values in parentheses represent strong but not significant correlations. Significant relationships (P<0.05) are in bold.

Figure 4. Six mammals found in the montane habitats of Mt. Kilimanjaro, Tanzania: *Praomys taitae* **(top left);** *Crocidura monax* **(top right);** *Myosorex zinki* **(middle left);** *Otomys tropicalis* **(middle right);** *Grammomys dolichurus* **(bottom left);** *Crocidura allex* **(bottom right); all photographs by W. Stanley.**

was a lack of capture independence among traps and buckets across the entire transect and at each site. Stanley & Hutterer [8] documented similar results in the Udzungwa Mountains, and hypothesized that multiple captures are influenced by the placement of individual traps and buckets. More specifically, while traps cannot catch more than one animal, generally, buckets can capture more than one on a given bucket-night. The possibility exists that the presence of a captured animal in a bucket may attract other animals into that bucket.

The only endemic mammal, as currently understood, on Mt. Kilimanjaro is *Myosorex zinki* [44]. Until this survey, and that of Mulungu *et al.* [15], *M. zinki* was only known from a few specimens captured in the moorland habitats above tree line [14]. This species was documented across the elevational range of 2500 to 4000 m in this survey [45] and between 2500 to 2600 m by Mulungu *et al.* [15]. Thus, this endemic shrew extends across several different habitats on the mountain. *Myosorex zinki* was not observed at our lowest sampling site (2000 m). Stanley *et al.* [45] suggest that human impact on the forests at 2000 m on the Maua route may be responsible for the absence of this endemic mammal, but this hypothesis has not been adequately tested.

Three species (2 shrews and 1 rodent) were found at all sites sampled and occur across the range from roughly 2000 to 4000 m: *Crocidura allex*, *Sylvisorex granti*, and *Dendromus insignis*. The latter was found in the highest trap set in the survey (3° 6.481' S, 37° 25.312' E, 4240 m, on the ridge leading to West Lava Hill), and four individuals of this species were collected in this single trap (a Museum Special). How high small mammals extend on Kilimanjaro remains unanswered. Wild dog (*Lycaon pictus*) is the only mammal (other than *Homo sapiens*) recorded at the summit (5895 m; [46]). However no small mammal surveys have been conducted above 4000 m, and such efforts would help elucidate the upper ranges of shrews and rodents on this unique and iconic mountain, and would further our understanding of its faunistic dynamics.

Acknowledgments

We are grateful to the Wildlife Conservation Society (Tanzania) for logistical support and to the Tanzania Commission for Science and Technology, Ministry of Natural Resources and Tanzanian National Parks for permission to conduct this research. E. Gereta, Mr. Lejora, N. Mafuru, and L. M. Ole Moirana, were particularly supportive. O. Mathayo and S. Temu all aided in fieldwork. R. Banasiak provided important assistance with the figures. S. Goodman and K. Helgen offered valuable editorial advice.

Author Contributions

Conceived and designed the experiments: WTS MAR PMK MJM. Performed the experiments: WTS MAR PMK MJM. Analyzed the data: WTS MAR PMK. Wrote the paper: WTS MAR.

References

1. Patterson BD, Meserve PL, Lang BK (1989) Distribution and abundance of small mammals along an elevational transect in temperate rainforests of Chile. J Mammal 70: 67–78.
2. McCain CM (2004) The mid-domain effect applied to elevational gradients: species richness of small mammals in Costa Rica. J Biogeogr 31: 19–31.
3. Md Nor S (2001) Elevational diversity patterns of small mammals on Mount Kinabalu, Malaysia. Glob Ecol Biogeogr 10: 41–62.
4. Heideman PD, Heaney LR, Thomas RL, Erickson KR (1987) Patterns of faunal diversity and abundance of non-volant small mammals on Negros Island, Philippines. J Mammal 68: 884–888.
5. Heaney LR, Heideman PD, Rickart EA, Utzurrum RB, Klompen JSH (1989) Elevational zonation of mammals in the central Philippines. J Trop Ecol 5: 259–280.
6. Rickart EA, Heaney LR, Utzurrum RB (1991) Distribution and ecology of small mammals along an elevational transect in southeastern Luzon, Philippines. J Mammal 72: 458–469.
7. Yu HT (1994) Distribution and abundance of small mammals along a subtropical elevational gradient in central Taiwan. J Zool 234: 577–600.
8. Stanley WT, Hutterer R (2007) Differences in abundance and species richness between shrews and rodents along an elevational gradient in the Udzungwa Mountains, Tanzania. Acta Theriol 52: 261–275.
9. Goodman SM, Ganzhorn JU, Rakotondravony D (2003) Introduction to the Mammals. In: Goodman, SM, Benstead, JP, editors. The Natural History of Madagascar. Chicago: The University of Chicago Press. Pp. 1159–1186.
10. McCain CM (2005) Elevational gradients in diversity of small mammals. Ecology 86: 366–372.
11. Moritz C, Patton JL, Conroy CJ, Parra JL, White GC, et al. (2008) Impact of a Century of Climate Change on Small-Mammal Communities in Yosemite National Park, USA. Science 322: 261–264.
12. Newmark WD (1991) Priorities for the conservation of Mount Kilimanjaro. In: Newmark, WD, editor. The Conservation of Mount Kilimanjaro. IUCN, Gland, Switzerland and Cambridge, UK. pp. 111–114.
13. Grimshaw J, Cordeiro N, Foley C (1995) The mammals of Kilimanjaro. J East Afr Nat Hist 84: 105–139.
14. Shore RF, Garbett SD (1991) Notes on the small mammals of the Shira Plateau, Mt. Kilimanjaro. Mammalia 55: 601–607.
15. Mulungu LS, Makundi RH, Massawe AW, Machang'u RS, Mbije NE (2008) Diversity and distribution of rodent and shrew species associated with variations in altitude on Mount Kilimanjaro, Tanzania. Mammalia 72: 178–185.
16. Thompson LG, Brecher HH, Mosley-Thompson E, Hardy DR, Mark BG (2009) Glacier loss on Kilimanjaro continues unabated. Proc Natl Acad Sci USA 106: 19770–19775.
17. Goodman SM, Newmark WD, Stanley WT, Howell KM (1995) The Ambangulu Forest, West Usambara Mountains, Tanzania: a threatened Eastern Arc forest. Oryx 29: 212–214.
18. Stanley WT, Kihaule PM, Howell KM, Hutterer R (1998) Small mammals of the Eastern Arc Mountains, Tanzania. J East Afr Nat Hist 87: 91–100.
19. Stanley WT, Goodman SM (2011) Small mammal inventories in the East and West Usambara Mountains, Tanzania. 4. Rodentia. In: Stanley WT, editor. Studies of Montane Vertebrates of Tanzania. Fieldiana Life Earth Sci 4: 53–73.
20. Sampson DN (1965) The geology, volcanology and glaciology of Kilimanjaro. Tanzan Notes Rec 64: 118–124.
21. Hanby J (1987) Kilimanjaro National Park. Tanzania National Parks/Africa Wildlife Foundation.
22. Mwasaga BC (1991) The natural forest of Mount Kilimanjaro. In: Newmark, WD, editor. The Conservation of Mount Kilimanjaro. IUCN, Gland, Switzerland and Cambridge, UK. pp. 111–114.
23. Stanley WT, Goodman SM, Hutterer R (1996) Notes on the insectivores and elephant shrews of the Chome Forest, South Pare Mountains, Tanzania (Mammalia: Insectivora and Macroscelidea). Zool Abh 49: 131–148.
24. Stanley WT, Goodman SM, Kihaule PM (1998) Results of two surveys of rodents in the Chome Forest Reserve, South Pare Mountains, Tanzania (Mammalia: Rodentia). Zool Abh 50: 145–160.
25. Stanley WT, Goodman SM, Kihaule PM, Howell KM (2000) A survey of the small mammals of the Gonja Forest Reserve, Tanzania. J East Afr Nat Hist 89: 73–83.
26. Stanley WT, Nikundiwe AM, Mturi FA, Kihaule PM, Moehlman PD (2005) Small mammals collected in the Udzungwa Mountains National Park, Tanzania. J East Afr Nat Hist 94: 203–212.
27. Stanley WT, Rogers MA, Howell KM, Msuya CA (2005) Results of a survey of small mammals in the Kwamgumi Forest Reserve, East Usambara Mountains, Tanzania. J East Afr Nat Hist 94: 223–230.
28. Stanley WT, Goodman SM, Hutterer R (2011) Small mammal inventories in the East and West Usambara Mountains, Tanzania. 2. Families Soricidae (Shrews) and Macroscelididae (Elephant Shrews). In: Stanley WT, editor. Studies of Montane Vertebrates of Tanzania. Fieldiana Life Earth Sci 4: 18–33.
29. Stanley WT, Goodman SM, Newmark WD 2011) Small mammal inventories in the East and West Usambara Mountains, Tanzania. 1. Study areas, methodologies, and general results. In: Stanley WT, ed. Studies of Montane Vertebrates of Tanzania. Fieldiana Life Earth Sci 4: 1–17.
30. Carleton MD, Stanley WT (2012) Species limits within the *Praomys delectorum* group (Rodentia: Muridae: Murinae) of East Africa: A morphometric reassessment and biogeographic implications. Zool J Linn Soc 165: 420–469.
31. Holden ME (2005) Family Gliridae. In: Wilson DE, Reeder DM, editors. Mammal species of the world: A taxonomic and geographic reference, Third Edition. Baltimore: Johns Hopkins University Press, pp. 819–841.
32. Hutterer R (2005) Order Soricomorpha. In: Wilson DE, Reeder DM, editors. Mammal species of the world: A taxonomic and geographic reference, Third Edition. Baltimore: Johns Hopkins University Press, pp. 220–299.
33. Musser GG, Carleton MD (2005) Superfamily Muroidea. In: Wilson DE, Reeder DM, editors. Mammal species of the world: A taxonomic and geographic reference, Third Edition. Baltimore: Johns Hopkins University Press, pp. 894–1531.
34. Sikes RS, Gannon WL, Animal Care and Use Committee of the American Society of Mammalogists (2011) Guidelines of the American Society of

Mammalogists for the use of wild mammals in research. J Mammal 92: 235–253.
35. Carleton MD, Schaefer Byrne E (2006) The status of *Otomys orestes dollmani* Heller, 1912 (Muridae: Otomyinae), a rodent described from the Mathews Range, central Kenya. Proc Biol Soc Wash 119: 477–515.
36. Swynnerton GH, Hayman RW (1951) A checklist of the land mammals of the Tanganyika Territory and Zanzibar Protectorate. J East Afr Nat Hist 20: 274–392.
37. Child GS (1965) Some notes on the mammals of Kilimanjaro. Tanzan Notes Rec 64: 77–89.
38. Carleton MD, Stanley WT (2005) Review of the *Hylomyscus denniae* complex in Tanzania, with description of a new species. Proc Biol Soc Wash 118: 619–646.
39. Stanley WT, Kihaule PM, Munissi MJ (2007) Small mammals of two forest reserves in the North Pare Mountains, Tanzania. J East Afr Nat Hist 96: 215–226.
40. Thomas O (1909) New African small Mammals in the British Museum Collection. Ann Mag Nat Hist 8: 98–112.
41. Dieterlen F (1979) Der früheste Fund der afrikanischen Kleinen Hamsterratte (*Beamys hindei*) (Cricetomyinae; Cricetidae; Rodentia) Stuttgarter Beitr zur Naturk ser. A 330: 1–3.
42. Demeter A, Hutterer R (1986) Small mammals from Mt. Meru and its environs (Northern Tanzania). Cimbebasia 8: 199–207.
43. Bishop IR (1979) Notes on *Praomys* (*Hylomyscus*) in eastern Africa. Mammalia 43: 521–530.
44. Stanley WT, Rogers MA, Hutterer R (2005) A morphological assessment of *Myosorex zinki*, an endemic shrew on Mount Kilimanjaro. Belg J Zool 135 Supplement: 141–144.
45. Stanley WT, Kihaule PM, Rogers MA, Makwetta AJ (2003) Elevational distribution and ecology of the endemic *Myosorex zinki* (Insectivora: Soricidae) along the southeastern slope of Mount Kilimanjaro, Tanzania. Tanzania Wildlife Research Institute: Proceedings of the third annual scientific conference, Arusha, Dec 3–5, 2002.
46. Thesiger W (1970) Wild Dog at 5894 metres. E Afr Wildl J 8: 202.

PERMISSIONS

All chapters in this book were first published in PLOS ONE, by The Public Library of Science; hereby published with permission under the Creative Commons Attribution License or equivalent. Every chapter published in this book has been scrutinized by our experts. Their significance has been extensively debated. The topics covered herein carry significant findings which will fuel the growth of the discipline. They may even be implemented as practical applications or may be referred to as a beginning point for another development.

The contributors of this book come from diverse backgrounds, making this book a truly international effort. This book will bring forth new frontiers with its revolutionizing research information and detailed analysis of the nascent developments around the world.

We would like to thank all the contributing authors for lending their expertise to make the book truly unique. They have played a crucial role in the development of this book. Without their invaluable contributions this book wouldn't have been possible. They have made vital efforts to compile up to date information on the varied aspects of this subject to make this book a valuable addition to the collection of many professionals and students.

This book was conceptualized with the vision of imparting up-to-date information and advanced data in this field. To ensure the same, a matchless editorial board was set up. Every individual on the board went through rigorous rounds of assessment to prove their worth. After which they invested a large part of their time researching and compiling the most relevant data for our readers.

The editorial board has been involved in producing this book since its inception. They have spent rigorous hours researching and exploring the diverse topics which have resulted in the successful publishing of this book. They have passed on their knowledge of decades through this book. To expedite this challenging task, the publisher supported the team at every step. A small team of assistant editors was also appointed to further simplify the editing procedure and attain best results for the readers.

Apart from the editorial board, the designing team has also invested a significant amount of their time in understanding the subject and creating the most relevant covers. They scrutinized every image to scout for the most suitable representation of the subject and create an appropriate cover for the book.

The publishing team has been an ardent support to the editorial, designing and production team. Their endless efforts to recruit the best for this project, has resulted in the accomplishment of this book. They are a veteran in the field of academics and their pool of knowledge is as vast as their experience in printing. Their expertise and guidance has proved useful at every step. Their uncompromising quality standards have made this book an exceptional effort. Their encouragement from time to time has been an inspiration for everyone.

The publisher and the editorial board hope that this book will prove to be a valuable piece of knowledge for researchers, students, practitioners and scholars across the globe.

LIST OF CONTRIBUTORS

Brandon Lee Drake, Wirt H. Wills, Marian I. Hamilton and Wetherbee Dorshow
Department of Anthropology, University of New Mexico, Albuquerque, New Mexico, United States of America

Luis Llaneza and Emilio J. García
A.RE.NA. Asesores en Recursos Naturales, S.L., Lugo, Spain

JoséVicente López-Bao
Research Unit of Biodiversity (UO/CSIC/PA), Oviedo University, Mieres, Spain
GrimsöWildlife Research Station, Swedish University of Agricultural Sciences (SLU), Riddarhyttan, Sweden

Kevin R. Ford, Ailene K. Ettinger and Janneke Hille Ris Lambers
Department of Biology, University of Washington, Seattle, Washington, United States of America

Jessica D. Lundquist and Mark S. Raleigh
Department of Civil and Environmental Engineering, University of Washington, Seattle, Washington, United States of America

Mikkel Willemoes, Anders P. Tøttrup and Kasper Thorup
Center for Macroecology, Evolution and Climate, Natural History Museum of Denmark, University of Copenhagen, Copenhagen, Denmark

Roine Strandberg, Yannis Vardanis and Thomas Alerstam
Department of Biology, Lund University, Lund, Sweden

Raymond H. G. Klaassen
Department of Biology, Lund University, Lund, Sweden
Dutch Montagu's Harrier foundation and Animal Ecology Group, Centre for Ecological and Evolutionary Studies, Groningen University, Groningen, the Netherlands

Paul W. Howey
Microwave Telemetry Inc, Columbia Maryland, United States of America

Martin Wikelski
Max Planck Institute for Ornithology, Department of Migration & Immuno-ecology, Radolfzell am Bodensee, Germany
Department of Biology, University of Konstanz, Konstanz, Germany

Logan Volkmann
Department of Biological Sciences, Simon Fraser University, Burnaby, British Columbia, Canada

Arne O. Mooers
Department of Biological Sciences, Simon Fraser University, Burnaby, British Columbia, Canada
Interdisciplinary Research in the Mathematical and Computational
Sciences (IRMACS) Centre, Simon Fraser University, Burnaby, British Columbia, Canada

Iain Martyn
Interdisciplinary Research in the Mathematical and Computational
Sciences (IRMACS) Centre, Simon Fraser University, Burnaby, British Columbia, Canada

Vincent Moulton
School of Computing Sciences, University of East Anglia, Norwich, United Kingdom

Andreas Spillner
Department of Mathematics and Computer Science, University of Greifswald, Greifswald, Germany

Yiping Fang
Institute of Mountain Hazards and Environment, Chinese Academy of Sciences, Chengdu, Sichuan, P.R. China

Yanqiang Wei
Institute of Mountain Hazards and Environment, Chinese Academy of Sciences, Chengdu, Sichuan, P.R. China
University of Chinese Academy of Sciences, Beijing, P.R. China

Heidi Nistelberger and J. Dale Roberts
School of Animal Biology, The University of Western Australia, Crawley, Western Australia, Australia

Margaret Byrne and David Coates
Science Division, Department of Parks and Wildlife, Bentley Delivery Centre, Bentley, Western Australia, Australia

Xiao-Xin Wei and Xiao-Quan Wang
State Key Laboratory of Systematic and Evolutionary Botany, Institute of Botany, Chinese Academy of Sciences, Beijing, China

Liu Liu Zhen-Zhen Hao and Yan-Yan Liu
State Key Laboratory of Systematic and Evolutionary Botany, Institute of Botany, Chinese Academy of Sciences, Beijing, China
Graduate University of the Chinese Academy of Sciences, Beijing, China

Yu-Zhi Cun
Dali University, Dali, Yunnan, China

Jutta Beher
ARC Centre of Excellence for Environmental Decisions, University of Queensland, Brisbane, Queensland, Australia

Junli Xu, Shiyin Liu, Wanqin Guo and Jian Wang
State Key Laboratory of Cryosphere Science, Cold and Arid Regions Environmental and Engineering Research Institute, Chinese Academy of Sciences, Lanzhou, PR China

Shiqiang Zhang
The Land in the Earth System, Max Planck Institute for Meteorology, Hamburg, Germany

Koustubh Sharma and Charudutt Mishra
Snow Leopard Trust, Seattle, Washington, United States of America
Nature Conservation Foundation, Mysore, Karnataka, India

Orjan Johansson
Snow Leopard Trust, Seattle, Washington, United States of America
Panthera, New York, New York, United States of America
Grimso Wildlife Research Station, Swedish University of Agricultural Sciences, Uppsala, Sweden

Lkhagvasumberel Tumursukh and Purevsuren Sevger
Snow Leopard Trust, Seattle, Washington, United States of America
Snow Leopard Conservation Foundation, Ulaan Baatar, Mongolia

Rana Bayrakcismith and Tom McCarthy
Panthera, New York, New York, United States of America

Yiling Wang, Guiqin Yan
College of Life Sciences, Shanxi Normal University, Linfen, China

Sandhya Sekar and Praveen Karanth
Centre for Ecological Sciences, Indian Institute of Science, Bangalore, India

Simon A. Black and David L. Roberts
1 Durrell Institute of Conservation and Ecology, School of Anthropology and Conservation, University of Kent, Canterbury, Kent, United Kingdom

Amina Fellous
Agence Nationale pourla Conservation de la Nature, Algiers, Algeria

Nobuyuki Yamaguchi
Department of Biological and Environmental Sciences, University of Qatar, Doha, Qatar

Nina J. Laitinen1, Dennis K. Shiozawa1
WIDB 401, Department of Biology, Brigham Young University, Provo, Utah, United States of America

Peter J. Unmack
WIDB 401, Department of Biology, Brigham Young University, Provo, Utah, United States of America
National Evolutionary Synthesis Center, Durham, North Carolina, United States of America

Carol L. Sęcor
School of Life Sciences, Arizona State University, Tempe, Arizona, United States of America

Thomas E. Dowling
School of Life Sciences, Arizona State University, Tempe, Arizona, United States of America
Department of Biological Sciences, Wayne State University, Detroit, Michigan, United States of America

Richard L. Mayden
Department of Biology, Saint Louis University, St. Louis, Missouri, United States of America

Gerald R. Smith
Museum of Zoology, The University of Michigan, Ann Arbor, Michigan, United States of America

Ian-Huei Yau and Joan R. Davenport
Crop and Soil Sciences, Washington State University, Prosser, Washington, United States of America

Richard A. Rupp
Crop and Soil Sciences, Washington State University, Pullman, Washington, United States of America

Shengwei Zong, Zhengfang Wu, Jiawei Xu, Haibo Du and Lei Wang
School of geographical sciences, Northeast Normal University, Changchun, China

Hongshi He
School of geographical sciences, Northeast Normal University, Changchun, China
School of Natural Resources, University of Missouri-Colombia, Columbia, Missouri, United States of America

Ming Li
College of Urban and Environment Sciences, Shanxi Normal University, Shanxi,China

Xiaofeng Gao
School of geography, Beijing Normal University, Beijing, China

David K. Wright
Department of Archaeology and Art History, Seoul National University, Seoul, Republic of Korea

Scott MacEachern
Department of Anthropology, Bowdoin College, Brunswick, Maine, United States of America

Jaeyong Lee
Department of Statistics, Seoul National University, Seoul, Republic of Korea

William T. Stanley and Mary Anne Rogers
The Field Museum of Natural History, Department of Science and Education, Chicago, Illinois, United States of America

Philip M. Kihaule
University of Dar es Salaam, Department of Zoology, Dar es Salaam, Tanzania

Maiko J. Munissi
Southern Highlands Conservation Programme, Mbeya, Tanzania

Index